Balances

Erich Robens · Shanath Amarasiri A. Jayaweera ·
Susanne Kiefer

Balances

Instruments, Manufacturers, History

 Springer

Erich Robens
Uniwersytet Maria Curie-Skłodowska
Lublin, Poland

Battelle-Institut e.V.
Frankfurt am Main, Germany

Johannes Gutenberg-Universität
Mainz, Germany

Shanath Amarasiri A. Jayaweera
School of Science & Engineering
Teesside University
Middlesbrough, UK

Susanne Kiefer
Kern & Sohn GmbH
Balingen, Germany

Philipp-Matthäus-Hahn Museum
Albstadt-Onstmettingen, Germany

ISBN 978-3-642-36446-4   ISBN 978-3-642-36447-1 (eBook)
DOI 10.1007/978-3-642-36447-1
Springer Heidelberg New York Dordrecht London

Library of Congress Control Number: 2013950174

Printed on acid-free paper

Springer is part of Springer Science+Business Media (www.springer.com)

"I often say that when you can measure what you are speaking about and can express it in numbers, you know something about it. But when you cannot measure it, when you cannot express it in numbers, your knowledge is of a meagre and unsatisfactory kind."

Sir William J. Thomson (Lord Kelvin)

*This book is dedicated to the memory of
Hans Richard Jenemann and Inis Jenemann.
The content is based to a large extend on
their work and on their collection of
balances, photographs and publications.*

Hans Richard Jenemann

Irene Jenemann

Preface

The balance is one of the oldest measuring instruments of civilised societies. Its invention most probably was in connection with development of commerce, trade routes and the establishment of towns within the Neolithic era, that means about 5000 BC. The oldest findings of weights and parts of scales, however, are from about 3300 BC. Development of balances with regard to sensitivity, precision, and easy operation happened slowly within millennia, answering the prevailing needs and concerning the technical possibilities.

With fast development of techniques and creation of modern chemistry towards the end of the 18th century the demands on sensitive balances grew dramatically and resulted in a relative sensitivity (resolution/maximum load) up to 10^{-9}. Thus, up to the middle of the 20th century balances had been the most sensitive and precise instruments of sciences.

A basic change occurred when electricity was introduced in balance constructions. Now the weighing result is indicated digitally on a monitor or directly evaluated by means of a computer. Still main parts are levers and/or springs. However, using Lorentz force for a reset of the balance to its zero position, measuring the deflection of an elastic body or frequency shift of oscillations, only small paths are to be considered. Thus, the problems of knife edges vanished. In only few years such mechatronic balances superseded the pure mechanical types.

The constructions of mechanical balances is well documented, however reports are distributed in many scientific journals. Because mechanical scales are popular subjects of collectors, we have many collections and some museums specialise on balances. Furthermore often balances are exhibited in museum's covering general techniques. Nevertheless we think it would be useful to write a survey in a lexical form, covering the beginning of balance development up to the present time. That could facilitate finding of more detailed descriptions.

The book starts with a survey on the development of the concept 'mass' and of theories of gravity. Keppler's laws and fundamental forces between basic particles are explained. Mass units and conversion factors are given in tables. Development of weights and their standardisation from antiquity to the present time is described. Examples of weights destined not for weighing are mentioned. A survey on the different methods of weighing is presented. Adjustment and calibration, diminution of erroneous influences, damping and methods of speeding up weighing are discussed. Different extensive and detailed types of balances are described. One chapter deals with some important applications of gravimetric methods and the special types of balances for that purpose: laboratory scales, vacuum and thermo balances, magnetic suspension balances, tensiometer, gravimeter, suspended particles, density, particle size, and magnetic susceptibility determination, post office and coin scales, bathroom and animal scales. Mass determination in astronomy is shortly described and some curiosities are reported.

Often the balance is used as symbolic instrument, e.g. in mythology, religion and esoteric context. In astronomy, commerce and in law it is used as an icon. Sometimes it is depicted in paintings or used as a component of statues. We find the balance in literature, and some poems are set to music.

In tabular form the history of the balance is presented as well as the museums in which balances are exhibited. History and profiles of some important balance manufacturers are reported and the addresses of many companies are listed. The appendix contains tables of prefixes and of recent literature.

Several hundred photographs, reproductions and drawings show instruments and their uses. These include commercial weighing instruments for merchandise and raw materials in workshops, as well as symbolic weighing in the ancient Egyptian's ceremony of 'Weighing of the Heart', the Greek fate balance, the Roman Justitia, Juno Moneta and Middle Ages scenes of the Last Judgement with Jesus or St. Michael, and of modern balances. The photographs are selected from the slide-archives of the late Richard Vieweg (1896–1972) (former President of the Physikalisch-Technische Bundesanstalt, Braunschweig, Germany), of the late Hans R. Jenemann (1920–1996) (former head of the Analytical Laboratory of Schott & Gen., Mainz, Germany) and of his wife Irene (1933–2008) and of Erich Robens.

Friedrichsdorf, Germany   Erich Robens
Middlesbrough, UK   Shanath Amarasiri A. Jayaweera
Messstetten, Germany   Susanne Kiefer

Contents

Authors

Ing. Erich Robens, Schlesiertstr. 5, 61381 Friedrichsdorf, Germany. erich.robens@t-online.de

*13.5.1924 at Mainz, Germany. Married with the chemist Gertrud Weingartz; 2 adopted children. After three years war service in Ukraine and Poland, studies of physics at the Johannes Gutenberg-Universität, Mainz, with unsuccessful result. Assistant at the Lycée Français at Mainz, than 27 years work as a physical engineer in the department Interface Science of Battelle-Institut e.V., Frankfurt am Main. Member of the board of directors of the Bausparkasse Mainz AG, and of the Mainzer Haus Vertriebs GmbH. Member of the German standardisation committee (DIN) and of the expert council Particulate Materials of the European Communities Bureau of Reference, Brussels, Belgium. Chairman of the working group Surface Measurement of DIN. Guest scientist at the Institute for Inorganic and Analytical Chemistry of the Johannes Gutenberg-Universität, Mainz; and of the Maria Curie-Skłodowska-Universytet, Lublin, Poland. Data protector for a youth's drug abuse society, Frankfurt am Main. Consultant of the Deutsche Gesellschaft für Luft- & Raumfahrt for a Mars experiment and NASA principal investigator of a research program Lunar samples. Chairman of the International Conferences of Vacuum Microbalance Techniques. In 2004 appointment as honorary professor of Maria Curie-Skłodowska-Universytet, Lublin, Poland. More than 500 scientific publications. 8 books, 60 patent applications.

Dr. Shanath Amarasiri A. Jayaweera, 42 Tanton road, Stokesley, Middlesbrough TS9 5HR, UK. saajayaweera@yahoo.co.uk

*20 September 1938. Nationality—dual: British (by naturalisation), Sri Lankan (by birth).

Academic Qualifications: BSc (First class Honours) University of Ceylon, MSc/PhD in chemistry University of London.

Professional Qualifications: Member of the Royal Society of Chemistry, Fellow of the Royal Society of Chemistry.

Employments in Sri Lanka: S Thomas' College, Gurutalawa, University of Ceylon.

Employments in UK: Plymouth College of Technology, Plymouth Polytechnic, Teesside Polytechnic/University, Robert Montefiore Secondary School, London, College of St Mark & St John, Middlesbrough College.

Examiner of theses: Curtin University of Technology, Perth, Australia, Université Mohammed V, Rabat, Morocco, University of Newcastle.

Academic awards and ~80 research publications.

Susanne Kiefer, Wacholderweg 32, 72469 Messstetten, Germany. skiefer@posteo.de

*1967. Studies of history and politics at the University of Konstanz, Germany; with qualification M.A. Since 1993 work on the history of precision balance production in the region of the Suabian Alb. 2002–2008 curator of the Matthäus Hahn-Museum at Albstadt-Onstmettingen. Since 2008 secretary of Martin Sauter (managing director) and marketing assistant of Kern & Sohn GmbH, Balingen, Germany.

Main publications: „Mechanische Kunstwerke Philipp Matthäus Hahns: Die Neigungswaage und die Allgemeine Hydrostatische Waage", 2002. „Lust zu allen Künsten: Philipp Gottfried Schaudt von Onstmettingen: Schulmeister, Uhrmacher und Mechanicus", 2003.

Chapter 1
Mass and Gravity

Besides length and volume, the mass, earlier designated as weight, is an important parameter characterising a body. It was recognised that there is proportionality between the ponderosity of a body, the resistance against moving it when it is at rest and stopping it when it is in motion. Nevertheless, mass as a physical quantity is a relatively new concept, defined mathematically first by Isaac Newton in 1697 [1, 2]. Since that time we distinguish between mass as an extensive property of a body and weight as a force exerted by the gravitational field of the earth on that body.

Xenia 165. Die Möglichkeit

Liegt der Irrtum nur erst,
wie ein Grundstein unten im Boden,
Immer baut man darauf,
nimmermehr kommt er an Tag.

Goethe/Schiller: Xenien [3]

The possibility

If a mistake is made at the beginning
And used in subsequent work
It will never be revealed later
Like building over a foundation stone.

Translation: S.A.A. Jayaweera

This xenia (host gift) most probably was written by Goethe against Newton. Goethe detested Newton's method of physical investigation. In his opinion Newton, therefore, made many mistakes.

In the International System of Units (SI) mass is one of the basic quantities. Today, due to the Theory of Relativity and Quantum Mechanics the concept mass becomes again fundamentally unclear, but not in conventional practice.

1.1 The Concept of Mass

Today, the aim of weighing is clearly the determination of the mass of a body (Table 1.12) or of a material sample; but earlier it was determination of its value [4]. In earlier times there were weight units with identical designation but different masses for different materials [5]. We still have some identical designations of weights and coins. Of course, the difference between weight and value of body was recognised in

E. Robens et al., *Balances*, DOI 10.1007/978-3-642-36447-1_1,
© Springer-Verlag Berlin Heidelberg 2014

earlier times. In the Old Kingdom the Egyptians had already distinguished between concrete 'weight' from abstract 'value', using different vocabulary for each. The unit of weight was Deben whereas of value was Sna (vocalised as *shena*), perhaps written during Dynasty 19 as *sniw*. We have no feeling of sense for mass but have for weight which is the force of a body in the Earth's gravitational field.

The word mass is derived from the Latin *massa* which means lumps of a homogeneous material without specific form, but also a conglomerate of bodies [6]. The idea of the 'quantitas materiae' and its conservation goes back to Greek philosophers. Anaxagoras (499–428 BC) argued "nothing comes into being or ceases to exist" [7]. Democritus (460–371 BC) said: "from nothing comes nothing and can only become nothing".

In the middle ages some alchemists doubted this concept and hoped to create noble metals from earthy matter, accompanied by weight increase. Furthermore, on religious reports like the creation of a woman from one rib of the man [8] or of the transubstantiation of bread and wine into Jesus' flesh and blood, difficulties arose in understanding mass as an invariable quantity [9].

Antoine Lavoisier (1743–1794) deduced a general conservation of mass from the weight ratios in chemical reactions [10]. John Dalton (1766–1844) established the atomistic theory in chemistry and made a first table of relative atomic masses [11]. The principle of conservation of mass was formulated finally by Isaac Newton (1643–1717) [1]. In classical physics mass is defined as an inertial property of matter. Nevertheless, what mass really is had not yet been explained reliantly. On the contrary, the concept mass is even today subject to continual changes. Today mass is a basic unit of our international system of units.

In the Newtonian mechanics, two types of mass were distinguished: inert mass and gravitational mass. Bondi [12] distinguishes three types of mass: inert mass, active gravitational mass and passive gravitational mass. A universal proportionality between those different types of mass was assumed and confirmed by many so-called Eötvös experiments (see Chap. 4). For Albert Einstein (1879–1955) that 'Weak Equivalence Principle' gave the basis for his theory of relativity. Furthermore, mass can be regarded as a special type of energy [13].

It is assumed that in the Universe most of mass is connected with invisible 'dark matter' and furthermore not obvious sources of 'dark energy' exist. On Earth we have to do in practice with the visible matter. This includes also transparent substances like gases, glass or water which can be made visible by electromagnetic radiation of certain frequencies.

1.2 Mass and Weight

The hypothesis of Leucippus (500–450 BC) and Democritus (460–371 BC) held everything to be composed of atoms, which are physically, but not geometrically, indivisible; that between atoms lays empty space; that atoms are indestructible; have always been, and always will be, in motion; that there are an infinite number of

Fig. 1.1   Freely adapted from Drew. LHC = Large Hadron Collider, CERN, Genève, Switzerland, FAIR = Facility for Antiproton Research in Europe, Darmstadt, Germany

atoms, and kinds of atoms, which differ in shape, size, and temperature. Of the weight of atoms, Democritus said "The more any indivisible exceeds, the heavier it is." But their exact position on weight of atoms is disputed [14, 15].

In daily life, in trade and commerce as well as in many physical and chemical processes mass is an important parameter [16, 17]. For more than 5000 years balances have been used and for centuries we find balances in every factory and in households. Nevertheless the parameter "mass" is relatively new and still under discussion. When weighing we measure not the mass but the force exerted on the sample by the gravitational field (Fig. 1.1). That force is directly proportional to the mass of the sample. On Earth, however, gravitational acceleration depends on the geographical location of the balance. On account of the elliptic shape of Earth, in equatorial regions its radius is larger. Indeed the shape of the Earth differs significantly from a sphere or ellipsoid. On mountains the distance to the centre is greater and in addition the mass distribution within the Earth is not uniform. In order to formulate the laws of motion it is desirable to have mass as an invariant parameter of a material.

Mass is a measure of the amount of material in an object, being directly related to the number and type of atoms present in the object. We have no sense for the quantity mass; however we do for the force of an accelerated body or of a body in rest within a gravitational field. Newton's second law [1] states: A body of mass m subject to a net force F undergoes an acceleration a that has the same direction as the force and a magnitude that is directly proportional to the force and inversely proportional to the mass. "Mutationem motus proportionalem esse vi motrici impressae, et fieri secundum lineam rectam qua vis illa imprimitur."

$$\vec{F} = m\vec{a} \tag{1.1}$$

The corresponding SI units are Newton: $N = kg\,m/s^2$, kg, and m/s^2, respectively.

Weight, W, is defined as the force acting on a body with mass m under gravitational acceleration g

$$W = mg \tag{1.2}$$

On Earth the acceleration is given by the gravitational field due to the large mass of the Earth. Correspondingly the mass will be determined only indirectly by measuring the weight which is a force W due to the gravitational field:

$$m = \frac{W}{g} \tag{1.3}$$

where m denotes the mass to be determined. g is the acceleration due to gravity which varies, depending on the geographical location, between about 9.77–9.83 $m\,s^{-2}$. The standard acceleration of gravity (standard acceleration of free fall) g_n is defined to be 9.806 65 $m\,s^{-2}$. Thus whilst mass is an unchanging quantity in classical mechanics, weight changes with gravitational acceleration. On the moon and the other planets that value is quite different depending on the mass of that celestial body (Table 1.2).

1.3 Gravity

Equation (1.3) is a special case of Newton's universal law of gravity: Two bodies of masses m_1, m_2 attract each other and the force between F_g is given by:

$$F_g = G\frac{m_1 m_2}{d^2} \tag{1.4}$$

where d is the distance between the centres of gravity of the masses and G is the gravitational constant. Obviously

$$g_n = \frac{Gm_1}{d^2} \tag{1.5}$$

Newton started his investigation of the laws of gravity when observing an apple falling from a tree. William Stukeley [18] reported that in 1726 Newton told him that story which happened in his garden in the year 1660. Newton was fascinated from the fact that the apples fall always in direction to the Earth centre.

In comparison to other known forces of nature gravity is a weak force. Only the large mass of the Earth gives us the impression of its essential effect. Therefore, it is difficult to determine the gravitational constant [19]. For its measurement extremely well defined and homogeneous test samples are required and screened from disturbing strong forces (electric, magnetic) and influences from the environment. It's no wonder that the value of the gravitational constant is less accurate than all others; its uncertainty is 1×10^{-4}. That means that the accuracy of all formula including G is limited to the per mille range. Many calculations in geological, meteorological, astronomical as well as of space operations are burdened with a basic uncertainty of per mille. With regard to that uncertainty until now it was not possible to check whether G is really a universal constant and to detect violations of the Equivalence Principle.

Methods to determine the value of G include free fall experiments from a tower or along an inclined plane. Better results are obtained by the measurement of the attraction of highly homogeneous test samples of several kilograms by means of a torsion pendulum (Chap. 4, Sect. 4.5.2). Similar experiments can be carried out likewise with a conventional beam balance. New techniques to measure G using atom interferometry are currently under development. This method is also being developed to measure the local acceleration due to gravity g. Recent experiments combine two vertically separated atomic clouds forming an atom-interferometer-gravity-gradiometer that measures the change in the gravity gradient when a well characterised source mass is displaced [20]. In Table 1.1 recent determinations of the gravitational constant and the CODATA recommendation for its value are listed [21].

1.4 Gravity and Motion

Since the time of the Greek philosopher Aristotle [31] in the 4th century BC, there have been many attempts to understand and explain attraction of material objects. Aristotle hypothesised that there was no effect without a cause and that each motion was caused by a force. He believed that everything tries to move towards its proper place in the crystalline spheres of the heavens, and that physical bodies fell towards the centre of the Earth in proportion to their mass [32].

The universal validity of free fall [33] was stated by the Byzantine Johannes Philoponos (485–555) [6] and in 1553 by Giambattista Bendetti. Galileo Galilei (1564–1642) rejected Aristotle's assumption that the velocity of bodies in free fall is proportional to its mass and quantified the theorem that all falling bodies descend with equal velocity "if one could totally remove the resistance of the medium, all substances would fall at equal speeds". He demonstrated this by dropping balls from the Leaning Tower of Pisa. Suitable measurements in this way however had not been possible because no clocks existed at his time with the required precision. He made measurements with balls of different mass rolling down inclined planes and pendulum experiments. In this way he reduced the velocity so that he was able to make reasonable measurements of time even by means of the inaccurate clocks. He determined the mathematical law for acceleration: The total distance covered, starting from rest, is proportional to the square of the time [34]. This law holds correctly only in vacuum because under atmospheric conditions air resistance obstructs the motion and this depends on the body's volume. That effect he discussed in mental

Table 1.1 Measurements 1996–2006 of the Newtonian constant of gravitation and 2006 CODATA recommend value

Year	Investigator	Method	G $\mathrm{m^3\,kg^{-1}\,s^{-2}}$	u_r	Ref.
1996	Karagioz, Izmailov	Fibre torsion balance, dynamic mode	6.6729(5)	7.5×10^{-5}	[22, 23]
1997	Bagley, Luther	Fibre torsion balance, dynamic mode	6.6740(7)	1.0×10^{-7}	[24]
2000 2002	Gundlach, Merkowitz	Fibre torsion balance, dynamic compensation	6.674255(92)	1.4×10^{-5}	[25]
2001	Quinn, Speake, Richman, Davis, Picard	Strip torsion balance, compensation mode, static deflection	6.67559(27)	4.0×10^{-5}	[26]
2002	Kleinevoß	Suspended body, displacement	6.67422(98)	1.5×10^{-4}	[27]
2003	Armstrong, Fitzgerald	Strip torsion balance, compensation mode	6.67387(27)	4.0×10^{-5}	[28]
2005	Hu, Guo, Luo	Fibre torsion balance, dynamic mode	6.6723(9)	1.3×10^{-4}	[29]
2006	Schlamminger, Holzschuh, Kündig, Nolting, Pixley, Schurr, Staumann	Stationary body, weight change	6.67425(12)	1.9×10^{-5}	[30]
2006	CODATA	CODATA-06 adjustment	6.67428(67)	1.0×10^{-4}	[20]

Newtonian constant of gravitation $G = 6.67428 \times 10^{-11}$ $\mathrm{m^3\,kg^{-1}\,s^{-2}}$, Standard uncertainty 0.00067×10^{-11} $\mathrm{m^3\,kg^{-1}\,s^{-2}}$, Relative standard uncertainty $u_r\,1.0 \times 10^{-4}$, Concise form $6.67428(67) \times 10^{-11}$ $\mathrm{m^3\,kg^{-1}\,s^{-2}}$

experiments of bodies falling from a tower [35]. Dropping of two different bodies in a gravitational field, the "weak equivalence principle" states that the bodies fall with the same acceleration; this is usually termed as the universality of free fall.

In a document "Theoremata circa centrum gravitatis solidorum" [36] Galileo was the first to discuss in 1585–1586 first results on the mass of solid bodies. In his manuscript "La Bilancetta" [37] he improved Archimedes's method of density determination and designed a hydrostatic balance. Galileo also put forward the basic principle of relativity, that the laws of physics are the same in any system that is moving at a constant speed in a straight line, regardless of its particular speed or direction. Hence, there is no absolute motion or absolute rest. This principle provided the basic framework [33] for Newton's (1643–1727) laws of motion [1] and is the infinite speed of light approximation to Einstein's special theory of relativity.

Galileo's Principle of Inertia stated: "A body moving on a level surface will continue in the same direction at constant speed unless disturbed." This principle was incorporated into Newton's 1st law of motion. Lex I: Corpus omne perseverare in statu suo quiescendi vel movendi uniformiter in directum, nisi quatenus a viribus impressis cogitur statum illum mutare: An object at rest or travelling in uniform motion will remain at rest or travelling in uniform motion unless acted upon by a net force.

Additionally measurements within the gravitational field force of a moving body demonstrate Newton's 2nd law of motion or law of inertia: Lex II: Mutationem motus proportionalem esse vi motrici impressae, et fieri secundum lineam rectam qua vis illa imprimitur: The rate of change of momentum of a body is equal to the resultant force acting on the body and is in the same direction.

Newton's second law as originally stated in terms of momentum p is: An applied force F_I is equal to the rate of change of momentum.

$$F_I = \frac{\mathrm{d}p}{\mathrm{d}t} \tag{1.6}$$

where t stands for time. If the mass m of the object is constant the differentiation of the momentum becomes

$$F_I = ma = m\frac{\mathrm{d}v}{\mathrm{d}t} \tag{1.7}$$

where a is the acceleration and v the velocity of the object.

According to Newton's classical mechanics the gravitational force F_g results from a field between two bodies and allows the determination of the mass of a sample in the gravitational field of a body which is large in comparison with the sample, for example, the Earth. This mass is called 'inert mass'. The force F_I according to Newton's law results from the motion of a single sample. The gravitational constant has been defined to a value that the mass of a sample is the same if determined with both methods. The 'weak equivalence principle' states that the property of a body called 'mass' is proportional to the force 'weight'. "This quantity that I mean hereafter under the name of ... mass ... is known by the weight ... for it is proportional to the weight as I have found by experiments on pendulums, very accurately made ..."

Within a closed room with opaque walls we cannot decide without additional information whether a gravitational field is due to mass attraction or due to acceleration. So as the basis of the General Theory of Relativity Einstein's Equivalence principle expresses the identity of the gravitational field with that generated by motion [38]: the gravitational 'force' as experienced locally while standing on a massive body (such as the Earth) is actually the same as the pseudo-force experienced by an observer in a non-inertial (accelerated) frame of reference [39].

Günter Sauerbrey demonstrated that 'General Equivalence Principle' experimentally [40, 41]. With two quartz resonators he measured the mass of a deposited thin film using the harmonic inertial field created on the surface of these resonators. With a torsion balance he measured the mass of an equivalent deposited thin film using the gravitational field. In the first case he used a time related parameter (frequency) to evaluate the mass of the deposition. In the second case he used a space related parameter to evaluate the mass. Mass, space and time are closely related as revealed by Einstein's General Theory of Relativity.

Newtonian mechanics turned out to be a special case within the General Theory of Relativity, Newton's law of universal gravitation provides an excellent approximation in many cases and most non-relativistic gravitational calculations are based on it. Up to now there are no hints from experimental investigations for a 'Fifth forth' or of weakening of the gravitational force when passing through matter [42] which could necessitate corrections of Newton's laws or of the Theory of Relativity.

1.5 Gravitational Field

1.5.1 Earth Gravity

The magnitude of the gravitational field near the surface of the Earth is about $9.81 \, \text{m s}^{-2}$. That means that ignoring air resistance, the speed of an object falling freely near the Earth's surface increases every second by about 9.81 metres per second. The standard value is defined as $g_n = 9.80665 \, \text{m s}^{-2}$ or $9.80665 \, \text{N kg}^{-1}$. The vector, gravity, is the force that acts on a body at rest near the surface towards the centre of the Earth. However, the strength of Earth's gravity varies with latitude, altitude, local topography and geology.

Due to its rotation, the Earth is shaped as an ellipsoid with a smaller distance from surface to the centre at the poles. The mean diameter of the equator is about 42 km larger than the rotational axis through the pole. Furthermore, landmass and oceans are distributed non- uniformly and mountains and valleys of the sea bed are irregular. In addition, the density of the material varies. So the Earth is described more accurately as an irregular geoid rather than as a sphere. According to C.F. Gauss, the geoid is considered as the mathematical surface of the Earth, as opposed to the visible topographical surface. Since 2009 the European research satellite Goce measured gravity variations of Earth's surface by means of a gradiometer. This satellite orbits at a height of 265 km. It contains six weights each of 300 g of a platinum/rhodium alloy suspended in electric fields. The acceleration by gravity is measured. In this way the shape of the Earth can be determined to an accuracy of 2 cm. The deepest dent of the geod with -110 m was detected in the Pacific Ocean and the highest bump with $+80$ m below Iceland, the North Atlantic and at New Guinea.

Gravity provides centripetal force which is weakened by centrifugal force due to Earth's rotation. The maximum of the centrifugal force is at the equator. At the poles the centrifugal force is zero. That small centrifugal force is inseparably superimposed on the attraction and this effect is usually included in the local value of the gravitational force. In combination, the equatorial bulge and the effects of centrifugal force mean that sea-level gravitational acceleration increases from about $9.780 \, \text{m s}^{-2}$ at the equator to about $9.832 \, \text{m s}^{-2}$ at the poles, so an object will weigh about 0.5 % more at the poles than at the equator.

The net force exerted on an object by the Earth is called apparent or effective gravity and may be influenced by other forces. For most purposes, the Earth's gravitational field may be considered invariable in time. However, attraction of the Sun and moon superimpose periodically the values because of their changing distance from the Earth. These attractions act also indirectly by slightly deforming the Earth and shifting the waters of the oceans, so that the attracting terrestrial masses themselves are modified. The effects are well within the measuring accuracy of modern gravimeters but negligible for practical purposes.

Gravity decreases with altitude, since greater altitude means greater distance from the Earth's centre. Approximately we can use:

$$g_h = g_n \left(\frac{r_e}{r_e + h} \right)^2 \tag{1.8}$$

where g_h is the gravity at height h above sea level, r_e is the Earth's mean radius and g_n the standard gravity. An increase from sea level to the top of Mount Everest (8.8 km) causes a decrease of weight of only about 0.28 %. At an altitude of 400 kilometres, equivalent to a typical orbit of the Space Shuttle, gravity is still nearly 90 % as strong as at the Earth's surface [43]. Nevertheless, when calibrating sensitive balances, this effect of the altitude must be considered (see Sect. 3.4.3 in Chap. 3).

Satellite methods have enormously improved knowledge of the gravity field. The Gravity Information System of the Physikalisch-Technische Bundesanstalt, Braunschweig, Germany provides gravity information for nearly every location above sea level without claiming highest scientific accuracy [44]. The information portal is designed for interested users from physics and metrology whose mechanical applications are affected by the gravitational field of the Earth. The data are of importance for calibration of balances.

1.5.2 Planet Gravity

The magnitude of the gravitational field near the surface of the Sun and the planets depends of the composition and diameter of the bodies. A survey of relative weights and of gravity on the Earth, other planets and the moon is given in Table 1.2.

1.6 Planetary Motion

Johannes Kepler (1571–1630) working with data collected by Tycho de Brahe (1546–1601) [45] without the aid of a telescope developed three laws which described the motion of the planets across the sky [46, 47]. Kepler's laws were derived for orbits around the Sun, but they apply to satellite orbits as well.

- Kepler's First Law (law of orbits): All planets move in elliptical orbits, with the Sun at one focus.
- Kepler's Second Law (law of areas): A line that connects the planet to the Sun sweeps out equal areas in equal times.

Table 1.2 Relative weights and of gravity on the Earth, other planets and the moon

Body	Multiple of Earth gravity	Gravity m/s^2	Body	Multiple of Earth gravity	Gravity m/s^2
Sun	27.90	274.1	Jupiter	2.640	25.93
Mercury	0.3770	3.703	Saturn	1.139	11.19
Venus	0.9032	8.872	Uranus	0.917	9.01
Earth	1 (by definition)	9.8226[a]	Neptune	1.148	11.28
Moon	0.1655	1.625	Pluto	0.0621	0.610
Mars	0.3895	3.728			

[a]This value excludes the adjustment for centrifugal force due to Earth's rotation and is therefore greater than the 9.80665 m/s^2 value of standard gravity

- Kepler's Third Law (law of periods) [48]: The square of the period t_p of any planet is proportional to the cube of the semi-major axis d of its orbit. The third Kepler's law is valid exactly for the period of two bodies.

$$\left(\frac{t_{p1}}{t_{p2}}\right)^2 = \left(\frac{d_1}{d_2}\right)^3 \tag{1.9}$$

Kepler nominated a as mean distance of Earth to the Sun i.e. mean of perigee and apogee. Combined with the law of gravity Kepler's Third Law for two masses m_1, m_2 becomes:

$$t_p^2 = \frac{4\pi^2 d^3}{G(m_1 + m_2)} \tag{1.10a}$$

$$t_p^2 \approx \frac{4\pi^2 d^3}{Gm_1} \quad \text{if } m_1 \gg m_2 \tag{1.10b}$$

These equations allow the determination of the total mass of a binary star system from distance d and period t_p. The approximation can be applied e.g. in our Solar System. Although Kepler's laws are valid exactly only for the two-body problem they are a good approximation for describing planet movements. Deviations from that law are due to gravitational influences of other planets, fluctuations of the Sun movement, imperfect spherical shape of the revolving planets and relativistic effects. Such effects on the elliptical orbit are summarised as irregularities of the orbit. Indeed the computational orbit can be regarded as an attractor only for the irregular orbital movement of planets.

Taking into account the different masses m_{p1} and m_{p2} of two planets as a three-body problem the Third Kepler Law writes:

$$\left(\frac{t_{p_1}}{t_{p_2}}\right)^2 = \left(\frac{d_1}{d_2}\right)^3 \frac{m_S + m_{P2}}{m_S + m_{P1}} \tag{1.11}$$

This is of importance only in case the masses m_{P1}, m_{P2} of both planets differ widely and the mass of the central star m_S is similar to the mass of one planet.

Using Eq. (1.10b), for the rotation of the Earth around the Sun, the sidereal period
is given by

$$t_u \approx \sqrt{\frac{(1.5 \cdot 10^{11})^3 4\pi^2}{6.67 \cdot 10^{-11} \cdot 2 \cdot 10^{30}}} \ [s] = 31602834 \ [s] = 365 \ [d] \qquad (1.12)$$

Mass determination of astronomical objects is made by observation of the relative
motion under the influence of a neighbouring object and application of Kepler's
Laws.

Newton's theory enjoyed its greatest success when it was used to predict the
existence of Neptune based on motions of Uranus that could not be accounted by
the actions of the other planets. By the end of the 19th century, it was stated that the
orbit of Mercury could not be accounted for entirely under Newton's theory, and this
was a first proof for the Theory of Relativity. Nevertheless, in most cases Kepler's
Laws can be applied as a good approximation.

1.7 Expansion of the Universe

Gravity is responsible also for movement of galaxies, which—on account of large
distances—can be regarded as point shaped objects. However, these motions cannot
be explained quantitatively if no additional sources of gravity are introduced. Because the gravitational laws are regarded as valid within the whole Universe it was
necessary to assume an invisible matter. Recent observations confirm such 'dark
matter'.

According to Einstein's prediction the Universe expands. To explain this phenomenon the concept of 'dark energy' was introduced which works opposite to
gravity.

1.8 Mass and Theory of Relativity

In 1881 Michelson and Morley [49] detected the independence of the speed of light
from any motion of the light source or of its receiver. That was the basis for Einstein's Theory of Relativity [38]. As a consequence of that theory mass lost its feature as an independent parameter of a body on account of its dependence on its
velocity v [13]:

$$m = m_{rest} \frac{1}{\sqrt{1 - v^2/c_0^2}} \qquad (1.13)$$

where m_{rest} stands for the mass of the body in rest, c_0 is the velocity of electromagnetic radiation in a vacuum. At ordinary velocities of terrestrial objects, v is very
small compared with c, and $m \approx m_0$. However, in an electron beam (for example in
an electron microscope), where v is much higher, the difference is significant, and

the above relativistic correction is applicable: The mass of a particle increases when moving with a velocity near the speed of light.

According to the Theory of Relativity the velocity of electromagnetic radiation in a vacuum is a maximum and a universal constant. Also any matter cannot travel faster. A recent observation of neutrino's moving at a velocity greater than that of light has been found to be a measuring error and all the physicists are happy that Einstein is correct.

A fundamental principle is conversation of mass and energy which is believed to hold for the whole Universe. In 1905 Einstein formulated his famous equation [38, 50]:

$$E = m_{rest} c_0^2 \qquad (1.14)$$

where E denotes the energy, m the mass and c_0 the velocity of light in a vacuum. So mass can be regarded as a special type of energy. When calculating the energy loss of a body emitting electromagnetic radiation, its decrease of mass should be considered. It should be noted that already much earlier the correlation between mass, energy and velocity of light had been stated and that Henri Ponicaré in 1900 had already formulated that equation. However Einstein developed the Special Theory of Relativity in which that equation was a logical constituent. The equation has been verified recently by comparing the mass of the stable isotope ^{28}Si with that of the unstable isotope ^{29}Si as well as of ^{32}S and ^{33}S and measuring the energy of the γ-radiation set free during the decay of the unstable atoms. Molecules of both isotopes were captured in a Penning trap and stimulated to rotation and the difference of the rotational frequencies was measured.

General Relativity is a geometric model of gravitation, a more accurate description of gravity in terms of the geometry of a curved space-time [51]. According to the General Relativity Theory, gravity is not a force acting on material particles. Instead it is identified as a curvature in space-time geometry [52]. In the absence of forces particles, including the photons of light, travel in the most straight possible way in curved space-time. This path is called 'geodesic' (Fig. 1.2a). In the absence of gravity, space-time is flat and geodesics are straight lines travelling at constant velocity. Near a star the geodesics follow the curvature of space-time and define the shortest natural paths. The presence of a large mass deforms space-time in such a way that the paths taken by particles bend towards the mass. That curved path is longer than a straight line and photons need more time to pass in dependence of the distance to the mass. This has an effect of a collecting lens (Fig. 1.2). Alternative theories of gravitation imply deviations of these geodesics.

Einstein developed the Theory of Relativity on the basis of a Gedankenexperiment: In an elevator without view to outside and without knowledge of the environment it cannot be decided whether attraction from a large mass or acceleration of the elevator is the reason of the observed gravity. The Theory of Relativity embodies the Weak (Galilean-Newtonian) Equivalence Principle:

1. Local bodies fall identically, because
2. Gravitational mass (Eq. (1.4)) is indistinguishable from inertial mass (Eq. (1.1)),
3. Regardless of composition,

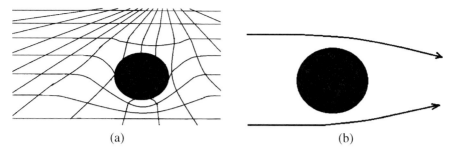

Fig. 1.2 (**a**) Schematic visualisation of geodesics in 4-dimensional space-time coordinates. In the absence of material objects space-time is flat and geodesics are straight lines. By a material accumulation geodesic will be buckled. A particle travelling along a geodesic will be deflected or even collide. (**b**) Deflection of a ray of light by a celestial body

4. Regardless of geometry (internal structure),
5. Regardless of amount of mass.

Inertial reference frames (coordinate systems) have constant relative velocity in a flat space-time manifold. Accelerating frames with consistent definitions of energy and momentum (or mass and angular momentum) require non-zero space-time curvature. Local space-time must have a unique curvature. Local test masses exhibiting non-parallel geodesic trajectories require simultaneous different values of local space-time curvature. Any paired (sets of) test masses violating the Equivalence Principle empirically falsify metric theories of gravitation at their founding postulate.

The Strong or Einstein Equivalence Principle states that all of the laws of physics (not just the laws of gravity) are the same in all small regions of space, regardless of their relative motion or acceleration.

6. Non-rotating free fall is locally indistinguishable from uniform motion absent gravitation. Linear acceleration relative to an inertial frame in Special Relativity is locally identical to being at rest in a gravitational field. A local reference frame always exists in which gravitation vanishes.
7. Local Lorentz invariance (absolute velocity does not exist) [53] and position invariance. All local free fall frames are equivalent.
8. The Strong Equivalence Principle embraces all laws of nature; all reference frames accelerated or not, in a gravitational field or not, rotating or not, anywhere at any time.

Gravity is still a central question of physics which may be expressed as "What gives an object mass (or inertia) so that it requires an effort to start it moving, and exactly the same effort to restore it to its original state?" [54]. A clarification of the existent theories could be found by search of violations of the Equivalence Principle and this requires extremely accurate measurement of the gravitational constant G in Eq. (1.4). Those which had been made are shown in Table 1.3.

Einstein predicted that ripples of low frequency in the space-time criss-cross the Universe. In order to examine the Universe by measuring intensity, properties and

Table 1.3 Test of the equivalence principle [55]

Year	Investigator	Method	Accuracy	Reference
~500	Philoponus	Drop Tower		[56]
1585	Stevin	Drop Tower	5×10^{-2}	[57, 58]
1600	Galileo	Balls on incline, pendulum	10^{-2}	[34, 59]
1680	Newton	Pendulum	10^{-3}	[1, 2]
1832	Bessel	Pendulum	2×10^{-5}	[60]
1910	Southerns	Pendulum	5×10^{-6}	[61]
1918	Zeeman	Torsion balance	3×10^{-8}	[62, 63]
1922	Eötvös	Torsion balance	5×10^{-9}	[64]
1923	Potter	Pendulum	3×10^{-6}	[65, 66]
1935	Renner	Torsion balance	2×10^{-9}	[67]
1964	Dicke, Roll, Krotkov	Torsion balance	3×10^{-11}	[68]
1972	Braginsky, Panov	Torsion balance	10^{-12}	[69, 70]
1976	Shapiro, Counselman, King	Lunar laser ranging	10^{-12}	[71]
1981	Keiser, Faller	Fluid support	4×10^{-11}	[72]
1987	Niebauer, McHugh, Faller	Drop tower	10^{-10}	[73]
1989	Heckel, Adelberger, Stubbs, Su, Swanson, Smith	Torsion balance	10^{-11}	[74]
1990	Adelberger, Stubbs, Heckel, Su, Swanson, Smith, Gundlach	Torsion balance	10^{-12}	[75]
1999	Baeßler, Heckel, Adelberger, Gundlach, Schmidt, Swanson	Torsion balance	5×10^{-13}	[76]
2008	Schlamminger, Choi, Wagner, Gundlach, Adelberger	Rotating cryogenic torsion balance	3×10^{-14}	[77]
2010	MiniSTEP	Earth orbit	10^{-17}	[78]

direction of these gravitational waves, and likewise the correctness of the General Theory of Relativity, can be checked. The concept of falling under the influence of gravity alone follows a geodesic in space-time is at the foundation of General Relativity, our best model of gravitation. Such an experiment is planned by the LISA Pathfinder mission (Laser Interferometry Space Antenna) which was scheduled for launch at the end of 2009. The LISA Pathfinder concept is to prove geodesic motion of two test-masses in a nearly perfect gravitational free-fall through laser interferometry. As gravity is a very weak interaction an extremely low level of nongravitational influences and parasitic accelerations is required. The test masses are cubes with 2 kg mass, 5 cm in diameter, made of a gold-platinum alloy and are surrounded with a gap of 4 mm to the electrode housing and thus can float free in the spacecraft and follow two parallel geodesics. The spacecraft is controlled by a micro-propulsion system and follows the test-masses with nanometer resolution. Violation of geodesic motion manifests itself as a relative acceleration of test-masses as measured by the laser interferometer with picometer distance resolution.

The graviton is a hypothetical elementary particle that mediates the force of gravity in the framework of quantum field theory. If it exists, the graviton must be massless because the gravitational force has unlimited range. Gravitons are postulated because of the great success of quantum field theory at modelling the behaviour of all other known forces of nature as being mediated by elementary particles. A quantum mass sometimes is postulated but never found.

Gravitation is weak but infinite ranging. Therefore aggregation of particles occurs in the huge and nearly empty space of the Universe. Then gravity + adhering forces may exceed dispersive forces and asteroids, planets and stars grow. By influence of gravity large bodies take a spherical shape, denser material concentrating in the interior.

Above a certain mass a star collapses to give a neutron star with mass of 1.35 to $2.1 \times m_{sun}$ and with diameter of 10 to 20 km [79]. In black holes the masses between 10 to $10^9 \times m_{sun}$ are concentrated. The respective diameter of such a spherical object is about 30 km for a mass of $10 \times m_{sun}$. Around a black hole there is an undetectable surface, called event horizon, which marks the point of no return. At the event horizon of a black hole, the deformation of space-time becomes so strong that all geodesic paths lead towards the black hole and not away from it. All the light is adsorbed, nothing reflected. From the interior nothing can escape, even not photons on account of its mass of motion. Karl Schwarzschild [80, 81] and independently Johannes Droste [82] gave a solution for the gravitational field of a point mass and a spherical mass. This solution has a singularity, the Schwarzschild radius, at which some of the terms in the Einstein equations become infinite.

Hess and Greiner [83, 84] made an extension of the theory of General Relativity based on pseudo-complex space-time coordinates. They constructed a pseudo-complex Schwarzschild solution, which does not suffer any more by a singularity. The solution indicates a minimal radius for a heavy mass object. As a result particles including photons may be repelled by an anti-gravitational effect when approaching closely another mass and then can leave the hole which is dark but no longer black.

1.9 Particle/Wave Duality

At the end of the 19th century the atomic theory was well established, namely, that much of nature was made of particles [85]. It was assumed that matter consists of particulate objects or atoms, and that electrons carry electric charge. Already Isaac Newton (1643–1727) argued that on account of its linear transmission light is composed of corpuscles and he could explain many optical effects by this assumption [86]. On the other hand Christiaan Huygens (1629–1695) proposed a wave theory of light [87] and later on by means of diffraction experiments the wave-like nature of light was clearly demonstrated. In the late 1800s, James Clerk Maxwell explained light as the propagation of electromagnetic waves according to the Maxwell equations [88]. By the turn of the 20th century, a duality of light was accepted.

Fig. 1.3 Light beam consisting of travelling wave packets (photons). Within the packet the oscillations have the frequency of the electromagnetic radiation. In vacuum the velocity of the photon
correspond to the speed of light as a maximum. Within the packet the velocity of oscillations may
be different so that the oscillations seem to move forwards or backwards, respectively. Rest mass
of a photon is zero

By the photoelectric effect electrons are emitted from matter after absorption
of energy from electromagnetic radiation such as X-rays or visible light. It was
observed that the energy of the emitted electrons did not depend on the intensity of
the incident radiation and this could not be explained by Maxwell's wave theory. In
1905 Albert Einstein described mathematically how the photoelectric effect could be
described by absorption of quanta of light (now called photons) [89]. The particle-
like behaviour of light was further confirmed with the discovery of the Compton
scattering in 1923 [90]. The Compton Effect is the decrease in energy and, hence,
increase in wavelength of an X-ray or gamma ray photon, when it interacts with
matter. Because of the change in photon energy, it is an inelastic scattering process
and it demonstrates that light cannot be explained purely as a wave phenomenon.

A ray of light may be regarded consisting of photons which are wave packets
travelling in vacuum with speed of light $c_0 = 299,792.458$ m s^{-1} (Fig. 1.3). The
velocity of the wave within the packet may be different. The light beam can be
affected by gravity. Photons, however, have no mass. Otherwise propagation with
speed of light would be impossible.

In 1924, Louis-Victor Pierre Raymond de Broglie claimed that all matter, not just
light, has a wave-like nature [91]. He related wavelength λ and momentum p:

$$\lambda = \frac{h}{p} \tag{1.15}$$

with the Planck constant $h = 6.626\,075\,5 \times 10^{-34}$ J s.

This is a generalisation of Einstein's equation (1.13). The wavelength of a moving body is given by:

$$\lambda = \frac{h}{mv} \tag{1.16}$$

where m = mass of particle, v = velocity of particle. When v is equal to the velocity of electromagnetic radiation c_0, and in combination with the Planck-Einstein
equation,

$$E = hv \tag{1.17}$$

with E = Energy of the photon, h = Plank constant and v = frequency, the Einstein
equation (1.13) can be derived.

Because h represents a very small value, the wavelength (de Broglie wavelength) of such matter-wave can be observed on sufficiently small particles as photons and electrons only. Nevertheless, experiments have been conducted also with neutrons, protons and molecules. In 1999, the diffraction of C_{60} fullerenes as the biggest particle so far was reported [92]. Fullerenes are comparatively large and massive objects, having an atomic mass of about 720 u. The de Broglie wavelength is 2.5 pm, whereas the diameter of the molecule is about 1 nm, about 400 times larger.

1.10 Fundamental Interactions

The so-called Standard Model of particle physics is a theory of three of the four known fundamental interactions and the elementary particles that take part in these interactions. These particles make up all visible matter in the Universe [93]. A fundamental interaction is a mechanism by which particles interact with each other, and which cannot be explained by another more fundamental interaction. Today four fundamental interactions are known: gravitation, electromagnetism, the weak nuclear interaction, and the strong nuclear interaction. Their magnitude and behaviour vary greatly. Grand unified theories seek to unify three of these interactions as manifestations of a single, more fundamental, interaction. The concept of quantum gravity aims to unify gravitation with the other three into an interaction that is completely universal.

Gravitation is by far the weakest fundamental interaction, about 10^{37} times smaller than the electromagnetic interaction. However, because it has an infinite range and because all masses are positive, it is nevertheless very important in the Universe. Because all masses are positive, large bodies such as planets, stars and galaxies have large total masses and therefore exert large gravitational forces. In comparison, the total electric charge of these bodies is zero because half of all charges are negative. This is similar for the weak and strong interactions. Unlike the other interactions, gravity works universally on all matter and energy. There are no objects that lack a gravitational 'charge'. Because of its long range, gravity is responsible for such large-scale phenomena such as the structure and action of galaxies. It is widely believed that in a theory of quantum gravity, gravity would be mediated by a particle which is known as the graviton. Gravitons are hypothetical particles not yet observed.

Although general relativity appears to present an accurate theory of gravity in the non-quantum mechanical limit, there are a number of alternate theories of gravity. Almost nothing is known about dark matter and dark energy and it is not clear whether it can be described completely by the present theories. Alternate theories under any serious consideration by the physics community all reduce to general relativity, and the focus of observational work is to establish limitations on what deviations from general relativity are possible.

Occasionally, physicists have postulated the existence of an additional fifth force in addition. Beyond the range of the force it is assumed to rapidly become insignificant. Many experiments have been undertaken to measure discrepancies in the measurement of the gravitational constant. This implies very sensitive measurements with pendulum [25] or weighing in the presence of varying large masses, e.g. on Earth's surface and in the shaft of a mine or at the wall of a barrage when the water level is changing [94]. Up to now there is no strong evidence for such a fifth force.

The 'standard model' describes the elementary particles and interactions between those particles, but not their mass. Peter Higgs (*1929) [95–97] and concurrently François Englert and Robert Brout [98], as well as Gerard Gouralnik, Karl Richard Hagen and Tom Kibble [99] developed field theories explaining the donation of mass to the particles. Higgs boson is a hypothetical massive scalar particle, a quantum that is postulated to be the carrier particle of the Higgs field, a theoretical field that permeates every place in the Universe at all times and endows all elementary subatomic particles with mass by interactions. Its existence is predicted by the Standard Model of particle physics and could explain how otherwise massless elementary particles still manage to construct mass in matter. In particular, it would explain the difference between the massless photon (travelling with light velocity) and other slower massive particles. On 4th July 2012 CERN informed that results of both, ATLAS and CMS experiments gave strong indications for the presence of a new particle, which could be the Higgs boson, in the mass region around 126 GeV $= 225 \times 10^{-27}$ kg. This corresponds to about 135 times the mass of a proton (hydrogen core). Such a heavy elementary particle decays very fast and it appears very seldom even under the experimental conditions which corresponded to the conditions existing after 10^{-9} s after the Big Bang.

About 99.9 % of the mass of the visible Universe is made up of protons and neutrons. Both particles are much heavier than their three quarks (about 5 %) and gluon constituents. Recently, it became possible to calculate the mass of such light hadrons using lattice quantum chromodynamics. 95 % of the mass is involved in the attractive forces between the quarks. A quantitative confirmation of this aspect of the Standard Model was represented with fully controlled uncertainties [100].

1.11 Theory of Everything

Already ancient philosophers have speculated that the apparent diversity of appearances conceals an underlying unity, and thus that the list of forces might be short, indeed might contain only a single entry [101]. Laplace formulated in 1814:

> "An intellect which at a certain moment would know all forces that set nature in motion, and all positions of all items of which nature is composed, if this intellect were also vast enough to submit these data to analysis, it would embrace in a single formula the movements of the greatest bodies of the Universe and those of the tiniest atom; for such an intellect nothing would be uncertain and the future just like the past would be present before its eyes" [102].

After Einstein's theory of gravity (general relativity) was published in 1915, the search for a unified field theory combining gravity with electromagnetism began, based on the assumption, that no other fundamental forces exist. Also Einstein in his later years was intensely occupied in finding such a unifying theory, without success.

For some decades theoretical physicists have looked for a so-called Theory of Everything. This ultimate theory should consist of a set of equations which give a connection between the currently known basic forces: gravitation, electromagnetism, the weak nuclear interaction, and the strong nuclear interaction. Unlike the point-shaped particles of quantum theory the basic particles are stretched strings, but include also more general items, called branes (derived from 'membrane') floating in space-time. Strings may be open-ended or in shape of a closed loop. It is speculated whether in a special 'supersymmetry' fermions exist as basic items of matter and bosons as particles that transmit a force to each fermion. Depending on size, shape and tension such objects form the particles known in the standard model. The average size of a string should be somewhere near the length scale of quantum gravity, called the Planck length, which is about 10^{-35} m. This means that strings are too small to be observed by today's methods and probably it will not possible at all. Although some critics concede that string theory is falsifiable in principle, they maintain that it is unfalsifiable for the foreseeable future, and so should not be called science.

Indeed such a Theory of Everything could not explain the complexity of the world and the existence of life, not even the occurrence of relatively simple inorganic structures.

1.12 International Unit of Mass

The international unit kilogram was determined as the mass of one cubic decimetre of distilled water at the temperature of its highest density at 4 °C (see Chap. 2). Kilogram weights were made of a platinum alloy and are used as prototypes (Fig. 1.4).

1.12.1 Definition of the Mass Prototype

In 1948 as a result of the General Conference on Weights and Measures it was published [103]:

Declaration on the unit of mass and on the definition weight; conventional value of g_n (CR, 70)

Taking into account the decision of the Comité International des Poids et Mesures of 15 October 1887, according to which the kilogram has been defined as unit of mass: Taking into account the decision contained in the sanction of the prototypes of the Metric System, unanimously accepted by the Conférence Générale des Poids et Mesures on 26 September 1889;

Considering the necessity to an end to the ambiguity which in current practice still exists on the meaning of the word *weight*, used sometimes for *mass*, sometimes for *mechanical force*;

The Conference declares:

Fig. 1.4 The International Prototype Kilogram (IPK) at Sèvres, France is made of an alloy of 90 % platinum and 10 % iridium (by mass) and is machined into a right-circular cylinder (height = diameter) of 39.17 mm to minimise its surface area. The IPK and its replicas are stored in air under two or more nested bell jars

1. The kilogram is the unit mass; it is equal to the mass of the international prototype of the kilogram;
2. The word 'weight' denotes a quantity of the same nature as a 'force': the weight of a body is the product of its mass and the acceleration due to gravity; in particular, the standard weight of a body is a product of its mass and the standard acceleration due to gravity;
3. The value adopted in the International Service of Weights and Measures for the standard acceleration due to gravity is 980.665 cm/s^2, a value already stated in the laws of some countries.

1.12.2 Alternative Mass Unit Definitions

Many units in the SI system are defined relative to the kilogram so its stability is important. Since the IPK and its replicas are stored in air (albeit under two or more nested bell jars), they adsorb atmospheric contamination onto their surfaces and gain mass. The relative change in mass and the instability in the IPK have prompted research into improved methods to obtain a smooth surface finish using diamond-turning on newly manufactured replicas and has intensified the search for a new definition of the kilogram. The kilogram is the last remaining base unit of the SI that is still defined by a material artefact. In 2005 the International Committee for Weights and Measures (CIPM) recommended that the kilogram be redefined in terms of fundamental constants of nature e.g. Avogadro constant.

An Avogadro constant-based approach attempts to define the kilogram as a quantity of silicon atoms. Silicon was chosen because the semiconductor industry knows processes for creating almost defect-free, ultra-pure monocrystalline silicon. To make a practical realisation of the kilogram, a one kilogram silicon globe is being produced from a rod-like, single-crystal ingot consisting almost of ^{28}Si calculated to eight decimal places. Natural silicon consists of 92 % of that isotope and some heavier atoms. By means of centrifuge that is refined to 99.99 % of ^{28}Si. The mass of the globe which is slightly more than 1 kg will be reduced by polishing down to the mass of the IPK. The surface of the ready-made globe is covered tightly with an oxide layer so that it can be handled without loss of surface molecules. The globe must be almost perfect in shape so that the volume can be determined exactly by means of an interferometer. Then the number of atoms can be calculated on the basis of the well-known value of the crystalline unit cell dimensions.

Another Avogadro-based approach, ion accumulation, would define and delineate the kilogram by creating new metal mass artefacts. It would do so by accumulating gold or bismuth ions (atoms stripped of an electron) and counting them by measuring the electrical current required to neutralise the ions. Gold and bismuth are used because they have the two greatest atomic masses of the chemical elements that have only one naturally occurring stable isotope. Because the production of enough deposits is time consuming the programme has just been cancelled.

Relativity and quantum mechanics show that even a single particle of mass m determines a Compton frequency $\omega_0 = mc^2/\hbar$, where c is the speed of light and $\hbar$ is the reduced Planck constant. A clock referenced to ω_0 would enable high-precision mass measurements and a fundamental definition of the second. By Holger Müller et al. [104] it was demonstrated that such a clock using an optical frequency comb to self-reference a Ramsey-Bordé atom interferometer and synchronise an oscillator at a subharmonic of ω_0. This directly demonstrates the connection between time and mass. It allows measurement of microscopic masses with 4×10^{-9} accuracy in the proposed revision to SI units. Together with the Avogadro project, it yields calibrated kilograms.

It is possible to compensate the force acting at a body in the gravitational field by means of the electromagnetic Lorentz force. That force acts on a point charge in an electromagnetic field. Both the electric field and magnetic field can be defined from the Lorentz force law:

$$\vec{F} = q(\vec{E} + v \times \vec{B}) \tag{1.18}$$

where F is the force [N], E is the electric field strength [V m^{-1}), B is the magnetic flux density [T = V s m^{-2}], q is the electric charge of the particle [C = A s], v is the instantaneous velocity of the particle [m s^{-1}]. Equation (1.18) is a vector equation and $\times$ means vector cross product. The electric force is straightforward, being in the direction of the electric field if the charge q is positive, and the direction of the magnetic field is perpendicular to it.

By compensation of weight forces by means of an electric current through a coil instead of a counterweight it is possible to substitute the unit mass by electrical units, supposed the geometry of the electromagnetic device is known. The Watt balance is a double-arm weighing scale with single pan for the kilogram test mass and a coil surrounding a permanent magnet at the opposite side (Fig. 1.5). The

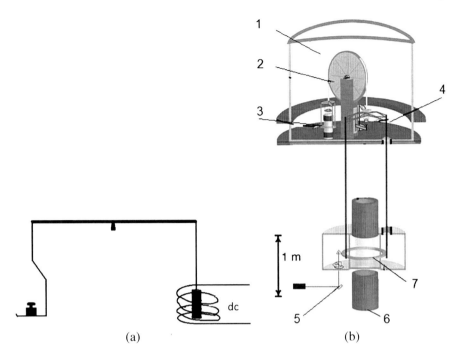

Fig. 1.5 (**a**) Diagram of an Ampère or Watt balance. (**b**) Watt balance of the UK National Physical Laboratory. *1* vacuum chamber, *2* balance wheel, *3* velocity drive coil, *4* reference mass, *5* interferometer (1 of 3), *6* superconducting solenoid, *7* induction coils. © NPL, Teddington, UK

electrical power necessary to oppose the weight of the test mass is measured as it is accelerated by gravity. However, because Eq. (1.17) contains two variables two experiments at the same balance are necessary. The Watt balance is a variation of an ampere balance in that it employs an extra calibration step that neutralises geometrical effects. Here the suspended coil is moved at defined velocity through the field of the fixed permanent magnet. In this way the electric potential in the watt balance is delineated by a Josephson voltage standard, which allows voltage to be linked to an invariant constant of nature with extremely high precision and stability. Its circuit resistance is calibrated against a quantum Hall resistance standard. The mass standard would be given by Planck's elementary quantum of action. The watt balance requires exquisitely precise measurement of gravity in the respective laboratory. Until now, the results of the various watt balances have wide scattering and the reasons of the deviations are still unclear. (See also Sect. 4.2 in Chap. 4.)

Another electrical method of defining the mass standard is the voltage balance [105–107]. Here an electrode is suspended movable at a balance and the force exerted by a fixed electrode is measured (Fig. 1.6). By application of high tension the force between the capacitor is measured using mechanical substitution balances. The balance and the capacitor are in an atmosphere of nitrogen gas.

Fig. 1.6 (**a**) Diagram of a
voltage balance. (**b**) Diagram
of the voltage balance of the
University of Zagreb, Croatia.
1 movable electrode,
2 high-voltage electrode,
3 auxiliary electrode,
4 column of the sliding unit.
5 carriage for the sliding unit,
6 drive, *7* interferometer for
measuring the displacement
path, *8* position sensor,
9 flexible bearings of the
balance, *10* balance beam,
11 counterweight, *12* mass
standard. © University of
Zagreb

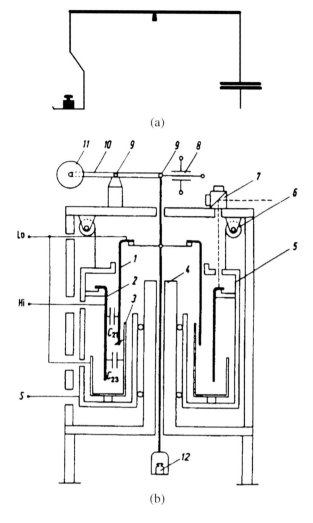

(a)

(b)

A liquid electrometer had been realised consisting of a horizontally arranged capacitor whereby the lower electrode is the surface of a mercury bath (Fig. 1.7) [108]. Voltage is applied between that surface and an electrode fixed at the distance d. The increase of the mercury level z is measured by means of a laser white-light interferometer in comparison to the distance of the mercury level to reference electrodes.

Proposals have been made for a new realisation of the static voltage balance for two dynamic methods: a drop method with a capacitor and an oscillation method [109].

There are also a number of proposals to replace the artificial mass standard by a natural factum. It is proposed to use the electron. However no experiment is known so far to determine the mass of an electron with the required accuracy. Furthermore it would be very difficult to extrapolate from those small mass weights for practical

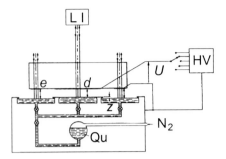

Fig. 1.7 Diagram of a liquid electrometer. d distance between electrodes, e_a, e_b distance between the reference electrodes, z change of the distance due to applied voltage U. The distance between electrodes is controlled by laser interferometer. *HV* high voltage source, *QU* mercury reservoir, *LI* laser with-light interferometer. © CSIRO, Australia

use. It was also suggested to use a mass derived from the Broglie wavelength [110] (see Sect. 1.10). Planck made proposals for defining mass, length and time by fundamental constants (see Sect. 1.12.2). The mass should be defined by:

$$m_{Pl} = \sqrt{\frac{\hbar c}{G}} = 2.17644(11) \times 10^{-8} \text{ kg} \approx 22 \text{ µg} \tag{1.19}$$

where G is the gravitational constant, $\hbar$ is the Planck constant, and c is the speed of light. However, for practical weighable bodies the Planck constant cannot be realised with the sufficient accuracy.

At the next session of the general BIPM conference in 2011 it should be decided whether the kilogram will get a new definition on the basis of one of these experiments

1.13 Mass Units

To describe physically the universe and the world we need a system of four basic quantities. From these independent quantities all others can be derived. However, for convenience most systems defined some more base quantities.

1.13.1 The International System of Units

The International System of Units (Système International d'Unités SI) is the modern form of the metric system. It is the world's most widely used system of units, both in everyday commerce and in science [111–113]. The metric system was conceived by a group of scientists (among them, Lavoisier) who had been commissioned by King Louis XVI of France to create a unified and rational system of measures. After the French Revolution, the system was adopted by the new government. On August 1,

Table 1.4 SI base units

Base quantity	Quantity symbol	Dimension symbol	SI base unit	Unit symbol
Length	l	L	Meter	m
Mass	m	M	Kilogram	kg
Time	t	T	Second	s
Electric current	I or i	I	Ampere	A
Thermodynamic temperature	T	Θ	Kelvin	K
Amount of substance	n	N	Mole	mol
Luminous intensity	I_v	J	Candela	cd

Fig. 1.8 Advertisement for the metric system

1793 the National Convention adopted the new decimal 'metre' with a provisional length as well as the other decimal units with preliminary definitions and terms. On April 7, 1795 the terms *gramme* and *kilogramme* replaced the former terms 'gravet' and 'grave' (Loi du 18 germinal, an III). The older metric system included several groupings of units. The SI was developed in 1960 from the metre-kilogram-second (mks) system. The SI introduced several newly named units. The SI is not static; it is a living set of standards where units are created and definitions are modified with international agreement as measurement technology progresses (Fig. 1.8).

In the SI system we have two basic quantities which are used to describe matter at rest: mass and length. The corresponding SI units are kilogram or kilogramme (kg) and meter (m). Mass is a relatively new and abstract quantity which can be measured by observing effects within gravitational or acceleration fields [103]. It is an inertial property; that is, the tendency of the object to remain at constant velocity unless acted upon by an outside force. The other parameter, volume (m^3), based on length, describes the extension of a structure of matter within the three-dimensional space. It is measured usually by means of the interaction of matter with electromagnetic radiations by which the structure becomes visible.

Another useful parameter, density, is defined as the relation of mass to unit volume and the corresponding SI unit is $kg\,m^{-3}$. The exact definition is relatively new, probably of Leonhard Euler [9, 114]. It should be mentioned that the notation 'specific gravity' (density related to water) is much older. Methods of determining of specific gravity are described in the books Ayin-Akbari of Muhammad Ibn-Ahmad Al-Biruni (873–1084) [115, 116] and in the Book of the Balance of Wisdom of Abd al-Rahman al-Manzur Al-Chazini 1120 [117].

A quantity whose magnitude is additive for subsystems is called 'extensive', examples are mass and volume. A quantity whose magnitude is independent of the extent of the system is called 'intensive', examples are temperature and pressure. The adjective 'specific' before the name of an extensive quantity is often used to mean 'divided by mass' [118]. When the symbol for the extensive quantity is a capital letter, the symbol used for the specific quantity is often the corresponding lower case letter.

The term 'specific' should be used only if the quantity can be (and is) divided by the total mass or partial mass of the sample in question. If it is related to another item as in the case of atomic and molecular mass the respective notation is 'relative'.

Likewise the term 'molar' before the name of an extensive quantity is used to denote 'divided by amount in moles'. Thus the heat capacity of a given quantity of a substance is the energy required to raise the temperature of that quantity of the substance by 1 °C (or 1 Kelvin), measured in joules $(J\,K^{-1})$. The specific heat capacity is this quantity divided by its mass, measured in $J\,K^{-1}\,kg^{-1}$. The molar heat capacity is the quantity divided by relative molar mass, measured in $J\,K^{-1}\,mol^{-1}$.

1.13.2 Natural Units

The natural units or Planck units are physical units of measurement defined exclusively in terms of five universal physical constants listed in Table 1.6, in such a manner that these five physical constants take on the numerical value of 1 when expressed in terms of these natural units.

By setting to (dimensionless) 1 the five fundamental constants in Table 1.4, the base units of length, mass, time, charge, and temperature shown in Table 1.5 are also

Table 1.5 Fundamental physical constants

Constant	Symbol	Dimension	Value in SI units with uncertainties
Speed of light in vacuum	c	$L\,T^{-1}$	$2.99792458108 \times 10^8$ m s^{-1}
Gravitational constant	G	$L^3\,M^{-1}\,T^{-2}$	$6.67428(67) \times 10^{-11}$ m^3 kg^{-1} s^{-2}
Reduced Planck constant	$\hbar = h/2\pi$	$L^2\,M\,T^{-1}$	$1.054571628(53) \times 10^{-34}$ J s
Coulomb constant	$\frac{1}{4\pi\varepsilon_0}$	$L^3\,M\,T^{-2}\,Q^{-2}$	$8.9875517873681764 \times 10^9$ kg m^3 s^{-2} C^{-2}
Boltzmann constant	k_B	$L^2\,M\,T^{-2}\,\Theta^{-1}$	$1.3806504(24) \times 10^{-23}$ J K^{-1}

Where h = Planck constant, ε_0 = permittivity of free space, L = length, M = mass, T = time, Q = electric charge, Θ = temperature. Speed of light and Coulomb constant are exact values by definition

Table 1.6 Natural Planck base units

Name	Dimension	Expressions	SI equivalent with uncertainties	Eq.
Planck length	L	$l_P = \sqrt{\dfrac{\hbar G}{c^3}}$	$1.616252(81) \times 10^{-35}$ m	(1.20)
Planck mass	M	$m_P = \sqrt{\dfrac{\hbar c}{G}}$	$2.17644(11) \times 10^{-8}$ kg	(1.19)
Planck time	T	$t_P = \dfrac{l_P}{c} = \dfrac{\hbar}{m_P c^2} = \sqrt{\dfrac{\hbar G}{c^5}}$	$5.39124(27) \times 10^{-44}$ s	(1.21)
Planck charge	Q	$q_P = \sqrt{4\pi\varepsilon_0 \hbar c}$	$1.875545870(47) \times 10^{-18}$ C	(1.22)
Planck temperature	Θ	$T_P = \dfrac{m_P c^2}{k_B} = \sqrt{\dfrac{\hbar c^5}{G k_B^2}}$	$1.416785(71) \times 10^{32}$ K	(1.23)

set to (dimensionless) 1. Particle physicists and cosmologists often use the reduced Planck mass, which is

$$\sqrt{\frac{\hbar c}{8\pi G}} \approx 4.340 \times 10^{-6} \text{ g} = 2.43 \times 10^{18} \text{ GeV}/c^2 \tag{1.24}$$

Planck units elegantly simplify particular algebraic expressions appearing in physical laws. Planck units are considered unique in that these units are not based on properties of any prototype object, or particle (that would be arbitrarily chosen) but are based only on properties of free space. However, most Planck units are many orders of magnitude too large or too small to be of any empirical and practical use, so that Planck units as a system are really only relevant to theoretical physics. That is because 1 Planck unit is often the largest or smallest value of a physical quantity that makes sense given the current state of physical theory.

1.13.3 Conventional Mass Units

By means of a balance we measure the weight but the scaling indicates mass units. The gravitational acceleration depends on the geographical location, but differences in the value of weight are almost insignificant in customary weighing. Thus in daily life the notation 'mass' will replace 'weight' only slowly or never. Furthermore 'weight' denotes the weight piece, which is a calibrated mass.

On the other hand, in science mass is an important physical quantity, which must be clearly distinguished from weight. Nevertheless, looking into the literature we find often the notation weight. For objects which will never be weighed by a conventional balance in the gravitational field—atoms and molecules—this notation is still familiar. Indeed, the mass of such particles is determined either indirectly or by means of mass spectrometers. These are methods in which gravity and, hence, weight is not included.

Mass is the only SI base unit with an SI prefix as part of its name and it is caused by the historical development. However, the usual prefixes in the SI system are never used additionally with kilogram (Table 1.7). With few exceptions, the system is legally being used in every country in the world, and many countries do not maintain official definitions of other units. In the United States, industrial use of SI is increasing, but popular use is still limited. In the United Kingdom, conversion to metric units is official policy but not yet complete. Those countries that still recognise non-SI units have redefined their traditional non-SI units in terms of SI units. There the avoirdupois pound is used as a unit of mass and its related unit of force is the pound-force. The European Union has a directive as a result of which non-SI markings will be banned after 31 December 2009 on any goods imported into the European Union [119]. This applies to all markings on products, enclosed directions and papers, packaging, and advertisements. Nevertheless, until now, consequences of that definition have not been completely accepted. A survey on other mass units is given in Chap. 2.

The practice of using the abbreviation 'mcg' rather than the SI symbol 'μg' was formally mandated for medical practitioners in 2004 by the Joint Commission on Accreditation of Healthcare Organisations (JCAHO) in their 'Do Not Use' List: Abbreviations, Acronyms, and Symbols because hand-written expressions of 'μg' can be confused with 'mg', resulting in a thousand-fold overdosing. The mandate was also adopted by the Institute for Safe Medication Practices. The metric carat or Karat is a unit of mass used for measuring gems and pearls. 1 ct = 200 mg.

Both in the British imperial system and U.S. customary units the avoirdupois pound is still used. 1 avoirdupois pound (lb) is defined as exactly 0.45359237 kg, making one kilogram approximately equal to 2.205 avoirdupois pounds.

A tonne (t) or metric ton, also referred to as a metric tonne or tonne de metrice, is a measurement of mass equal to 1000 kilograms (Table 1.8). It is not a SI unit but tolerated. Still in use are long and short tons. One long ton (2240 lb) which is 101.605 % of a tonne; one short ton (2000 lb) is 90.72 % of a tonne.

Occasionally ton describes the percentage of a metal within an alloy. The tonne of trinitrotoluene (TNT) is used as a proxy for energy of explosives (explosive force),

Table 1.7 SI multiples for gram. (Prefixes are never used additionally with kilogram!)

Value	Symbol	Name	Value	Symbol	Name
10^{-1} g	dg	decigram	10^0 g	g	gram
10^{-2} g	cg	centigram	10^1 g	dag	decagram
10^{-3} g	mg	milligram	10^2 g	hg	hectogram
10^{-6} g	μg	microgram (mcg)	10^3 g	kg	kilogram
10^{-9} g	ng	nanogram	10^6 g	Mg	megagram = 1 tonne
10^{-12} g	pg	picogram	10^9 g	Gg	gigagram
10^{-15} g	fg	femtogram	10^{12} g	Tg	teragram
10^{-18} g	ag	attogram	10^{15} g	Pg	petagram
10^{-21} g	zg	zeptogram	10^{18} g	Eg	exagram
10^{-24} g	yg	yoctogram	10^{21} g	Zg	zettagram
			10^{24} g	Yg	yottagram

Table 1.8 Multiples for tonne

Value t	Symbol	Name	Value g	SI value kg	Symbol	Name
10^0 t	t	ton, tonne	10^6 g	10^3	Mg	megagram
10^1 t	dat	decaton	10^7 g	10^4		
10^2 t	ht	hectoton	10^8 g	10^5		
10^3 t	kt	kiloton	10^9 g	10^6	Gg	gigagram
10^6 t	Mt	megaton	10^{12} g	10^9	Tg	teragram
10^9 t	Gt	gigaton	10^{15} g	10^{12}	Pg	petagram
10^{12} t	Tt	teraton	10^{18} g	10^{15}	Eg	exagram

referred to as 4.184 GJ (gigajoules). Ton is used also used as volume unit for ships: tonnage, displacement tonnage, gross tonnage the brutto register ton BRT: 1 BRT = 100 cubic foot = 2,83164 m^3.

Obsolete is a mass unit defined in the former cgs-system: 1 kp s^2 m^{-1} = 9.8 kg.

A survey on mass units accepted by the Bureau International des Poids et Mesures (BIPM) [120] and of related units and constants is given in Table 1.9.

1.13.4 Atomic Mass

An important unit for chemists is amount of atomic or sub-atomic species expressed in moles, based on Avogadro constant. Furthermore, it is based on the number of constituent particles (protons, electrons and neutrons) of an atom of carbon-12. The mole was confirmed in 1969 as a basic SI unit.

Table 1.9 Mass and mass related units

Name	Symbol	Definition	SI unit
Number of entities (e.g. molecules, atoms, ions, formula units)	N		–
Amount of substance, amount (chemical amount)	n	$n_B = N_B/L$	mol
Avogadro constant	L, N_A	$6.02214179 \pm 0.00000030 \times 10^{23}$	mol^{-1}
Planck constant$/2\pi$	$\hbar = h/2\pi$	1.0546×10^{-34}	J s
Planck mass	m_{Pl}	$2.17645(16) \times 10^{-8}$	kg
Mass of atom, atomic mass	m_a, m		kg
Mass of entity (molecule, formula unit)	m, m_f		kg
Atomic mass constant	m_u	$m_u = m_a(^{12}\text{C})/12 = 1\,\text{u}$	kg
Dalton, unified atomic mass unit	Da, u	$1\,\text{Da} = 1\,\text{u} = 1.66053886(28) \times 10^{-27}$	kg
Molar mass	M	$M_B = m/n_B$	kg mol^{-1}
Molar mass constant	M_u	$M_u = m_u N_A$	kg mol^{-1}
Relative molecular mass (relative molar mass, molecular weight)	M_r	$M_r = m_f/m_u$	–
Relative atomic mass (atomic weight)	A_r	$A_r = m_a/m_u$	–
Molar volume	V_m	$V_{m,B} = V/n_B$	$\text{m}^3\,\text{mol}^{-1}$
Mass fraction	w	$w_B = m_B/\sum_i m_i$	–
Electronvolt	eV	$1\,\text{eV} = 1.60217653(14) \times 10^{-19}$	J
Electronvolt$/c^2$	eV/c^2	1.783×10^{-36}	kg

In the book "The International System of Units (SI)" of the Bureau International des Poids et Mesures (BIPM) [113] we read:

"'Atomic weights' and 'molecular weights' ... are in fact relative masses. Physicists and chemists have ever since agreed to assign the value 12, exactly, to the so-called atomic weight of the isotope carbon with mass number 12 (carbon-12, ^{12}C), correctly called the relative atomic mass A_r (^{12}C). The unified scale thus obtained gives the relative atomic and molecular masses, also known as the atomic and molecular weights, respectively."

In the IUPAC manual 'Quantities, Units and Symbols in Physical Chemistry' [118], the so-called 'Green book', it is concluded:

"For historical reasons the terms 'molecular weight' and 'atomic weight' are still used. For molecules M_r is the 'relative molecular mass' or 'molecular weight'. For atoms M_r is the relative atomic mass or 'atomic weight', and the symbol A_r may be used. M_r may also be called the relative molar mass, $M_{r,B} = M_B/M_u$, where M_u = mass (in grams) of 1 mole of carbon-12."

The term 'atomic weight' is being phased out slowly and being replaced by 'relative atomic mass', however 'standard atomic weights' have maintained their name [121].

Relative atomic masses are listed in Table 1.7; a periodic table with standard atomic weights is depicted in Tables 1.10a and 1.10b.

The molecular mass of a substance, formerly also called molecular weight and abbreviated as MW, is the mass of one molecule of that substance, relative to the unified atomic mass unit u (equal to $1/12$ of the mass of one atom of carbon-12). This is distinct from the relative molecular mass of a molecule M_r, which is the ratio of the mass of that molecule to $1/12$ of the mass of carbon-12 and is a dimensionless number. Molar masses are almost never measured directly. They may be calculated as the sum of the standard atomic weights included in the chemical formula. The molar mass of carbon C is 12.0107 g/mol. Salts consist of ions and therefore the value is notified as formula mass instead of a molecular mass.

The unified atomic mass unit (u), or dalton (Da), is a small unit of mass used to express atomic and molecular masses. It is defined to be one-twelfth of the mass of an unbound atom of ^{12}C at rest and in its ground state. The unit is convenient because one hydrogen atom has a mass of approximately 1 u, and more generally an atom or molecule that contains N_p protons and N_n neutrons will have a mass approximately equal to $(N_p + N_n) \cdot u$. Atomic masses are often written without any unit and then the unified atomic mass unit is implied. In biochemistry, particularly in reference to proteins, the term 'dalton' is often used. Because proteins are large molecules, they are typically referred to in kilodaltons, or 'kDa' The unified atomic mass unit, or dalton, is not an SI unit of mass, although it is accepted for use with SI under either name. The symbol amu for atomic mass unit is obsolete.

The Avogadro constant $N_A \approx 6.022 \times 10^{23}$ mol^{-1} is the number of 'entities' (usually, atoms or molecules) in one mole. (The pure value 6.022×10^{23} is called Avogadro's number.) The Avogadro constant and the mole are defined so that one mole of a substance with atomic or molecular mass 1 u have a mass of 1 g. For example, the molecular mass of a water molecule containing one ^{16}O isotope and two ^{1}H isotopes is 18.0106 u, and this means that one mole of this monoisotopic water has a mass of 18.0106 grams. Water and most molecules consist of a mixture of molecular masses due to naturally occurring isotopes. For this reason these sorts of comparisons are more meaningful and practical using molar masses which are generally expressed in g/mol, not u. The one-to-one relationship between daltons and g/mol is true but in order to be used accurately calculations must be performed with isotopically pure substances or involve much more complicated statistical averaging of multiple isotopic compositions [122].

Molar mass, symbol m_m is the mass of one mole of a substance (chemical element or chemical compound). Molar masses are almost quoted in grams per mole (g mol^{-1}). Mole is defined as the molecular mass in grams:

$$m_m = \frac{m_u}{N_A} \tag{1.25}$$

where m_u is the mass in atomic mass units and N_A is Avogadro's number.

Table 1.10a Relative atomic masses related to $^{12}C = 12,0000$ [123]. The value enclosed in brackets, indicates the mass number of the most stable isotope of the element

Atomic number	Symbol	Name	Relative atomic mass
1	H	Hydrogen	1.00794(7)
2	He	Helium	4.002602(2)
3	Li	Lithium	6.941(2)
4	Be	Beryllium	9.012182(3)
5	B	Boron	10.811(7)
6	C	Carbon	12.0107(8)
7	N	Nitrogen	14.0067(2)
8	O	Oxygen	15.9994(3)
9	F	Fluorine	18.9984032(5)
10	Ne	Neon	20.1797(6)
11	Na	Sodium	22.98976928(2)
12	Mg	Magnesium	24.3050(6)
13	Al	Aluminium	26.9815386(8)
14	Si	Silicon	28.0855(3)
15	P	Phosphorus	30.973762(2)
16	S	Sulfur	32.065(5)
17	Cl	Chlorine	35.453(2)
18	Ar	Argon	39.948(1)
19	K	Potassium	39.0983(1)
20	Ca	Calcium	40.078(4)
21	Sc	Scandium	44.955912(6)
22	Ti	Titanium	47.867(1)
23	V	Vanadium	50.9415(1)
24	Cr	Chromium	51.9961(6)
25	Mn	Manganese	54.938045(5)
26	Fe	Iron	55.845(2)
27	Co	Cobalt	58.933195(5)
28	Ni	Nickel	58.6934(4)
29	Cu	Copper	63.546(3)2
30	Zn	Zinc	65.38(2)
31	Ga	Gallium	69.723(1)
32	Ge	Germanium	72.64(1)
33	As	Arsenic	74.92160(2)
34	Se	Selenium	78.96(3)
35	Br	Bromine	79.904(1)
36	Kr	Krypton	83.798(2)
37	Rb	Rubidium	85.4678(3)
38	Sr	Strontium	87.62(1)
39	Y	Yttrium	88.90585(2)

Table 1.10a (Continued)

Atomic number	Symbol	Name	Relative atomic mass
40	Zr	Zirconium	91.224(2)
41	Nb	Niobium	92.90638(2)
42	Mo	Molybdenum	95.96(2)
43	Tc	Technetium	[98]
44	Ru	Ruthenium	101.07(2)
45	Rh	Rhodium	102.90550(2)
46	Pd	Palladium	106.42(1)
47	Ag	Silver	107.8682(2)
48	Cd	Cadmium	112.411(8)
49	In	Indium	114.818(3)
50	Sn	Tin	118.710(7)
51	Sb	Antimony	121.760(1)
52	Te	Tellurium	127.60(3)
53	I	Iodine	126.90447(3)
54	Xe	Xenon	131.293(6)
55	Cs	Caesium	132.9054519(2)
56	Ba	Barium	137.327(7)
57	La	Lanthanum	138.90547(7)
58	Ce	Cerium	140.116(1)
59	Pr	Praseodymium	140.90765(2)
60	Nd	Neodymium	144.242(3)
61	Pm	Promethium	[145]
62	Sm	Samarium	150.36(2)
63	Eu	Europium	151.964(1)
64	Gd	Gadolinium	157.25(3)
65	Tb	Terbium	158.92535(2)
66	Dy	Dysprosium	162.500(1)
67	Ho	Holmium	164.93032(2)
68	Er	Erbium	167.259(3)
69	Tm	Thulium	168.93421(2)
70	Yb	Ytterbium	173.054(5)
71	Lu	Lutetium	174.9668(1)
72	Hf	Hafnium	178.49(2)
73	Ta	Tantalum	180.94788(2)
74	W	Tungsten	183.84(1)
75	Re	Rhenium	186.207(1)
76	Os	Osmium	190.23(3)1
77	Ir	Iridium	192.217(3)
78	Pt	Platinum	195.084(9)

Table 1.10a (Continued)

Atomic number	Symbol	Name	Relative atomic mass
79	Au	Gold	196.966569(4)
80	Hg	Mercury	200.59(2)
81	Tl	Thallium	204.3833(2)
82	Pb	Lead	207.2(1)
83	Bi	Bismuth	208.98040(1)
84	Po	Polonium	[209]
85	At	Astatine	[210]
86	Rn	Radon	[222]
87	Fr	Francium	[223]
88	Ra	Radium	[226]
89	Ac	Actinium	[227]
90	Th	Thorium	232.03806(2)
91	Pa	Protactinium	231.03588(2)
92	U	Uranium	238.02891(3)
93	Np	Neptunium	[237]
94	Pu	Plutonium	[244]
95	Am	Americium	[243]
96	Cm	Curium	[247]
97	Bk	Berkelium	[247]
98	Cf	Californium	[251]
99	Es	Einsteinium	[252]
100	Fm	Fermium	[257]
101	Md	Mendelevium	[258]
102	No	Nobelium	[259]
103	Lr	Lawrencium	[262]
104	Rf	Rutherfordium	[267]
105	Db	Dubnium	[268]
106	Sg	Seaborgium	[271]
107	Bh	Bohrium	[272]
108	Hs	Hassium	[270]
109	Mt	Meitnerium	[276]
110	Ds	Darmstadtium	[281]
111	Rg	Roentgenium	[280]
112	Cp	Copernicum	[285]
113	Uut	Ununtrium	[284]
114	Uuq	Ununquadium	[289]
115	Uup	Ununpentium	[288]
116	Uuh	Ununhexium	[293]
118	Uuo	Ununoctium	[294]

Table 1.10b Periodic table with standard atomic weights

G→ / P↓	1	2	3	4	5	6	7	8	9	10	11	12	13	14	15	16	17	18
1	H 1.008																	He 4.003
2	Li 6.941	Be 9.012											B 10.81	C 12.01	N 14.01	O 16.00	F 19.00	Ne 20.18
3	Na 22.99	Mg 24.31											Al 26.98	Si 28.09	P 30.97	S 32.07	Cl 35.45	Ar 39.95
4	K 39.10	Ca 40.08	Sc 44.96	Ti 47.87	V 50.94	Cr 52.00	Mn 54.94	Fe 55.84	Co 58.93	Ni 58.69	Cu 63.55	Zn 65.39	Ga 69.72	Ge 72.61	As 74.92	Se 78.96	Br 79.90	Kr 83.80
5	Rb 85.74	Sr 87.62	Y 88.91	Zr 91.22	Nb 82.91	Mo 95.94	Tc [99]	Ru 101.1	Rh 102.9	Pd 106.4	Ag 107.9	Cd 112.4	In 114.8	Sn 118.7	Sb 121.8	Te 127.6	I 126.9	Xe 131.3
6	Cs 132.9	Ba 137.3	*	Hf 178.5	Ta 180.9	W 183.8	Re 186.2	Os 190.2	Ir 192.2	Pt 195.1	Au 197.0	Hg 200.6	Tl 204.4	Pb 207.2	Bi 209.0	Po [209]	At [210]	Rn [222]
7	Fr [223]	Ra [224]	**	Rf [263]	Db [262]	Sg [266]	Bh [264]	Hs [269]	Mt [268]	Ds [272]	Rg [272]	Uub [277]	Uut [284]	Uuq [289]	Uup [288]	Uuh [292]	Uus [291]	Uuo [293]

*Lanthanides	La 138.9	Ce 140.1	Pr 140.9	Nd 144.2	Pm [145]	Sm 150.4	Eu 152.0	Gd 157.3	Tb 158.9	Dy 162.5	Ho 164.9	Er 167.3	Tm 168.93	Yb 173.0	Lu 175.0
**Actinides	Ac [227]	Th 232.0	Pa 231.0	U 238.0	Np [237]	Pu [244]	Am [243]	Cm [247]	Bk [247]	Cf [251]	Es [252]	Fm [257]	Md [258]	No [259]	Lr [262]

P = period, G = group

Fig. 1.9 Definition electronvolt

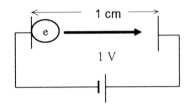

Table 1.11 Multiples for electronvolt/c^2

Symbol	Value
1 EeV/c^2	1.783×10^{-18} kg
1 PeV/c^2	1.783×10^{-21} kg
1 TeV/c^2	1.783×10^{-24} kg
1 GeV/c^2	1.783×10^{-27} kg
1 MeV/c^2	1.783×10^{-30} kg
1 keV/c^2	1.783×10^{-33} kg
1 eV/c^2	1.783×10^{-36} kg

Table 1.12 Mass of some objects in rest

Object	Mass	Mass kg	Object	Mass kg
Photon	0	0	Sun	1.989×10^{30}
Electron	$510\,998.9$ eV $c^{-2} =$ $5.485\,799\,110(12) \times$ 10^{-4} u	$91.09\,381\,88(72) \times$ 10^{-30}	Moon	7.349×10^{22}
			Mercury	3.302×10^{23}
			Venus	4.869×10^{24}
Quark Up	1.5–4.0 MeV c^{-2}	2.7–7×10^{-30}	Earth	5.9736×10^{24}
Quark Down	4–8 MeV c^{-2}	7–14×10^{-30}	Mars	6.419×10^{23}
Quark Strange	80–130 MeV c^{-2}	0.14–0.23×10^{-27}	Jupiter	1.899×10^{27}
Quark Charm	1150–1350 MeV c^{-2}	2–2.41×10^{-27}	Saturn	5.685×10^{26}
Quark Bottom	4100–4400 MeV c^{-2}	7.3–7.8×10^{-27}	Uranus	8.683×10^{25}
Quark Top	170900 ± 1800 MeV c^{-2}	134×10^{-27}	Neptun	1.0243×10^{26}
Higgs particle	126 GeV c^{-2}	225×10^{-27}	Neutron star	2.7–4.2×10^{30}
Proton	$1.007\,276\,466\,88(13)$ u	$1.672\,621\,58(13) \times$ 10^{-27}	Black holes	$\sim 5 \times 10^{30}$–10^{40}
Mist particle	~ 0.01–0.3 g	~ 0.01–0.3×10^{-3}		
Rain droplet	~ 0.05 g	$\sim 0.05 \times 10^{-3}$	Galaxy Milky Way	3.6×10^{41}
1 l Water at 4 °C		1	Universe (visible)	10^{52}–10^{54}
Water on Earth		1.384×10^{21} kg		

Nearly 99.9 % of the mass of the solar system is concentrated in the Sun. The solar mass $M_{Sun} = 1.989 \times 10^{30}$ kg is used as an astronomical unit. Occasionally also the Earth mass is used as a reference, $M_{Sun} = 332\,946\,M_{Earth}$. Whereas the mass of the Sun decreases almost imperceptibly due to radiation, the mass of the Earth increases slightly because of the cosmic debris (dust, meteorites) 10^6–10^8 kg/day

1.13.5 Electronvolt

One eV is a very small amount of energy widely used in solid state, nuclear and particle physics. For large energies, the million electronvolt (MeV) is used. One electronvolt is based on a notional experiment as the amount of energy that could be gained by a single unbound electron when it would be accelerated through an electrostatic potential difference of one volt, in vacuo (Fig. 1.9, Table 1.11) [124]. It is equal to one volt times the (unsigned) charge of a single electron $C_{el.}$:

$$1 \text{ eV} = 1 \text{ JC}^{-1} \times C_{el.} = 1.602\,176\,53(14) \times 10^{-19} \text{ J} \approx 0.160 \text{ aJ} \quad (1.26)$$

The unit electronvolt is accepted (but not encouraged) for use with SI (Table 1.4). A single atom is such a small entity that to talk about its energy in joules would be inconvenient. But instead of taking the appropriate SI unit attojoule physicists (and chemists) have unfortunately chosen, arbitrarily, a non-conformist unit called an electronvolt (eV). In mitigation, one can acknowledge the convenience of this unit when it comes to experimental measurement of ionisation energies of elements, an important parameter for chemists. The experimentally measurable parameter is the potential difference (measured in volts) required to dislodge the electron from the atom, leading to the erroneous description of the energy as the 'ionisation potential'. Whereas ionisation energies in eV are convenient numerical quantities, the energies for dislodging electrons from single atoms are very small. This can be overcome if ionisation energies are expressed in kJ mole^{-1}, the recommended unit.

According to Einstein's relation [13] mass can converted into energy (Eq. (1.13)) and

$$E/m = c^2 = (299\,792\,458 \text{ m/s})^2 = 89.875 \times 10^{15} \text{ J/kg} = 89.875 \text{ PJ/kg} \quad (1.27)$$

So one gram of mass is equivalent to the energy of 89.9 terajoules (TJ).

It is common in particle physics, where mass and energy are often interchanged, to use eV/c^2 as a unit of mass.

$$1 \text{ eV}/c^2 = 1.783 \times 10^{-36} \text{ kg} \quad (1.28)$$

Obsolete: Bevatron (BeV) was used for 'billion-electron-volt; occasionally; it is equivalent to the GeV.

1.13.6 Summary

The mass as an inertial quantity of a body should be clearly distinguished from its weight, which is the product of mass and the acceleration due to gravity. Therefore, use of the terms wt%, atomic and molecular weight should be avoided and instead mass %, relative atomic mass, relative molecular mass or relative molar mass and formula mass should be used [125]. If electronvolt is used it would be appropriate to mention in addition the respective value in SI units because that would facilitate further assessments and calculations.

Table 1.13 Symbols used

Symbol	Explanation	SI unit
a	Acceleration	$\mathrm{m\,s^{-2}}$
d	Distance	m
E	Energy	$\mathrm{J} = \mathrm{kg\,m^2\,s^{-2}}$
W	Weight = force due to the gravitational field.	N
F_I	Applied force	$\mathrm{N} = \mathrm{kg\,m\,s^{-2}}$
g	Acceleration due to gravity	$\mathrm{m\,s^{-2}}$
g_n	Standard acceleration due to gravity = 9.80665	$\mathrm{m\,s^{-2}}$
G	Gravitational constant = $6.672\,59(85) \times 10^{-11}$	$\mathrm{kg^{-1}\,m^3\,s^{-2}}$
h	Planck constant = $6.626\,075\,5 \times 10^{-34}$	$\mathrm{J\,s}$
m	Mass	kg
m_p	Mass of a planet	kg
m_S	Mass of the Sun or a star	kg
m_0	Mass of a body at rest	kg
N_A	Avogadro constant = $6.022\,141\,79 \times 10^{23}$	$\mathrm{mol^{-1}}$
p	Momentum	$\mathrm{kg\,m\,s^{-1}}$
t	Time	s
t_u	Time of a sidereal period	s
c_0	Speed of light in vacuum = 2.99792458×10^8	$\mathrm{m\,s^{-1}}$
v	Velocity	$\mathrm{m\,s^{-1}}$
λ	Wavelength	m
ν	Frequency	$\mathrm{s^{-1}}$

References

1. I. Newton, *Philosophiae Naturalis Principia Mathematica* (London, 1687)
2. I. Newton, *The Principia: Mathematical Principles of Natural Philosophy Trans. I* (University of California Press, Berkeley, 1999)
3. J.W.v. Goethe, F. Schiller, *Xenien*. Schillers Musenalmanach (1797)
4. E. Pritchard (ed.), *Measurement of Mass* (Royal Society of Chemistry, Cambridge, 2003). (brochure with CD)
5. H.R. Jenemann, The development of the determination of mass, in *Comprehensive Mass Metrology*, ed. by M. Kochsiek, M. Gläser (Wiley/VCH, Berlin, 2000), pp. 119–163
6. M. Gläser, The concept of mass, in *Comprehensive Mass Metrology*, ed. by M. Kochsiek, M. Gläser (Wiley/VCH, Berlin, 2000), pp. 16–47
7. W. Capelle, *Die Vorsokratiker. Die Fragmente und Quellenberichte* (Kröner, Stuttgart, 1968)
8. Bible, *Genesis*, in *Bible*, p. 22
9. M. Jammer, *Der Begriff der Masse in der Physik* (Wissenschaftliche Buchgesellschaft, Darmstadt, 1974)
10. A.L. de Lavoisier, *Traité élémentaire de Chimie* (Paris, 1789)
11. J. Dalton, W.H. Wollaston, *Die Grundlagen der Atomtheorie* (Akademische Verlagsgesellschaft Geest & Portig, Leipzig, 1983)
12. H. Bondi, Negative mass in general relativity. Rev. Mod. Phys. **29**, 423–428 (1957)
13. A. Einstein, Zur Elektrodynamik bewegter Körper. Ann. Phys. **17**, 891–921 (1905)

14. N. Bakalis, *Handbook of Greek Philosophy: From Thales to the Stoics: Analysis and Fragments* (Trafford Publishing, 2005)
15. Wikipedia, *Democritus*. http://en.wikipedia.org/wiki/Democritus (2009)
16. M. Kochsiek, M. Gläser (eds.), *Comprehensive Mass Metrology* (Wiley/VCH, Berlin, 2000)
17. M. Kochsiek (ed.), *Handbuch des Wägens*, 2nd edn. (Vieweg, Braunschweig, 1985)
18. W. Stukeley, *Memoirs of Sir Isaac Newtons Life* (Royal Society, London, 1752)
19. M. Engel, *Die Gravitation im Test – Gravitationskonstante und Äquivalenzprinzip*. http://www.pi5.uni-stuttgart.de/lehre/hauptseminar2001/Gravitationskonstante/Gravitation_2ndversion.htm, (2002).
20. P.J. Mohr, B.N. Taylor, D.B. Newell, CODATA recommended values of the fundamental physical constants: 2006. Rev. Mod. Phys. **80**, 633–730 (2008)
21. CODATA, *The Committee on Data for Science and Technology, 5 rue Auguste Vacquerie, F-75016 Paris, France*. http://www.codata.org/ (2010)
22. O.V. Karagioz, V.P. Izmailov, Izmer. Tekh. **39**(10), 3 (1996)
23. O.V. Karagioz, V.P. Izmailov, Meas. Tech. **39**, 979 (1996)
24. C.H. Bagley, G.G. Luther, Phys. Rev. Lett. **78**, 3047 (1997)
25. J.H. Gundlach, S.M. Merkowitz, Measurement of Newton's constant using a torsion balance with angular acceleration feedback. Phys. Rev. Lett. **85**, 2869 (2000)
26. T.J. Quinn et al., Phys. Rev. Lett. **87**, 111101 (2001)
27. U. Kleinevoß, *Bestimmung der Newtonschen Gravitationskonstanten G* (Universität Wuppertal, Wuppertal, 2002)
28. T.R. Armstrong, M.P. Fitzgerald, Phys. Rev. Lett. **91**, 201101 (2003)
29. Z.-K. Hu, J.-Q. Guo, J. Luo, Phys. Rev. **D71**, 127505 (2005)
30. S. Schlamminger et al., Phys. Rev. **D74**, 082001 (2006)
31. Aristoteles, a.t., ed., Questiones mechanicae, in *Kleine Schriften zur Physik und Metaphysik*, ed. by P. Gohlke (Paderborn 1957)
32. *Gravitation*. http://en.wikipedia.org/wiki/Gravitation (2007)
33. *Galileo Galilei*. http://en.wikipedia.org/wiki/Galileo_Galilei; http://de.wikipedia.org/wiki/Galileo_Galilei (2006)
34. G. Galilei, *De motu antiquiora* (Pisa, 1890)
35. A. Mudry, *Galileo Galilei: Schriften-Briefe-Dokumente 1586–1638* (Rütten & Loening, Berlin, 1985)
36. G. Galilei, *Theoremata circa centrum gravitatis solidorum* (Pisa, 1585/86)
37. G. Galilei, *La Bilancetta* (Pisa, 1588)
38. A. Einstein, Die Grundlagen der allgemeinen Relativitätstheorie. Ann. Phys. **49**, 769–822 (1916)
39. *Equivalence principle*. http://en.wikipedia.org/wiki/Equivalence_principle (2007)
40. G. Sauerbrey, Verwendung von Schwingquarzen zur Wägung dünner Schichten und zur Mikrowägung. Z. Phys. **155**, 206–222 (1959)
41. E. Robens, V.M. Mecea, In memoriam prof. dr. Günter Sauerbrey. J. Therm. Anal. Calorim. **86**, 7 (2006)
42. G. Dragoni, G. Maltese, La misure di massa de Quirino Majorana nella ricerca sull'absorbimento della gravità, in *La Massa e la Sua Misura—Mass and Its Measurement*, ed. by L. Grossi (CLUEB, Bologna, 1995), pp. 66–72
43. Wikipedia, *Earth's gravity*. http://en.wikipedia.org/wiki/Earth%27s_gravity (2010)
44. A. Lindau, *Gravity Information System of PTB*. http://www.ptb.de/cartoweb3/SISproject.php (2007), Physikalisch-Technische Bundesanstalt, Braunschweig
45. T. de Brahe, *Opera Omnia. Sive astronomiae instauratae progymnasmata. In duas partes distributa, quorum prima de restitutione motuum* (Olms, Frankfurt/Hildesheim, 1648/2001)
46. J. Kepler, Astronomia Nova, in *Bibliothek des verloren gegangenen Wissens*, vol. 58, ed. by F. Krafft (Marixverlag, 1609/2005)
47. J. Kepler, in *Harmonices Mundi libri V*, ed. by F. Krafft (Marixverlag, 1619/2005)
48. J. Kepler, *Epitome Astronomiae Copernicanae* (Frankfurt, 1609)

49. A.A. Michelson, E.W. Morley, On the relative motion of the Earth and the luminiferous ether. Am. J. Sci. **34**(208), 333–345 (1887)
50. A. Einstein, Ist die Trägheit eines Körpers von seinem Energieinhalt abhängig? Ann. Phys. **18**, 639–641 (1905)
51. A.M. Schwartz, *Novel equivalence principle tests*. http://www.mazepath.com/uncleal/eotvos.htm (2007)
52. S. Vitale et al., *The science case for LISA Pathfinder*, vol. 1 pp. 1–25 (ESA-SCI, 2007)
53. Wikipedia, *Lorentz covariance*. http://en.wikipedia.org/wiki/Lorentz_covariance (2010)
54. B. Haisch, A. Rueda, H.E. Puthoff, Inertia as a zero-point field Lorentz force. Phys. Rev. A **49**(2), 678–694 (1994)
55. I. Ciufolini, J.A. Wheeler, in *Gravitation and Inertia*, ed. by P.W. Anderson, A.S. Wightman, S.B. Treiman (Princeton University Press, Princeton, 1995)
56. J. Philoponus, *Corollaries on Place and Void* (Cornell University Press, Ithaca, 1987)
57. S. Stevin, *De Beghinselen der Weeghconst (Principles of the Art of Weighing)* (Leyden, 1586)
58. E.J. Dijksterhuis, *The Principal Works of Simon Stevin* (Swets & Zeitlinger, Amsterdam, 1955)
59. G. Galilei, *Sidereus Nuncius* (Thomam Baglionum, Venezia, 1609)
60. F.W. Bessel, Versuche über die Kraft mit welcher die Erde Körper von verschiedener Beschaffenheit anzieht. Ann. Phys. Chem. (Poggendorff) **25**, 401 (1832)
61. L. Southerns, Proc. R. Soc. Lond. **84**, 325 (1910)
62. P. Zeeman, Proc. K. Ned. Akad. Wet. **20**, 542 (1917)
63. P. Zeeman, Proc. K. Akad. Amsterdam **20**(4), 542 (1918)
64. L.R.v. Eötvös, D. Pekár, E. Fekete, Beiträge zum Gesetze der Proportionalität von Trägheit und Gravität. Ann. Phys. (Leipz.) **373**(9), 11–66 (1922)
65. H.H. Potter, Proc. R. Soc. Lond. **104**, 588 (1923)
66. H.H. Potter, On the proportionality of mass and weight. Proc. R. Soc. Lond. Ser. A, Math. Phys. Sci. **113**(765), 731–732 (1927)
67. J. Renner, Mat. és természettudományi ertsitö **53**, 542 (1935)
68. P.G. Roll, R. Krotkov, R.H. Dicke, Ann. Phys. (NY) **26**, 442 (1964)
69. V.B. Braginsky, V.I. Panov, Zh. Eksp. Teor. Fiz. **61**, 873 (1971)
70. V.B. Braginsky, V.I. Panov, Sov. Phys. JETP **34**(3), 463 (1972)
71. I.I. Shapiro, C.C. Counselman III., R.W. King, Verification of the principle of equivalence for massive bodies. Phys. Rev. Lett. **36**, 555 (1976)
72. G.M. Keiser, J.E. Faller, Eötvös experiment with a fluid fiber. Bull. Am. Phys. Soc. **24**, 579 (1979)
73. T.M. Niebauer, M.P. McHugh, J.E. Faller, Galilean test for the fifth force. Phys. Rev. Lett. **59**(6), 609–612 (1987)
74. B.R. Heckel et al., Experimental bounds on interactions mediated by ultralow-mass bosons. Phys. Rev. Lett. **63**(25), 2705–2708 (1989)
75. E.G. Adelberger et al., Testing the equivalence principle in the field of the Earth: particle physics at masses below 1 μeV? Phys. Rev. D **42**(10), 3267–3292 (1990)
76. S. Baeßler et al., Improved test of the equivalence principle for gravitational self-energy. Phys. Rev. Lett. **83**(18), 3585 (1999)
77. S. Schlamminger et al., Test of the equivalence principle using a rotating torsion balance. Phys. Rev. Lett. **100**(4), 041101 (2008)
78. F. Everitt, P. Worden, *Satellite Test of the Equivalence Principle (STEP). A Cultural History of Gravity and the Equivalence Principle*. http://einstein.stanford.edu/STEP/ (Stanford University, 2010)
79. B. Knispel, Pulsare mit dem Heimcomputer entdeckt. Spektrum Wiss. **12**, 78–82 (2010)
80. K. Schwarzschild, *Über das Gravitationsfeld eines Massenpunktes nach der Einsteinschen Theorie*. Sitzungsber. Preuss. Akad. d. Wiss. (1916), pp. 189–196
81. K. Schwarzschild, *Über das Gravitationsfeld eines Kugel aus inkompressibler Flüssigkeit nach der Einsteinschen Theorie*. Sitzungsber. Preuss. Akad. d. Wiss. (1916), pp. 424–434

82. J. Droste, On the field of a single centre in Einstein's theory of gravitation. Proc. K. Ned. Akad. Wet. **17**(3), 998–1011 (1915)
83. P.O. Hess, W. Greiner, Pseudo-complex general relativity. Int. J. Modern Phys. E (2009)
84. W. Greiner, *Es gibt keine schwarzen Löcher – Von Einstein zu Zweistein, in FIAS* (Frankfurt am Main, 2010)
85. Wikipedia, *Wave-particle duality.* http://en.wikipedia.org/wiki/Wave%E2%80%93particle_duality, (2009)
86. I. Newton, *Optics* (London, 1704)
87. C. Huygens, *Traité de la lumière* (Leiden, 1690)
88. J.C. Maxwell, A dynamical theory of the electromagnetic field. Philos. Trans. R. Soc. Lond. **155**, 459–512 (1865)
89. A. Einstein, Über einen die Erzeugung und Verwandlung des Lichtes betreffenden heuristischen Gesichtspunkt. Ann. Phys. **322**(6), 132–148 (1905)
90. A.H. Compton, A quantum theory of the scattering of X-rays by light elements. Phys. Rev. **21**(5), 483–502 (1923)
91. L.V.P.R. de Broglie, *Recherches sur la théorie des quanta* (Paris, 1924)
92. M. Arndt et al., Wave–particle duality of C60. Nature **401**, 680–682 (1999)
93. Wikipedia, *Standard Model.* http://en.wikipedia.org/wiki/Standard_Model (2008)
94. E. Adelberger, B. Heckel, C.D. Hoyle, Testing the gravitational inverse-square law. Phys. World **18**, 41–45 (2005)
95. P. Higgs, Broken symmetries, massless particles and gauge fields. Phys. Lett. **12**, 132 (1964)
96. P. Higgs, Broken symmetries and the masses of gauge bosons. Phys. Rev. Lett. **13**, 508 (1964)
97. P. Higgs, Spontaneous symmetry breakdown without massless bosons. Phys. Rev. **145**, 1156 (1966)
98. F. Englert, R. Brout, Broken symmetry and the mass of gauge vector mesons. Phys. Rev. Lett. **13**, 321 (1964)
99. G.S. Guralnik, C.R. Hagen, T.W.B. Kibble, Global conservation laws and massless particles. Phys. Rev. Lett. **13**, 585 (1964)
100. S. Dürr et al., Ab initio determination of light hadron masses. Science **322**, 1224 (2008)
101. Wikipedia, *Theory of everything.* http://en.wikipedia.org/wiki/Theory_of_everything (2008)
102. P.S.d. Laplace, *Essai philosophique sur les probabilités.* 18, Paris
103. F. Cardarelli, *Encyclopaedia of Scientific Units, Weights and Measures, Their SI Equivalences and Origins* (Springer, Heidelberg, 2003)
104. S.-Y. Lan et al., *A clock directly linking time to a particle's mass.* http://www.sciencemag.org/content/ea...cience.1230767, in Science journal. 2013-01-10
105. T. Funck, V. Sienknecht, Determination of the volt with the improved PTB voltage balance. IEEE Trans. Instrum. Meas. **40**, 158–161 (1991)
106. V. Bego, Determination of the volt by means of voltage balances. Metrologia **25**, 127–133 (1988)
107. V. Bego et al., Progress in measurement with ETF voltage balances. IEEE Trans. Instrum. Meas. **42**, 335–337 (1993)
108. W.K. Clothier et al., Determination of the volt. Metrologia **26**, 9–46 (1989)
109. F. Cabiati, Alternative methods for relating electrical to mechanical quantities through a power equation. IEEE Trans. Instrum. Meas. **40**, 110–114 (1991)
110. J.W.G. Wignall, Proposal for an absolute, atomic definition of mass. Phys. Rev. Lett. **68**, 5–8 (1992)
111. I. Mills et al., *Quantities, Units and Symbols in Physical Chemistry*, 2nd edn. (Blackwell Science, Oxford, 1993)
112. *International System of Units.* http://en.wikipedia.org/wiki/SI (2006)
113. E. Göbel, I.M. Mills, A.J. Wallard (eds.), *The International System of Units (SI)*, 8th edn. (Bureau International des Poids et Mesures, Sèvres, 2006)
114. L. Euler, *Disquisitio de Billancibus.* Commentari Acadimiae Scientiarium Imperialis Petropolitanae, 1738/1747. **10**

115. J.J. Clement-Mullet, Pesanteur spécifique de divers substances minérales (Excerpt de Ayin-Akbari). J. Asiat. **11**, 379–406 (1858)

116. E. Wiedemann, Über die Kenntnisse der Muslime auf dem Gebiete der Mechanik und Hydrostatik. Arch. Gesch. Nat.wiss. **2**, 394–398 (1910)

117. Al-Chazini, *Buch der Waage der Weisheit.* 1120, Merw

118. IUPAC, *Quantities, Units and Symbols in Physical Chemistry*, 3rd edn. (RSC Publishing, Cambridge, 2007) (prepared by E.R. Cohen et al.)

119. Council-EEC, *Council Directive 80/181/EEC of 20 December 1979 on the approximation of the laws of the Member States relating to units of measurement and on the repeal of Directive 71/354/EEC, as amended with Directive 89/617/EEC (which changed the cutoff date in article 3.2to31December1999) and Directive 1999/103/EC* 1979–2006, Brussels, Eur-Lex

120. BIPM, *Bureau International des Poids et Mesures, Pavillon de Breteuil, F-92312 Sèvres cedex, France.* http://www.bipm.org/

121. P. de Bievre, H.S. Peiser, Atomic weight—the name, its history, definition, and units. Pure Appl. Chem. **64**(10), 1535–1543 (1992)

122. Wikipedia, *Atomic mass unit.* http://en.wikipedia.org/wiki/Atomic_mass_unit (2008)

123. M.E. Wieser, Atomic weights. Pure Appl. Chem. **78**, 2051–2066 (2006

124. Wikipedia, Electronvolt. http://en.wikipedia.org/wiki/Electronvolt (2008)

125. E. Robens et al., On the terms mass and weight. J. Therm, Anal. Cal. (2008)

Chapter 2
Weights

The term 'weight' denotes the force acting on the mass of a body in the gravitational field. Furthermore it designates an object of known mass which is in conformation with a system of units. Weights are used for comparison with the mass of the weighing items. Originally weights were made of stone or earthenware. For light weights other natural material like seeds and glass had also been used. Later on these materials were largely replaced by metal. Also the kilogram prototypes are weights. There are many forms of weight, which fall into three major categories: geometric (Fig. 2.1), figurative (Figs. 2.2) and symbolic (Fig. 2.3) [1]. Often weights have inscriptions. Many Greek weights show the inscription 'Zeus' which should obviously demonstrate the right weight (Fig. 2.1). Similar seems to be the case for weights shaped as head of a god (Fig. 2.4). Often—today always—figures are engraved giving the value of the weight.

Also for non-metrological purposes weights are used in order to fix light objects like textiles or paper, table-cloth, carpets, fur skins or material of tents. Those weights occasionally were designed artistically and they may have been confused with real weight stones.

2.1 The Origin of Weights

Trade is based in principle on the value of goods to be exchanged. In addition to the 'inner value' of the object the amount as an extensive quantity determines the value. Therefore soon for exchange of larger amounts the measurement of quantities was required. First the volumetric measurement was used in particular after the invention of ceramic beakers. The origin of gravimetry is obscure but doubtless it was at a later period. Volume and mass units often had the same designations and represented more a value rather than a physical property. Much later when coins had been introduced, these also took such similar designations.

The discovery of weights is the earliest indication of weighing. Of course it is often difficult to identify an object like a worked up stone as a weight or

E. Robens et al., *Balances*, DOI 10.1007/978-3-642-36447-1_2,
© Springer-Verlag Berlin Heidelberg 2014

Fig. 2.1 Greek weight with inscription 'Zeus'

to distinguish a set of small weights from game figures. So to identify weight stones it is important to find series of weight stones and in particular objects with inscriptions. An antique series is shown in Fig. 2.4a and modern examples in Fig. 2.4b.

The balance at first was used to determine the value of a commodity rather than its mass. This becomes obvious when recognising that

- Some weights were made in the form of the goods they represented (Figs. 2.2a, b,d).
- We find equal designations for weights with different masses for different materials with regard to the value of the material. E.g. in the Egyptian Middle Kingdom the Gold Deben corresponds to 91 g and this was equivalent to 2.5 Silver Deben or 200 Copper Deben [2].
- Even today we have 'small' and 'large' or 'short' and 'long' units which consider parts of the weighed bodies of less value or which indicate the differences of similar weight systems in different countries.
- Initially we had identical units of money and weights. Examples are the Babylonian Royal Mina (~1.01 kg), the Greek Drachma (~10 g). The Roman Pondus (0,328 kg) survived as British or American Pound and German Pfund (0.35–0.56 kg), and the Pound Sterling is still British currency. In addition to the double meaning the Babylonian Talent (29 kg) [3] has the symbolic significance of aptitude.
- Identical naming of the instrument and the commercial calculating sheet 'balance' demonstrate the equivalence of mass and value in old times.

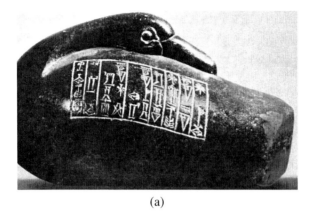

(a)

(b)

Fig. 2.2 (**a**) Hethytic weight in the form of a duck. (**b**) Weights of Sumer, Akkad, Assur, Babylon. Some in form of grain. Pfundsmuseum, Kleinsassen, Germany. (**c**) Weight collection of the Egyptian Museum, Cairo, Egypt. Weights of geometric shape. (**d**) Weight collection of the Egyptian Museum, Cairo, Egypt. Weights of different shape, one in form of a sheep's head

2.1.1   Development of the Weight Systems from the Weight of Grains

There are reasons to assume that first the weighings were for the measurement of amounts of grain and seed because there are some advantages in comparison with volumetric measurements. So the basic unit of weight may have been originally the mass of a grain of corn, wheat (Fig. 2.5), barley, mustard seed, or carob (Fig. 2.6) or of a number of these [4]. Indeed the notations 'gran' and 'carat' (or equivalent words in other languages) have been used for weights from antiquity until now.

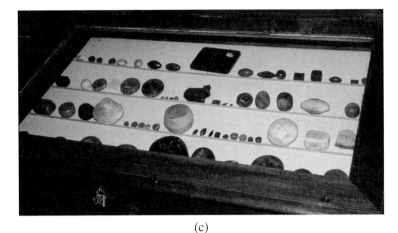

(c)

(d)

Fig. 2.2 (Continued)

The original 'grain' weight found in sets of weights in the Mediterranean area may be based upon a wheat seed [2, 5]. Antique wheat is different from modern varieties, it includes the species einkorn (Triticum monococcum), emmer (Triticum dicoccum), hard corn (Triticum durum), and kamut (Triticum turgidum). The masses

Fig. 2.3 Steelyard with
sliding weight shaped as head
of a goddess. Bronce, Rome
2nd century. The Czartoryski
Collection, Krakow, Poland.
AD

of such grains differ and it was standardised in different regions with figures varying between 52 and 73 mg. As an approximate value 65 mg is a reasonable estimate. (The exact value the Troy grain is indicated in Table 2.5.) The Roman
granum = 1/20 scrupulum = 1/60 drachma = 1/480 uncia were used until recently
as a weight in the drug and cosmetic market as well as in medical and pharmaceutical sciences.

Counts of barleycorns (Hordeum vulgare) were also used as a reference to the
grain weight of the Hebrew gold shekel (she = wheat, kel = bushel). Dried carob
beans (Ceratonia siliqua) became the medium, at some latter era, for the jeweller's
weights, the carat (1 ct = 0.2 g), in weighing out quantities of precious gems and
gold. It was assumed that carob beans are very uniform in mass. This is not the case,
but it is possible indeed by visual inspection to find out easily beans of a standard
mass value.

2.1.2 Development of Weights for Coarse Materials

Besides that grain system for weighing small quantities, a weight system had probably been developed independently for comparison and control of larger quantities
of goods of high value like pieces of meat or metal and for weighing tithes and tributes. Because the use of a balance replaced barter, such weights occasionally were

(a)

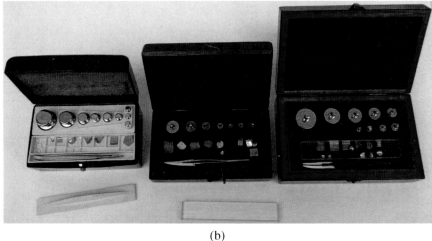

(b)

Fig. 2.4 (**a**) Assyrian set of weights in form of lions. (**b**) Modern and historical sets of weights. Weights above 1 g have geometric shape of cylinder with a handle allowing operation with tweezers. Weights below 1 g are geometric shaped plates. Its form symbolises its mass and allows for easy identification. Gottlieb Kern & Sohn, Albstadt-Ebingen, Germany & Sartorius, Göttingen, Germany

shaped in the form of the goods they represented, e.g. animals (Fig. 2.2a) or parts of animals (Fig. 2.2c). Weights from Sumer, Akkad, Assur Babylon often are shaped in form of grains (Fig. 2.2b).

Fig. 2.5 (**a**) Wheat ear.
(**b**) Wheat grains

(a)   (b)

Fig. 2.6 Carob capsicum
with seed

2.2 Antique Weight Systems

Weights of stone and of metal were used to determine the value of commodities. The oldest weights from before 3000 BC were found widely in the area of eastern Mediterranean up to the Indus valley and some hundred years later in the Huanghe valley (China) [4]. Different weight systems have been developed at different places with regard to the local need, and subsequently efforts of having a unified system had been undertaken. Furthermore small and big weights were found widely and most probably there had also been attempts at unifying the wide measuring range of balances. It is, however, hardly believable that anyone in antiquity counted several thousand grains in order to define a pound. Most probably only a loose connection existed between small and heavy weights.

Most findings of weight stones can be classified according to the different counting systems which existed in ancient times, the basis of which included 6, 10, 12, 24 and 60. Nevertheless it is almost impossible to find exact values for antique standard weights. On the contrary the values vary much more than to be expected by the relative sensitivity of the balances of those times which were in the per mille range. There are several reasons:

- There are few findings which can be identified clearly as weight prototypes.
- Weights found always show detritions, corrosion or damage and thus need corrections in their mass value.
- Derivations from the prototype and corresponding copies show deviations.
- In a wide region, local differences are to be expected despite standardisation. In adjacent regions divergences of the units are often observed.
- Up to now in Mediterranean countries it was usual to give an accurate weight for purchased goods (or to cheat with somewhat less). Therefore in the Bible as well as in the Koran we find many recommendations of correct weighing and the use of fair weight. However, this does not imply using the exact standard weight.

- Divergences between standard masses in adjacent regions facilitate bargaining in commerce.

Balances of that time usually had a relative sensitivity in the per mille range. This is in accordance with the needs of those times. Therefore we can assume likewise that the mass of a standard prototype in antiquity was not realised and determined better.

Because of the lack of prototypes the standard values of antique weight systems are determined usually by averaging the strongly deviating values of many findings. By such a mathematical operation a large number of decimal digits may be obtained. Such figures, however, are insignificant and should be rounded up to avoid misinterpretation. It is misleading to determine today the value of antique mass units more accurately than it was possible in those times. Therefore in the survey given in Table 2.1 and following ones figures are presented only to three decimal places.

2.2.1 Egypt

More than 4000 old Egyptian weights have been found, in particular from the excavations of W. Flinders Petrie [6].

> "In several groves at Naqadeh were cylindroid stones with domed ends. They were never worn, and had no use as implements. On comparing the weights of them they all agree within limits of variation of the gold standard, *nub*, the *beqa* of Palestine, which was certainly known in the 4th dynasty, by the weight of Khufu, and in the 1st dynasty by the gold bar of Aha."

Other objects assumed as weights are in the form of cylinders with flat or rounded base, rounded oblongs, rings, cones, or double cones. The materials are limestone, porphyre, breccia, alabaster, and syenite. Some are coloured or decorated with line ornaments. The weights range up to about 8 kg; the unit seems to be 13 g and the weights have multiples of $\frac{1}{2}$, 6, 20, 30, 40. Most of them are in the Petrie Museum and in the Science Museum in London or in the Egyptian Museum, Cairo [4]. Many of these weights can be identified by line markings or inscriptions from which the respective basic unit could be recognised. Nevertheless it is difficult to derive weight units because of abrasion, corrosion or damage to the objects. Furthermore, regarding the wider region, and the fact that Upper and Lower Egypt were not always unified, local differences are to be expected despite standardisation. The shapes of the weights differ widely—geometrical like cubes (rectangular or rounded), cylinders, spheres, cones, while others are artificially worked up.

In Egypt we have pre-dynastical findings but it is speculative to assume the same system of units as in the dynastic era. In every dynastical period the unit seems to have been called dbn which is pronounced usually as deben. Weight findings of the Old Kingdom show high discrepancies. Usually the gold deben is regarded as the leading standard. Dbn means 'ring' but we have no ring-shaped weights. However raw materials of precious metals were stored in the shape of rings and

Table 2.1 Ancient weight systems. Usually commercial balances up to the end of the middle ages had a relative sensitivity in the per mille range. Thus we give conversion factors with five digits as a maximum

Country	Period (approx. data)	Unit	Equivalents	Mass (approx. values)
Britannia		1 avoidupois pound	1 lb = 16 oz. = 256 dr. = 7000 gr	453.59273 g
United States		1 troy pound	12 oz.tr. = 96 dr.ap. = 5760 gran	373.24 g
		1 st. (stone)	175 troy pound	6.35029 kg
		144 pound	20 l.ctw. = 160 st. = 2240 lb	65.3173 kg
		1 long ton	20 sh.ctw. = 2000 lb	1016.05 kg
		1 short ton		907.18 kg
Burma, Pegu		Abucco		
		biza	4 agiros	196 g
		agiro	2 abbuci	
		abacco	$12\frac{1}{2}$ teccalis	
China		Shek, Stone	120 catty	72.6 kg
East Asia		Candareen, fēn; fàn	10 cash; 10 li, 100 yuan	378 mg
		mace, qian, qián, tsin	10 candareen	3.8 g
		Catti, kan, kati, Jin		500–600 g
		Picul, tam, dàn	100 catty	60.5 kg
		Tael, leun, liang	1/16 catty, 10 mace	38 g
Egypt, Old Kingdom	2025 BC– 1700 BC	dbn (deben)	1 gold dbn	12–14 g
			1 copper dbn	27 g
Egypt, New Kingdom	1700 BC– ~0	dbn (deben)	1 gold dbn = $2\frac{1}{2}$ silver dbn	91 g
			1 gold dbn = 200 copper dbn	
			1/6 silver dbn = 1/3 lead dbn	
		qdt (quedet, kite) shaty	1 qdt = 0.1 dbn	9 g
Greek		Talent	60 Mina	
		Mina	100 Drachma	
		Stater	2 Drachma	9–16 g
		Electrontrite	1/3 Stater	4.71 g
		Drachma	1/100 Silver mina	3.4 g
		Obol	1/6 Drachma	
Indus valley (Harrapan)	2600 BC			26–28 g
Malta		Ottav	1/8 uqija	3.31 g
		Quart	1/4 uqija	6.61 g
		Uquija		26.5 g

Table 2.1   (Continued)

Country	Period (approx. data)	Unit	Equivalents	Mass (approx. values)
		Ratal		794 g
		Qsima		992 g
		Wiżna		3.97 kg
		Quantar		79.4 kg
		Peżata		238 kg
Palestine		nub, beqa		13 g
		heavy talent		58.9 kg
Persia [21, 22]		Babylonian talent	7 Babylonian talents = 10 Attic talents	29 kg
Sumeria	3000 BC	Shekal	180 Se	8.36 g
		Mina	60 Shekal	500 g
		Talent	60 Mina	30 kg
		Se		0.047 g
Tartar		Öleş (доня)		44.4 mg
		Mısqal (золотник)		4.26 g
		Lot (лот)	3 mısqal	12.8 g
		Qadaq or göränkä (фунт)	32 lot	410 g
		Pot (пуд)	40 qadaq	16.4 kg
		Qantar	$2\frac{1}{2}$ pot	41.0 kg
		Berkovets (берковец)	10 pot	164 kg
Transylvania		Dram		3.2 g
		Font		0.5 kg
Vedic	1500–400 BC	Drone	4 Akada = 16 Prastha = 64 Kudova = 256 Pala = 1024 Karsa = 16 384 Masa	18 kg
		Aksa	1 Karsa (gold) = 16 Masa (gold)	17.6 g
		Dharna	10 Pala = Masaka (silver) = 1 Purana (silver) = 19 Sispava = 2/5 Kara = 24 Ratikitika	
Yemen		Ratl kabir	24 Uqiyya	673.6 g
		Ratl sana'ani	20 Uqiyya = 20 × 28.1 g	561.3 g
		Ratl saghir	16 Uqiyya	449.6 g
			1 Ratl = 2 nuss = 3 thilith = 4 rub'r = 8 thumun	

may be from that the notation of the standard weight was derived. A late Middle Kingdom account (Papyrus Boulaq 18 [7, 8]) refers to 'small' and 'large' deben. Other sources refer to a gold deben (12–14 g) and a copper deben (27 g) (Table 2.1). In the New Kingdom (about 1550–1069 BC) the system changed. Egypt favoured a decimal system and the new deben of 91 grams was divided into 10 qdt of about 9 g (pronounced as qedet or kite). Weights of $\frac{1}{2}$, 1, $2\frac{1}{2}$, 5, 10 and 50 qedets were used. Documents and inscriptions state 60 quedets being 1 Babylonian silver mina [9]. The deben-qedet system continued in use to the Late Period.

It should be noted that the Egyptians distinguished concrete weight from value, using different vocabulary for each. From the Old Kingdom to some point in the New Kingdom, the unit of value was Sna (pronounced as shena), perhaps written in Dynasty 19 as sniw. This time span is the same as that for the gold/copper deben system of 12–14 g and 27 g, and presumably there was some correlation between the two systems, the one for weight, the other for value. In the Rhind Mathematical Papyrus, about 1550 BC, Problem 62 deals with calculations of value for equal quantities of gold, silver and lead in one bag: the workings for the Problem have been taken to indicate that 1 deben is equal to 12 shena.

2.2.2 Near East

The measuring systems in the Mediterranean area were developed from those of Egypt, Sumeria and Babylon. The oriental (light) royal mina with about 500 g mass formed the basis of the Greek system of weights, with 100 drachma = 1 mina and 6 obolus = 1 drachma. The invention of minted coins removed the identity between amount/mass and value of trading goods. Nevertheless a close connection of the weight and currency system remained because the basis of the value of coins was governed by their material and weight. However, whereas the large currency units like mina, libra, uncia served as an operand only, weights of any size were in use and most probably standard weights were stored in the temples in the safekeeping of priests.

In the Near East 8 g and multiples are prominent but with ancient margins of error, it is difficult to determine whether a weight belongs to the 9 g Egyptian system, or to an 8 g Near Eastern system. In Babylon, as in Egypt, a large and a small unit of mass were in use. The mass of the large royal mine was about 1010 g and the small one about 505 g. In Mesopotamia and neighbouring countries likewise a large number of weights has been found [10]. In these countries including, at first in Sumeria and later in Babylon, Persia, Attica and the Aegean region, a mixed sexagesimal/decimal system was used. In Sumeria a standard shekel of 8.4 g existed. From the Shekel multiplied by 60 the Sumerian mina was derived corresponding to about 500 g. Multiplying again by 60 gives the talent corresponding to about 30 kg (Babylonian talent). Dividing the shekel by 180 results in the se (corn or grain of corn) corresponding to about 0.047 g. Because such a mass could hardly be measured by scales at that time this value may have served indeed only as an operand.

The talent is based on the volume of an amphora. It was the mass of water required to fill an amphora, approximately 30 kg. A Greek (or Attic) talent, was 26 kilograms, a Roman talent was 32 kilograms, an Egyptian talent was 27 kilograms. Ancient Israel, and other Levantine countries, adopted the Babylonian talent, but later revised the mass. The heavy common talent, about after the year 0, was 58.9 kilograms.

In about the 5th century BC the Libra was introduced, and its value had been constant for a long period. According to the Greek system 1 Libra = 12 Unciae = 48 Sicilici = 72 Sextulae = 98 Drachmay = 2888 Scripula = 576 Oboli = 1728 Siliquae.

Due to its geographic situation Yemen was a crossing point of important trade routes between Eastern Africa, Asia, Arabia and towards the Mediterranean regions. So until today we find weights and coins of different kind [11, 12]. Nevertheless a special weight system had been standardised.

2.2.3 India

In the Indus river valley region (now Pakistan) weights of the bronze era [13] since about 2600 BC have been found. Uniform, polished stone cubes discovered in early settlements were probably used as weight-setting stones in balance scales [14]. Although the cubes bear no markings, their weights are multiples of a common denominator based on multiples of 0.05, 0.1, 1.2, 5, 10, 20, 50, 100, 200 and 500, where the base unit was about 28 g. It is interesting to note that the cubes are made of many different kinds of stones with varying densities.

In India, in Vedic times (1500–400 BC) different weights and nomenclatures for its use of different merchandise existed [15]. For food, grains and similar items the basic unit was the drone ≈ 18 kg. Each successive unit is in multiples of four of its predecessor.

2.2.4 China

In China the common weights and measures were called 'du', 'linag', 'heng', meaning length, capacity or volume, and weight, respectively [16]. Wu Chengluo's book "A History of Weights and Measures in China" of 1937 [17] may be considered as the first monograph from that country. This book, however, suffers from the absence of profound research supported by any ancient artefacts. A modern and critical survey is given by the abundantly illustrated book "Investigations on Weights and Measures through the Ages of China" by Qiu Guangming [16] and a survey on the literature is presented by a recent bibliography [18].

In China during the Zhou Dynasty, 1122 to 255 BC [19], there had been a direct connection between the 'liang as a weight' (12 grams when applied to coinage) and the liang as a denomination of coinage (Fig. 2.7). About the time the Chin Dynasty established control over China (and possibly a little earlier), the pan liang

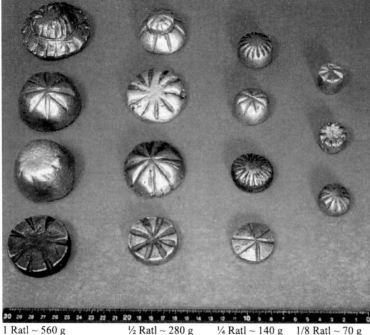

1 Ratl ~ 560 g   ½ Ratl ~ 280 g   ¼ Ratl ~ 140 g   1/8 Ratl ~ 70 g

(a)

(b)

Fig. 2.7 (a) Jemen Ratl weights 2004, © Dr. Theuerkauf. (b) Antique Chinese money/weight

(or 1/2 liang) coinage was introduced as this weight standard (about 6 g), but very quickly the connection between the weight and the monetary unit ceased to apply. Findings of specimens within a single hoard occasionally will be of uniform diameter but the weight can vary significantly between 2 to 18 g and diameters from 14 to over 34 mm. So the definition of weights in China at that time seems to be very doubtful.

In the Ch'in period 221 BC–206 BC defence was strengthened by creating the Great Wall. Furthermore, the lawmakers strengthened state power and control over the people. Weights and measures, and the Chinese writing system were unified.

In more recent years, one Jin was equal to 500 grams or about 1.1 pounds; it used to be divided into 16 liang, but now 10 liang. A qian is one-tenth of a liang [20].

2.2.5 Rome

The measuring system of the Roman republic (including Byzantium) was derived from the old East-Mediterranean systems [23]. Basic units of the Roman weight system were libra (*LB* or *lb*; libra = scales or balance) and uncia [9, 24]. Uncia (Table 2.7) means 1/12 and so we find this notation likewise for time, length and volume measures. Scrupulum, a tiny stone (from *scrupus* sharp stone), indicates a weight of 1/24 of an uncia. The standard weight may be connected first to the coinage in the silver drachme [25]. Many attempts have been made to find exact values by averaging the values of findings (taking into account wear and corrosion). Nevertheless resulting mean values have a scatter more than could be expected by the limits of sensitivity of the calibration procedure possible in those times. Standard values recommended recently are between 322.5 and 327.5 g. Considering the large area covered by the Roman Empire and its long era, obvious deviations of the standards existed.

2.3 Middle Ages Weight Systems

In the Middle Ages units of weights and coins that had developed form antiquity were still closely linked. Pursuing his father's Pippin reforms, Charlemagne ∼ AD 793/94 replaced both the coin and the weight system defining the silver pondus caroli (Pfund, Karlspfund, livre carolienne) [26] equal to 20 solidus (later shilling) or 15 unciae ∼ 407 g as the standard, from which 240 denarii (denier, later: Pfennige, penny) were minted. This is documented by a contemporary manuscript [27] as well as by the reports of the Frankfurt synode 794 [28]. The pondus caroli was the basis of both, the weight and the coinage system. It is disputed whether a deviating weight standard existed, a pound of 16 unciae or 256 denari ∼ 434 g [29]. The Carolingian denarius ∼ 1.7 g, the predecessor of the Pfennig, becomes the basis of weights and currency in the Middle Ages whereas for coinage the pound for currency was an operand only.

As a result of weakening of central power, mediaeval European weight and currency systems evolved due to the agriculture-intensive way of life. These systems may also be referred to as feudal weight systems, the units of which differed in the different states. Variations in the mass value resulted in complicated conversions

Table 2.2 Roman weights and coins

Weight	Equivalent	Mass
granum	1/3 chalkus = 1/72 drachma	47 mg
chalcus	1/8 obolus = 1/48 drachma	71 mg
siliqua	1/3 obolus = 1/18 drachma	188 mg
lupinus	1/96 uncia	282 mg
obolus	1/48 uncia = 1/2 scrupulum = 1/6 drachma	564 mg
scrupulum	1/24 uncia = 1/3 drachma	1.13 g
dimidia sextula	1/12 uncia	2.3 g
drachma	1/8 uncia = 1/2 shekel	3.39 g
sextula	1/6 uncia	4.55 g
sicilicus	1/4 uncia = 1 shekel = 2 drachma	6.77 g
banie sextulae	1/3 uncia	9 g
semuncia	1/2 uncia	14 g
uncia	4 shekel	27.1 g
sescuncia	$1\frac{1}{2}$ unciae	40.5 g
sextans	2 unciae	54.2 g
quadrans	3 unciae	81.3 g
trians	4 unciae	108 g
quincunx	5 unciae	136 g
semis	6 unciae	163 g
septunx	7 unciae	190 g
bes	8 unciae	217 g
dodrans	9 unciae	244 g
dextans	10 unciae	271 g
deunx	11 unciae	298 g
libra = as	12 unciae	327 g
mina		434 g

and control. As a system commercial weights (Krämergewicht) were established: 1 pound = 16 ounces = 32 lot = 128 Quentchen = 512 Pfennige = 1024 Heller. The lot was both, a volume and mass unit. However, almost every town had their own basis of measurement. Whereas in Nuremberg the pound equalled 510 g, in Berlin it was 467 g only. After 1200 the Kölner Mark (Cologne mark) weighing 234 g got supranational importance as a basis of currency [30]. As a derivate of the Cologne mark in the 13th century Alfonso III of Portugal defined the marco with 229g = 8 ounces as the national standard [31]. The marco become the first world standard of weights because Portuguese sailors and businessmen used it in world-wide trade. For weighing the Cologne pound with twice the mass (468 g) was used as a weight in commerce. In 1838 the Zollvereinsmark was defined with 234 g and in 1854 the Deutsche Zollverein (German customs-union) fixed the pound (Zollpfund) to be equal exactly to 500 g. Since 1884 use of the pound is forbidden in Germany for official measurements and in trade, but has survived unofficially until now as half a kilogram.

In the 12th century in France several variants of Pondus Caroli developed—in the reign of Louis VI, the Libra Parisi, and with the beginning of the 13th century the Livre Tournois. Simultaneously the Livre de Troyes was defined as the livre de poids-de-marc. In the late 17th century, the French emperor Philippe-Auguste replaced the existing unit of weight (the pound) with the marc, a unit of weight which remained in use until 1795 and which was used by most European countries but with different values.

In 1555 a system of pharmaceutical weights prevailed based on weights manufactured in red copper foundries at Nuremberg. The standard weight was the silver ounce of about 30 g. 12 times of its mass gave the pharmaceutical pound of about 360 g. 1 ounce = 8 drachme (3.75 g) = 24 scruples (1.25 g) = 480 granum (0.06 g). The granum was of corn or pepper. Sets of weights were sold in wooden cases as 'tant von Nurenberch'. This system was replaced in Germany first in 1870 when the metric decimal system became official.

A survey on the differing values of the pound is given in Table 2.3 and of the ounce in Table 2.4. As mentioned, uncia means 1/12 and so we find this notation occasionally for time and length. Still in use is the liquid ounce as a volume measure for perfume and for liquid drugs, though not allowed in the European Union in trade or in pharmacy.

2.4 Modern Ounce Systems

The industrial development in England during the 19th century resulted in the country coming to a leading position in commerce and trade. This and the establishment of the British Empire made the British version of the Middle Ages measuring system dominant. It is still in use in the United Kingdom, in the Commonwealth and in the United States. In the UK we have still remnants of the corresponding coinage system.

There are two variants of the British weight system based on different size of the pound and the ounce: the older troy system and the more recent avoirdupois system. Both are used simultaneously but for different materials. They have been reformed and a fixed relation to the metric system has been standardised. Officially those weight systems are forbidden for use in trade and commerce in the European Union. However, they are only slowly replaced by the metric system.

As late as the beginning of our millennium, two green grocers Neil Herron and Steve Thoburn from Sunderland in the UK refused to mark the price of his goods in metric measurements, using only pounds. (UK law requires all weights to be in grammes or kg, but allows other units also to be displayed in the price). Many supermarkets and shops display prices in pounds and kg. The green grocers were taken to court and charged and convicted. They still refused to change their prices from only

Table 2.3 Pound weights

Country	Pound unit	Equivalents	Mass g
Roman Empire	Pondus		327
Roman/German Empire	Pondus Caroli (Karlspfund)	20 solidus = 15 uncia ~ 240 denare	404–408
	Krämergewicht (Commercial pound)	16 Unzen = 32 Loth = 128 Quentchen = 512 Pfennige = 1024 Heller	467–510
	Apothekerpfund Nürnberger Pfund Pharnmaceutical pound	12 unzen = 96 Drachme =288 Scrupula = 5760 Granum	351–358
	Wiener Pfund	2 Wiener Mark	561
	Kölner Pfund	2 Kölner Mark	468
Austria, Hungary	Pfund	32 Lot = 128 Quentchen = 256 Achtel = 512 Sechzehntel = 512 Pfennige	560
Denmark	Pund	32 Lod = 100 Kvint = 1000 Ort	500
France	Livre Tournois		367
	Libra Parisi		459
	Livre de Troyes	2 Marcs = 16 Onces = 123 Gros = 384 Deniers	490
	Livre usuelle	4 Quarterons = 16 Onces = 128 Gros	
Italy (Rome)	Libra	12 Once	339
Portugal	Libra, Arratel	4 Quartas = 16 Onças = 128 Oitavas = 1024 Scropulos	459
Prussia	Pfund	32 Loth = 128 Quentchen	468
Germany (after 1854)	Pfund, Zollpfund		500.0
Russia	Funt	96 Solotniki = 9216 Doli	410
Sweden	Skålpund	100 Ort = 1000 Korn = 32 Lod = 128 Qvintin = 8848 Ass	425
Switzerland	Pfund	16 Onzas = 128 Drachmas	500.0
Spain	Libra	16 Onzas = 128 Drachmas = 384 Escrupulos	460
United Kingdom, USA	Pound	Avoirdupois pound lb	453.59237
		Troy pound	373.24

pounds. They came to be known as the 'Metric Martyrs' and got a lot of support from local people, who had sentimental and nationalistic feelings to the Imperial System of measurements. Later the EU and/or the UK government allowed other measurements units to be used in addition to the metric measurements. Meanwhile Steve Thoburn died (of natural causes!).

Table 2.4 Ounce units

Ounce unit	Country	Value
Uncia	Roman Empire	27.1 g
Onza	Bolivia	28.1 g
Onza	Argentinia, Mexico	28.8 g
Onza	Chile, Guatemala, Peru	28.7 g
Onza	Haiti, Honduras	30.5 g
Ons	Netherlands	100 g
Onza	Portugal	28.7 g
Uqiyya	Yemen	28.1 g
Maria Theresia Taler	Austria, Hungary	28.0688 g
Liquid ounce (fl. oz)	United Kingdom	28.413 ml
Liquid ounce (fl. oz)	USA	29.5735295625 ml
Shoe ounce	United Kingdom	1/64 inch = 0.396875 mm
Time ounce	United Kingdom	1/12 moment = 7.5 s
Troy ounze (ozt, oz.tr.) (Apothecaries Ounce ℥ U + 2125, *oz. ap.*, Feinunze)	United Kingdom, USA	31.1034768 g
Avoirdupois ounce (oz)	United Kingdom, USA	28.349523125 g

2.4.1 The Troy System

Based on the standard pound of the French town Troyes the troy system of weights was established in medieval times and applied throughout in the British Empire. In 1758 the troy pound was defined. One cubic inch (16.39×10^{-6} m^3) of distilled water, at 62 °F (16.67 °C), and at a barometric pressure of 30 inches of mercury (101.305 kPa), was defined as the weight of 252.458 troy grains (gr), each 64.79891 mg. The troy ounce (31.1034768 g) is 480 grains and is 1/12 of the troy pound ($\sim$373.242 g).

The troy ounce (Feinunze) (ozt, oz.tr.) troy ounze (ozt, oz.tr.) (*Apothecaries Ounce* ℥ U + 2125, *oz. ap.*, Feinunze) $\sim$31.1 g is about 10 percent heavier than the avoirdupois ounce. It was the unit in the apothecary system in the Middle Ages until recently. It is the only ounce customarily used in the pricing of precious metals, gold, platinum, silver, and gemstones on an international level.

It designates only the proportion of the precious metal; foreign matter and impurities are subtracted. The grain, which is identical in both the troy and avoirdupois systems, is used to measure arrow and arrowhead weights in archery and bullets and powder weights in ballistics. Grains were long used in medicine but have been largely replaced by milligrams.

In Scotland the Incorporation of Goldsmiths of the City of Edinburgh used a system in multiples of sixteen. Thus there were 16 drops to the troy ounce, 16 ounces to the troy pound, and 16 pounds to the troy stone.

Table 2.5 Troy system

Unit	Equivalent	Mass
grain (gr)	1/5760 ozt	64.79891 mg
ounce (ozt)	1/12 lb troy, 8 dr.ap. = 480 gr = 20 pennyweights	31.1035 g
pennyweight	24 gr	1.555 g
pound (lb troy)	12 ozt = 5 760 gr	373.2417216 g
stone (st)	14 lb = 1/2 qtr	∼6.35 kg
quarter (qtr)	28 lb = 2 st	∼12.7 kg
hundredweight (cwt)	112 lb = 4 qtr	∼50.8 kg
ton (t)	2 240 lb = 20 cwt	∼1 016 kg

2.4.2 The Avoirdupois System

The English mercantile and Tower systems are considered more modern. The avoirdupois pound is $14\frac{7}{12}$ ($\sim$14.583) troy ounces, since troy ounces are larger than avoirdupois ounces. 192 oz. = 175 oz.tr. = 5.44310844 kg. When the British began to use this system they included the stone, which was eventually defined as fourteen avoirdupois pounds. The quarter, hundredweight, and ton were altered, respectively, to 28 lb, 112 lb, and 2,240 lb for easier conversion between the units.

The avoirdupois system of weights is based on exactly 7 000 gr per pound which, in turn, was divided into sixteen ounces. It is the everyday system of weights used in the United States. It is still widely used by many people in Canada and the United Kingdom despite the official adoption of the metric system, including the compulsory introduction of metric units in shops. The word *avoirdupois* is from French *avoir de pois*, 'goods of weight' or 'goods sold by weight'. This term originally referred to a class of merchandise: *aveir de peis*, 'goods of weight', things that were sold in bulk and were weighed on large steelyards or balances. Only later did it become identified with a particular system of units used to weigh such merchandise.

The thirteen British colonies in North America (not including those that formed Canada), however, adopted the French system as it was. In the United States, quarters, hundredweights, and tons remain defined as 25, 100, and 2,000 lb respectively. The quarter is now virtually unused, as is the hundredweight outside of agriculture and commodities. If no ambiguity is required then they are referred to as the 'short' units, as opposed to the British 'long' units.

In the avoirdupois system, all units are multiples or fractions of the pound, which is now defined as 0.45359237 kg in most of the English-speaking world since 1959 [32].

Table 2.6 Avoirdupois system

Unit	British (long units)		USA customary units (short units)	
	Equivalent	Mass	Equivalent	Mass
grain (gr)	1/7000 lb	64.79891 mg		64.79891 mg
dram or drachm (dr)	1/256 lb = 1/16 oz	~1.772 g	1/256 lb = 1/16 oz	~1.772 g
ounce (oz)	1/16 lb = 16 dr = 437.5 gr	28.349523125 g	1/16 lb = 16 dr	~28.35 g
pound (lb)	16 oz = 7000 gr	453.59237 g	16 oz	~453.6 g
stone (st)	14 lb = 1/2 qtr	~6.35 kg		
quarter (qtr)	28 lb = 2 st	~12.7 kg	25 lb = 1/4 cwt	~11.34 kg
hundredweight (cwt)	112 lb = 4 qtr	~50.8 kg	100 lb = 4 qtr	~45.36 kg
ton (t)	2 240 lb = 20 cwt	~1 016 kg	2 000 lb = 20 cwt	~907.2 kg

2.4.3 The Fluid Ounce

The fluid ounce (*fl oz*) is not a mass unit but a volume unit used for liquids. It is still used today for perfume in the United States and the Commonwealth. 1 fl.oz. equals 1/160 imperial gallon ~ 28.4131 ml. In the United States it is 1/128 of a US gallon ~ 29.5735 ml.

2.5 The International Prototype Kilogram

Louis XVI charged a group of savants including Lavoisier to develop a new system of measurement. Their work laid the foundation for the 'decimal metric system', which has evolved into the modern SI. The original idea of the king's commission was to create a unit of mass that would be known as the 'grave'. By definition it would be the mass of a litre of water at the ice point (i.e. essentially 1 kg) "la pesanteur de l'eau, prise au terme de la glace et pesée dans le vide". Indeed, weighing in vacuum results in the *poids absolu*, the true mass. The definition was to be embodied in an artefact mass standard.

Weighing of an open vessel of water in vacuo, however, is impossible. In order to realise the new unit mass, therefore, hydrostatical weighings were performed, using a cylinder made by N. Fortin. First, A.L. Lavoisier and R.J. Haüy were charged with the measurements. In 1793, however, during the French revolution, the commission was 'cleaned' and Lavoisier was executed.

After the Revolution, the new Republican government took over the idea of the metric system Based on the law of the 18th Germinal of the year 3 (7 April 1795) the work was continued by a new commission which made some significant changes. The designation grave was changed to kilogramme. It was decreed "*Gramme*, le poids absolu d'un volume d'eau pure égal au cube de la centième partie du mètre, et à la température de la glace fondante." "the absolute weight of a volume of water equal to the cube of the hundredth part of the meter, and the temperature of melt-ing ice" [33–35]. However, since many mass measurements of the time concerned masses much smaller than the kilogram, they decided that the unit of mass should be the 'gramme' with 1/1000 of the mass of 1 kg. Nevertheless, since a one-gramme standard would have been difficult to use as well as to establish, Lefèvre-Gineau realised a mass of 1000 gram. For that purpose, work was also commissioned to precisely determine the mass of a cubic decimeter of water. Furthermore in 1799 the commission redefined the standard to water's most stable density point: the tem-perature at which water reaches maximum density, which was measured at the time as 4 °C. They concluded that one cubic decimeter of water at its maximum density was equal to 99.92072 % of the mass of the provisional kilogram made four years earlier. The results of mass determination were not reduced to vacuum conditions and differing air pressure was not regarded in their initial measurements [35].

That same year, 1799, an all-platinum kilogram prototype was fabricated with the objective that it would equal as close to, as was scientifically feasible for the day, the mass of a cubic decimeter of water at 4 °C. The prototype was presented to the Archives of the Republic in June and on 10 December 1799, the prototype was formally ratified as the Kilogramme des Archive and the kilogram was defined as being equal to its mass [35]. By 1875 the unit of mass had been redefined as the 'kilogram', embodied by a new artefact whose mass was essentially the same as the kilogram of the archives.

Since 1889, the SI system defines the magnitude of the kilogram to be equal to the mass of the International Prototype Kilogram (IPK) (Fig. 1.4). The IPK is made of an alloy of 90 % platinum and 10 % iridium (by mass) and is machined into a right-circular cylinder (height = diameter) of 39.17 mm to minimise its sur-face area. The addition of 10 % iridium improved on the all-platinum Kilogram of the Archives by greatly increasing hardness while still retaining platinum's many virtues: extreme resistance to oxidation, extremely high density, satisfactory elec-trical and thermal conductivities, and low magnetic susceptibility. The IPK and its six sister replicas are stored in an environmentally monitored safe in the lower vault located in the basement of the Bureau International des Poids et Mesures at Sèvres, France (BIPM) at Breteuil in Sèvres on the outskirts of Paris. Three independently controlled keys are required to open the vault. In addition, more than eighty copies of 1 kg prototypes in Pt/Ir have been manufactured by the BIPM for use as national prototypes and also the BIPM maintains a number of 1 kg copies for current use. These are compared with the IPK roughly every 50 years.

The IPK is one of three cylinders made in 1879. In 1883, it was found to be indis-tinguishable from the mass of the Kilogram of the Archives made eighty-four years prior, and was formally ratified as *the* kilogram by the 1st CGPM in 1889. Modern measurements of the density of Vienna Standard Mean Ocean Water—purified water that has a carefully controlled isotopic composition—show that a cubic decimeter of water at its point of maximum density, 3.984 °C, has a mass that is 25.05 parts per million less than the kilogram. This small, 25 ppm difference, and the fact that the mass of the IPK was indistinguishable from the mass of the Kilogram of the Archives, speak volumes for the scientists' skills over 209 years ago when making their measurements of water's properties and in manufacturing the Kilogram of the Archives.

Meanwhile it was detected that the IPK lost mass of about 0.05 mg in comparison with the copies. One reason may be more often cleaning procedures. IPK are cleaned by rubbing the surface quite hard with chamois leather. Initially the leather is cleaned several times by soaking in a mixture of ethanol and ether and wrung out. After that the surface is treated by a water steam jet and dried by touching with filter paper. It is recommended that a platinum-iridium standard should be cleaned and washed twice before comparison with another standard. It has been found that there was no significant change in mass after the second cleaning.

2.6 Standardisation of Weights and Calibration

2.6.1 Early Antiquity

In local regions as well as in the large empires of ancient times (Egypt, Rome, China, etc.) weight systems had been standardised and in the process of time also reforms occurred. In addition, in order to facilitate trade, efforts always had been undertaken to make the own system compatible with those of neighbouring empires.

Sumer is widely considered to be the earliest settled society in the world to have manifested all the features needed to qualify fully as a civilisation, dating back approximately to 7300 years ago. Most probably during that time standardisation of measures began and was taken over by the Babylonians. In the Babylonian Empire (~3000–539 BC) a state-supervised system of metrology existed. Besides weights, bearing the mark of the ruler to attest their correctness, measures of length chiselled into statues have also been found. The standard measures were deposited in temples in the safekeeping of priests who most probably were also empowered to adjust and control balances and weights. It should be noted that a temple in antiquity had not only a religious, but also a cultural and an economic significance. Large temples included workshops, commercial shops, education centres and served in the Greek era as banks. The oldest document of weight standardisation is a tablet in the tomb of Hesire in Saqqara (Egypt) (~2620 BC) which shows two balance beams and sets of weights and cylinders (Fig. 2.8) [10].

2.6.2 Antiquity

In the Roman Empire the most accurate standards, the exagia, were kept in Rome in the Capitol [9] probably in the temple of Juno Moneta. Juno Moneta was the goddess of the imperial mint situated nearby her temple and she is depicted as Juno Moneta Aequitas usually with scales in her hand. A close network of bases was distributed in every corner of territory. In the provinces, every legion received copies of the

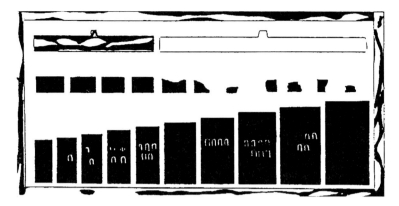

Fig. 2.8 Tablet showing two balance beams, weights and standard cylinders. Tomb of Hesire, Saqquara, Upper Egypt

exagia and in turn the cohorts received utility standards with which the measuring devices of the tradesmen and craftsmen could be compared.

Since the oldest times there was a close connection between volumes for the measurement of seed and weights and until now we have ton units both for volume and mass. Agricola assumed that already in Greece and then in the Roman Empire the units for volume and mass were fixed by special vessels 'sextarius' of calibrated volume to be filled with oil and the mass of oil giving the weight standard [36–38]. Already the Greek medical Galen should have reported on a calibration procedure by means of a special vessel 'libra horn'. The content of oil of that horn was 12 unciae corresponding to 10 Roman weight ounces. However, theories that the Roman weight standards are based on volume measurements could not be verified by findings of such vessels until now.

Solidi exagia are among the rarer and more interesting artifacts associated with Roman coins. They began to be used in 363, during the reign of Julian II, who instituted a policy of having an official coin weigher, a zygostates, in each municipality (Codex Theodosianus 12.7.2) [39]. This was meant to restore confidence in the solidus, which had become suspect due to widespread clipping and forgery. To make this task efficient and verifiable, weighing was performed with base metal exagia that were the same weight as an unadulterated solidus. Most exagia in the East were round, and seemingly were struck at Constantinople, since the mintmark CONS begins to appear on them late in the reign of Theodosius I. In the West, exagia tend to be square or rectangular (Fig. 2.9), and it is not known where they were produced [40]; Rome has been suggested, but three have been found in Trier, and it is possible that each mint produced their own. In the 6th century, Byzantine exagia came to be made of glass, which gave rise to that rather prolific tradition in the Muslim world.

Fig. 2.9 Exagium solidi 379–423, 4.34 g. Facing bust of Theodosius I, diademed, draped and cuirassed, in the centre, flanked by Honorius, diademed, draped and cuirassed on left, and Arcadius, diademed, draped and cuirassed on right. Rev. AV/GGG within wreath. Hunterian, vol. V, p. 424, 1, pl. 86, CW. This exagium is of the same type (and is struck from the same obverse die) as the piece illustrated in Hunterian, vol. V, pl. 86, C.W.; it shows three diademed, draped, Imperial busts, two being larger and one smaller. Since they represent Theodosius I in the center flanked by Arcadius on the right and Honorius on the left, we may narrow the timeframe for its issue to January, 393 through January, 395

2.6.3 Middle Ages

After the fall of the Roman Empire standardisation followed varying paths in the different states of the former empire. Increasing commercial relationships required a lively conversion of weight and coin values and strong supervisions at trading centres. An example may be given with the market of Yemen's capital Sana'a [12]. The market law (Quanun) was established by the imam Mutawakkil al-Qasim (1716–1727) on the basis of regulations from pre-Islamic times. In the 19th century the market master (Shaykh al Mashayikh) together with his assistants produced weights to be used and controlled the market. Weights and scales had been inspected and controlled by means of standard weights, and faulty samples were confiscated.

2.6.4 Mendenhall Order

The Mendenhall Order marked a decision to change the fundamental standards of length and mass of the United States from the customary standards based on those of England to metric standards. It was issued on April 5, 1893 by Thomas Corwin Mendenhall, superintendent of the US Coast and Geodetic Survey, with the approval of the United States Secretary of the Treasury, John Griffin Carlisle. Mendenhall ordered that the standards used for the most accurate length and mass comparison change from certain yard and pound objects to certain meter and kilogram objects, but did not require anyone outside of the Office of Weights and Measures to change from the customary units to the metric system. In the Mendenhall Order for references [32], all units of the avoirdupois are multiples or fractions of the pound, which is now defined as 0.45359237 kg in most of the English-speaking world since 1959. Whereas before it was a system of weights now it indicated also the mass of weighing goods.

In 1866 the Congress passed a law which allowed, but did not require, the use of the metric system. The US Coast and Geodetic Survey Office of Weights and Measures had on hand a number of metric standards, and selected the iron 'Committee Meter' and the platinum 'Arago Kilogram' to be the national standards for metric measurement; the standard yard and pound previously mentioned continued to be the standards for customary measurements. The conversions were 1 yard = 3600/3937 meter and 1 pound = 0.4535924277 kilogram.

A series of conferences in France between 1870 and 1875 lead to the signing of the 'Metric Convention' in 1875, and to the permanent establishment of the International Bureau of Weights and Measures (Table 2.7). The BIPM made meter and kilogram standards for all the countries that signed the treaty; the two meters and two kilograms allocated to the United States arrived in 1890, and were adopted as national standards.

The definitions of 1893 remained unchanged for 58 years, but increasing precision in measurements gradually made the differences in the standards in use in English-speaking countries important. Effective from 1st July, 1959, Australia, Canada, New Zealand, South Africa, the United Kingdom, and the United States agreed that 1 yard = 0.9144 meter and that 1 avoirdupois pound = 0.45359237 kilogram.

2.6.5 Introducing the Metric System

A month after the coup of 18 Brumaire, the metric system was definitively adopted in France by the First Consul Bonaparte (later Napoleon I) on December 10, 1799. During the history of the metric system a number of variations have evolved and their use spread around the world replacing many traditional measurement systems. Karl Friedrich Gauß (1777–1855) proposed the three-dimensional MKS-system with length [m], mass [kg] and time [s]. Wilhelm Eduard Weber (1804–1891 proposed the three-dimensional coherent CGS-system with units of length. The FPS-system based on length [imperial yard], mass [pound avoirdupois] and time [s] is still being used in Great Britain and in the USA. Besides four-dimensional variations of unities had been developed.

By the end of World War II it was recognised that additional steps were needed to promote a worldwide measurement system. As a result the 9th General Conference on Weights and Measures (CGPM), in 1948, asked the International Committee for Weights and Measures (CIPM) to conduct an international study of the measurement needs of the scientific, technical, and educational communities.

Based on the findings of this study, the 10th CGPM in 1954 decided that an international system should be derived from six base units to provide for the measurement of temperature and optical radiation in addition to mechanical and electromagnetic quantities. The six base units recommended were the metre, kilogram, second, ampere, Kelvin degree (later renamed the Kelvin), and the candela. In 1960, the 11th CGPM named the system the *International System of Units*, abbreviated SI

Table 2.7 Organisations and authorities engaged in metrological work concerning mass

AGME	Arbeitsgemeinschaft Mess- und Eichwesen. (Organisation of the standardisation authorities of the German counties—Landesämter für Mess- und Eichwesen)	www.agme.de
ASTM international	(Originally American Society for Testing and Materials) 100 Barr Harbor Drive, West Conshohocken, PA, 19428-2959, USA	www.astm.org
BAM	Bundesanstalt für Materialforschung und -prüfung. Unter den Eichen 87, D-12205 Berlin, Germany	www.bam.de
BIPM	Bureau International des Poids et Mesures, Pavillon de Breteuil, F-92312 Sèvres, France	www.bipm.org
CCM	Consultative Committee for Mass and related Quantities—Comité consultatif pour la masse et les grandeurs apparantées	www.bipm.org
CCQM	Consultative Committee for amount of Substance: Metrology in Chemistry—Comité consultatif pour la quantité de matière: Métrologie en chimie	www.bipm.org
CCU	Consultative Committee for Units—Comité consultatif des unites	www.bipm.org
CGPM	Conférence Génerale des Poids et Mesures	www.bipm.org
CIPM	Comité international des poids et mesures. International Bureau of Weights and Measures. Pavillon de Breteuil, F-93212 Sèvres, Haute-de-Seine, France	www.bipm.org
CNAM	Conservatoire National des Arts et Metiers 292 Rue Saint-Martin, 75003 Paris, France	www.cnam.fr
CODATA	Committee on Data for Science and Technology, Paris, France	www.codata.org
CSIRO	Commenwealth Scientific and Industrial Research Organisation, Sidney: Lindfield, NSW, Australia	www.csiro.au
DAM	Deutsche Akademie für Metrologie, Bayerisches Landesamt für Maß und Gewicht, Franz-Schrank-Str. 9D-80638 München	www.dam-germany.de
DIN	Deutsches Institut für Normung e.V., Burggrafenstraße 6, D-10772 Berlin, Germany	www.din.de
DKD	Deutscher Kalibrierdienst, Bundesallee 100, D-38116 Braunschweig, Germany	www.dkd.eu/; www.kalibrieren.de

Table 2.7 (Continued)

EA	European Co-operation for Accreditation, 75 avenue Parmentier, F-75544 Paris	www.european-accreditation.org
IRMM	Institute for Reference Materials and Measurements. Retieseweg 111, B-2440 Geel, Belgium	irmm.jrc.ec.europa.eu
INM LNE	Institut National de Métrologie—Laboratoire national de métrologie et d'essais France; LNE-SYRTE, 61, avenue de l'Observatoire, F-75014 Paris, France	lne-syrte.obspm.fr
ISO	International Organisation for Standardisation. 1, ch. de la Voie-Creuse, CH-1211 Geneva 20, Switzerland	www.iso.org
IUPAC	International Union of Pure and Applied Chemistry, Research Triangle Park, NC 27709-3757, USA	www.iupac.org
IUPAP	International Union of Pure and Applied Physics, Institute of Physics, 76 Portland Place, London W1B 1NT, UK	www.iupap.org
LCIE	Laboratoire Centrale des Industries Electriques, 33, avenue du Général Leclerc F-92260 Fontenay-aux-Roses, France	
NIST NBS	National Institute of Standards and Technology, Gaithersburg, USA	www.nist.gov
NPL	National Physical Laboratory, Hampton Road Teddington, Middlesex TW11 0LW	www.npl.co.uk
NRC	National Research Council of Canada, 1200 Montreal Road, M-58 Ottawa, Ontario, K1A 0R6 Canada	www.nrc-cnrc.gc.ca
NRLM	National Research Laboratory of Metrology, Tsukuba Research Center of AIST, Tsukuba, Ibaraki, Japan	www.aist.go.jp
OIML	Organisation International de Métrologie Légale—International Organisation of Legal Metrology, Paris, France	www.oiml.org
OFMET METAS	Bundesamt für Meteorologie und Klimatologie—Office féderale de métrologie, Lindenweg 50, CH-3003 Bern-Wabern, Switzerland	www.meteoschweiz.admin.ch www.metas.ch
PTB	Physikalisch-Technische Bundesanstalt, Bundesallee 100, D-38116 Braunschweig, Germany	www.ptb.de
SMU	Slovenský Metrologický Ùstav, Karloveská 63, 842 55 Bratislava, Slovakia	www.smu.sk

from the French name: *Le Système International d'Unités*. The seventh base unit,
the mole, was added in 1970 by the 14th CGPM.

2.6.6 *Weight Calibration and Control*

From antiquity we have knowledge of mint control by means of special coin scales
and of administrative measures against forgery of coins. Certainly also control of
weights took place in trade and in the market but there are only few reports of
such procedures in daily life. In ancient times calibration of scales and weights
most probably was performed or supervised by priests in the temples. In the Roman
Empire this task was taken over by the administration which, in the colonies, was
associated with the military. Since the Middle Ages we have standard offices or
departments, Eichämter, calibration services etc.

The International Standards Organisation (ISO) was established in 1947 and
is an association of over 150 countries, each represented by a National Standards
Body. ISO's headquarters are based in Geneva, Switzerland. ASTM International,
formerly known as the American Society for Testing and Materials, is a globally
recognised leader in the development and delivery of international voluntary con-
sensus standards. NIST is a non-regulatory federal agency within the U.S. Department of Commerce. NIST's mission is to promote U.S. innovation and industrial
competitiveness by advancing measurement science, standards, and technology in
ways that enhance economic security and improve our quality of life. The National
Physical Laboratory (NPL) is one of the UK's leading science and research facilities. It is a world-leading centre of excellence in developing and applying the most
accurate standards, science and technology available. In Germany the Physikalisch
Technische Bundesanstalt (PTB) performs fundamental research and development
work in the field of metrology as a basis for all the tasks entrusted to it in the areas
concerning the determination of fundamental and natural constants, the realisation,
maintenance and dissemination of the legal units of the SI, safety engineering, services and metrology for the area regulated by law and for industry, as well as technology transfer. Central tasks of the Deutsche Akademie für Metrologie (DAM) are
basic and advanced training in the field of legal metrology of technical staff from
all Federal States (Bundesländer) of Germany and of experts in state accredited test
centres run by private enterprises. The 'Arbeitsgemeinschaft Mess- und Eichwesen
(AGME)' coordinates the standardisation authorities in the German states. Similar
institutions are found in all industrialised states in the world. The work of a weights
and measures inspector is insistently described by Joseph Roth in his famous novel
"Weights and Measures" [41, 42].

2.7 Coins

There is a close connection between money and weight, and occasionally coins are
used as weights. Initially weighing was made to determine the value of goods rather

Fig. 2.10 Electrum
$(4.71 \text{ g}) = 1/12$ stater, about
610–560 BC. © rjohara.net

than their mass. First there may be comparison of similar amounts of goods but later on the weighing item was counterbalanced by a weight which represented the value of the goods. Roman coins had intrinsic value: the mass of a metal coin corresponds roughly to the value of the (precious) metal or alloy, respectively, from which the coin was made. The basis of mint was a given amount of metal (mina, libra) from which a set of coins was minted. The weight of the coin was controlled by means of special coin balances in order to detect faked pieces or worn down coins.

2.7.1 Early Antiquity

It is assumed that bronze figurines of domestic animals from 2000 BC which were found in the Mediterranean area indeed served as the first metallic money. Thousands of years later between 650 and 600 BC in Lydia in the western part of Asia Minor pieces of molten electron or electrum without marks came into use. Electron is a naturally occurring gold-silver alloy. During the government of Alyattes II (605–561 BC) or somewhat earlier minting of coins was started. In the Artemis temple of Ephesus egg-shaped electron coins with lion-head stamping were found [43, 44] (Fig. 2.10). Better known is his son Kroísos, whose stamp is often found in early coins. Afterwards coins were made also of gold and—at first in about 550 BC at the Ägina island—of silver, later on small coins were made of copper. In Greece until 400 BC mint commerce slowly replaced barter and the attic tetradrachmon of 17 g mass, the obolos and multiples became standard. After Alexander the Great (356–323 BC) it became customary to show the portrait of the emperor [45].

In the coastal regions of the Indian Ocean and the Pacific including East Africa, Australia and China the first means of payment was shell money the Cypraea moneta, or money cowry.

During the Shang dynasty in China (1600–1100 BC), pieces of bronze were used as currency. Later on, during the Zhou Dynasty (1122 to 255 BC) coins were shaped as knives or spades. The first emperor Qin Shi Huangdi in 221 BC normalised currency to round and punched copper coins (Fig. 2.7). The inscription included the mass of the coin, e.g. banliang $\sim$ 15 g [46]. That shape of coins was kept for 2000 years.

Some believe that India's Karshapana was the world's first coin. The first Indian coins were minted around the 6th century BC by the Mahajanapadas of the Ind-Gangetic plain. The coins of this period were punch marked coins called Puranas, Karshapanas or Pana.

2.7.2 Rome

The first currency of Rome consisted of large irregular, unworked lumps of bronze known as aes rude. The earliest surviving piece of aes rude dates from the early BC 200s and has a mass of 10.9 g.

The imperial mint of Rome was situated near the temple of Juno Moneta on the Arx, a height on the Capitoline Hill. Juno was an important goddess of the Romans. Her epithet Moneta is usually assumed to derive from the Latin word monere. = warn. This goes back to an occurrence when in 390 BC, the Gauls attacked the Capitol, but the geese of Juno started a honking ruckus and the citizens were alerted and thus able to drive off the Gauls. Alternatively Cicero reports that it was a warning before an earthquake by a voice from the temple. Furthermore, Moneta could be derived from mons = hill, mountain. Questionable is the identification of Juno Moneta with Mnemosyne, the Greek Goddess of Memory and mother of the Muses and the idea that this could be the origin of that epithet.

The Temple of Juno Moneta, built as the result of a vow taken by L. Furius Camillus during the war against the Auruncii, was built on the Arx in 344 BC. Suidas reports that during a war against Tarent the Romans had a lack of money. On an advice of Juno in 273 BC the Romans established therefore a mint near Juno Moneta's temple, and for the next four hundred years all the silver coins of Rome were made there, up to the time of the Emperor Augustus in the 1st century AD. The three mint masters, triumviri monetales, were called auro, argento, and aeri flando feriundo. Juno Moneta then accordingly became associated with money and minting coins, and the English words 'money' and 'mint', the French monnaie, the German Münze and Moneten and the Italian moneta trace back to Juno's epithet Moneta. Oddly enough for a temple of such importance and fame, no trace of it survives and no one is even sure now where it was located.

In antiquity mintage was personalised: Aeskulanus was the god of copper coins, his son Argentinus the god of silver coins and Aurinus (?) that of gold. In Rome copper coins first appeared, starting in 286 BC silver coins appeared and since beginning of the empire 27 BC also gold. The three types of coinage usually were represented by three women each equipped with a horn of plenty and with equal armed scales in equilibrium. The goddess Juno-Moneta-Aequitas with scales and cornucopia is often depicted on the reverse of mints.

Also in other places of the large Roman Empire coins were produced by the local administration, sometimes different in size and of minor quality with regard to the content of precious metal. Such money actually had only local significance. Nevertheless such mints diffused occasionally. Besides, forgery had been practised.

Table 2.8   Relative coin values of the Roman empire [47]

Relative Early Republic values (after. 211 BC)

Denarius	Sestertius	Dupondius	As	Semis	Quincunx	Triens	Quadrans
1	4	5	10	20	24	30	40

Relative Augustan values (27 BC–301 AD)

Aureus	Quinarius Aureus	Denarius	Quinarius	Sestertius	Dupondius	As	Quadrans
1	2	25	50	100	200	400	1600

Relative Diocletian values (301–305 AD)

Solidus	Argenteus	Nummus	Radiate	Laureate (coin)	Denarius
1	10	40	200	500	1000

Relative Late Empire coin values (301–305 AD)

Solidus	Miliarense	Siliqua	Follis	Nummus
1	12	24	180	7200

Although the denarius remained the backbone of the Roman economy, from its introduction in 211 BC until it ceased to be normally minted in the middle of the third century, the purity and weight of the coin slowly, but inexorably, decreased. The problem of debasement in the Roman economy appears to be pervasive, although the severity of the debasement often parallelled the strength or weakness of the Empire. When introduced, the denarius contained nearly pure silver at a theoretical weight of approximately 4.5 g. The coinage of the Julio-Claudians remained stable at 4 grams of silver, until the debasement of Nero in 64 AD, when the silver content was reduced to 3.8 grams, perhaps due to the cost of rebuilding the city after fire consumed a considerable portion of Rome. The denarius continued to decline slowly in purity. The exact reason that Roman coinage sustained constant debasement is not known, but the most common theories involve inflation, trade with India, which drained silver from the Mediterranean world, and inadequacies in state finances.

2.7.3   Middle and Modern Ages

In the Middle Ages units of weights and coins that had developed form antiquity were still closely linked. Lacking adequate own gold sources the Merowingians, and after them the Carolingians, confined themselves to silver coins that had been used by the Romans together with gold coins. Pursuing his father's Pippin reforms, in 793/94 AD Charlemagne introduced a new system for weights and coins. The basis gave the *pondus Caroli* (Karlspfund, livre carolienne), consisting of 406 g silver, equal to 20 solidus (later shilling) or 15 unciae $\sim$ 407 g as the standard, from which denarii (denier) (equal to 1/240 pondus) was minted. The new silver coinage was in a fixed relationship to that of the Romans [48]. During this period for coinage, libra (livre) and solidus (sou) were only units in accounting, whereas Denare was a real coin. The Carolingian Denarius ∼ 1.7 g, the predecessor of the Pfennig, became the basis of weights and currency. The Karolingian coin system survived in its basic relations in the British system until 1971. In the reign of the Salier, (beginning in 1024) the Kölner Mark (Cologne Mark) weighing 234 g got supranational importance as a basis of currency [30].

Whereas the equal designations for coin and weight units remained the masses of these were different. In France Livre = 20 Francs was used. The French troy system was the basis for the pre-decimalisation of the British system of coinage introduced by Henry II of England, in which the penny was literally one pennyweight of silver. One pound sterling silver was equal to twenty shillings, with each shilling equal to twelve pennyweights or pennies. The term is derived from the fact that, in about the year of 775, silver coins known as 'sterlings' were issued in the Saxon kingdoms, 240 of them being minted from a pound of silver, the weight of which was probably about equal to the later troy pound. Sterling silver is an alloy of 925 parts of sliver and 75 parts of copper and for 200 years there was no change. The origin of the word sterling is unclear. It could be derived from the Greek statera = balance or from star because of some Norwegian coins which had some stars minted. It could be also a shortening of easterling which denoted coins from Northern Germany. The usual designation become 'pounds of sterlings' with the currency sign £ (Libra) originally with two cross-bars, then later £ with a single cross-bar. After the Norman Conquest the pound was divided for simplicity of accounting into 20 shillings (solidi) and into 240 pennies or pence (denari). The abbreviation LSD—librae, solidi, denarii was used for the pounds, shillings and pence from of the original duodecimal currency system. That pound sterling system with 1 £ = 20 shilling = 240 pence remained until 1971 in the UK.

Maria Theresa Thalers (MTT) were silver-alloy coins, with varying impression of the empress Maria-Theresia (1717–1780) on the obverse and the Habsburgian double eagle on the reverse (Fig. 2.11). Inscription: **M**(aria) **Theresia D**(ei) **G**(ratia) **R**(omanorum) **IMP**(peratrix) **HU**(ngaria) **BO**(hemia) **REG**(ina), and **ARCHID**(dux) **AUST**(riae) **DUX BURG**(undiae) **CO**(mes) **TYR**(rolis). MTT were first minted in 1741 as currency of the Habsburgian Empire, and had been valid until 1858 [49]. It was accepted in German countries and had been very important for trade with the Levant (parts of Turkey, Lebanon, Syria). Over time, MTT became the unofficial currency in some areas of Africa and Asia. In 1857, Emperor Franz Joseph of Austria declared the Maria Theresa Thaler to be an official trade coinage. For that purpose MTT has been restruck at Vienna, Austria, and at many other mints including Bombay with the 1780 date frozen in time. During the Japanese occupation of Indonesia in World War II, USA created counterfeit MTTs for use by resistance forces. Until today, nearly 400 million MTTs were minted.

Acceptance was based on confidence; faked items could be easily detected by inspection of MTT's appearance and weight. It was hard to strike faked coins with regard to small details. Lettering near of the outer rim did not allow filing off some metal, and the weight allowed assessing the silver content. Today MTTs serve as

Fig. 2.11 Maria Theresia Taler. © Münze Österreich AG

(a)   (b)

Fig. 2.12 (**a**) Stolberger Alte Münze (Old Mint) museum, mint workshop built 1535, Stolberg, Germany. (**b**) Alte Münze, Schleiz

raw silver for silversmiths, as investment and jewellery. However at many markets and in many countries they are still accepted as a basic currency unit. Because the mass of the coin corresponds to that of one ounce MTT is used likewise as a weight.

In Berlin and Köln (Cologne, Germany) until 1980 the designation 1 Pfund = 20 DM was in use and vanished with the introduction of the Euro.

Everywhere in important old European towns we have buildings of the mint in which coinage was produced (Fig. 2.12).

Fig. 2.13 (a) Paper weight of
the administration of Mainz,
Germany. (b) Paper glass
weights

(a)

(b)

2.8 Weights not for Scales

2.8.1 Loading Weights

Weights have been used in order to fix light objects like textiles or paper, furs,
furskins or cloth of tents, and also today we have paperweights (Fig. 2.13) and car-
pet weights. Recently Yigael Yadin wrote that he found in Palace 6000 at Megiddo
from the level dated 11th century BC that a 'bronze weight' of a monkey and another animal in a bag containing loom weights and spindles; he believed these were
the contents of a woman's bag [50–52].

Weight-lifting is a sport. By 2011 world records for men were 215 kg (snatch)
and 263 kg (clean and jerk), for women 147/187 kg. (Fig. 2.14). Lifting weights can
help to develop muscles and to increase the energy burned while walking, but it is
far safer to just walk further instead. Weights at the ankles, feet, wrists and hands
increase the risk of injury to the knees, shoulders, muscles, tendons and repetitive
motion injury. A weighted backpack can cause strain if the weight is not distributed
across the hips. Only a weighted vest or hip belt would be recommended for training.

Counterweights and balance weights are used for tensing ropes, threads, yarn
e.g. in weaver's loom (Fig. 2.16). Weaver's weights are found everywhere as relics
of antique civilisations; recently in Germany weights from the Celts 1000 BC were
found, made of clay [53].

Fig. 2.14
(**a**) Counterweights in a force
training apparatus. Viversani
Sport & Therapiezentrum,
Friedrichsdorf, Germany.
(**b**) Lifter weights.
(**c**) Dumbbells.
Krankengymnastik
Bodenbender-Müller,
Friedrichsdorf, Germany

(a)

Large weights are used to counterbalance lever arms e.g. of a crane or of tractors (Fig. 2.17).

Pendulum clocks may be equipped with two types of weights: Gravity weights to drive the clockwork and a weight of the pendulum (Fig. 2.18). Whereas the weight of the drive unit may consist of any metal or even stone the pendulum weight mostly consists of lead within a brass shell, the rod going through the middle. Except for tubular bell, the time weight is the same as the strike weight if a wood stick pendulum is used and the same as the chime weight on a Westminster if a lyre pendulum is used.

Fig. 2.15 Continued

(b)

(c)

In gravity clocks the mass of the clock is used to drive the clock as it descends a rack, cord, incline, etc.

Gravity light is a new approach to storing energy and creating illumination, developed especially for countries in which some regions not dispose on electricity. The light is lifted and descends on a rope, creating 30 minutes of light on its descent.

Fig. 2.16 Weaver's loom
with draw-weights.
Hirtenmuseum (Shepheard
museum) Hersbruck,
Germany

Fig. 2.17 Counterbalance
weight for tractors (Baywa,
Hersbruck, Germany)

Deadweights were earlier used as loads for the safety valve of steam engines
(Fig. 2.19). Today they are replaced by springs. A counterweight is applied on the
pickup of the turntable in order to reduce the weight acting on the record, likewise
the beam at level-crossing gates. Heavy weights are used to operate or counterbal-

Fig. 2.18 (a) Wooden pendulum clock with stone weight drive and horizontal pendulum loaded with two metal weights. © Bert Neeb. (b) Metal pendulum clock with metal weight drive and wood stick/metal weight pendulum. © Hiltrud Müller

(a)   (b)

Fig. 2.19 Pressure relief valve, safety valve with weight. © Bengs Modellbau, Im Kirchfelde 6, D-31675 Bückeburg, Germany

ance gates and elevator cabins. In old forges (Fig. 2.20) as well as in old organs bellows had been loaded with weights.

Fig. 2.20 Forge, bellows
with loading weights,
Hirtenmuseum Hersbruck,
Germany

2.8.2 Wheel Weights

Automobile wheels, tyres and the combination of both may have a mass imbalance.
Those asymmetries of mass cause the wheel to wobble when accelerated. This wob-
bling can give rise to driving disturbances, usually vertical and lateral vibrations.
The disturbance due to imbalance usually increases with speed. One can distinguish between two effects: static balance and dynamic balance. The static balance
is measured at rest by placing the object in its vertical axis on a non-rotating spindle
tool and determining the spot on the tyre where the mass is greatest is acted on by
gravity to deflect the tooling downward. Dynamic balance describes the forces generated by asymmetric mass distribution when the tyre is rotated, usually at a high
speed (≥ 300 rpm). Sensors measure the forces, analyse and compare them with tol-
erance values. Strictly speaking, it is mathematically impossible to exactly align the
principal axis of the inertia tensor with the axle by adding a finite number of small
masses. But in practice, this is exactly done.

By means of wheel weights (wheel balancing weights) (Fig. 2.21) the combined
effect of the tyre and wheel imbalance is counteracted. Most are of metal (usually
lead with 3 % antimony for hardening, zinc-alloy or steel-alloy) and are clipped to
the wheel rim. Wheel weights come in various sizes and styles, and must be properly

Fig. 2.21 Wheel weights

attached to the rim so they don't move or fall off. Different styles of clips are available for various types of rims. Self-adhesive stick-on weights are also available that are mounted on the inside face of alloy wheels. According to the US Environmental Protection Agency worldwide every year millions of small weights are attached to tyres balancing them, including about 70 000 tons of lead.

2.8.3 Water Weights

Occasionally a weight is made of a hollow receptacle filled with sand or a liquid. The 'Nerobergbahn' at Wiesbaden, Germany, is the only existing mountain railway driven by water ballast (Fig. 2.22). It enables ascending the Neroberg (260 m above sea level) by covering a height difference of 83 m at a distance of 438 m with a gradient up to 26 %. The maximum speed is 7.3 km h^{-1}. This is a funicular on two rails; a third rail in the middle serves as a brake only. A 7 m^3 container below the upper carriage is filled with water corresponding to the weight of passengers in the lower carriage. Water released at the tracking station is pumped to a reservoir on the hill.

Fig. 2.22 Nerobergbahn, Wiesbaden, Germany

References

1. M.A. Crawforth, *Handbook of Old Weighing Instruments* (Chicago, International Society of Antique Scale Collectors, 1984)
2. H.G. Wiedemann, G. Bayer, Events in the historical development of weights and money, in *Materials Issues in Art and Archaeology II*, ed. by P.B. Vandiver, J.R. Druzik, G. Wheeler, (Materials Research Society, 1991)
3. G. Hellwig, *Lexikon der Maße und Gewichte* (Bertlsmann/Lexikothek, Gütersloh, 1983)
4. H.R. Jenemann, The development of the determination of mass, in *Comprehensive Mass Metrology*, ed. by M. Kochsiek, M. Gläser (Wiley/VCH, Berlin, 2000), pp. 119–163
5. A. Eran, Samen in der Metrologie, in *Die historische Metrologie in den Wissenschaften*, ed. by A. Witthöft (St. Katharinen, 1986), pp. 248–261
6. W.M.F. Petrie, *Ancient Weights and Measures* (London, 1926)
7. A. Mariette, *Les papyrus egyptiens du Musee de Boulaq*, vol. II (Paris, 1872)
8. A. Scharff, Ein Rechnungsbuch des königlichen Hofes aus der 13. Dynastie. Z. ägypt. Sprache Altert. Kd. **57**, 51–68 (1922)
9. M. Kochsiek, R. Schwartz, The unit of mass, in *Comprehensive Mass Metrology*, ed. by M. Kochsiek, M. Gläser (Wiley/VCH, Weinheim, 2000), pp. 484–579
10. F.G. Skinner, *Weights and Measures—Their Ancient Origins and Their Development in Great Britain up to AD 1855* (Science Museum, London, 1967)
11. G. Theuerkauf, Arabisch-jementische Waagen und Gewichte – Maß und Symbol. Teil 1: Metrologische Betrachtungen aus kulturhistorischer Sicht. Jem.-Rep. **35**(2), 7–10 (2004)
12. G. Theuerkauf, Arabisch-jementische Waagen und Gewichte – Maß und Symbol. Teil 2: Waagen und Gewichte aus dem Suq von Sana'a. Jem.-Rep. **36**(1), 15–21 (2005)
13. Wikipedia, Indus-Kultur. http://de.wikipedia.org/wiki/Indus-Kultur#Wissenschaft (2007)
14. K.M. Petruso, Early weights and weighing in Egypt and the Indus Valley. MFA Bull. **79**, 44–51 (1981)

15. S.V. Gupta, in *Units of measurement*, ed. by R. Hull et al. Springer Series in Materials Science, vol. 122 (Springer, Heidelberg, 2010)

16. Q. Guangming, *Research on Weights and Measures Through the Ages in China* (Kexue chubanshe, Beijing, 1992), p. 520

17. W. Chengluo, *A History of Weights and Measures in China* (Shangwu Yinshuguan, Commercial Press, Shanghai, 1937)

18. U. Theobald, H.U. Vogel, Chinese, Japanese and Western Research in Chinese Historical Metrology. A classified bibliography (1925–2002), http://www.sino.uni-heidelberg.de/staff/janku/hilfsmittel/materials/bibliography_metrology.pd. Universität Tübingen (2004)

19. R. Kokotailo, Chinese cast coins. http://www.calgarycoin.com/ (2008), Calgary Coin

20. China-Window, Chinese Steelyard-Gancheng. http://www.china-window.com/china_culture/china_culture_essentials/chinese-steelyard-ganchen.shtml (2008)

21. M.I. Marcinkowski, *Measures and Weights in the Islamic World* (ISTAC, Kuala Lumpur, 2002)

22. W. Hinz, Islamische Maße und Gewichte, in *Handbuch der Orientalistik, erste Abteilung, Ergänzungsband I, Heft 1* (Brill, Leiden, 1970)

23. Ancient Roman units of measurement. http://en.wikipedia.org/wiki/Ancient_Roman_units_of_measurement#Weight (2006)

24. W. Trapp, *Kleines Handbuch der Maße, Zahlen, Gewichte und der Zeitrechnung* (Reclam, Stuttgart, 1992)

25. M.H. Crawford, *Roman Republican Coinage* (Cambridge University Press, Cambridge, 1974)

26. Karlspfund. http://de.wikipedia.org/wiki/Karlspfund, Wikipedia (2008)

27. *Captiulare episcoporum CCVI*, vol. 4635, fol. 43v. 793, Paris: Bibliothèque Nationale Lat., pp. 70–92

28. Frankfurter Synode 794. http://de.wikipedia.org/wiki/Synode_von_Frankfurt, Wikipedia (2008)

29. H. Witthöfft, *Münzfuß, Kleingewichte, Pondus Caroli und die Grundlegung des nordeuropäischen Maß- und Gewichtswesens in fränkischer Zeit* (Scripta Mercaturae, St. Katharinen, 1984)

30. H. Ziegler, Die Kölner Mark in neuem Licht. Hans. Geschichtsbl. **98**, 39–60 (1980)

31. A. Cruz, *Weights and Measures in Portugal* (Instituto Português da Qualidade, Caparica, 2007)

32. T.C. Mendenhall, Fundamental standards of length and mass, in *Weights and Measures Standards of the United States: A Brief History*. http://physics.nist.gov/Pubs/SP447/, L.E. Barbrow and L.V. Judson, Editors. 1893/2006 NBS Superintendent of Documents: Washington D.C., pp. 28–29

33. Kilogram, http://en.wikipedia.org/wiki/Kilogram, Wikipedia (2008)

34. T.R. Quinn, The kilogram: the present state of our knowledge. IEEE Trans. Instrum. Meas. **40**(4), 81–85 (1991)

35. H.R. Jenemann, Das Kilogramm der Archive vom 4. Messidor des Jahres 7: Konform mit dem Gesetz vom 18. Germinal des Jahres 3? in *Genauigkeit und Präzision*, ed. by D. Hoffmann, H. Witthöfft (Physikalisch-Technische Bundesanstalt, Braunschweig, 1996), pp. 183–213

36. G. Fraustadt, W. Weber (eds.), *Agricola, Georgius: Schriften über Maße und Gewichte (Metrologie)*. Georgius Agricola: Ausgwählte Werke, ed. by H. Prescher (Berlin, 1959)

37. H. Witthöfft, *Georgius Agricola über Maß und Gewicht* (1998). www.unze.de

38. G. Agricola, *De mensuris et ponderibus* (Basel, 1550)

39. I. Gothofredus, *Codex Theodosianus 16,8,1-29. Europäische Hochschulschriften, Reihe III*, B. 453, ed. by R. Frohne (Bern, 1991)

40. Hunterian

41. J. Roth, Weights and Measures. Everyman's Classics S. 1983: EBLA (UK)

42. J. Roth, *Das falsche Gewicht. Die Geschichte eines Eichmeisters*. Romane, vol. 3 (Köln, 1999), pp. 135–231

43. R.J. O'Hara, Ancient Greek Coins of Miletus. http://rjohara.net/coins/lydia-electrum/ (2009)

44. L. Weidauer, *Probleme der frühen Elektronprägung*. Monographien zur antiken Numismatik, I (Office du Livre, Fribourg, 1975)

45. M. Miller, *Münzen des Altertums* (Braunschweig, 1963)
46. N. Ferguson, *The Ascent of Money: A Financial History of the World* (Allen Lane, London, 2008)
47. Roman currency. http://en.wikipedia.org/wiki/Roman_currency, Wikipedia (2008)
48. H. Witthöfft, Die Rechnung und Zahlung mit Gold und Silber nach Zeugnissen des 6. bis 9 und 13/14 Jahrhunderts. Hamburger Beiträge zur Numismatik, 1985. **30/32** (1976/78), pp. 9–36
49. J. Renger, Silbermünzen im Jmen – Der Maria Theresia Thaler. Jem.-Rep. **39**(2), 13–16 (2008)
50. Y. Yadin, Megiddo of the Kings of Israel. Bibl. Archaeol. **3**(33), 66–95 (1970)
51. Y. Yadin, *Biblical Megiddo*. In L.I., in *Jerusalem Cathedra*, ed. by L.I. Levine (Jerusalem), pp. 120–181
52. M. Feinberg Vamosh, Monkey weights. Personal communication (2008)
53. Anonymus, Zur Keltenzeit Weber in Eschborn. Frankfurter Allgemeine Zeitung, 2012(138), p. 57

Chapter 3
Weighing

Today, to our mind, the aim of weighing is determination of the mass of a sample (Fig. 3.1).

Indeed a balance in the gravitational field measures a force which is expressed in the SI system in newtons. A conventional double armed two-pan balance compares forces due to gravity of the weighing article and of the counterweight. A single pan balance could be regarded as a mass sensor which measures the attractive force between the earth's mass and the mass of the weighing article. Indeed here also a comparison is made between the force exerted on the sample mass and those on hidden counterweights (Fig. 3.2). Using one of Eqs. (1.1)–(1.3) we can calculate the mass. Because there is only a constant ratio between force and mass we can standardise balances and weights in mass units. Balances can be used also for other force measurements, e.g. as so-called gravimeter to determine the gravitational field of the earth.

This book is concerned in particular with measuring methods which are aimed for direct mass determination. We should consider, however, that indirect methods also are in use, e.g. in chemistry when observing reactions or when determining the molecular mass, in particle physics and in astronomy. Furthermore, we experience that it is impossible to measure the mass alone. Using a gravitational balance in our environment, in fact we measure the sum of gravitational force and buoyancy. Likewise in all other methods we have to realise other influences. Nevertheless, weighing always was and still is one of the most sensitive measuring methods and the measuring range of balances is extraordinarily large.

A balance is a sophisticated instrument. In antiquity weighing needed special knowledge and its operators enjoyed a high rank in society. Still in the middle of the 20th century precision weighing in laboratories was highly ritualised, time consuming, and performed in a separate cabinet. Today the balance has became a black box with one single weighing tablet and a digital output. It can be placed everywhere and operated at the touch of a button by anyone. Nevertheless, the physical process in the background should be regarded and the various influences which can cause disturbances, in particular, in sensitive weighing. Regular correcting adjustment should never be neglected.

E. Robens et al., *Balances*, DOI 10.1007/978-3-642-36447-1_3,
© Springer-Verlag Berlin Heidelberg 2014

3.1 Weighing Methods

Weighing is determination of the mass of a sample by comparison of the force exerted on that mass with a test force in Earth's gravitational field, in a centrifugal acceleration field or in alternating gravitational acceleration field. Two basic interactions are applied: gravitation and electromagnetism.

- Gravitational lever balance = comparison of two gravitational forces.
- Gravitational spring balance = comparison of the gravitational force of the sample with elastic force of a test body, that is basically the force of physico-chemical bonding of molecules.
- Gravitational electromagnetic or electrostatic balance = comparison with macroscopic electromagnetic or electrostatic forces.

  Three methods of direct weighing may be distinguished:

- Weighing in the gravitational field.
- Measurement of the impulse of an accelerated sample.
- Measurement of the frequency shift of an oscillating sample.

According to Einstein's equivalence principle, the difference between gravitational force and the force of an accelerated mass cannot be distinguished [1, 2] and thus, all three methods are equivalent and give the same result. However, when weighing in the gravitational field of the earth the test sample is always in a local constant field and the constant force is measured. With impulse measurements

Fig. 3.2 View in an opened mechanical single pan balance. One side of the balance beam carries the sample pan and is equipped with a mechanical loading device for substituent weights. The other side has a fixed counterweight.
© Sartorius, Göttingen, Germany

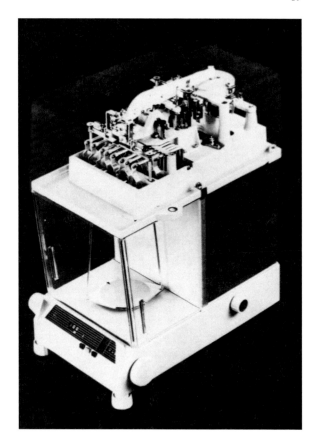

the sample is once accelerated—e.g. by falling through a defined distance—and the peak force subjected to an obstacle is measured. In the third method the sample is subjected to an oscillating acceleration field which can be varied experimentally. The sample is connected firmly to the generator of that field. In general its mass is determined by the feedback to the oscillator.

During the latter half of the 20th century measuring methods turned from analogue to digital. However, we should distinguish between measuring principle, indication and treatment of results. Only when weighing in the quantum region, digital principles are used. Of course, such weighing may get importance for practical purposes in future. Today all weighings are performed by analogue instruments. Those analogue weighing methods had been investigated exhaustively by Theodor Gast (1916–2010) [3]. Classical weighing is designed to the measurement of a digital equilibrium value, either by placing of discrete weights on scale or by indication of discrete values. More recent investigations include measurements of dynamic chemical and physical processes, e.g. in thermogravimetry. For visual assessment resulting diagrams are helpful, but for data processing digital values are used and likewise their corrections are made using algebraic methods.

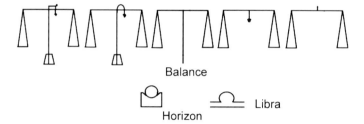

Fig. 3.3 Hieroglyphs denoting 'balance' (michAt). The hieroglyphs bottom show sunrise or sunset and stand for horizon (Akhet). Today the hieroglyph bottom right is used for 'Libra'

In Old Egypt the balance (michAt) had special hieroglyphs (Fig. 3.3). The notion of balance came from the Latin bilanx = bivalvous (bi-, bis = twice, and plax = (scale-) pan). In the Middle Ages, the word 'balance' (which means scales) entered the French language. The original notation for the steelyard in Rom was statera (stater = weight, also an antique weight unit). Another Latin word for the balance was libra which is included in the word equilibrium.

Figure 3.4 shows the principle of several variants of weighing in the gravitational field: the beam balances which compare the mass of the sample with that of the counterweight and a force sensor which measures just the force of the sample due to gravity. The balance can be operated either by observing the changed output signal of the sensor e.g. the deflection of the beam or by compensating the force acting on the sensor and measuring the compensating force. The following considerations concern the conventional beam balance but may be applied likewise to other types of balances.

3.2 Deflection, Compensation and Substitution Weighing

In an equal-armed beam balance, the sample mass is tared until the balance beam is at zero position (Fig. 3.5a,b). Alternatively the main part is tared and the rest and likewise mass variations are evaluated from its inclination (Fig. 3.5c). Its inclination is an indication of the measured weight. Optical observation by means of pointer and microscope with micrometer eyepiece or mirror and autocollimator or using photoelectric devices does not cause reaction forces, whereas inductive and capacitive position sensors have an influence on the weighing.

The range of measurement can be extended by removing weights from the load side (substitution) or adding them to the opposite side (compensation). Substitution weighing has the advantage that the total load of the beam at the zero point is always brought back to the same value, so the bending and the sensitivity remain constant.

Because the measurement range Δm_{max} and sensitivity s are connected by the angle of deflection φ_{max} in the deflection method:

$$\Delta m_{max} = \frac{\varphi_{max}}{s} \tag{3.1}$$

Fig. 3.4 Basic types of lever balances © Hans Jenemann. (**a**) Symmetrical beam balance. (**b**) Unsymmetrical beam balance with sliding counterweight (steel yard). The balance arm is marked with mass values. Alternatively the counterweight is fixed and the balance beam is shifted within a bearing (besemer). (**c**) The single pan balance with a fixed counterweight can be regarded as a force sensor. When a small additional mass m_φ is deposited on the balance, the beam turns through an angle φ

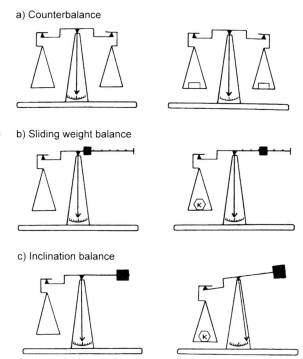

a) Counterbalance

b) Sliding weight balance

c) Inclination balance

the high sensitivity desired for the balance's beam restricts the range of measurement. There is no such disadvantage with the null method. Here, the beam is brought back by a measurable, external effect to the position of equilibrium which can be observed with high sensitivity. According to various physical principles, the required forces or moments stand in a reproducible association with generating quantities which are indicated or registered with a high useful resolution. If the beam's deflection itself produces the required effect, this is known as automatic compensation. It takes place in a servo-loop comprising the balance beam with bearings, a position sensor, a servo-control and a force element.

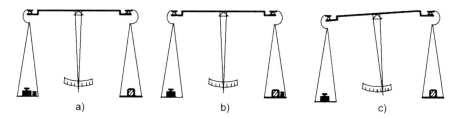

a)   b)   c)

Fig. 3.5 Modes of the lever balance: (**a**) Complete compensation towards equilibrium of the sample mass by weights on the compensation pan, (**b**) by tare weights on the load side. (**c**) Evaluation of the angular deflection after compensation of a major part of the sample mass

3.3 Characteristics of a Balance

The features characterising scales have been standardised worldwide for a long time
[4–8]. The importance of those parameters changed when new balance types had
been offered or new applications were introduced. Basic requirements are

- maximum load
- measuring range
- resolution.

    Quality characteristics are

- accuracy
- relative resolution or relative sensitivity
- reproducibility.

Such requirements are expressed in parameters such as linearity and stability
of the value indicated, discrimination and reproducibility. More practical features
are indifference against shocks, inclination of the stand and of the situation of the
weighing object in the pan. Today a balance is characterised mostly by its maximum
capacity, the resolution, given by the digits indicated, and by the weighing range,
also given by the number of digits indicated.

Standardised definitions of parameters are designated to characterise the features of commercially manufactured instruments in their regulated uses. In general,
balances for commerce, laboratories etc. are used for single mass determinations
of a sample in one particular weighing procedure. In industrial research weighing
of series of samples require automatic sample changing. Thermobalances, vacuum
balances, layer thickness monitors etc. are applied to record mass variations as a
function of time, whereas the mass of the sample under investigation is determined
separately using a laboratory balance. In the case of a symmetric balance the counterweight is used exclusively to counterbalance the sample mass, to set the zero
position and to suppress buoyancy. Furthermore, mass and force sensors are used to
monitor and to control production processes.

With microchemical balances the actual mass of a sample should be determined
most accurately in a discrete weighing procedure. For this purpose an adequate characterising of the balance can be made using sensitivity, discrimination and the maximum capacity. The graphical representation of sensitivity should result in a straight
line of 45° and standard deviations from that line characterise the quality of the
balance.

With vacuum and thermo microbalances the sample mass is mechanically compensated using counterweights and mass variations are measured via electrodynamic
compensation. Usually, the sample mass is measured using an analytical balance.
Then the sample is placed on the microbalance and compensated by a counterweight
of the same density. Mass variations as a consequence of physical or chemical processes are observed as a function of time. Thus, the rough range of the balance is of
minor interest and the need of a definition of the sensitivity exists only for the electrodynamically compensated range. Such balances can be characterised adequately

using bench-marks like discrimination, maximum capacity and the longtime stability of the zero point and of measuring values. With regard to a representative sampling it is often irrelevant whether a 1 mg or 10 g sample is investigated. Thus, the ratio discrimination to maximum capacity CDR or relative resolution or relative sensitivity best characterises the balance.

3.3.1 Maximum Load

Following Czanderna [9] we propose that the definition of the maximum load of a balance is the maximum sample mass which can be placed without damage to the balance or affecting its operation. The maximum load may be far above the measuring range.

3.3.2 Capacity, Measuring Range

The maximum capacity (Höchstlast, portée maximale) is the upper limit of the specified measuring range not considering the additive tare mass [1–3]. For special research balances the tare (balance pan, crucible, suspension) can be varied so that approximately the whole capacity is available for the sample. If necessary the nominal load is exceeded and the maximum load is determined empirically.

The minimum capacity is the lower limit of the measuring range. Below this, measuring results are burdened with a too large relative error. The specified measuring range is the region between minimum and maximum capacity [1–3].

3.3.3 Sensitivity

In reports and in descriptions of instruments we are very often confronted with the incorrect use of parameters characterising the features of a balance. In particular the notation 'sensitivity (Empfindlichkeit, sensibilité)' seems to be unclear, though it is well defined by the Organisation International de Métrologie Légale (OIML) [10] as well as by national and international standards [4–8, 11]. According to DIN/ISO, sensitivity is defined as the response of a measuring instrument divided by the corresponding change in the stimulus. The stimulus of a balance is the mass m placed on it [10, 12]. The sensitivity is often not constant in the whole measuring range of the balance and depends on the value of the mass placed on it. The sensitivity is the magnitude of the reversible displacement of an indication of the balance.

When both sides of the balance are loaded with equal masses and a small additional mass m_φ is deposited on the balance in equilibrium, the beam turns by an

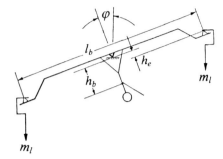

Fig. 3.6 Diagram of a symmetrical beam balance. The beam with a length l_b and a mass m_b is in equilibrium when the end knife-edges are loaded with equal masses m_l. h_e is the distance of the pivot from the line connecting the end knife-edges and h_b is the distance of the pivot from the centre of gravity. When a small additional mass $m_{\varphi\varphi}$ is deposited on the balance, the beam turns by an angle φ

angle φ. The sensitivity of a beam balance may be defined as the change of the beam's deflection $\delta\varphi$ divided by the difference in the mass δm_φ to be weighed:

$$s = \frac{\partial\varphi}{\partial m_\varphi} \tag{3.2a}$$

The constancy of its sensitivity is one of the characteristics of the quality of a balance. Therefore, the beam of a beam balance should be sufficiently rigid. For a beam balance the beam's pivot and the bearings (e.g. knife edges) of the pan suspensions should be on one level.

The sensitivity may be appropriate to characterise usual balances. It is the quotient of the observed variation of the variable l and the corresponding variation of the measured mass m at a given mass value [1–3]

$$s = \partial l / \partial m \tag{3.2b}$$

For balances with digitised output the analogue variable l is replaced by the appropriate digital value. The sensitivity may be appropriate to characterise ordinary balances.

The sensitivity of a beam balance (Fig. 3.6) may be defined as the change of the beam's deflection $\delta\varphi$ divided by the difference in the mass δm_φ to be weighed according to Eq. (3.1).

If the pivots are exactly centred in the balance the unloaded beam is in indifferent equilibrium. To obtain a balance with stable equilibrium (beam in horizontal position) it is necessary to add a mass below the lower beam's centre or to arrange the central pivot somewhat above the line connecting the pan pivots. That additional mass will draw back (in equilibrium) the beam into the horizontal position when deflected. The centre of gravity is somewhat below the pivot and by reduction of the distance the sensitivity can be increased within limits.

If the beam of a symmetric balance is sufficiently rigid, the bending is negligible and no elastic restoring force is acting on it, in equilibrium and small angular deflections the following is valid:

$$m_\alpha g \frac{l_b}{2} = (m_b h_b + 2m_l h_e) g \Delta\varphi \tag{3.3}$$

The sensitivity then results to

$$s = \frac{l_b}{2(m_b h_b + 2m_l h_e)} \tag{3.4}$$

One expects a linear relation between $1/s$ and m_l.

$$\frac{1}{s} = \frac{2}{l_b}(h_b m_b + h_e m_l) \tag{3.5a}$$

If the beam's pivot and the suspension device are on one level, the formula is simplified to

$$s = \frac{l_b}{2h_b m_b} \tag{3.5b}$$

If $(m_b h_{b+2m_l}, h_e) > 0$, the beam is fundamentally stable. To determine h_b and h_e, after equilibrium has been established the deflection occurring when the balance is successively loaded with small additional masses is measured against a fixed yardstick for various loads. For small angles the following is valid approximately:

$$\frac{2l_\varphi}{l_b} = \sin\varphi \approx \frac{\varphi}{\text{rad}} \approx \tan\varphi \tag{3.5c}$$

The sensitivity of other balances with an analogue scale display is equal to the quotient of the longitudinal shift δl_φ of a mark observed on the scale and the change in load δm_φ producing it [6]:

$$s = \frac{\partial l_\varphi}{\partial m_\varphi} \tag{3.6}$$

The particular indicating display is included in this definition, as it is in the definition of sensitivity for balances with numerical displays with the numerical step $\delta Z =$ digit variation:

$$s = \frac{\partial Z}{\partial m_\varphi} \tag{3.7}$$

3.3.4 Linearity

Sensitivity is not a suitable criterion for judging balances and mass sensors, as it can be expanded by means of lever transmission, luminous pointers and amplifiers. Proof of the indicator's linearity is more important, i.e. a graph in which the response of the uncorrected sensor is shown against mass variations.

3.3.5 Resolution

Resolution is the quantitative expression of the ability of an indicating device to distinguish meaningfully between closely adjacent values of the quantity indicated [4]. Alternatively the expression discrimination (Ansprechvermögen, mobilité) is used as the ability of a measuring instrument to respond to small changes in the value of the stimulus. Discrimination threshold (Ansprechschwelle, seuil de mobilité) is the smallest change in the response of a measuring instrument. This may be defined as follows: Discrimination is the minimum variation in mass that can be observed experimentally in a reproducible manner.

3.3.6 Stability, Drift

Stability (Meßbeständigkeit, constance) is the ability of a measuring instrument to maintain constant its metrological characteristics. It is usual to consider stability with respect to time [4]. If mass variations are observed as a function of time, the fluctuation of the readings, and in particular the variation of the zero point, should be limited. The spontaneous variation of the indication at constant sample weight and constant environmental conditions within a given period of time is an important characteristic of a balance. Often a slow drift (Meßgerätedrift, dérive) is observed, caused by the irreversible change of the components of the instrument (e.g. fatigue of the material, wear). Furthermore, slow variations caused by environmental variations cannot be excluded completely. The sum of the variations may be quantified by the mean square deviation within an extended period of time.

3.3.7 Relative Resolution

In many applications the mass of the sample is not the important factor, the relative change of the object's mass being of greater interest. Consequently, the product maximum capacity × sensitivity (LEP) [8], maximum capacity to discrimination threshold [9], or the relative resolution (= sensitivity divided by the maximum capacity) [13–16] are the essential criteria for judging a balance. Basedow and Jenemann [13] use the reciprocal relative resolution (Auflösungsverhältnis). The relative resolution gives no information about the accuracy [17] of weighing but indicates the overall state of the art of such instruments.

3.3.8 Reproducibility

Reproducibility is the ability of the balance to indicate identical results for repeated measurements of one sample under uniform conditions.

3.4 Errors and Influences

Balances are sensible instruments which require a careful installation. For precision balances a heavy stand is used. This may be a rigid table in which a heavy stone is suspended. Mechanical dampers may be used as shock-resistors to protect the stand of the balance from spurious vibrations of the bottom.

For accurate weighing corrections of the measured values are required, accounting for deficiencies of the instrument, environmental influences and principle errors of the weighing method. The sensitivity of a balance is limited by the precision of its mechanical construction and the stability of the electronic components. For vacuum microbalances and thermobalances it is chiefly ambient influences which restrict their useful sensitivity.

Besides effects observed with conventional balances operated at ambient conditions Table 3.1 includes also effects observed when weighing at reduced pressure and when large temperature differences exist between the sample and the measuring system. The values 'maximum effect' are observed with a vacuum microbalance in its typical use: pressure range $0-10^5$ Pa, temperature range 70–1000 K [3]. The 'relative uncertainties' are obtained mainly with metrological balances [18]. The values give a hint to the limits of the application of beam balances. Further problems with vacuum, hydrostatic and thermo balances are discussed in the respective

Table 3.1 Effects governing accuracy of beam balances. For explanations see text

Effect	Maximum effect µg	Relative uncertainties
Uneven thermal expansion of the balance beam	0.6	
Inelasticity of flexural strips		$<4 \times 10^{-12}$
Error of the suspension bearing	10	
Brownian motion	0.006	1×10^{-13}
Thermal gas flows (Knudsen forces)	300	
Cavity forces	2	
Convection	100	5×10^{-11}
Magnetic fields	0.1	
Electric fields	10	
Light pressure	1	
Building vibration	0.3	$10^{-12}-10^{-10}$
Buoyancy of the specimen	4000	
Buoyancy of the adsorbate layer	3	
Adsorption on contamination layers	0.1	
Water sorption on a fingerprint	400	
Noise from the electronic circuit	<0.01	$<4 \times 10^{-12}$
Quantum mechanical limit: Heisenbergs time/energy uncertainty relation		1×10^{-29}

chapters. Some recent literature on the calculation of errors and uncertainties of measurements may be found in the references [19–24].

3.4.1 Mechanical Effects

A few mechanical errors are governed by the construction of the various types of balances: these are unequal arm-length of the beam, displacement of the central bearing under load, changes in the ratio of arms in taut-band suspensions, bending of the beam. Effects of this kind may have an indirect influence on the indicator system. These balance errors should be negligible for commercial instruments within the stated measurement range, and should be tested by adjustment and calibration procedures.

3.4.2 Buoyancy Effects

3.4.2.1 Buoyancy Under Atmospheric Conditions

Buoyancy in a liquid was first described by Archimedes [25] in a general, theoretical treatment. In weighing (Fig. 3.7) under atmospheric conditions the article weighed displaces the respective volume of air and therefore the weight w_s^* indicated by the balance is diminished by the weight of the volume of displaced air w_g [26, 27]:

$$w_s^* = w_s - w_g = w_s - \rho_g V_s g_n \tag{3.8}$$

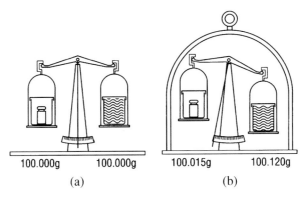

100.000g   100.000g   100.015g   100.120g

(a)   (b)

Fig. 3.7 Buoyancy action. (**a**) A balance loaded with a vessel containing water indicates in equilibrium 100.000 g. © Mettler, Greifensee, Switzerland. (**b**) At reduced pressure a lower mass of air is pushed away, buoyancy is reduced and therefore a higher mass is indicated. In vacuum the balance indicates 100.120 g. © Mettler, Greifensee, Switzerland

where ρ_g is the density of air, V_s the volume of the weighed object and g_n the gravitational acceleration. The density of the air depends on the air pressure and the relative humidity and may be calculated approximately using

$$\rho_g = \frac{0.348444 \, p_g/\text{hPa} - (0.00252 T_g/°C - 0.02082)\Phi_g/\%}{273.15 + T_g/°C} \, \text{kg m}^{-3} \qquad (3.9)$$

where p_g is the air pressure /hPa, T_g the ambient temperature /°C, Φ_g the relative humidity /%. The corrected weight, w_s, is given by:

$$w_s = w_s^* + w_g = w_s^* \left(1 + \frac{\rho_g/\rho_s}{1 - \rho_g/\rho_s}\right) \qquad (3.10)$$

The buoyancy error is much less if in an equal armed balance a mass comparison is made using a counterweight of similar size w_c and of density ρ_c:

$$\frac{w_s}{w_c} = \frac{w_s^*}{w_c^*}\left(1 + \frac{\rho_g}{\rho_c}\frac{\rho_c - \rho_s}{\rho_s - \rho_g}\right) \qquad (3.11)$$

Usually laboratory scales are equipped with a calibration weight of an alloy with density of about 8 kg dm^{-3}.

3.4.2.2 Buoyancy of Vacuum Balances

In the use of vacuum balances, the pressure, temperature and kind of gas used are often changed a great deal, with the result that the buoyancy effect may exceed the changes in mass to be measured by orders of magnitude [28].

The apparent change in mass m_A of a body with the volume V_s corresponds to the mass of the gases displaced by it. If the ideal gas equation is applicable without correction to the gas, it is calculated as:

$$m_A = \frac{p V_s M}{RT} \qquad (3.12)$$

where M is the molecular mass of the gas, R the molar gas constant and T the temperature. In tests with non-ideal gases or in the proximity of the condensation curve, modified gas equations should be used, although these second-order corrections are generally obscured by other experimental influences.

In the standard use of vacuum microbalances, the buoyancy of every individual component with a different temperature must be separately determined and calculated: the balance beam or coil, the suspension device, the pans with sample and counter weight. The resultant buoyancy turns out to be the sum of the partial buoyancy. In the case of equal-armed balances the buoyancy of the beam should be eliminated by means of gold and aluminium wires [29]. The balance is loaded on both sides with gold weights of typical sample mass and the buoyancy registered in large pressure ranges between, for example, 103 and 105 Pa in nitrogen in approximately equal steps. The results are plotted in a diagram m_A/p. The buoyancy m_A referred to the unit of pressure $[\text{kg Pa}^{-1}]$ is calculated in a regression calculation. The mass of the wires is calculated as:

$$m_{Au} = m_{Al} = \frac{\rho_{Au}\rho_{Al}m_A 10^5}{\rho_{N_2}(\rho_{Au} - \rho_{Al})} \qquad (3.13)$$

with $\rho_{Au} = 19.3 \times 10^3$ kg m^{-3} the density of gold, $\rho_{Al} = 2.7 \times 10^3$ kg m^{-3} the density of aluminium, and $\rho_{N2} = 1.25$ kg m^{-3} the density of the nitrogen in gas form under standard conditions.

Subsequently, the balance is loaded with the sample and the measurement repeated. The sample's buoyancy is compensated by a counter weight of the same density. Densities between 2.2 and 19.3×10^3 kg m^{-3} are obtained with gold and quartz crystal pieces. The composition of the compound is calculated with the following equation:

$$m_Q = m_s \frac{\rho_Q(\rho_{Au} - \rho_s)}{\rho_s(\rho_{Au} - \rho_Q)} \qquad (3.14)$$

$$m_{Au} = m_s - m_Q \qquad (3.15)$$

with m_s the sample mass, ρ_s the sample density, m_{Au} the mass of gold, $\rho_{Au} = 19.3 \times 10^3$ kg m^{-3} the density of gold, m_Q the mass of quartz and $\rho_Q = 2.2 \times 10^3$ kg m^{-3} the density of quartz. Counter weights with densities below these values can be made with hollow quartz balls. The remaining buoyancy is either corrected or taken into account by calculation. With unequal-armed balances or spring balances, only a calculated correction according to the following equation can be used:

$$m_{Probe} = m^* + m_A p \qquad (3.16)$$

where m^* is the mass value reading. If the buoyancy was measured at 293 K but the measurement carried out at 77 K, the term $m_A\rho$ should still be multiplied by the temperature ratio $293/77$. If the reaction gas is exchanged, the ratio of the molecular weights of both gases must be taken into account.

Specific problems occur with suspension balances because the suspended magnet has a large volume. Its temperature should be kept constant, that is, it should be registered for the buoyancy correction. Furthermore, the buoyancy of the balance with the carrying magnet, which is subjected to changing air pressure in the atmosphere, must also be taken into account. It might be practical to enclose it in a container under constant inert gas pressure.

In traditional weighing in the gravitational field it is always only the reduced mass of the commodity around the buoyancy which is measured in a gas atmosphere. In sorption measurement it is therefore not the adsorbate mass m_a which is measured, but the quantity:

$$m^* = m_a - \rho_g V_a \qquad (3.17)$$

where ρ_g is the density of the gaseous sorptive. The volume V_a of the adsorbate is unknown and must be determined indirectly. As with rotary pendulums the amount of non-adsorbed sorptive moves with it and so is also weighed, here too a volume correction is necessary in a determination of the adsorbate mass [30] in layers close to the surface (but usually only in the first layer). The density of the adsorbate may deviate substantially from the density of the liquid, being larger, smaller or the same, depending on the temperature [31].

In a thermal analysis the sample volume at increasing temperature is reduced by the reaction, with a simultaneous change in density. On the other hand, in chemical reactions the volume may also increase. The resulting change in buoyancy should be measured by density determinations at the beginning and at the end of the test, and in a first approximation be taken into account by a correction proportional to the pressure.

3.4.3 Gravitational Effects

The weight force measured depends on the local strength of the gravitational field. Values can be obtained e.g. by the Gravity Information Systems of the Physikalisch-Technische Bundesanstalt, Braunschweig [32]. Usually a balance will be adjusted when placed in a specific location initially. However, if the instrument is moved into another floor of the building this factor should be considered. If a sample indicates a kilogram by the balance in the first floor (1.000 000 kg) the balance indicates in

$$1.000000 \frac{(radius_{earth})^2}{(radius_{earth} + 10 \text{ m})^2} = 1 \left(\frac{6370000 \text{ m}}{6370010 \text{ m}} \right)^2 = 0.9999968 \text{ kg} \qquad (3.18)$$

Therefore, the balance should be adjusted again at its new location. On the other hand we see that this is necessary only in case when we need a sensitivity of more than 6 digits.

3.4.4 Thermal Effects and Adsorption

Temperature variations in the weighing room may effect weighing. Whereas influences on the electronic parts in general are compensated mechanical parts will vary in length and may influence the sensitivity. The linear dependence of the balance with temperature is characterised usually by the temperature coefficient $C_T/\% \text{ }^\circ\text{C}^{-1}$. For highly accurate measurements such an error should be avoided by placing the scales in a thermostated room. Scales should be shielded from the sun or light irradiation which may produce temperature differences. The human body irradiates heat of about 80 watt. In all cases body effects of the operator (temperature, breathing) should be prevented by suitable shielding.

Another temperature influence occurs when the temperature of the sample is not in equilibrium with that of the environment. In this case air currents are produced (Fig. 3.8). If the sample is warmer, then the upwards air flow along the flask causes an upwards force at the surface and a decrease in the indicated mass occurs [27]. The effect is reversed for a colder sample.

Fig. 3.8 (**a**) Air current
produced by a warm sample
flask: decreasing weight
indication. © Mettler,
Greifensee, Switzerland.
(**b**) Air current produced by a
cold sample flask: increasing
weight indication. © Mettler,
Greifensee, Switzerland

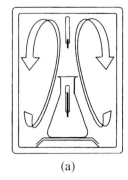 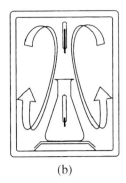

(a)   (b)

At ambient conditions the sample and the sample vessel are covered with a water
film which varies with temperature. By adsorption of water a cold body appears
heavier and a warm body lighter than when under equilibrium due to adsorption or
desorption, respectively.

In many measurement tasks involving microbalances, the sample is cooled with
liquid gas or freezing mixtures in an open Dewar vessel. Because the tank's temperature is air-pressure dependent, it must be monitored. An even temperature is
obtained by stirring or gently heating the bottom layer. If liquid nitrogen is used,
it should be noted that atmospheric oxygen dissolves, leading to a gradual increase
in temperature. A few cm^3 of liquid oxygen can be added to the counter weight's
nitrogen tank to prevent condensation occurring there. When liquid oxygen is used
it should be borne in mind that this represents a fire hazard. In tests on condensable vapours, such as water vapour at room temperature, the balance must be kept
at a higher temperature to avoid unwanted condensation. An example of this kind of
apparatus is shown in Fig. 3.9 [33].

Thermogravimetric measurements usually are made at elevated temperature;
sorption measurements at lower temperature. As the specimen is not in direct contact with the thermostat, sufficiently large areas of constant temperature must be
produced in the field surrounding it. In vacuum, heat is transferred almost exclusively by radiation. The irradiation from the balance chamber, generally maintained
at room temperature, must be shut off by screens (Fig. 3.10). It is advisable to measure the temperature in the specimen, and there are various systems which enable
the temperature signal to be externally transmitted. Otherwise, a thermo-element
should be arranged close to the specimen, or the temperature checked by measurement. Errors on account of thermal effects should be avoided by waiting for thermal
equilibrium: this may well last for hours.

3.4.5 Electrostatic and Magnetic Effects

Both electric and magnetic fields will influence the action of scales. There are even
reports of faking weighing by means of electromagnets or electric tension below

Fig. 3.9 Gravimetric apparatus of Willems for the measurement of water vapour sorption isotherms.
1 microbalance, *2* vacuum chamber, *3* diaphragm pressure gauge, *4* water cooler, *5* heater, *6* to temperature control and record, *7* to pressure control and record, *8* Pirani vacuum gauge, *9* heat insulated case, *10* to balance control and mass record, *11* thermostat control, *12* to temperature record, *13* solenoid valve, *14* metering valve, *15* main suction valve, *16* suction port, *17* thermostat (25.0 °C), *18* thermostatted water bulb for pressure control, *19* thermostat.
© H.H. Willems

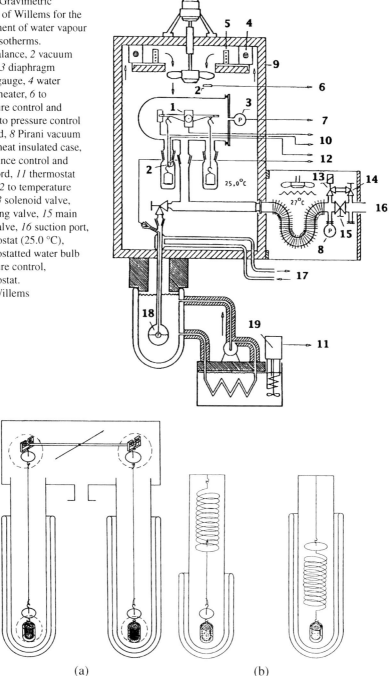

Fig. 3.10 (**a**) Vacuum microbalance with heat shieldings. (**b**) Vacuum spring balance with heat shieldings

Fig. 3.11 Electrical
thermostat with bifilar
winding for temperatures up
to 600 °C for a vacuum
microbalance. *1* cover (2 mm
nonmagnetic steel), *2* sleeve,
3 Fibrefrax heat insulation,
4 bifilar winding (Megapyr),
5 outer tube (5 mm
nonmagnetic steel),
6 fibreglass, *7* inner tube
(light metal), *8* plug (light
metal), *9* ceramic heat
insulation, *10* thermocouple
for temperature measurement,
11 thermocouple for
temperature control,
12/13 electrical connections

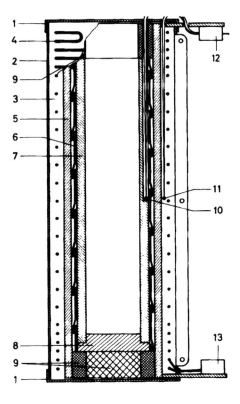

Fig. 3.12 Influence of
electric charge on weighing

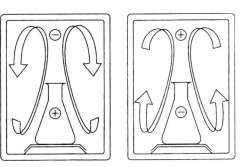

the weighing pan. Of course scales must be carefully shielded from strong fields
that may be produced by electrical apparatus. In thermogravimetry the sample is
surrounded by an oven, usually electrically heated. By bifilar winding a magnetic
field can be avoided (Fig. 3.11).

Sample containers consist often of non-conducting material which can be easily
charged by rubbing (Fig. 3.12). When a powdered sample is poured the particles
may be charged. Such charges should be removed by placing the sample into a
metallic container and/or by touching with an earthed device.

In a balance in a closed container/vacuum vessel charges may be introduced to the sample by gas flow or irradiation. Such charges may be annulled by means of a small radioactive sample e.g. of Americium.

3.4.6 Effects of the Brownian Motion

Brownian motion is a fundamental phenomenon in nature. Depending on the temperature, gas molecules move irregularly and collide with movable parts of the balance, causing a stochastic fluctuation of the emitted signal [34, 35]. The balance represents a system with only one degree of freedom (disregarding the lateral movements of the suspension device). From the principle of the equipartition of energy, the disturbing mass effect for an automatic compensating balance is calculated as:

$$\Delta m_B = \frac{4}{l_b g} \sqrt{\frac{JkT}{\tau_i \tau_D}} \tag{3.19}$$

For an estimate we select as the length of the balance beam, 0.2 m, the moment of inertia, $J = 2 \times 10^{-4}$ kg m^2, $T = 300$ K, the integration time $\tau i = 1$ s, the damping time $\tau D = 10$ s, and for the Boltzmann constant $k = 1.4 \times 10^{-23}$ J K^{-1}. We then obtain $\Delta m = 6$ ng. This value represents a practical limit for the resolution capacity of a balance of such dimensions. In the nanogram range and below, an increasing noise level can be expected.

3.5 Adjustment of the Balance and Error Estimate

Today we can expect that commercially available scales are not burdened with deficiencies of their sensitivity within the given measuring range. Constructional shortcomings are compensated by technical measures or computationally corrected to a large extent. It is, however, advisable to look occasionally into the correcting black box. A balance must be calibrated at regular intervals. For official and commercial use, authorised adjustment of scales and weights is required by national regulations, and for laboratory scales calibration services are available commercially. Balances are calibrated by means of sets of standard weights. Precision balances are often equipped with an internal standard weight allowing an adjustment of the instrument. Furthermore, balances are checked either with transposition or with substitution weighing [36].

Weighing results of equal armed balances exhibiting a specific sensitivity of 10^{-3} as used in trade need corrections of systematic errors neither of unequal arm length of the balance beam, nor of buoyancy. Even in the 18th century, however, the specific sensitivity of balances was improved to about 10^{-6}. Scientists became aware of such systematic errors and they started to develop methods to diminish the influence of unsymmetries of the symmetric balance.

Fig. 3.13 Transposition
weighing (Gauss)

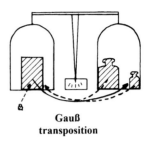

**Gauß
transposition**

3.5.1 Transposition Weighing

In the early days of trade it was already known how to check the correctness of
scales. Commodities and weights were exchanged on the two pans and if the pointer
changed its position the scales were deemed faulty [37] (Fig. 3.13). From this custom resulted transposition or exchange weighing. After the pans' contents had been
exchanged, weighing was repeated. For the weight force w the geometric mean of
the two weighings w_1, w_2 gives the correct result:

$$w_{geometric} = \sqrt{w_1 w_2} \tag{3.20}$$

For small differences the arithmetical mean gives a sufficient approximation:

$$w_{arithmetic} = \frac{w_1 + w_2}{2} \tag{3.21}$$

A text book on docimasy from 1739 indicates that in the first half of the 18th century
essayists used the transposition method for the precise analysis of ores and metals:
"Extrahatur tandem numeri sic producti radex quadrata, quae indicabit verum pondus rei" "Extract the square root of the product of the numbers to obtain the true
weight" [38]. A few decades later, Lavoisier (1743–1794) also used this method in
his work on the nature of water [39]. Later, transposition weighing was named after
Carl Friedrich Gauß who in 1836 recommended it as best method for metrological
weighing [40].

3.5.2 Substitution Weighing

The substitution method (Fig. 3.14) is named after Jean Charles de Borda who about
1791 recommended this method for standardising the kilogram [41]. The sample
was first counterbalanced and in a second step the sample was substituted by a
weight of similar size. Substitution of the sample by weights was applied already in
antiquity to check the mass of coins and later on for many other purposes [40]. Based
on Gauss' judgement the substitution method was superseded by the transposition
method and fell into oblivion until the 1950's.

   In substitution weighing the commodity and the weights are weighed subsequently each against a comparison mass on the opposite side of the balance. This is

Fig. 3.14 Substitution weighing (Borda)

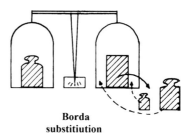

Borda substitiution

according to the mathematical statement that two quantities are equal to each other if each equals a third one. In this way it is not necessary that the balance is exactly symmetric; an arm length error has no influence on the result. There are two variants of substitution weighing, one in which the load is held constant, the other with a changing load. The term 'double weighing' is often used for these, and even more frequently the ambiguous expression is also used for transposition weighing [42]. Because the three axles of the balance are in one plane connecting them is often not fulfilled. In the variation with changing load, the balance's sensitivity has a tendency to decline in higher ranges [43, 44]. However, in weighing with constant load the sensitivity remains constant in all ranges for any type of balances.

3.5.3 Calculation of the Mean Value

Carl Friedrich Gauss (1777–1855) developed a method of error calculation based on the theory of least error square [45]. The standard method is as follows: If in a series of measurements N scattering values m_i result, a mean value m_m can be calculated using:

$$m_m = \frac{1}{N} \sum_{i=1}^{N} m_i \qquad (3.22)$$

The mean square deviation s results:

$$s = \sqrt{\frac{1}{N-1} \sum_{i=1}^{N} (m_i - m_m)^2} \qquad (3.23)$$

3.5.4 Accuracy

Measurement accuracy or trueness is the closeness of agreement between the average of an infinite number of replicate measured quantity values and a reference quantity value.

3.5.5 Correct Weighing

In its history the capabilities of the balances was adapted to the prevailing requirements at this time. In recent centuries the balance was developed to become the most
sensitive instrument of science. In addition methods were developed for highly accurate readings: A typical procedure was observing the amplitudes of the pointer
and arithmetical calculation of the mean value in order to obtain its equilibrium deflection. Later on effective damping methods were introduced. Today such methods
are supplemented by an automatic digital output of a (apparently) stable value, calculated electronically using given criteria. The result is often much more precise
than required in the special task.

In physical equations mass may be only one variable besides others. The accuracy
of the final result depends on the less correct values of such variables. It is advisable
to perform an error calculation for both: assessment of the required accuracy of
measurements of the variables $A, B, C, \ldots$ and of characterising the accuracy of the
final result X. In the most general case, if the value in question X is a function of
some variables: $X = f(A, B, C, \ldots)$, the error of X is determined by the following
procedure [46]:

$$\Delta X = \sqrt{\left(\frac{\partial X}{\partial A}\right)^2 (\Delta A)^2 + \left(\frac{\partial X}{\partial B}\right)^2 (\Delta B)^2 + \left(\frac{\partial X}{\partial C}\right)^2 (\Delta C)^2 + \cdots} \qquad (3.24)$$

Furthermore it should be mentioned that correctness of a value cannot be improved
by mathematical operations. It is misleading to increase the number of digits of a
measured value e.g. after multiplication or division. In this case the resulting value
should be rounded to the appropriate number of significant figures.

3.5.6 Trickery

Unfortunately, deception is not a curiosity but often executed in commerce. So the
oldest reports of weighing from Egypt, in the *Bible* and in the *Koran* are warnings
and exhortations to weigh honestly (see next chapter). Therefore, already in the
Old Empire in Egypt we had a standardisation of balances and weights. The wall
painting in the tomb of Hesire in Saqqara shows two balance beams and sets of
weights and cylinders (Fig. 2.8) [47]. Hesire (Fig. 3.37) was the most important
scientist and highest dentist under the 3rd Egyptian dynasty's pharaoh, Netjenkhet
Djoser (2630–2611 BC) (see Chap. 1). Weights and balances had been kept under
surveillance of priests of the temples in Egypt, Greece and Rome.

Besides altering of the adjusting screw, old-fashioned beam scales could be ma-
nipulated by means of an underfloor magnet below the balance pan. One of the
authors knew a salesman who coughed on a pan when weighing. A mobile green-
grocer, who used a punched disc of a weight-lifter with 5 kg weight inscribed, was
known to one of the authors. Its real weight was 4.6 kg. For modern retail scales adjustment devices are sealed. Typical deceptions are made by not taring the balance

of the weight of wrapping paper, ends of sausages, and by leaning of e.g. a fork on the scale (Fig. 5.128).

More often fraud in the mass of an article is made by addition of worthless filling material, in particular water in food.

3.6 Speeding up Measurements

Correct weighing with the symmetrical beam balance is time consuming. Since the beginnings of the use of balances, methods of braking and calming down the eddy movements have been applied. Fast balances are required for observing the mass change occurring in a fast process or using the mass as a criterion for selection or control. The present chapter deals with methods of shortening the measuring time [48, 49].

Conventional balances, including both beam or spring type balances and also other force sensors, are systems which can easily be stimulated to mechanical oscillations. Enhancing the sensitivity of a balance intensifies simultaneously the tendency to oscillate and extends the time of settling equilibrium [50]. Two types of mechanical oscillations can be observed: Oscillation of the balance beam around its swivelling axis or in direction of the helical spring axis, respectively (one degree of freedom) and oscillations of the pans (two degrees of freedom). Both modes of motions are connected and stimulate each other alternately. This leads to complex indications [51]. Oscillations may be stimulated when placing sample or counterweight on the pans, when sudden variations of the sample mass occur due to reactions with the environment or by transference of external vibrations. After an extended period of time movements decline and the balance comes to rest due to friction in the bearings or in the spring material, respectively, and due to friction in the surrounding air (Fig. 3.15, graph b). Insufficiently damped balances in vacuo may oscillate for an indefinite period of time because always some stimulating disturbances are acting and friction by the residual gas is insignificant (Fig. 3.15, graph a).

A variety of methods have been developed for decreasing the measuring time. In principle weighing time can be shortened in two ways:

- appeasement of the balance and observation of the equilibrium value at rest or
- observation of the oscillating output signal and calculation of the equilibrium value [52].

3.6.1 Capturing Methods

3.6.1.1 Locking of the Balance

The simplest method to quieten down the movements consists of arresting the balance by hand. This was practised already in Old Egypt: with one hand the public

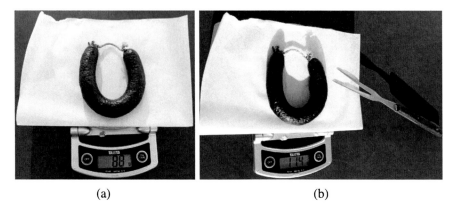

(a)   (b)

Fig. 3.15 Fraudulent weighing:

Weight indication	114 g
Wrapping paper	3 g
Ends of sausage	3 g
Leaning fork	26 g
Sausage	82 g

Fig. 3.16 Undamped (*a*) and damped (*b*) harmonic oscillations and aperiodic limit (*c*) of a balance. The diagram shows the course of angular deflection as a function of time t, simplified for the case of pure beam oscillation and calculated using Eq. 3.1. The *bold line* denotes the evaluated part

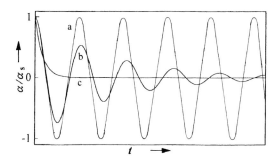

weigher catches the hangdown cords (Fig. 3.16) of the weighing scale or calms down the beam (Fig. 3.17). This was necessary because the beam was suspended in such a way that it could swing not only in vertical direction but also rotate horizontally around its suspension. With his other hand he touches the dangling plummet which was a part of the indicator system [53]. Since then, oscillations of hand scales were stopped by putting down the instrument so that one pan touches the desk. Further development resulted in devices with levers and gears or tackle line, which allowed setting of the balance (Fig. 3.18). Later on two discs were installed which could be lifted until they touch the pans (Fig. 3.19).

Sometimes the balance tube of a vacuum microbalance is equipped with a clamp, operated from outside by means of which oscillations of the hangdown wire can be stopped and simultaneously the sample electrically grounded. Oscillations may be broken by inclining the balance tube a little within the flange and touching the sample pan. Also the compensation current of electrobalances may be decreased so

(a)

(b)

Fig. 3.17 (**a**) Weighing of precious metals. Tomb of Mencheperreseneb, Schech abd el Gurna, 18th Dynasty. (**b**) Weighing of precious metals. Tomb of Rechmere

that the sample pan descends and touches the bottom. In this way also a thermal contact to the pan is realised and temperature equilibrium is speeded up [54].

For analytical balances the locking mechanism is often combined with levers lifting the weighing scales and the beam in order to relieve the knife-edges. Today, even kilogram prototype comparators of extreme sensitivity with knife-edges are operated without locking and relieving [55, 56]. Microbalances are suspended mostly in taut bands or springs and relieving of the beam is of minor importance and made for transport only.

3.6.1.2 Limitation of the Oscillation Amplitude

In order to curtail weighings, the Romans fastened a rectangular frame at a bracket at some distance from the scale beam axis to one or both ends. The beam is fed through

Fig. 3.18 Putting down the balance beam by means of a tackle line, pans touching the desk

Fig. 3.19 Sartorius DP 3. Two discs operated by a rotary knob below the ground plate are lifted until they touch the pans and move the beam into the arrest position. Pan suspensions are equipped with air damping devices

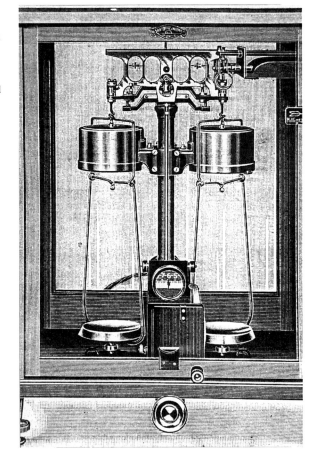

Fig. 3.20 Relief on a Roman public weigher's gravestone, ~200, showing a steelyard with sliding weight. The beam is fed through a frame fastened at a bracket. In this way the beam motion is limited (Landesmuseum Trier, Germany)

this frame and in this way its motion is limited. An example with a steelyard is shown in the relief on a gravestone of a public weigher (Fig. 3.20). At the beginning of the 19th century high-performance balances were equipped with adjustable bars which gave some resistance to the swinging beam [57] (Fig. 3.21). In this way fast corrections of the counterweight are possible. Also modern electronic balances have delimiters for the beam; however, this is not to decrease weighing time, but to protect the beam and its suspension. Delimiter devices may be regarded as a precursor of half-arresting, a method which came into use at the end of the 18th century. Here the arresting lever was operated in such a way that the beam was allowed to swing only with restricted amplitude.

3.6.2 Damping Methods

We assume that the balance is well situated on a heavy stand and protected by dampers from shocks and vibrations from outside. Then by handling, oscillations of the beam and additionally of the suspended sample and counterweight are stimulated. In the following discussion we restrict ourselves to the movements of the beam, assuming suspended parts being in rest with respect to the beam [58]. Because of friction within the bearings or in the spring material and by the surrounding air, the oscillations are damped. After a transient effect the angular movement

Fig. 3.21 Adjustable beam
delimiters of the long-armed
precision balance of Nathan
Mendelssohn, 1808

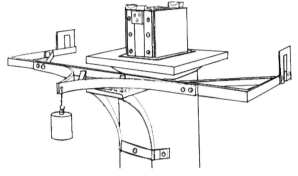

of the beam around the new equilibrium can be described by the following equation
of motion (Fig. 3.15):

$$T = J\ddot{\alpha} + k\dot{\alpha} + C\alpha \tag{3.25}$$

where J is the moment of inertia, k is the damping constant and C is the constant
of the restoring moment, α is the angular deflection and T is the turning moment.
In order to obtain maximum reproducibility, friction in the bearings is minimised as
far as possible and air damping is low. Thus k is very small for balances without
special damping devices. Damping may be defined as application of reacting forces
which are synchronous and proportional to spurious oscillations and vanish as soon
as the interfering signal becomes zero. Thus, damping in principle has no influence
on the weighing result. However, a damping device itself may be subject to environmental influences and in this way the weighing can be disturbed. Therefore, for high
sensitive comparison of mass standards no additional damping devices are applied.

In the 19th century various damping methods for engine vibrations and for indicating instruments were already known and could be adapted for the special requirements of balances. Fast acting damping of the balance beam oscillation by means
of air, liquid or eddy current allowed for faster weighings and direct readings of the
result and enabled combinations with other measuring techniques.

Fig. 3.22 The balance of
Plattner, 1833, could be
lowered by means of a tackle
line. With the brush
movements of the balance
were appeased. Movements
of the weighing scales were
stopped by touching the desk

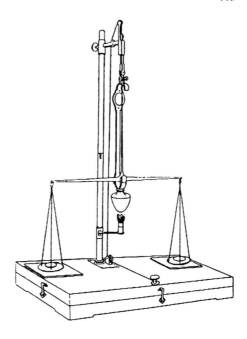

3.6.2.1 Dry Frictional Damping

The oscillations of a balance can be stopped by dry boundary friction of a sliding
mechanical contact to a fixed surface. However, that brings the balance at rest at an
uncertain position on account of the static friction effect.

Widely used within the 19th century was arresting of the balance by means of a
brush. As soon as the oscillations were nearly zero, the brush was removed and the
beam was allowed to swing freely and to find its rest position. This method may be
regarded as a combination of half arresting and damping using dry friction. In the
balance of Plattner [59] damping of the balance by means of a brush was combined
with arresting of the weighing scales by contact with the desk (Fig. 3.22).

3.6.2.2 Fluid Damping

The first real damping device for a balance, air damping, was designed by Arzberger
[60, 61] 1875 (Fig. 3.23). It consists of a plate suspended within the frame of the
support of one weighing scale which moved in a cylinder fastened to the balance casing. Further developments resulted in two cups packed into each other with a small
gap only between the walls of the cylinders [62]. A variety of technical realisations
of that was successfully used to damp mechanical analytical balances (Figs. 3.24,
3.25 and 3.26).

Friction by the residual gas in a vacuum is insignificant. The suspended balance
pans may oscillate for an indefinite period of time, whereas oscillations of the beam
are eliminated electrically. Kuhn et al. [65] recommended the introduction of some

Fig. 3.23 Air damping
according to Arzberger, 1875.
At one side of the symmetric
balance a plate is fastened at
the pan support, which moves
in a cylinder fastened at the
casing

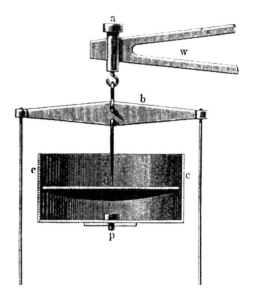

Fig. 3.24 Air damping
according to P. Curie, 1889
with two cylinders at bottom
of both scales. The counter
cylinders are fastened at the
desk

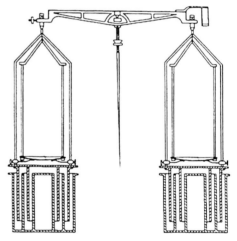

helium into the vessel at the beginning of adsorption experiments. That gas provides
both: damping of the oscillations and faster temperature equilibration of the scales.
Helium is a low adsorbing gas and can quickly be pumped off as soon as the oscillations end. By addition of helium it is also possible to perform measurements in the
pressure region of maximal Knudsen forces.

A special case of oil damping by means of rotating vanes in connection with
a balance is reported by Poynting (Fig. 3.27). Damping by means of a vertical arranged vane moving through oil is widely applied for large mechanical balances and
for rough weighing. Because of disturbing wetting effects and variations in buoyancy this method is not suitable for precision weighing. However, Tripp et al. [66]
immersed the bottom of a special tare pan of a Cahn electrobalance in vacuum oil

Fig. 3.25 Analytical balance
DA 201 of Draegerwerk,
Lübeck, Germany 1949, with
air dampers above the beam
[63]

Fig. 3.26 Analytical balance
of Spoerhase, Gießen,
Germany 1932, with a
horizontal air cylinder
arranged at the pointer [64]

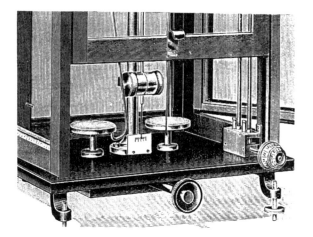

in order to dampen vibrations. Since the beam of the compensating microbalance is
always at null position, they observed no errors due to buoyancy and no decrease in
sensitivity.

Fig. 3.27 Oil damping of a
gravimetric instrument for the
determination of the mean
density of the Earth according
to J.H. Poynting, 1891. At the
right hand side the end of a
0.6 m long pointer which
drives a bracket equipped
with a mirror for microscopic
observation. The mirror is
damped by means of crossed
vanes turning in oil

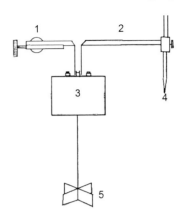

Fig. 3.28 (**a**) Spiral spring
equilibration of Marek.
(**b**) Laboratory balance of
W. Marek, produced by
Nemetz Wien, 1906, with
permanent magnet/eddy
current damper

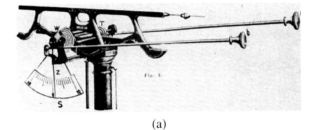

(a)

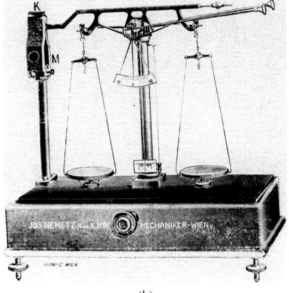

(b)

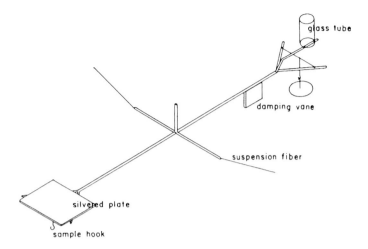

Fig. 3.29 Quartz beam microbalance according to Kolenkow and Zitzewitz, with copper vane as conductor for the eddy current generated by an external magnet

3.6.2.3 Eddy Current Damping

In very early times, eddy current damping was applied for mirror galvanometers [67]. First, Marek [68] developed an analytical beam balance with eddy current damping (Fig. 3.28). Then, in the 1930s, in the USA, laboratory balances were equipped with such devices whereas in Europe air damping was preferred.

Walker [69] damped beam oscillation of a Gulbransen balance [70] by means of a thin plate of pure aluminium suspended on the aluminium suspension wire for the counterweight. Outside the balance tube an external alnico permanent magnet was arranged, the poles of which encircled the aluminium plate. Kolenkow and Zitzewitz [71] fastened a copper vane on the beam (Fig. 3.29). Mayer et al. [72] combined an eddy current device with automatic control of the balance (Fig. 3.30). Damping of the torsion UHV microbalance was accomplished in two ways—eddy currents and delayed circuit by means of a permanent magnet plunging into a copper cylinder with coil.

3.6.3 Electronic Methods

3.6.3.1 Electronic Damping

In electronic balances lag, lead and filter elements are used in the control loop and absorptive attenuators are included in the indicating circuit. Electronic compensating balances [73, 74] use a feedback loop in which an error signal is amplified, producing directly the compensating force. In this case, one should consider not only the mechanical self-oscillations of the balance but also those of the control loop [75].

Fig. 3.30 Quartz torsion microbalance according to Mayer et al. with means for eddy current and electronic damping

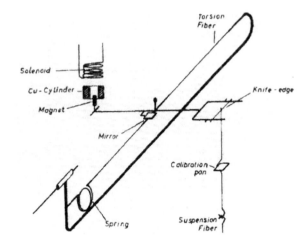

Fig. 3.31 Diagram of a filter circuit in order to damp the output signal of a Cahn balance

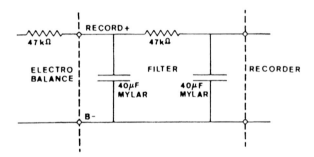

Both oscillations can be influenced by means of lag, lead and filter elements in the control loop. Furthermore, absorptive dampers can be included in the indicating circuit. Mauer [76] stopped beam oscillations of an automatic analytical balance by phasing the beam position signal with a velocity damping signal obtained by differentiating the beam position signal. Cahn [77] equipped his electrobalance with an adjustable shunted capacitor in the indicating circuit in order to damp oscillations of the signal [78] (Fig. 3.31). For his suspension balance, Gast [79] inserted an RC circuit between discriminator and the output amplifier which feeds the magnetic coil controlling the distance of the suspension magnet.

3.6.3.2 Digital Methods

For digital balances fast reacting nullification of eddy signals is applied with uncoupling of the indicated signal as soon as a reliable equilibrium value is calculated [49]. The digital indication can be disconnected and stopped when either for a given period of time the mass value does not vary or as soon as a reliable means value can be extrapolated.

3.6.4 Fast Weighing

In industrial applications weighing is often a bottleneck in the production line. Therefore, in sorting devices many balances are often arranged in parallel. In this section we describe methods of avoiding this expensive technical expenditure.

As can be seen in Fig. 3.15 the undamped signal is always faster than any damped one. So damping should be avoided. Already the initial part of the undamped harmonic oscillation curve includes all information about the equilibrium value. Instead of reading the equilibrium value, the equation of motion of a balance Eq. (3.16) can be used to calculate the unknown torque T and to measure, for that purpose, all the other quantities in the equation: the angle of deflection of the beam α, the moment of inertia J, the damping constant k and the constant of the restoring moment C [34, 80].

When working in the field of magnetism, we applied this method to read a very slow balance in a reasonable time [80]. Later on we discussed whether this method could be used also for fast balances and we considered several sources of error relevant for this method [81]. We found out that using this method the balance reading could be at least ten times faster as waiting for equilibrium [82, 83] Similar results were reported by Horn for load cells [84, 85] Weighing errors which are either intrinsic to the instruments or are due to environmental influences, can be smaller than those when waiting for the equilibrium position [86]. Therefore, using this method, microbalances could be applied better for the monitoring of fast chemical and thermal processes [87, 88].

3.6.4.1 Concept of a Fast Balance

The following considerations are made for a balance operated electrodynamically. Optical or electromagnetic sensing of the balance arm deflection provides high precision and sensitivity. We shall, for the time being, suppose that there are no limits to either of these features. This assumption leads at first sight to the conclusion that no damping or electrical feedback should ever be used as this could only involve extra errors. On the other hand, the use of such a completely undamped balance would very quickly cause unacceptable deflections and velocities of the balance beam.

Poulis, Massen, and Robens suggested separating measurement and feedback in time. So we think about periods of measurement (lengths t_m) alternating with periods of adjustment of the current through the coil (lengths t_a). During t_m, we suggest reading deflection (at constant current), differentiating once and twice, and using the equation of motion:

$$T_x + T_c = J\frac{d^2\alpha}{dt^2} + k\frac{d\alpha}{dt} + C\alpha \qquad (3.26)$$

In this equation T_x stands for the torque to be measured, T_c for the compensating torque (which we assume to be constant during t_m). We assume we know the values of J, k, and C. When the deflection becomes too large we stop the period of measurement (Fig. 3.32).

Fig. 3.32 The three pulses of the compensating torque. Designations see text

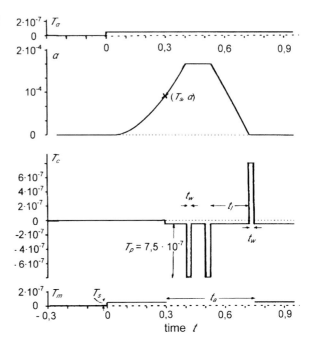

In the following period t_a we:

(1) adjust T_c
(2) apply a peak in the current to reduce the beam velocity
(3) apply a combination of a positive pulse and a negative one (of equal magnitude) to reduce the beam deflection.

In order to assess the behaviour of a balance in such a procedure we use data derived from an electrodynamic beam microbalance made by C.I. Electronics (Fig. 3.33, Table 3.2). At the centre of the beam a coil is attached which moves in the field of a fixed permanent magnet stator to provide for the compensating torque T_c.

The method described requires that the current pulses through the coil should be generated quickly, within t_a. However, this is limited by the self-induction L and the resistance R of the coil. The angular frequency $\omega = R/L$ is therefore crucial. This angular frequency of the C.I. Instruments balance amounts to 11500 rad s^{-1} ($f = 1700$ s^{-1}).

Therefore, it is reasonable to assume that we can pulse the current I through the coil during an interval t_p of 10^{-2} s. Suppose, for estimation purposes, that I is constant during such intervals, we get for the compensation torque T_c the following expression:

$$T_c = n\frac{1}{2}b2lBI = 0.1I \qquad (3.27)$$

where n is the number of turns of the moving coil, b and l are the dimension of the coil and B is the magnetic induction in the air gap.

Table 3.2 Data of the CI Electronics balance and material parameters used

Symbol	Quantity/Parameter	Value
l_{beam}	Total length of the beam	2×0.04 m
J	Moment of inertia of the beam, with no suspensions attached	10^{-7} kg m^2
f_0	Mechanical resonance of beam without electrical feedback	1 Hz[a,b]
$l \cdot b \cdot d$	Dimensions of the coil	21 mm $\times$ 15.5 mm $\times$ 5 mm
n	Number of turns of the moving coil	1400
R	Resistance of the coil	2.3 kΩ
L	Coil inductance (measured with the coil mounted in the magnetic circuit)	200 mH
B	Magnetic induction in the air gap	0.23 T
$c_{v\,copper}$	Specific heat of copper	4×10^3 J kg^{-1} K^{-1}
ρ_{copper}	Resistivity of copper	1.7×10^{-8} Ω m
d_{copper}	Density of copper	8×10^3 kg m^{-3}
$\alpha_{lin\,copper}$	Linear expansion coefficient of copper	1.7×10^{-6} K^{-1}

[a] Approximate

[b] Much higher with electrical feed-back

Fig. 3.33 Electrodynamic compensating beam microbalance of C.I. Electronics. *1* balance beam (metal), *2* bearing for sample/counterweight hangdown wire, *3* permanent magnet with fixed coil, *4* moving coil, *5* glass casing, *6* ground-in connection, *7* metal flange with electrical connection

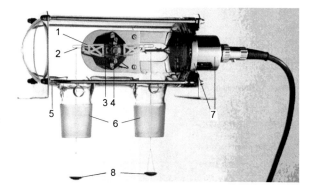

Such a peak causes a variation $\Delta\left(\frac{d\alpha}{dt}\right)$ of $\frac{d\alpha}{dt}$ which satisfies:

$$T_c \Delta t = J \Delta\left(\frac{d\alpha}{dt}\right) \tag{3.28}$$

resulting in

$$\Delta\left(\frac{d\alpha}{dt}\right) = 10^4 I \tag{3.29}$$

One of the effects, which limit the current through the coil, is the heat Q that this current develops:

$$Q = RI^2 t_p = 2 \cdot 10^{-5} \left[\Delta \left(\frac{d\alpha}{dt} \right) \right]^2 \tag{3.30}$$

Using the data of Table 3.2 the mass of the coil m_{coil} is estimated to be 7×10^{-4} kg.

The temperature increase $\Delta\Theta$ due to the heat caused by the peak of the current is:

$$\Delta\Theta = \frac{Q}{m_{copper} C_{v copper}} \tag{3.31}$$

$$\Delta\Theta = 7 \cdot 10^5 \left[\Delta \left(\frac{d\alpha}{dt} \right) \right]^2 \text{[K]} \tag{3.32}$$

This temperature increase leads to an increase in width b of the coil. Thus according to Eq. (3.6) the relative error $\Delta_{error} T_c / T_c$ satisfies:

$$\frac{\Delta_{error} T_c}{T_c} = \alpha_{copper} \Delta\Theta = 10^{-12} \left[\Delta \left(\frac{d\alpha}{dt} \right) \right]^2 \tag{3.33}$$

where α_{copper} is the linear coefficient of expansion.

In the above-suggested procedure, in order to demonstrate that the temperature effect is more serious than in the case of a conventional automatic balance, let us consider the following example. We let $t = 0$ coincide with the beginning of a measurement period which therefore lies between $t = 0$ and $t = t_m$. We take the example where both T_x and T_c are 0 between $t = 0$ and $t = \frac{1}{2} t_m$ and where T_x jumps to T_0 at $t = \frac{1}{2} t_m$. For a conventional automated balance the value of T_c will become $-\frac{1}{2} T_0$ at $t = \frac{1}{2} t_m$.

Using $-I_0$ to denote the compensating current, which is required to achieve the torque $T_c = -T_0$, the heat P_{aut} developed in the coil during a complete cycle $t_m + t_a$ amounts to:

$$P_{aut} = \left(\frac{1}{2} t_m + t_a \right) R I_0^2 \tag{3.34}$$

for the case of a conventional balance.

To discuss the heat developed by our procedure introduced above we calculate the angular velocity at $t = t_m$:

$$\left(\frac{d\alpha}{dt} \right)_{t=t_m} = \frac{T_0}{J} \frac{1}{2} t_m \tag{3.35}$$

To eliminate this velocity we use a rectangular peak: $T_c(t) = T_p$ during t_p at $t - t_m$ where T_p satisfies:

$$T_p t_p = -T_0 \frac{1}{2} t_m \tag{3.36}$$

The necessary peak in the current I_p satisfies:

$$I_p = -\frac{I_0}{T_0} T_0 \frac{1}{2} \frac{t_m}{t_p} \qquad (3.37)$$

The heat P involved satisfies:

$$P = I_p^2 R t_p = \left(\frac{1}{2} \frac{t_m^2}{t_p} + t_a \right) R I_0^2 \qquad (3.38)$$

If we compare Eqs. (3.10) and (3.14) we see that the reduction of t_p aimed for involves an extra heating of the coil.

It has to be mentioned that the choice on the basis of the example given above was a worst case one. When at $t = 0$ the torque T_c and T_x are not taken to be zero, the heat effect will be less serious.

3.6.4.2 Extrapolation of Weighing Results

The mass change during chemical or physical processes often occurs slowly so that the response time of the balance is sufficient but a result of the observation of the process cannot be attained in reasonable time. In such cases it is appropriate to extrapolate the result. Extending a resulting curve by visual estimate or fixing of a period in which no significant change of mass is observed is dangerous, because long ranging mass changes may be overlooked. It is reasonable to search for a mathematical law describing the process under investigation and to use the respective equation for extrapolation of the result. As an example we present a solution of Jäntti [89–91] for the measurement of adsorption processes which occur very slowly.

In 1970 Jäntti published a method to obtain equilibrium values at an early stage of gravimetric sorption measurements [89, 90]. He calculated the equilibrium value of the dynamic adsorption isotherm on the basis of a molecular model for the adsorption process by means of a computer [92]. He applied his method to a large number of series of gravimetric measurements, in particular of complete nitrogen sorption isotherms on activated carbons at 77 K and the determination of the specific surface area and the pore size distribution. He was able to reduce the measuring time by about 70 percent.

In earlier papers Poulis and co-workers criticised [93, 94] and extended that method by using different models for sorption processes [52, 95–107] and introduced functions which could be used in the evaluation of sorption parameters [108–111].

Important technical applications of Jäntti's method are the measurement of uptake and release of vapours, the observation of drying processes and the determination of the dry mass. For such processes, the establishment of equilibrium may need days and even weeks. Often the measurements are truncated after a given period of time or when the variation of the signal is below a given value. Obviously slow mass changes, which may lead to a different equilibrium value, can be overlooked. Therefore an early assessment of the equilibrium value in this way can only be applied where the sorption behaviour of the material or of the material group is well known. If this is not the case, under certain pre-conditions, truncation of the measurement may nevertheless be done and then the asymptotical equilibrium value calculated by means of extrapolation. An appropriate procedure provides Jäntti's method.

If at constant temperature a solid sample is exposed to a sorptive vapour at a varying and then constant partial pressure, uptake or release can be described in many cases approximately by a simple exponential law:

$$m_a(t) = m_{as}\left(1 - e^{-t/\tau}\right) \tag{3.39}$$

where m_{as} = asymptotical equilibrium value of the mass adsorbed, m_a = adsorbed mass, t = time, τ = characteristic time. Here the initial value of mass is set as zero. This law fits in with a simple molecular sorption model [89, 90, 110].

Jäntti used three consecutive values m_{a1}, m_{a2} and m_{a3} at times, t_1, t_2 and t_3 respectively and at equal time intervals, $\Delta t = t_2 - t_1 = t_3 - t_2$ and assessed the approximate equilibrium value m_{as} using:

$$m_{as} = \frac{m_{a2}^2 - m_{a1}m_{a3}}{2m_{a2} - m_{a1} - m_{a2}} \tag{3.40}$$

Poulis et al. proposed a differential form of equation in order to evaluate continuously the measured mass curve and to observe whether Eq. (3.31) holds:

$$J(t) = m_a - \frac{(dm_a/dt)^2}{d^2m_a/dt^2} \tag{3.41}$$

By differentiating the equation and inserting it is easily shown that $J(t)$ has a constant value which equals the asymptotical equilibrium value m_{as}:

$$J(t) = \text{const.} = m_{as}. \tag{3.42}$$

For the numerical calculation by using discrete measuring values should be applied repeatedly to obtain:

$$J(t) = \frac{m_{a2}^2 - m_{a1}m_{a3}}{2m_{a2} - m_{a1} - m_{a2}} \tag{3.43}$$

In principle, $J(t)$ should result in the equilibrium value m_{as} at a very early stage of the measurement. On account of disturbances which occur by the change of the gas pressure, the uncertainty of the signal may be large. Therefore the measurements should be repeated sequentially and recorded to give the Jäntti curve. Although Eq. (3.41) may not hold good in practice in a very early stage, the Jäntti curve approximates the asymptotical equilibrium value m_{as} much faster than the curve of the experimental mass values m_{ai} as is demonstrated in Fig. 3.34.

If $J(t)$ does not approximate to a constant value (Fig. 3.35) the adsorption process does not fit a simple model. Scattering of the Jäntti curve is an early warning that equilibrium may not be obtained in reasonable time. This will give a warning that the adsorption process is more complicated. In some cases a more complicated exponential function can be applied [99, 100, 107].

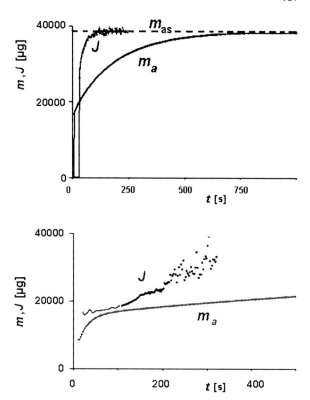

Fig. 3.34 Measured mass adsorbed $m_a(t)$ after an increase step of adsorptive pressure and curve calculated using Eq. (3.40): $J(t)$ with $\Delta t = 22, 5$ s, m_{as} – – –. The adsorption process is well described by Eq. (3.39) and the Jäntti curve $J(t)$ approximates the asymptotical equilibrium value m_{as} much faster than the curve of the experimental mass values $m_a(t)$

Fig. 3.35 Measured mass adsorbed $m(t)$ after an increase step of adsorptive pressure and curve calculated using Eq. (3.40): $J(t)$ with $\Delta t = 22.5$ s. The adsorption process fits not a simple model as described by Eq. (3.39). Scattering of the Jäntti curve warns early that equilibrium may not be obtained in reasonable time

3.7 Balance Standardisation

In Chap. 2 the standardisation of the system units and of the weights is reviewed. However, the balance itself and its characteristic qualities are also subject to standardisation. Furthermore there exist standardised instructions and recommendations for the operation of this sophisticated instrument.

Already in antiquity trade with faraway regions and chemical production e.g. in metallurgy and pharmacy, required standardisation of the balance and its use [112]. A tablet in the tomb of Hesire in Saqqara (Egypt) shows two balance beams and sets of weights and cylinders (Fig. 2.8) [47]. Hesire (Fig. 3.36) was the most important scientist and highest dentist under the 3rd Egyptian dynasty's pharaoh, Netjenkhet Djoser (2630–2611 BC). The balance beams depicted were of the same shape as those about a thousand year's older Naqada object. Such scales of technically unfavourable design could resolve mass differences of 1 g and had a relative sensitivity of about 10^{-2}. Within the 3000 years pharaonic period in Egypt the balance was improved to a relative sensitivity of 10^{-4}. During the New Kingdom a quite other type appeared characterised by four suspension cords coming out laterally at both ends of the wooden balance beam. Considering findings and many pictures we can assume that also this type was standardised and in addition it involved somewhat difficult operation.

Fig. 3.36 Hesire, High
officer under Pharaoh
Netjenkhet Djoser
(2630–2611 BC).Chief of the
royal writers, Governor of
Buto, Highest dentist and
physician, Highest of the Ten
of Upper Egypt

In the following millenniums new types of balances were developed, however reports on standardisation are seldom. In the two volumes 'Weseler Edikte' of 1324–1600 [113] more than 50 regulations are included concerning different balance types. In the 20th century efforts were made to define common parameters characterising the abilities of balances. This process is still in progress on account of new balance types and because the analogue indication is replaced more and more by digital. Table 2.7 includes institutions engaged in such standardisation work.

Recommendations for the use of balances are given by the manufacturers. To allow for an objective judgement of the capability and for comparison of competing instruments national authorities published definitions, standards and guidelines [5, 6, 8, 10, 13, 114, 115]. Common regulations concern the certification and installation and control of scales in public use. It should be noted that one should distinguish between adjustment and calibration. Adjustment means that the scaling of the balance is adapted to correct values regarding the special environmental conditions (gravity due to geographic position, temperature, etc.). Calibration means that the indication is controlled and corrected by means of standard weights. For example in Germany 'Deutscher Kalibrierdienst' (DKD) is an association of calibration laboratories of industrial firms, research institutes, technical authorities, inspection and testing institutes. These laboratories are accredited and supervised by the Accreditation Body of Deutscher Kalibrierdienst. They calibrate measuring instruments and material measures for measurands and measurement ranges specified within the framework of accreditation (DIN EN ISO 9001:2000).

Fig. 3.37 Range of
weighing. Range of mass
occurring in nature. Range of
weighing technology:
0.1 μg–100 t. © Wiley/VCH
[117]

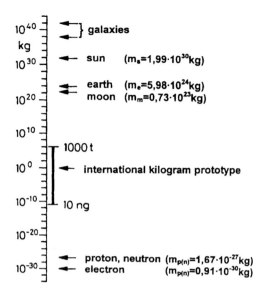

3.8 The Limits of Weighing

The typical range of weighing is between 0.1 μg and 1000 tons (Fig. 3.37). Upper limits of weighing are governed by practical considerations. Up to the present time, all objects could be weighed. For practical reasons large quantities of material often are weighed by proportioning. The Eiffel Tower in Paris has a mass of about 10 000 tons and deriving this figure could be based on hydraulic presses in the foundations which had been planned for the correction of the upright position of the tower [116]. Eiffel installed, at each of the 16 piers of the foundation, a permanent hydraulic press, forming part of the anchorage, by means of which the whole gigantic structure could be tilted bodily if it deviated measurably from the perpendicular. This deviation never occurred.

The lower limits of weighing have been decreased systematically according to the development of new techniques. Such developments may be expressed by the minimum detectable mass or the resolution of the balance and by the relative resolution which is the ratio of resolution to maximum load.

In Fig. 3.38 the oldest balance beam is depicted, made of limestone and about 85 mm in length (the ropes are added for demonstration). It was found in Upper Egypt and dated to pre-dynastic times [118, 119]. In the tomb of Hesire of the 3rd dynasty about 2650 BC a standardisation table is depicted containing cylinders, weights and two balance beams [47]. Such scales of technically unfavourable design could resolve mass differences of 1 g and had a relative sensitivity of about 10^{-3}. At about 2000 BC balances in ancient Egypt had a relative sensitivity down to 10^{-4} [120]. Childe [121] wrote: "The existence of purpose-made alloys of copper with lead for small ornaments and alloys of copper with varying amounts of tin for a wide variety of bronzes implies an ability to make accurate measurements with a weighing device ca. 3000 BC and perhaps earlier".

Fig. 3.38 Copy of the oldest
balance, Egypt, Science
Museum London

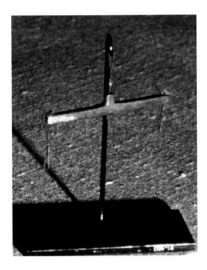

Fig. 3.39 Model of a large
Egyptian balance

There are many drawings of balances in Egyptian tombs of the 18th Dynasty $\sim$ 1567–1320 BC and later. The Egyptians believed that in a death tribunal the heart representing the soul of a deceased person would be weighed against an ostrich feather which was the symbol of truth [122]. Such a balance, of a man's height, was reconstructed in order to determine its sensitivity and to understand its operation (Fig. 3.39) [120]. The model had a capacity of about 30 kg and a resolution of 1 g. So indeed the Old Egyptians could weigh that feather in such scales (see Chap. 7).

In 350 BC Aristotle and about 250 BC Archimedes clarified the theory of the of lever [123]. Archimedes used the balance for the determinations of the relative density (specific gravity) of solids (Fig. 3.40) [25]. The masterpiece of a hydrological balance was Al Chazini's 'Balance of Wisdom' built about 1120 AD (Fig. 3.41) [124]. It achieved a relative resolution of 2×10^{-5}. Further improvement of the

Fig. 3.40 Diagram of
Archimedes's density balance

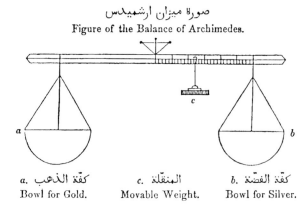

صورة ميزان ارشميدس

Figure of the Balance of Archimedes.

a. كفة الذهب   *c.* المنقلة   *b.* كفة الفضة
Bowl for Gold.   Movable Weight.   Bowl for Silver.

Fig. 3.41 Al Chazini's
'balance of wisdom'.
Reconstruction.
© Islamisches Institut der
Johann-Wolfgang-Goethe-U-
niversität, Frankfurt am Main,
Germany

balance progressed very slowly. Leonardo da Vinci at the end of the 15th century
designed several gravimetric hygrometers (Fig. 4.33) [125]. That means he measured water vapour adsorption. The balances were very poor; nevertheless he was
able to observe variations of their readings for weather forecasting.

Real progress took place when scientists like Lavoisier founded modern chemistry (Fig. 3.42). At that time, at the end of the 18th century, the French revolution

Fig. 3.42 Fortin's balance
for Lavoisier

took place. Lavoisier was beheaded, but his work survived including the introduction of the new metric system of units [126]. To weigh the mass prototypes highly sensitive balances operated *in vacuo* were needed. At the end of the 19th century metrological comparator balances for the kilogram prototypes achieved a relative resolution of 2.5×10^{-9}. The Bunge transposition balance was used for more than 80 years between 1879 and 1951 and is now in the museum of the Bureau International des Poids et Mésures at Sèvres, France [63, 127]. In many weighings by standardisation bureaus the mean uncertainty of measurement was 10^{-6}. The resolution of comparator balances had been improved by means of electromagnetic systems which compensate the deviations of the beam after course compensation by counterweights. In the National Laboratory of Metrology (NRLM), Japan, a relative standard deviation of $s_{relative}$ 3×10^{-10} has been reached using a mechanical equal-arm 1 kg metrological balance with resolution of 0.1 pg [55, 56]. At present, the most accurate balance is an equal-arm beam balance with maximum capacity of 2.3 kg at the Bureau International des Poids et Mesures (BIPM) (Fig. 3.43). That balance has flexural strips instead of bearings and is equipped with an automatic sample changer. A value for $s_{relative}$ equal to 3×10^{-12} has been achieved.

Real progress in this field was obtained by applying Lorentz forces to compensate the deflection of the balance beam [73]. Indeed the relative sensitivity of mechanical balances could not be improved; however the measuring range was extended down to the nanogram range. The principle of electromagnetic compensation today is used for most precision balances in all fields of application.

Using a 5 cm long beam made of a quartz wire framework, Hans Petterson [128–131] reported on a balance for samples of 0.1 to 0.2 g and a sensitivity of 0.1 ng, corresponding to a relative resolution of 10^{-9}. The assertion, however, is not substantiated by published measuring results. Czanderna and Rodder produced an ultrahigh vacuum ultramicrobalance equipped with a three-dimensional quartz wire framework beam [9]. The deflection was detected by means of photocells and compensated for electrodynamically. The maximum load was 20 g and the discrimination threshold 30–100 ng, so the relative resolution was 10^{-8} [132]. The signal was constant to 1–2 µg within a period of half a year. For typical commercial vacuum microbalances the relative resolution goes down to about 10^{-7}, and this is a realistic

Fig. 3.43 (**a**) Flexure-strip
comparator balance with
electromagnetic
compensation (FB-2)
designed and constructed at
the Bureau International des
Poids et Mesures at Sèvres,
© BIPM, (**b**) The beam and
servo-mechanism of the FB-2
balance. © BIPM

(a)

(b)

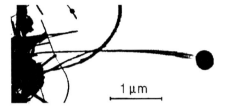

Fig. 3.44 Oscillating carbon nanotube with carbon particle as load. The mass of a graphite particle attached to the end of a oscillating carbon nanotube could be determined as 22 femtogram. © W. de Heer

figure regarding the environmental influences as predetermined by the experimental conditions.

In 1909 Ehrenhaft and Millikan [133] determined the relation electric charge to mass of suspended particles. That method modified by Straubel was used to measure water adsorption isotherms on suspended particles in the range of 0.1 nanogram [134].

In 1665 Robert Hooke invented the spring balance [135, 136]. The relative sensitivity of spring balances is poor in comparison with beam balances. However, its simple design allows the construction of cheap instruments. Spring balances with sensitivities of percent or permille ranges are used for many purposes. Quartz spring balances are favourably applied in sorption experiments in a corrosive atmosphere. In connection with strain gauges their relative sensitivity could be remarkably improved and such load cells today are used everywhere. A load cell based on this principle should be used as a first balance on Mars. The principal objectives are measurements of the surface structure of Martian soil and its sorption capacity [137]

In 1957 Sauerbrey invented the oscillating quartz crystal balance [138]. Using high frequencies, the method is restricted to measure mass changes of samples which are firmly connected to the sensor. However, oscillators produce an acceleration field which can be very much stronger than the gravitational field of the Earth. Thus measuring range and resolution of such a balance can be far higher than that of a conventional balance. An American group of scientists weighed the mass of 22 femtograms of a carbon particle by means of an oscillating carbon nanotube (Fig. 3.44) [139]. Observing the frequency shift of oscillating carbon nanotubes or of silica nanorods recently masses or mass changes in the attogram or zeptogram range have been observed [140–142]. Furthermore graphene layers are used for mass change measurements by observing either its deformation or the shift of resonance vibrations.

A nanomechanical resonator has been used to measure quantum effects of thermal Brownian motion [143, 144].

Recently the Physikalisch-Technische Bundesanstalt, Braunschweig (Germany) developed a nano-force sensor which measures simultaneously course and forth non-destructively [145]. The probe is fastened at a meander shaped spring. Forces between 1 μN and 500 μN are transferred capacitively.

With the atomic force microscope (AFM) it is possible to catch and displace single molecules or clusters. The manipulation allows simultaneously determining binding forces. So-called optical tweezers is a laser beam which can catch a particle of nano- or micrometer size in a 'trap'. If such a particle is moved by means of the laser beam to another object, it may be edged out of that trap and the forces exerted can be measured [146].

The exact statement of an atomic mass in the kilogram unit is possible still only by indirect methods because the mass of an atom is about 10^{-26} kg. The uncertainties for the differences between two atomic masses correspond to 10^{-36} kg.

References

1. A. Einstein, Die Grundlagen der allgemeinen Relativitätstheorie. Ann. Phys. **49**, 769–822 (1916)
2. Equivalence principle. http://enwikipedia.org/wiki/Equivalence_principle (2007)
3. T. Gast, T. Brokate, E. Robens, Vacuum weighing, in *Comprehensive Mass Metrology*, ed. by M. Kochsiek, M. Gläser (Wiley/VCH, Weinheim, 2000), pp. 296–399
4. DIN, *DIN 1319, Teil 2: Grundbegriffe der Meßtechnik. Begriffe für die Anwendung von Meßgeräten* (Beuth, Berlin, 1980)
5. DIN, *DIN 1319 Teil 3: Grundbegriffe der Meßtechnik. Begriffe für die Meßunsicherheit und für die Beurteilung von Meßgeräten und Meßeinrichtungen* (Beuth, Berlin, 1983)
6. DIN, *DIN 8120, Teil 3: Begriffe im Waagenbau* (Beuth, Berlin, 1981)
7. DIN, *DIN ISO: Internationales Wörterbuch der Metrologie. International Vocabulary of Basic and General Terms in Metrology* (Beuth, Berlin, 1984)
8. DIN, *DIN EN 45501: Metrologische Aspekte nichtselbsttätiger Waagen* (Beuth, Berlin, 1992)
9. A.W. Czanderna, S.P. Wolsky, *Microweighing in Vacuum and Controlled Environments* (Elsevier, Amsterdam, 1980)
10. OIML, OIML R 76-1 (E). Non-automatic weighing instruments. Part 1: Metrological and technical requirements—Test. OIML R 76-1 (F). Instruments de pesage à fonctionnement non automatique. Partie 1: Exigences métrologiques et techniques—Essais (OIML, Paris, 1992)
11. BIPM, International Vocabulary of Metrology—Basic and General Concepts and Associated Terms VIM, 4th edn. JCGM 200:2008, ISO Guide 99 by ISO (ISO/IEC Guide 99-12:2007) (BIPM Sevres, 2008)
12. EN, D., *DIN EN 45501: Metrologische Aspekte nichtselbsttätiger Waagen* (Beuth, Berlin, 1992)
13. A.M. Basedow, H.R. Jenemann, Waage und Wägung, in *Quantitative organische Elementaranalyse*, ed. by F. Ehrenberger (VCH, Weinheim, 1991), pp. 79–108
14. E. Robens, Bemerkungen zur Charakterisierung der Leistungsfähigkeit von Waagen. Wägen + Dos. **26**, 6 (1995)
15. E. Robens, The characterization of the capability of a balance. J. Therm. Anal. **47**, 619–622 (1996)
16. C. Berg, *Grundlagen der Wägetechnik* (Sartorius, Göttingen, 1995)
17. DIN, *DIN 55 350 Begriffe der Qualitätssicherung und Statistik; Teil 13: Begriffe der Qualitätssicherung; Genauigkeitsbegriffe* (Beuth, Berlin, 1987)
18. R. Schwartz, Mass determination with balances, in *Comprehensive Mass Metrology*, ed. by M. Kochsiek, M. Gläser (Wiley/VCH, Weinheim, 2000), pp. 232–295
19. M. Grabe, *Measurement Uncertainties in Science and Technology* (Springer, Heidelberg, 2010)
20. I.G. Hughes, T.P.A. Hase, *Measurements and Their Uncertainties: A Practical Guide to Modern Error Analysis* (Oxford University Press, Oxford, 2010)
21. H. Günzler, P. De Bièvre, *Measurement Uncertainty in Chemical Analysis* (Springer, Heidelberg, 2010)

22. S.G. Rabinovich, *Measurement Errors and Uncertainties: Theory and Practice* (Springer, Heidelberg, 2010)
23. B. Pesch, *Messunsicherheit: Basiswissen für Einsteiger und Anwender* (Books on Demand Verlag, Norderstedt, 2010)
24. G. Genta, *Methods for Uncertainty Evaluation in Measurement: Statistical Issues in Metrology* (VDM Verlag Dr. Müller, Saarbrücken, 2010)
25. Archimedes, *The Works of Archimedes, §7. About Swimming Bodies*, vol. 1 (Wissenschaftliche Verlagsbuchhandlung, Frankfurt am Main, 1987)
26. M. Kochsiek (ed.), *Handbuch des Wägens*, 2nd edn. (Vieweg, Braunschweig, 1985)
27. Mettler-Toledo, *Mettler Wägefibel* (Mettler, Greifensee, 1989)
28. E. Robens, T. Gast, Errors due to zero uncertainties and buoyancy in the gravimetric measurements of sorption isotherms. J. Vac. Sci. Technol. **15**(2), 805–809 (1978)
29. E. Robens et al., Auftriebsanwendungen und Auftriebsfehler. Vak.-Tech. **36**(5), 139–147 (1987)
30. E. Robens et al., Determination of the amount adsorbed from the gas phase in a porous solid, in *Proceedings of the XXVIth International Conference on Vacuum Microbalance Techniques*, ed. by M.b.B. Chanaa (Université Cadi Ayyad, Marrakesh, 1995), pp. 219–224
31. R.A. Pierotti, Gas-solid interactions and buoyancy, in *Vacuum Microbalnce Techniques*, ed. by A.W. Czanderna (Plenum, New York, 1967), pp. 1–16
32. A. Lindau, Gravity Information System of PTB. http://www.ptb.de/cartoweb3/SISproject. php. Physikalisch-Technische Bundesanstalt, Braunschweig (2007)
33. H.H. Willems, *Creep Behaviour and Microstructure of Hardened Cement Pastes* (Technische Universiteit, Eindhoven, 1985)
34. J.A. Poulis, J.M. Thomas, Sensitivity of analytical balances and relevance of fluctuation theory, in *Vacuum Microbalance Techniques*, vol. 3, ed. by K.H. Behrndt (Plenum, New York, 1963), pp. 1–14
35. C.H. Massen, J.A. Poulis, Sources of error in microweighing, in *Microweighing in Vacuum and Controlled Environments*, ed. by A.W. Czanderna, S.P. Wolsky (Elsevier, Amsterdam, 1980), pp. 95–123
36. D. Morse, *Lab Balance Handbook* (IES Corporation, Portland, 2011). www.iescorp.com/HB.pdf
37. H.R. Jenemann, The development of the determination of mass, in *Comprehensive Mass Metrology*, ed. by M. Kochsiek, M. Gläser (Wiley/VCH, Berlin, 2000), pp. 119–163
38. J.A. Cramer, Elementa artis docimasticae 1739/1744, Lugdunum Batavorum = Leyden
39. A.L. de Lavoisier, Second mémoire sur la nature d'eau. Histoire de l'Académie Royale des Sciences—avec les Mémoires de Mathématique et de Physique, 1770/3, pp. 90–107
40. H.R. Jenemann, Zur Geschichte der Substitutionswägung und der Substitutionswaage. Technikgeschichte **49**, 89–131 (1982)
41. A.L. de Lavoisier, R.J. Hauy, Expériences de Lavoisier et Hauy, in *Oevres de Lavoisier* (Paris, 1793), pp. 683–685
42. C.F. Gauß, H.C. Schumacher, Briefwechsel, ed. by C.H.F. Peters (Altona, 1861), pp. 34, 61
43. J. Leupold, Theatrum staticum – das ist: Schauplatz der Gewichtskunst (Theatrum staticum universale, Leipzig, 1726)
44. L. Euler, Disquisitio de Billancibus. Commentari Acadimiae Scientiarium Imperialis Petropolitanae 1738/1747. **10**
45. C.F. Gauß, *Abhandlungen zur Methode der kleinsten Quadrate*, ed. by A. Börsch, P. Simon (Vaduz, Berlin, 1886/7)
46. Error calculation (Universität Basel, 2009)
47. F.G. Skinner, *Weights and Measures—Their Ancient Origins and Their Development in Great Britain up to AD 1855* (Science Museum, London, 1967)
48. H.R. Jenemann, Die Geschichte der Dämpfung an der Laboratoriumswaage. Ber. Wiss.gesch. **20**, 1–17 (1997)
49. T. Gast, H.R. Jenemann, E. Robens, The damping of balances. J. Therm. Anal. Calorim. **55**(2), 347–355 (1999)

50. W. Felgenträger, *Feine Waagen, Wägungen und Gewichte* (Springer, Berlin, 1932)
51. T. Gast, T. Brokate, E. Robens, Vakuumwägung, in *Massebestimmung*, ed. by M. Kochsiek, M. Gläser (VCH, Weinheim, 1996), pp. 294–399
52. E. Robens, High-speed weighing. J. Therm. Anal. Calorim. **55**, 455–460 (1999)
53. H.R. Jenemann, E. Robens, Indicator system and suspension of the old Egyptian scales. Thermochim. Acta **152**, 249–258 (1989)
54. C.J. Williams, A new lease on life for gravimetric adsorption. American Laboratory 1969(6)
55. Y. Kobayashi et al., Prototype kilogram balance II of NRLM. Bull. NRLM **33**(2), 7–18 (1984)
56. Y. Kobayashi et al., Prototype kilogram balance II of NRLM. Bull. NRLM **35**(2), 143–158 (1986)
57. N. Mendelssohn, Beschreibung einer großen und sehr genauen Wage zum Gebrauch für Physiker und Chemiker. Gilberts Ann. Phys. **29**, 153–161 (1808)
58. M. Gläser, Massekomparatoren, in *Massebestimmung*, ed. by M. Kochsiek, M. Gläser (VCH, Weinheim, 1997), pp. 442–479
59. C.F. Plattner, *Die Probirkunst mit dem Lötrohre* (Leipzig, 1833)
60. F. Arzberger, Luftdämpfung für analytische Waagen. Liebigs Ann. Chem. **178**, 382–384 (1875)
61. H.R. Jenemann, Zur Geschichte der Herstellung von Präzisionswaagen hoher Leistung in Wien. Bl. Tech.gesch. **49**, 7–85 (1987)
62. P. Curie, Sur une balance de précision et à lecture directe des derniers poids. C. R. Hebd. Séances Acad. Sci. **108**, 663–666 (1889)
63. H.R. Jenemann, Paul Bunge und die Fertigung wissenschaftlicher Waagen in Hamburg. Z. Unternehm.gesch. **31**, 117–140 (1986), see also 165–183
64. H.R. Jenemann, Zur Geschichte der Präzisionsmechanik und der Herstellung feiner Waagen in Gießen. Mitt. Oberhess. Geschichtsver Gießen **66**, 5–54 (1981)
65. W. Kuhn et al., Methods of the elimination of weighing troubles due to convection in a microbalance, in *Vacuum Microbalance Techniques*, ed. by C.H. Massen, H.J. van Beckum (Plenum, New York, 1961), pp. 1–21
66. W.C. Tripp, R.W. Vest, N.M. Tallan, System for measuring microgram weight changes under controlled oxygen partial pressure to 1800°C, in *Vacuum Microbalance Techniques*, ed. by P.M. Waters (Plenum, New York, 1965), pp. 141–157
67. O. Frölich, *Die Entwickelung der elektrischen Messung* (Braunschweig, 1905)
68. W. Marek, Aperiodische Wage mit Hilfsfedern. Österreichische Zentral-Zeitung für Optik und Mechanik **1**, 5–7 (1906)
69. R.F. Walker, Microbalance techniques for high temperature applications, in *Vacuum Microbalance Techniques*, ed. by M.J. Katz (Plenum, New York, 1961), pp. 87–110
70. E.A. Gulbransen, K.F. Andrew, An enclosed physical chemistry laboratory: the vacuum microbalance, in *Vacuum Microbalance Techniques*, ed. by M.J. Katz (Plenum, New York, 1961), pp. 1–21
71. R.J. Kolenkow, P.W. Zitzewitz, A microbalance for magnetic susceptibility measurements, in *Vacuum Microbalance Techniques*, ed. by P.M. Waters (Plenum, New York, 1965), pp. 195–208
72. H. Mayer et al., On some modifications of a torsion microbalance for use in ultrahigh vacuum, in *Vacuum Microbalance Techniques*, ed. by K.H. Behrndt (Plenum, New York, 1963), pp. 75–84
73. H.R. Jenemann, The early history of balances based on electromagnetic and elektrodynamic force compensation, in *Microbalance Techniques*, ed. by J.U. Keller, E. Robens (Multi-Science Publishing, Brentwood, 1994), pp. 25–53
74. H.R. Jenemann, *Die Waage des Chemikers—The Chemist's Balance* (DECHEMA, Frankfurt am Main, 1997)
75. R.L. Schwoebel, Beam microbalance design, construction and operation, in *Microweighing in Vacuum and Controlled Environments*, ed. by A.W. Czanderna, S.P. Wolsky (Elsevier, Amsterdam, 1980), pp. 59–93

76. F.A. Mauer, Analytical balance for rapid changes in weight. Rev. Sci. Instrum. **25**, 598–602 (1954)

77. L. Cahn, H.R. Schultz, The Cahn recording GRAM electrobalance, in *Vacuum Microbalance Techniques*, ed. by K.H. Behrndt (Plenum, New York, 1963), pp. 29–44

78. W.E. Boggs, The adaption of the Cahn electrobalance control system to the automatic operation of a quartz-beam vacuum microbalance, in *Vacuum Microbalance Techniques*, ed. by A.W. Czanderna (Plenum, New York, 1967), pp. 45–58

79. T. Gast, Microweighing in vacuo with a magnetic suspension balance, in *Vacuum Microbalance Techniques*, ed. by K.H. Behrndt (Plenum, New York, 1963), pp. 45–54

80. C.H. Massen et al., Automated balances of the second generation. Thermochim. Acta **103**, 1–4 (1986)

81. C.H. Massen et al., Computer simulation of balance handling. J. Therm. Anal. Calorim. **55**(2), 367–370 (1999)

82. J.A. Poulis, C.H. Massen, E. Robens, Verfahren zur Ablesung einer schwingenden Anzeige, insbesondere von Waagen, in *Offenlegungsschrift DE 197 30 070 A 1*, Germany (1997)

83. J.A. Poulis, C.H. Massen, E. Robens, Verfahren zur Ablesung einer schwingenden Anzeige, insbesondere von Waagen, in *Offenlegungsschrift DE 198 26 438 A 1*. Germany (1998)

84. K. Horn, To make load-cell-scales settle faster, in *XI IMEKO Congress*, vol. 2 (1988), pp. 217–243

85. K. Horn, Verfahren und Meßeinrichtung zur Bestimmung mechanischer Meßgrößen, insbesondere eines unbekannten Gewichts, in *Patentschrift DE 3743897 C2*. Germany (1987)

86. C.H. Massen et al., Optimizing of balances of the second generation. J. Therm. Anal. Calorim. **55**(2), 449–454 (1999)

87. C.H. Massen et al., Fast electromagnetic balance. J. Therm. Anal. Calorim. **71**(1), 47–51 (2003)

88. J.A. Poulis et al., Schnelle Wägung. Wägen Dos. Misch. **3**, 11–13 (2003)

89. O. Jäntti, J. Junttila, E. Yrjänheikki, Mikropunnitusajan Lyhentämisestä Ekstrapolaatiomenetelmällä. (On curtailing the microweighing time by an extrapolation method). Suom. Kemistil., A **43**, 214–218 (1970)

90. O. Jäntti, J. Junttila, E. Yrjänheikki, On curtailing the micro-weighing time by an extrapolation method, in *Progress in Vacuum Microbalance Techniques*, ed. by T. Gast, E. Robens (Heyden, London, 1972), pp. 345–353

91. E. Robens, C.H. Massen, On the applicability of Jäntti's method of shortening sorption measurements. J. Therm. Anal. Calorim. **94**(3), 711–714 (2008)

92. O. Jäntti, E. Robens, Computerised reduction of the weighing time required for the determination of adsorption isotherms on activated carbons. Thermochim. Acta **51**, 67–75 (1981)

93. C.H. Massen, J.A. Poulis, E. Robens, Criticism on Jäntti's three point method on curtailing gas adsorption measurements. Adsorption **6**, 229–232 (2000)

94. E. Robens, J.A. Poulis, C.H. Massen, Fast measurements, fast evaluation, fast results, fast richtig. GIT Z. Labortechnik **46**(5), 556–559 (2002)

95. C.H. Massen et al., Extension of the applicability of Jäntti's method for fast calculation of desorption data. Adsorp. Sci. Technol. **18**(10), 853–856 (2000)

96. J.A. Poulis et al., Fast adsorption measurements on silicium dioxide, in *Theoretical and Experimental Studies of Interfacial Phenomena and Their Technological Applications: Book of Abstracts, VIII Ukrainian-Polish Symposium*, ed. by Y. Tarasevich et al., September 19–24 2004, Sergijiwka–Odessa, Ukraine (SCSEIO, Odessa, 2004), pp. 258–263

97. J.A. Poulis, C.H. Massen, E. Robens, Measurement of gas adsorption with Jäntti's method using continuously increasing pressure. J. Therm. Anal. Calorim. **68**(2), 719–725 (2002)

98. J.A. Poulis, C.H. Massen, E. Robens, The Jäntti approach using a two-layers model. J. Therm. Anal. Calorim. **71**(1), 61–66 (2003)

99. J.A. Poulis et al., The application of Jäntti's method for the fast calculation of equilibrium in the case of multilayer adsorption, in *Characterization of Porous Solids VI*, ed. by F. Rodríguez-Reinoso et al. (Elsevier, Amsterdam, 2002), pp. 761–767

100. J.A. Poulis et al., Jäntti's method for the fast measurement of adsorption combined with diffusion. Z. Phys. Chem. **218**, 245–254 (2004)
101. J.A. Poulis et al., A fast two-point method for gas adsorption measurements, in *Characterisation of Porous Solids*, ed. by K.K. Unger, G. Kreysa, J.P. Baselt (Elsevier, Amsterdam, 1999), pp. 151–154
102. J.A. Poulis et al., The possible use of Jäntti's method for the explanation of adsorption onto rough surfaces, in *Analytical Forum, Book of Abstracts* (Warszawa, 2004)
103. J.A. Poulis et al., The application of Jäntti's method for the explanation of adsorption onto rough surfaces. J. Therm. Anal. Calorim., 39–42 (2006)
104. J.A. Poulis et al., Application of Jäntti's method to volumetric adsorption measurements. J. Therm. Anal. Calorim. **76**(2), 579–582 (2004)
105. J.A. Poulis et al., Evaluation of the applicability of Jäntti's method to volumetric sorption measurements. J. Therm. Anal. Calorim. **86**(1), 43–45 (2006)
106. E. Robens et al., Fast measurements of adsorption on porous materials using Jäntti's method. Adsorp. Sci. Technol. **17**(10), 801–804 (1999)
107. E. Robens, J.A. Poulis, C.H. Massen, Fast gas adsorption measurements for complicated adsorption mechanisms. J. Therm. Anal. Calorim. **62**, 429–433 (2000)
108. J.A. Poulis et al., Introduction of new functions to speed up sorption measurements. Adsorption **12**(3), 213–217 (2006)
109. J.A. Poulis et al., General application of Jäntti's method for the fast calculation of sorption equilibrium. J. Therm. Anal. Calorim. **76**(2), 583–592 (2004)
110. J.A. Poulis et al., A Jäntti approach for quick calculations of sorption equilibria. Z. Phys. Chem. **216**, 1123–1135 (2002)
111. J.A. Poulis, C.H. Massen, E. Robens, Saving time when measuring BET isotherms. J. Colloid Interface Sci **311**(2), 391–393 (2007)
112. I. Jenemann, S. Kiefer, E. Robens, Some intriguing items in the history of scientific balances. J. Therm. Anal. Calorim. (2008)
113. Anonymus, *Weseler Edikte* (Wesel, 1324–1600)
114. DIN (ed.), *Internationales Wörterbuch der Metrologie* (Beuth, Berlin, 1984)
115. DIN, *DIN 51006: Thermische Analyse (TA) – Thermogravimetrie (TG) – Grundlagen* (Beuth, Berlin, 2005)
116. E. Schneider, Daten zu Eiffelturm. http://www.baufachinformation.de/denkmalpflege.jsp?md=1988017121187 (2010)
117. M. Kochsiek, M. Gläser (eds.), *Comprehensive Mass Metrology* (Wiley/VCH, Berlin, 2000)
118. W.M.F. Petrie, *A Season in Egypt* (London, 1888)
119. W.M.F. Petrie, *Ancient Weights and Measures* (London, 1926)
120. C.H. Massen et al., Investigation on a model for a large balance of the XVIII Egyptian dynasty, in *Microbalance Techniques*, ed. by J.U. Keller, E. Robens (Multi-Science Publishing, Brentwood, 1994), pp. 5–12
121. G.V. Childe, The prehistory of science: archaeological documents, Part 1, in *The Evolution of Science, Readings from the History of Mankind*, ed. by G.S. Metraux, F. Crouzet (New American Library/Mentor Books, New York, 1963), pp. 66–67
122. C. Seeber, in *Untersuchungen zur Darstellung des Totengerichts im Alten Ägypten*, ed. by H.W. Müller, Münchner Ägyptologische Studien, vol. 35 (Deutscher Kunstverlag, München, 1976)
123. Aristoteles, a.t. (ed.), Questiones mechanicae, ed. by P. Gohlke (Kleine Schriften zur Physik und Metaphysik, Paderborn, 1957)
124. Al-Chazini, *Buch der Waage der Weisheit* (Merw, 1120)
125. L. da Vinci, *Codex atlanticus—Saggio del Codice atlantico*, ed. by Aretin, vol. fol. 249 verso-a + fol. 8 verso-b (Milano, 1872)
126. H.R. Jenemann, Das Kilogramm der Archive vom 4. Messidor des Jahres 7: Konform mit dem Gesetz vom 18. Germinal des Jahres 3? in *Genauigkeit und Präzision*, ed. by D. Hoffmann, H. Witthöfft (Physikalisch-Technische Bundesanstalt, Braunschweig, 1996), pp. 183–213

127. H.R. Jenemann, Die Werkstatt von Paul Bunge: 100 Jahre Präzisionswaagenherstellung in Hamburg. Beitr. Dtsch. Volks- Altert.kd. **26**, 169–188 (1988/1991)

128. H. Pettersson, A new micro-balance and its use. Diss. Stockholm, Göteborg's Vet. of Vitterh. Samhalle's Handlinger (Göteborg, 1914)

129. H. Pettersson, Experiments with a new micro-balance. Proc. Phys. Soc. Lond. **32**, 209–221 (1919)

130. R. Strömberg, Adsorptionsmessungen mit einer verbesserten Mikrowaage. K. Sven. Vetensk.akad. Handl. III **6**(2), 33–122 (1928)

131. R. Strömberg, Adsorption measurements with an improved microbalance. K. Sven. Vetensk.akad. Handl. III **6**(2), 1–12 (1928)

132. A.W. Czanderna et al., Photoelectrically automated, bakeable, high-load ultramicrobalance. J. Vac. Sci. Technol. **13**, 556–559 (1976)

133. U. Kilian, C. Weber, *Lexikon der Physik*, vol. 4 (Spektrum Akademischer Verlag, Heidelberg, 2000)

134. G. Böhme et al., Determination of relative weight changes of electrostatically suspended particles in the sub-microgram range, in *Progress in Vacuum Microbalance Techniques*, ed. by S.C. Bevan, S.J. Gregg, N.D. Parkyns (Heyden, London, 1973), pp. 169–174

135. R. Hooke, De Potentia Restitutiva or of Spring, explaining the Power of Springing Bodies. Lectiones Cutleriana or a collection of Lectures: Physical, Mechanical, Geographical, & Astronautical. Early Science in Oxford, vol. VIII (Gunther, London, 1678/9)

136. H.R. Jenemann, Robert Hooke und die frühe Geschichte der Federwaage. Ber. Wiss.gesch. **8**, 121–130 (1985)

137. E. Robens, D. Möhlmann, Planning of gravimetric investigations on Mars. J. Therm. Anal. Calorim. **76**(2), 671–675 (2004)

138. G. Sauerbrey, Verwendung von Schwingquarzen zur Wägung dünner Schichten und zur Mikrowägung. Z. Phys. **155**, 206–222 (1959)

139. P. Poncheral et al., Electrostatic deflections and electromechanical resonances of carbon nanotubes. Science **283**, 1513–1516 (1999)

140. B. Ilic et al., Attogram detection using nanoelectromechanical oscillators. J. Appl. Phys. **95**(7), 3694–3703 (2004)

141. S. Gupta, G. Morell, B.R. Weiner, Electron field-emission mechanism in nanostructured carbon films: a quest. J. Appl. Phys. **95**(12), 8314–8320 (2004)

142. J. Wood, Mass detection finds new resonance. Mater. Today **4**, 20 (2004)

143. LaHaye et al., Science **304**, 74 (2004)

144. C. Sealy, Probing the quantum world with uncertainty. Mater. Today **6**, 9 (2004)

145. U. Brand, Nanosensor misst gleichzeitig Kraft und Weg. Nachrichten und Presseeldungen aus Labor und Analytik (2013)

146. S. Arnold, E.-C. Reiff, Kraftmessung in nanoskopischen Dimensionen. LaborPraxis **7/8**, 54–56 (2010)

Chapter 4
Balances

The book is concerned in particular with measuring methods which are aimed at direct mass determination. The respective mass sensors are scales or balances. Apart from simple rulers like yardsticks and cups, the balance is the oldest real measuring instrument, invented more than 5000 years ago [1–3]. Besides watches, rulers and measuring beakers the balance is still one of the most widely used instruments in the world. It is used in the house, in industry, in commerce and of course in every laboratory.

In balances the force of the mass of a sample in a gravitational or acceleration field is compared with a counterforce. For a long period of time compensation of the sample mass by counterweights was the most accurate measuring method and yielded an unrivalled large measuring range. Positioning of the weights, however, exerts forces on the balance and the knife edges resulting in damages and disturbances and the change of weights is a time consuming procedure. Therefore, various methods have been tested to facilitate and to speed up the equilibration. Restoring of the balance to its initial situation after a change of load can be effected by combination of different forces; typically mechanical and electromagnetic. Mechanical forces can be caused by the elastic response e.g. of a torsion wire or discontinuously by position of counterweights. Elastic forces as applied in spring balances strongly depend on the position of the balance beam, which must be accurately observed, and strain gauges as well as piezo-electric sensors exhibit comparatively small measuring ranges. The most successful methods, however, are electromagnetic compensation which was introduced very early [4] and electrodynamic restoring of the beam. Lorentz forces may be actuated continuously or pulsed.

In this chapter balances are classified with regard to the type of the dominant counterforce applied.

4.1 Classification of Balances

Types of balances may be distinguished basically according to the applied method of weighing [5, 6]:

E. Robens et al., *Balances*, DOI 10.1007/978-3-642-36447-1_4,
© Springer-Verlag Berlin Heidelberg 2014

- Gravitational balances
- Momentum measuring systems
- Oscillating weighing systems
- Indirect weighing systems

Until recently the most widely used balances were gravitational balances and there are a variety of such types. We can classify gravitational balances according to their working principle. In the conventional lever or beam type balance the weight force of the sample is counterbalanced by a counterweight; besides many other types of forces are applied for counterbalancing, in particular elastic and electromagnetic forces. Often several methods are combined. In this chapter balances are classified according to their counterbalancing force. In mechanical balances the counterweight may be differently arranged and we have:

- Symmetrical counterweight balances
- Asymmetrical counterweight balances

  – Steelyard
  – Besemer
  – Deflection balance

- Symmetrical counterweight balances with double beam

  – Roberval balance
  – Béranger balance
  – Pfanzeder balance

- Asymmetrical counterweight balances with double beam

  – Platform balance with sliding weight
  – Decimal balance
  – Platform deflection balance

Whereas the use of counterweight balances is restricted to measurements within gravitational fields and usually at a fixed place parallel to the ground, other types of counterforce can be used also for measurements in accelerated devices and in any position. For less accurate weighing, spring balances are widely used. Load cells with strain gauges achieve better sensitivity and can be applied in industry even in difficult situations. Today for accurate weighing electromagnetic counterbalancing is the most important technique used. Less known is the hydraulic balance in which either the sample exerts a pressure on a liquid in a closed vessel and that pressure is measured by means of manometer or the force is counterbalanced by buoyancy of vessels floating in a liquid or by the varying height of a liquid column. Gyro scales are somewhat exotic but applied for sensitive industrial mass determination. The following is a classification of counterforce type balances:

- Spring balances, load cells
- Balances with electric force compensation
- Balances with electromagnetic force compensation
- Hydraulic balances

- Buoyancy balances
- Gyro scales

The investigation of moving objects includes the determination of the inert mass using momentum measuring devices. Such instruments are used in industry for rough determination of the amount of flowing bulk material and liquids:

- Momentum belt weigher
- Momentum liquid flow weigher
- Deflector plate
- Measuring chute

Furthermore:

- Mass spectrometry
- Astronomical measurements using Kepler's Laws

Oscillating systems extended the application of mass determination into new areas (see Sect. 3.8 in Chap. 3). Their features, and hence applications, depend on the frequency range of oscillations. However, here we classify such instruments by the type of the oscillating element:

- Crystals oscillator
- Mechanical oscillator
- Pendulum

In indirect mass determination extensive quantities are measured, the values of which are a function of the sample mass. Furthermore, in chemical reactions, the mass of a compound or of a component can be determined if the reaction and the amount of the other component is known.

- Radiometric weigher
- Density measurement
- Chemical reaction

Alternative classifications may be based on sensitivity, capacity or measuring ranges, others on their applications or on additional equipment e.g. in order to control the environmental conditions.

Today, the term 'electronic balances' is used for a great variety of mass measuring instruments, based on different principles and used in various fields of application. Many balances are equipped with an electrical digital output which allows data processing, but its operation principle is hidden. Indeed, the term 'electronic balances' does not describe the weighing principle and is therefore misleading [7].

In many cases operation principles are combined. The sensitivity of mechanical balances can be improved by means of a weak spring attached to the beam or with an electromagnetic or electrodynamic device. On the other hand the measuring range of an electronic balance can be extended by put on weights to its lever system. Such balances indeed are hybrids.

4.2 Gravitational Balances

Gravitational balances make use of the advantages of the locally highly constant gravitational field of the Earth (Fig. 4.1). Each mass is attracted in that field in a direction to the centre of Earth mass. We measure not the mass directly but the weight force on the sample due to gravity. That force divided by the local value of the acceleration due to gravity using Eqs. (1.1–1.3) gives the mass. Indeed the scaling of balances is given in mass units because there is only a locally constant factor. Gravitational balances work either on the counterweight principle or by the weight force compensated by any other type of force: elastic, electric, electromagnetic, hydrostatic or gyroscopic.

Very sensitive measurements are possible by comparing sample and counterweight using a lever supported on a central pivot. Because the gravitational field depends on the distance to the centre of the Earth and on the distribution of the minerals of different density in Earth's crust, a gravitational balance must be adjusted when installed. Furthermore, unless weighing is made in vacuum, it is not the weight force that is determined but the difference of weight force and buoyancy of sample and counterweight. Whereas buoyancy of parts of the balance should be compensated or corrected by adjustment, differences of buoyancy due to differences in density of sample and counterweight must be considered in sensitive weighing.

Until the beginning of the last century mechanical load receptors were used where the weight is transmitted and spread by the lever arm and the deflection was indicated by a pointer. The mass of the article to be weighed was either compensated or the deflection was related to the mass. This kind of instrument has fallen from favour as it requires qualified technicians to adjust the system and is somewhat difficult and slow in operation. For less accurate measurements spring balances were used. Mechanical lever and spring type scales remain in use because they need no power supply. They are superseded today by sensor types with electrical digital output. As sensor elements and calculator chips become cheaper than mechanical systems, the devices can be made smaller and the results can be recorded and evaluated by means of a computer.

4.2.1 Parts of a Balance

Each gravitational balance consists of several parts and the main important ones are the following:

- sample container, usually sample pan
- connecting parts: suspensions, hinges, lever
- deflection sensor
- counteracting system
- indicator system

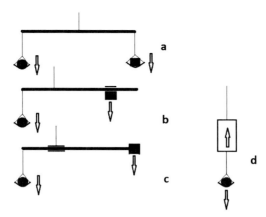

Fig. 4.1 Gravitational balances use of the advantages of the locally highly constant gravitational field of the Earth. (**a**)–(**c**) counterweight balances, a symmetrical beam balance, (**b**) asymmetrical beam balance with sliding weight, (**c**) asymmetrical balance with sliding beam: besemer, (**d**) sensor balance, ▬ balance beam, □ force sensor, ↓↑ force vector (not to scale), ● weighing article, ■ counterweight

Since the development of these components is strongly connected with the development of lever balances, its discussion is centred on that type. In most other balance types the mass of the weighing article is connected to the counterbalancing force by means of levers and so far we have to do mostly with mechanical hybrids. In the following sections, parts are described which are common to the mechanical and mechanical hybrid balances.

4.2.1.1 Balance Beam and Lever

The balance beam has several tasks: as a girder it is a supporting and connecting component and as a lever it is a module which transforms the weight forces exerted from sample and counterweight and is responsible for the sensitivity of the instrument. A symmetrical counterweight balance is composed of three pivots: two as suspension of the weighing article and the counterweight, respectively and one central as a support for the beam (Fig. 4.1a). In asymmetrical beam balances the counterweight may be fixed to the beam and the beam support is not central but at a place in which the balance is in horizontal position (Fig. 4.1b). Whereas in counterweight balances the beam support always divides the beam into two opposite parts, the bearing of a force transforming lever in sensor type balances can be arranged or placed at one end of that part. In this case we have a unequal-armed balance, the counterforce acting in opposite direction to the load (Fig. 4.1c). The shape of such beams may be highly complicated, sometimes put together of several parts or folded up. In highly accurate scales today mono-block designs are widely used: metal blocks in which a complicated twisted 'beam' is cut by laser treatment including thin sections as bearings (Figs. 4.2).

Fig. 4.2 (**a**) Monobloc
balance system. © Mettler.
(**b**) Monobloc balance
system. © Mettler.
(**c**) Monobloc balance system
© Sartorius

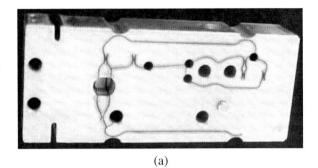

(a)

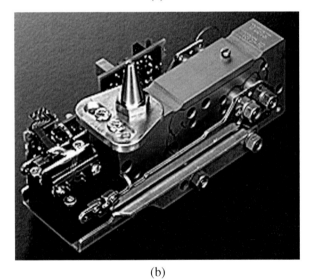

(b)

(c)

The sensitivity of beam balances depends on the beam length. In symmetrical balances the central beam pivot and the bearings of the pan suspensions should be in one line and at equal height. Extension of the beam length increases the sensitivity; on the other hand the beam should not bend when loaded. These conflicting requirements lead to a relative short beam [8]. With symmetrical beam balances the thermal expansion of the object is compensated to the largest possible extent by using counter weights of similar density and ensuring that load and counter weight are at the same temperature.

Above all, constancy of lever ratios is required of balance beams, which indicates a simple construction using materials with a low coefficient of expansion and small internal tensions. The beam should be as light as possible to avoid unduly loading the suspension, and the dynamics make a small moment of inertia desirable. A high degree of flexural strength is important for constant sensitivity with changing load in the deflection method and the speed of adjustment in automatic compensation weighing. It increases in relation to mass from the rod over the tube to the framework. The effects on sensitivity of changes in geometry due to load and temperature are easier to recognise, and the effects on zero stability are often more apparent if the beam is symmetrical. For microbalances, vacuum balances and thermobalances, quartz glass is popular as beam material as a result of its low coefficient of expansion, its deformation resistance, its excellent chemical stability, its weldability and its bakeability. One disadvantage is its brittleness. Stainless steel, invar, glass and glass ceramics are also used. For analytical balances and scales for household, industry and trade, metal framework or even massive connecting rods of brass, stainless steel or light metal alloys are used.

Also in sensor type balances mostly levers are found as connecting parts and as mass transformers. Similar considerations lead to short, sturdy but light parts.

4.2.1.2 Hinges and Bearings

For beam balances the bearings are essential because they fix the length of the lever as defined in Eqs. (4.1), (4.2) or (4.9) and thus also accuracy and sensitivity of the balance [9]. In ancient times and also in old-fashioned scales today the rope connection to the beam served as the flexible points of the balance. With the introduction of metal scales ring shaped hooks on pivots and knife edges were used. In the 19th century a balance with flexible bearings was presented and toe bearings were applied first in the 20th century for vacuum microbalances.

Today the problem of friction in general has lost its importance, because for sensitive balances the deflection of the beam is reduced to rather small angles by application of electrodynamic counterforce. By means of additional elastic levers the knife edges can be centred in a well defined working position and in many cases knife edges could be avoided using taut bands, flexural pivots or similar elastic devices. Even for sensitive balances, thin flexible metal strips can form the turning points of the lever system. Electrodynamical control systems allow the operation of balances near its null position. So the strips can be chosen sturdy enough allowing

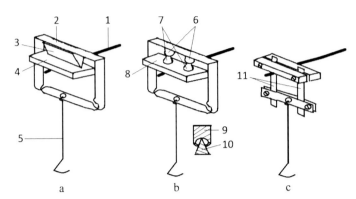

Fig. 4.3 Bearing types: (**a**) knife-edge bearing, (**b**) toe bearing, (**c**) flexible strip (taut band) bearing. *1* balance beam, *2* suspension frame, *3* knife edge, *4* sapphire sheet, *5* suspension with pan, *6, 9* sapphire calottes, *7, 10* diamond pins, *8* pin support, *9–10* calotte/pin cross-section, *11* taut bands

nevertheless heavy loads. Today metal strip bearings are applied even for metrological comparators.

Knife-Edge Bearings   In the 18th century the standard deviations referred to the maximum load of balances was improved to about 10^{-6}. When in 1791 work was started in France to introduce the metric system, even more accurate balances were required to define the kilogram prototype and to compare secondary standard weights with them [10]. To minimise random errors by friction, the knife edge bearings were improved with regard to material, shape and mounting. Diminishing the radius of curvature increases the stress which endangers the brittle materials especially in arresting and unloading of the balance. By securing permanent contact between knife edges and blocks in substitution and transposition weighing, the specific sensitivity of comparator balances could be improved to 10^{-9}.

Knife-edge bearings are distinguished by a good relative resolution (Fig. 4.3a). Usually they are made of a hard metal and the counterpart is V-shaped which provide an automatic centring of the knife edges. For analytical balances knife-edges attached to the beam and the static plane bearing stone have been made usually of sapphire. The advantages of sapphire are its hardness, chemical stability and low sorption of water vapour. Its coefficient of expansion is smaller than that of metal but higher than that of quartz glass, and must be taken into account when it is combined with other constructional components. Knife-edge bearings are also used for the end joints. V-shaped blocks in combination with plane demarcation plates function without exact guidance by an arrestment mechanism. The radius of friction is minimised by slanting both end surfaces of the knife-edges.

Toe Bearings   While a knife-edge transfers the load on a length of a few mm to the stone bearing, the diameter of the contact surface of a toe (Fig. 4.3b) is of an order of magnitude of a few μm. This means that the knife-edge bearing can transfer fundamentally greater forces than the toe bearing.

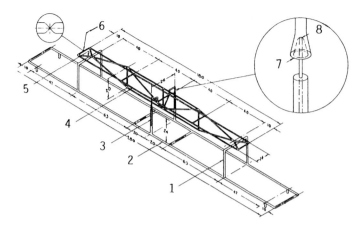

Fig. 4.4 Beam of the Czanderna balance, consisting of a three-dimensional quartz wire framework. *1* beam arrestment, *2* frame, *3* 20 mm tungsten wire, *4* three-dimensional reinforced beam, *5* yoke, *6* constricted quartz thread as carrier for the suspension hooks, *7* quartz block, *8* tungsten toe. All measurements in mm

The Czanderna and Honig balance (Fig. 4.4) [11] described above with a framework beam mounted on two tungsten wire toes in the parabolic cavities of quartz cups fixed to the frame has a reproducibility of 10 ng at a total load of 6 g. Czanderna has proved that a single bearing of this kind can take a load of up to 10 g without any significant increase above 1 μg of the balance's standard deviation. Poulis et al. [12] have investigated a combination of a pick-up and diamond toes, with a radius of curvature of 18 μm and sapphire bearing shells. Loads of up to 5 g showed no lasting impressions. For an almost equal-armed balance beam with an arm-length l_{arm} mounted with a sphere of the radius r on a level plate, and which shows a difference in arm length ΔL_{arm}, the authors obtained:

$$m = m_c - \frac{m_c \Delta l_{arm}}{l_{arm}} + \frac{m_b h}{l_{arm}} \varphi - \frac{r}{l_{arm}}(2m_c + m_b)\varphi \qquad (4.1)$$

Here, m_c is the counterweight, φ the deflection of the beam and φ_s the inclination of the surface of the stone bearing which causes an additional error. Compared with a knife-edge bearing, the toe bearing is much more strongly influenced by the angle γ. With a radius of curvature of 18 μm, an arm-length of 90 mm, and assuming that $2m_c = m_b$, this must be in an order of magnitude of 10^{-5} in order for the relative error to be $\frac{\delta m}{m} \leq 10^{-8}$.

Regarding the usefulness of a toe bearing as an end joint, the authors have calculated that an offset of the plane bearing stone of δl in the direction of the beam causes the stone to turn at an angle of $\delta l/h$, h being the distance between the centre of gravity of the combination load + bearing stone and the plane face supported by it. The relative error is:

$$\frac{\delta m}{m} = \frac{r}{l_{arm}} \frac{\delta l}{h} \qquad (4.2)$$

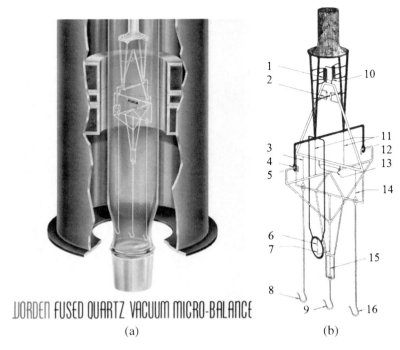

JORDEN FUSED QUARTZ VACUUM MICRO-BALANCE
(a)   (b)

Fig. 4.5 (**a/b**) Worden quartz beam vacuum microbalance. *1* damping magnet, *2* palladium stops, *3* sample suspension, *4* protection loop, *5* approximate centre of mass, *6* reference pointer, *7* beam pointer, *8* sample hook, *9* sensitivity adjustment hook, *10* aluminium damping vane, *11* beam suspension, *12* permanent magnet for electromagnetic balancing, *13* tare weight suspension, *14* sensitivity adjustment hook suspension, *15* tare hook

Assuming that $r = 18$ μm, $h = 2$ cm, $l_{arm} = 9$ cm and with the requirement $\delta m/m = 10^{-8}$, the uncertainty Δl of the support must be restricted to 10^{-4} mm. This condition is difficult to fulfil. The error is avoided by effecting a concave curvature of the bearing stone in such a way that the centre of curvature coincides with the centre of gravity of the load + bearing stone. In a state of equilibrium the point of contact and the centre of gravity of load and bearing stone are always perpendicular through the centre of the toe, and no arm-length error occurs.

Suspension Wire   Vertical suspension wires are used in the Worden fused quartz vacuum microbalance (Fig. 4.5). The balance beam of such microbalances is supported by the freely tautened torsion thread, providing a largely hysteresis-free bearing and, if necessary, generating the righting moment in the turning or load moment in the compensation process. It may be in the form of a wire (Fig. 4.6), band (Fig. 4.7) or fibre. In a cruciform cross section the torsion constant is given by [13]:

$$K_T = \frac{\pi G_S r^4}{2l_0} \tag{4.3}$$

Fig. 4.6 (a/b) Components
and housing of Rodder's
quartz balance. *1* sample tube,
2 window, *3* flag, *4* magnet,
5 vacuum flange, *6* ring seal

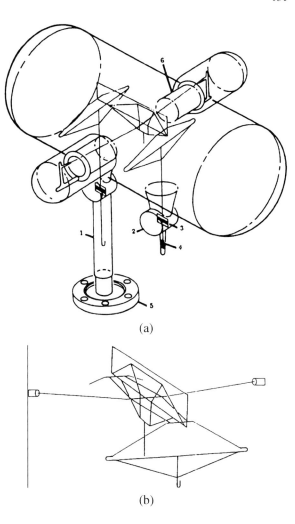

(a)

(b)

when G_S is the shear modulus, r the radius and l_0 the length. The torsion moment
of the taut band suspension depends on the properties of the material and the dimension [14]. It is proportional to small angles of deflection φ, with larger angles
it increases disproportionately. It increases with width b and thickness h, and decreases in indirect proportion to the length l. Particularly with thin bands

$$\left(v = \frac{b}{h} \gg 1 \right)$$

the tensile load L_Z results in an increase of the moment. The following is then valid:

$$M_\varphi = \frac{\varphi}{l} G_S b^4 f_1(v) + \frac{\varphi}{l} L_Z b^2 f_2(v) + \frac{\varphi^3}{l^3} E b^6 f_3(v) \tag{4.4}$$

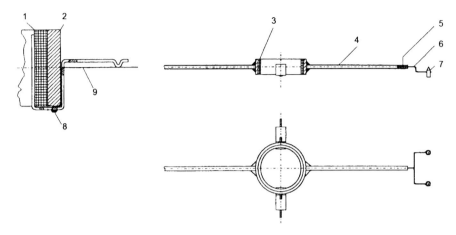

Fig. 4.7 Beam of the Sartorius vacuum balance designed by Th. Gast. *1* coil, *2* quartz ring, *3* teflonised wire coil, *4* quartz tube, *5* araldit glue, *6* sapphire toe, *7* plumb, *6* taut band

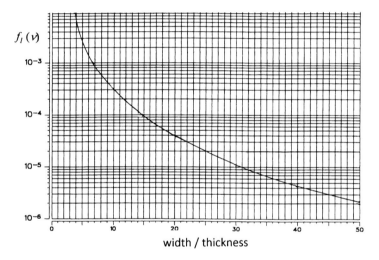

Fig. 4.8 Diagram for the calculation of torsion threads for the balance beam suspension. $f_1(n)$ is a function of the ratio of width to thickness n

where E is Young's modulus. The dependence on the width-thickness ratio is given by the functions f_1, f_2, f_3. For $v > 6$ it is $f_2 \approx 1/12$. As the third term is usually negligible, the simplified equation suffices:

$$M_\varphi = \frac{\varphi}{l}G_s b^4 f_1(v) + \frac{1}{12}\frac{\varphi}{l}L_z b^2 \tag{4.5}$$

f_1 is shown in Fig. 4.8.

Fig. 4.9 Suspension of the balance beam on a vertical fibre. j horizontal swing, d offset, M_j moment, L load

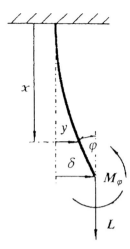

Vertical Fibre and Band Suspension   The fibre or band is integrated into the suspension of the beam as a link [15], and this bears a load L (Fig. 4.9):

$$L = g(m_b + 2m) \tag{4.6}$$

and undergoes a moment M_φ which produces at the lower end an inclination δ and a horizontal swing φ. At a distance x from the restraint the localised moment M_y for small inclinations is:

$$M_y = M_\varphi - L(\delta - y) \tag{4.7}$$

The following differential equation is therefore valid:

$$E J_A \frac{d^2 y}{dx^2} = M_\varphi - L(\delta - y) \tag{4.8}$$

Here, E is Young's modulus and J_A the geometrical moment of inertia. With the boundary conditions:

$$y(0) = 0, \quad y(l) = \delta, \quad \frac{dy}{dx} = 0 \quad \text{for } x = 0$$

the solution is:

$$y = \frac{M_\varphi}{L \cosh\left(l\sqrt{\frac{L}{E J_A}}\right)} \left\{ \cosh\left(x\sqrt{\frac{L}{E J_A}}\right) - 1 \right\} \tag{4.9}$$

where l is the length of the vertical fibre. This yields an effective torsion constant:

$$K_T = \sqrt{E J_A L} \coth\left(l\sqrt{\frac{L}{E J_A}}\right) \tag{4.10}$$

4.2.1.3 Indicator Systems

Deflection Sensors   The deflection sensor of a mechanical balance is a pointer connected to the balance beam. It may be connected directly or via level or wheels e.g. For sensitive balances light pointers have been used whereby a mirror is fastened at the balance beam. Today mostly the value of deflection or of the compensating force is transferred into an electric signal and the output is digitally indicated. Such transducer may operate according to various physical effects: magneto-elastic, capacitive, inductive, piezo-electric, resistive or electromagnetic.

Photosensors   A mirror fixed to the beam influences the direction, or an associated screen the cross section of a bundle of rays emitted from a radiation source to a radiation receiver. Light emitting diodes (LED) are chiefly used as radiation sources because they are reasonably priced, give long service and are spectrally well-adapted to the receiver. On account of its high efficiency only little waste heat is produced. Red and near infrared are the spectral ranges used for preference, and their good modulation capability and small spectral band width are useful for suppressing extraneous light.

Various photosensors are available. A photoelectric device can be either intrinsic or extrinsic. An intrinsic semiconductor has its own charge carriers and is not an efficient semiconductor, e.g. silicon. In intrinsic devices the only available electrons are in the valence band, and hence the photon must have enough energy to excite the electron across the entire bandgap. Photocells based on the external photo effect are of little significance today. It is the semiconductor component elements based on the internal photo effect upon which attention is now focused.

The highly sensitive, depletion layer-free photoresistors or Light Dependent Resistors (LDR) or photoconductor change their ohmic resistance according to light incident intensity by forming electron-hole pairs. The time required at 10^3 lux is in the millisecond range; at 1 lux it is several seconds. This would be a disadvantage for self-compensating balances. Between a p- and an n-conducting region of a photo element, due to the recombination of defective and conducting electrons, there develops a depletion layer free of charge carriers. The uncompensated negative charge of the acceptors remains on the p-conducting side, and the corresponding positive charge of the donors on the n-conducting side. An electric field thus arises in the depletion layer (or space charge region). Incident light quanta generate electron-hole pairs which are separated by the space charge region. In an open circuit, a voltage proportional to the logarithm of the illumination occurs between the functions of the p and n regions and in short circuit operation it flows to the proportional current. The short circuit operation (element operation) can be realised in approximation with an operational pre-amplifier as shown in Fig. 4.10a. The circuit is low noise as there is no dark current flow. Lead sulfide (PbS) and indium antimonide (InSb) LDRs are used for the mid infrared spectral region. GeCu photoconductors are among the best far-infrared detectors available. Inexpensive cadmium sulfide cells can be found in many consumer items.

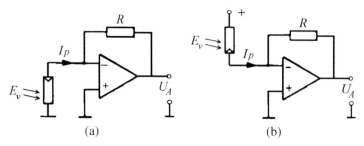

Fig. 4.10 (a) Photo element operation. © Theodor Gast. **(b)** Photodiode operation with 'short circuit operation'. © Theodor Gast

Fig. 4.11 Phototransistor with operational amplifier. Phototransistor with operation amplifier. © Theodor Gast

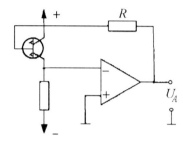

A photodiode is a type of photo detector capable of converting light into either current or voltage, depending upon the mode of operation. It can operate in photo-voltaic mode or photoconductive mode. If the photodiode is operated in the blocking direction with additional voltage, at incident light a current proportional to the intensity flows, even with a relatively high load resistance. Figure 4.10b shows diode operation with bias voltage. LEDs reverse-biased can also be used as photodiode light sensors.

Phototransistors act like amplifying photodiodes: a photodiode lies parallel to the base collector transition. Its photoelectric current affects an augmented collector current around the transistor's current amplification. Both hybrid and integrated construction is possible. In the circuit shown in Fig. 4.11 the potentials on the transistor are constant. In this way higher frequencies are reached than with load resistance in the emitter or collector conduction.

For application of the push-pull principle, the optical scanning device has a radiation source and two receivers upon which variable luminous fluxes impinge in opposite directions when the balance beam inclines. In the circuit illustrated in Fig. 4.12, a voltage proportional to the deflection occurs. The push-pull circuit eliminates zero-point fluctuations resulting from ageing of the component parts and regulates the steady-state characteristic. Multiplicative errors do not arise in the formation of the quotients from the difference and total of the photoelectric current.

The deflection of the balance beam affects the amplitude of an alternating voltage. In crossover, the phase jumps by 2π. Phase-dependent modulation and low-pass filtering bring about a constant voltage proportional to the deflection [16]. This

Fig. 4.12 Push-pull circuit.
© Theodor Gast

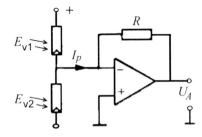

has a low noise level and extremely high zero point accuracy. A diagrammatic view
is given in Fig. 4.13.

Inductive Sensors   In the differential choke, a soft ferromagnetic cylinder integrated into the cavity of two cylindrical coils moves in the load suspension of the
balance as shown in Fig. 4.14, thereby reversing their inductance. Together with
two fixed resistors, the coils form a Wheatstone bridge as illustrated in Fig. 4.15,
topped by an amplifier, a phase-sensitive rectifier and a low-pass filter. The steady-
state characteristic is clearly indicated by the phase-sensitive rectification. A usable
resolution of 10 nm is reached [17].

The differential transformer (Fig. 4.16a) contains a primary winding excited by
alternating current the coupling of which with two coaxial, anti-serially connected

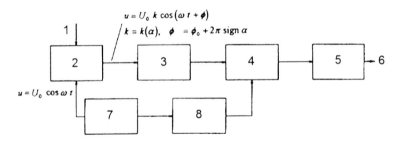

Fig. 4.13 Carrier frequency set-up of measuring instruments. *1* measuring quantity α, *2* sensor,
3 amplifier, *4* phase dependent rectifier, *5* low pass, *6* output signal $U \approx \alpha$, *7* oscillator, *8* phase
shifter. © Theodor Gast

Fig. 4.14 Differential choke.
© Theodor Gast

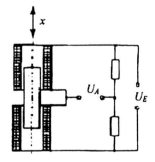

Fig. 4.15 Bridge circuit with opposed variable inductances. © Theodor Gast

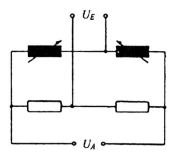

coils is variably reversed by shifting a ferromagnetic core. With the core in neutral position, the output voltage goes with a phase jump from 180 degrees through zero. Phase-dependent rectification results in a clear, linear steady-state characteristic (Fig. 4.16b). The usable resolution is 10 nm. With eddy current transducers, the self-inductance of a coil is reduced by the proximity of a conductor as a result of flux displacement. In a symmetrical arrangement with two coils and between them a plate moving in an axial direction, with an induction comparison in a bridge circuit the temperature drift of the zero point is eliminated and the repulsive forces are compensated in the neutral position [18, 19]. The screening effect of a metal plate in the opening between two coaxial coils acting as the primary and secondary windings of a transformer is also suitable to detect the path or angle of a balance [20]. A moving coil used in a homogeneous high-frequency field [21] to translate the angle of deflection has the advantage of being independent of the translation, and with modulation and selective amplification, it is extraordinarily sensitive.

Capacitive Sensors The variation of distance in plate capacitors results, according to the following equation:

$$C = \frac{\varepsilon \varepsilon_0 A}{d} \tag{4.11}$$

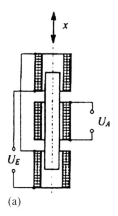

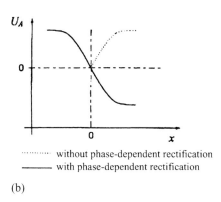

(a)   (b)

Fig. 4.16 Differential transformer and its steady-state characteristic. © Theodor Gast

(ε_0 absolute permittivity of vacuum, ε relative permittivity of vacuum of the liquid in the capacitor, A plate area, d plate distance) together with a hyperbolic steady-state characteristic, in a carrier-frequency evaluative change in capacity. In the bridge circuit of differential capacitors, the steady-state characteristic is linearised. The usable resolution is 0.1 mm. Mayer and co-workers have developed an ultra-microbalance with a capacitive indicator [22] which also compensates electrostatically. When a dielectric plate is inserted, the capacity C of a plate capacitor is increased according to the relation:

$$\Delta C = \varepsilon_0 b\left(\frac{x}{d_1 + d_2/\varepsilon_2} - \frac{x}{d_1 + d_2}\right) \tag{4.12}$$

where, as shown in Fig. 4.17a, b is the width of the electrode and the plate, d_2, ε_2 their thickness and relative permittivity of vacuum (DC), d_1 the dual cycle, x the insertion depth, and $\varepsilon_1 = 1$:

$$\Delta C = \varepsilon_0 b\left(\frac{x}{d_1 + d_2/\varepsilon_2} - \frac{x}{d_0}\right) \tag{4.13}$$

for $d_1 = 0$ the equation is simplified to:

$$\frac{dC}{dx} = (\varepsilon - 1)\frac{\varepsilon_0 b}{d_0} \tag{4.14}$$

Figure 4.17b shows a differential capacitor with an agitated dielectric. The advantage here is that there is no movable supply line, insensitive to displacement in a normal direction. Figure 4.17e shows a capacitive indicator with a movable terminal without fixed potential. Here, the following is valid:

$$C_s = \frac{C_1 C_2}{C_1 + C_2} = \frac{C_1}{2} \tag{4.15}$$

In the sensor shown in Fig. 4.17c, the capacity between two fixed terminals is controlled by the proximity of a movable screen. Finally, Fig. 4.17e shows a differential capacitor without a supply line to the movable terminal, in which a suspension control is used [23].

Capacitive sensors as well as inductive ones can be integrated into resonant circuits and used in the conversion of paths and angles in analogue frequencies. From these, direct-current signals are derived by means of discriminators [24].

Magnetic Sensors   The application of Hall probes and photo resistance cells will be discussed in the section on suspension balances.

4.2.2 Separation of the Parts of a Balance

In mass determination with the aid of mechanical balances, the weight of the object is compared with a counterforce produced by elasticity or gravitation. The mechanical parts, transferring, converting and transmitting the forces or indicating force

Fig. 4.17 (**a**) Capacitive
sensor. Beam deflection
detected by insertion of a
dielectric plate. © Theodor
Gast. (**b**) Capacitive sensor
with inserted dielectric.
(**c**) Distance detection.
© Theodor Gast. (**d**) Direct
capacitance. © Theodor Gast.
(**e**) Differential capacitor with
movable terminal without
potential. © Theodor Gast

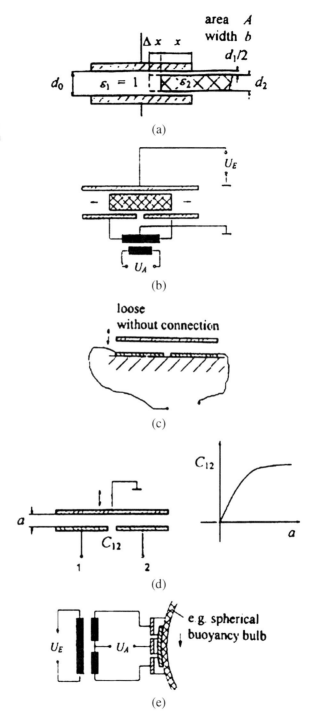

Table 4.1 Effects used to replace mechanical joints in mass and force measurement

Force field	Application
hydrostatic	density measurement (areometer), counterforce, hydrostatic balance
fluid dynamic	flow meter, fluid bearing
acoustic	sample positioning, density measurement
electrostatic	sample positioning, e/m measurement, counterforce, indication
electromagnetic	sample suspension, suspension balance, counterforce
electrodynamic	counterforce, indication
photons	light pointer, counterforce

differences are usually connected by various kinds of joints. These joints cause disturbances especially when utilising the instrument up to its limits. Hence, disjunction can possibly reduce random errors. Separation of the sample from the balance is beneficial in other respects. Independent of the environment, the sample can be held in ultrahigh vacuum or in a corrosive atmosphere. In the first case, no vapours desorbed from the balance can contaminate the sample. On the other hand, the delicate instrument is protected against corrosive gases or vapours contained in the reaction chamber for treatment of the sample or produced by a chemical reaction, respectively by thermolysis. Furthermore, using a hermetically closed sample chamber facilitates the handling of dangerous materials. In the following parts a disconnection is possible:

– sample suspension,
– counterforce support,
– hinges between load suspension and beam, and the beam suspension,
– transmission of the output signal.

In Table 4.1 the effects are summarised which have been applied for that purpose.

4.2.2.1 Bearings

Magnetic bearings on account of their weight and the insufficient definition of the rotational axis are not suitable for this purpose. A rotational gas bearing with axial limitation is serviceable as central fulcrum of a balance beam, if laminar flow is secured by using sintered metal for the bearing faces (Fig. 4.18) [25, 26]. Rotational gas bearings are applied in precision dynamic torque measurement [27]. For the moving parts of a self-compensating balance a linear gas bearing was developed [28]. It carries the balance pan, a capacitive position sensor, and a moving coil.

4.2.2.2 Counterforce

Several kinds of counterforces may be introduced touch-free to the balance beam. Zinnow and Dybwad [29] even used the pressure of light to counterbalance a highly

Fig. 4.18 Diagram of a rotational gas bearing for a balance

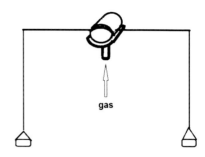

Fig. 4.19 Electromagnetic microbalance. *1* balance beam, *2* sample and counterweight pans, *3* suspended permanent magnet and stationary field coil

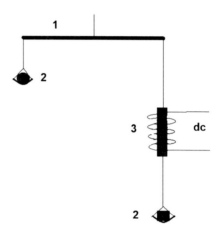

sensitive vacuum balance. An electrostatic field can be applied especially in high vacuum [22]. In gases difficulties are encountered because of the different dielectric properties [30]. On account of the high tensions needed, the gas between the electrodes is ionised producing a current or a flashover. The most successful methods, however, are based on electromagnetic compensation (Figs. 4.19, 4.20) [4] and restoring of the beam.

Buoyancy is mostly used to determine the density of solids and liquids. Figure 4.21 show various types of areometer [31]; the method has been described already in antiquity [32]. Rather simple balances were developed for this purpose. However, buoyancy has also been used as a source of constant upward force for a hydrostatical comparator balance of high accuracy. Likewise, the effect of buoyancy is used to measure the density of gases and their variations by means of a microbalance (Fig. 4.22a) and in free suspension (Fig. 4.22b). In special cases, buoyancy has been used to calibrate microbalances [33]. In this context the widely used flow meters should be mentioned in which the gravitational force of a floating body counteracts against the force of the streaming fluid.

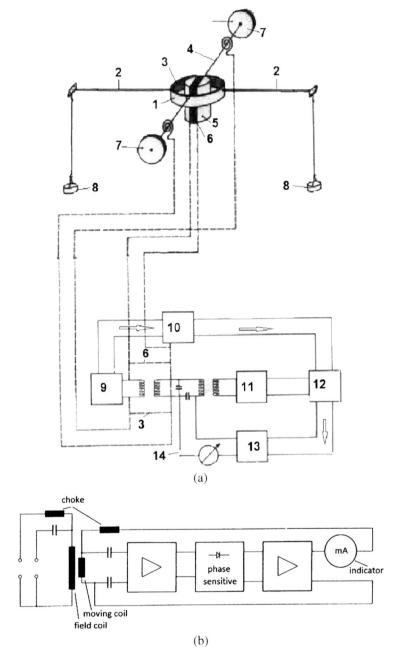

Fig. 4.20 (**a**) Electrodynamic microbalance. *1* quartz ring, *2* balance beam (quartz), *3* moving coil, *4* taut band, *5* permanent magnet, *6* fixed field coil, *7* turnable disk, *8* sample and counterweight pans, *9* oscillator, *10* phase corrector, *11* high frequency amplifier, *12* phase dependent rectifier, *13* dc amplifier, *14* monitor. (**b**) Block diagram of an electronic microbalance with electrodynamic restoring of the balance beam. © Theodor Gast

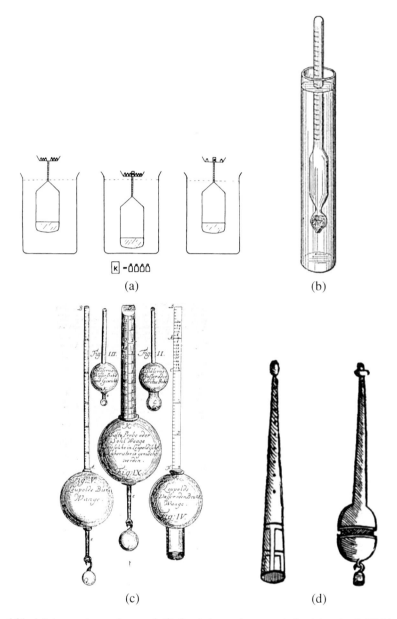

Fig. 4.21 (a) Areometer used as a substitution balance. k = mass to be determined. (b) Modern areometer. (c) Areometer of Jacob Leupold (1674–1724). (d) Areometer from 1603

4.2.2.3 Sample

Free suspension of a sample or a test object means that the sample or a test object hangs at the balance without mechanical connection to the sensor. Free suspension

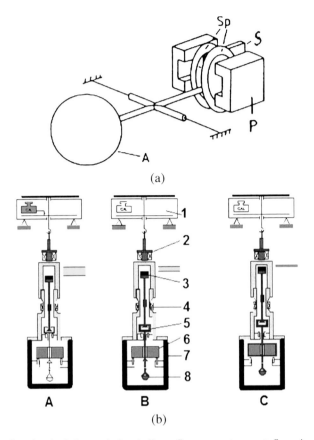

(a)

(b)

Fig. 4.22 (**a**) Gast density balance. *A* glass balloon, *P* permanent magnet, *S* moving coil, *Sp* stator coils. © Theodor Gast. (**b**) Operation of the suspension balance for density measurement. In position *A* (tare and calibration) only permanent magnet and stem below are floating whereas sample bucket and sinker rest on supports. In measuring position *B* (adsorption measurement) the bucket with sample is lifted. In position *C* (density measurement) both bucket with sample and sinker are lifted and weight *1* electronic laboratory balance, *2* electromagnet, *3* suspended permanent magnet, *4* position sensor, *5* load decoupling device, *6* sinker, *7* thermostat, *8* pan with sample © Rubotherm, Bochum, Germany

is applied when the sample is suspended in a liquid as in the case of a hydrometer or by electromagnetic or electrostatic attachment [34].

Separation of sample holder and balance protects the delicate instrument against corrosive gases or vapours contained in the reaction chamber for treatment of the sample or produced by a chemical reaction by thermolysis. Additionally it inversely prevents contamination of the sample by desorption of vapours from the balance, as required in ultrahigh vacuum. Furthermore, it facilitates safe weighing of dangerous samples e.g. in nuclear techniques. In general this measure extends remarkably the usability of the balance.

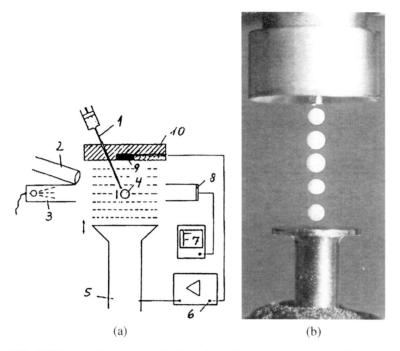

(a)   (b)

Fig. 4.23 (a) Scheme of acoustic positioning and density determination of droplets according to Heide. *1* pipette, *2* telescope, *3* lamp, *4* droplet or particle, *5* acoustic transmitter, *6* feed back generator, *7* recorder, *8* photoelectric cell, *9* acoustic receiver, *10* reflector. © W. Heide. (**b**) Styropor spheres levitated into pressure centres of a standing ultrasound wave. © Rudolf Tuckermann, Sigrid Bauerecker, Bernd Neidhart

The principle of coupling the sample held in a separate vessel to the balance is shown in Fig. 4.24 [35]. An insulating, non-magnetic window is required at the top of the sample vessel. A permanent magnet, which is attached to the sample, is kept in floating state by the attractive force of an electromagnet. The latter is fastened at the load suspension of an electromagnetic beam balance. The floating state is maintained by a position control circuit, consisting of an inductive vertical position detector, a controller and a control winding to generate the attractive force. If the controlling current is kept constant by a superimposed control loop, no position sensor is necessary [19]. In this case, the error signal can be derived from velocity of the magnet, using a voltage induced in the control winding. Moreover, the control winding cannot exert any force onto the suspended system in equilibrium. As a consequence, the winding can be attached to the frame of the instrument without errors in the indicated weight [36].

Furthermore the above mentioned methods of particle suspension within an electrical field (Fig. 4.23a) or an ultrasonic wave (Fig. 4.23a) can be applied. Such methods, however, are used only for the investigation of single particles as discussed later.

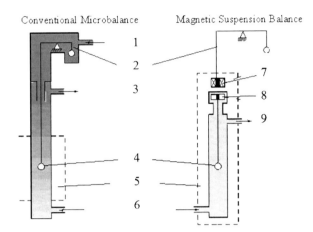

Fig. 4.24 Comparison of conventional microbalance with magnetic suspension balance. *1* inert flushing gas, *2* balance, *3* gas outlet, *4* sample, *5* thermostat, *6* vacuum line or measuring gas inlet, *7* holding electromagnet, *8* permanent suspension magnet, *9* measuring gas outlet.
© Rubotherm

4.3 Counterweight Balances

A lever balance consists of the sample pan which is connected to the balance beam. That beam is situated rotating about an axis arranged horizontally. The counterbalancing force is given by the gravitational force of a counterweight arranged on the opposite side of the balance beam. Mass comparison is carried out in the gravitational field, enabling the difference in the object's change in mass to be measured against the comparison mass [6].

4.3.1 Symmetrical Counterweight Balances

The oldest known balances are of lever type symmetrical beam balances. A symmetrical balance beam made of limestone older than 3000 BC was found in Egypt. Furthermore weight stones of those times indicate the existence of double-armed scales. The principle of such a balance is obvious: comparing the weight force or the mass, respectively, of the weighing article with an equal mass of the counterweight. Indeed by that comparative measurement the largest part of the weight force of the article is compensated by the weight force of the counterweight; and only small deviations are observed with high accuracy. That is an important advantage in comparison with other methods using a force sensor. Furthermore, using a symmetrical design and weighing in equilibrium needs no calculation of the weighing result other than adding the values of a set of known weights. Findings and drawings from Old Egypt always show highly symmetric constructions and careful weighing in equilibrium.

The balance beam of Old Egyptian balances was made mostly of wood. Metal pans were suspended either on ropes or on metal chains. Within a period of more than 2000 years the symmetric balance was improved with regard to the shape of the beam, aligning of the central pivot within the connection line of the edges where the

Fig. 4.25 Diagram of a symmetrical, equal armed beam balance. The beam with a length l_b and a mass m_b is in equilibrium when the end knife-edges are loaded with equal masses. If $m_t, m_c > m_l$, the beam turns by angle φ

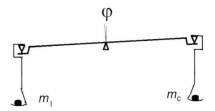

ropes of the pans were suspended, better definition of the bearings and a more so-phisticated indicator system [37]. This balance type was in use until recently. Further developments were possible only after the beginning of use of metal and by help of theoretical considerations as developed by Greek philosophers some hundred years BC [38]. The law of the lever was derived by Aristotle (384–322 BC) [39, 40] or one of his pupils and Archimedes (~287–212 BC) [41, 42]. From indications in the works of other authors it can be concluded that besides the surviving 'Deplanorum aequilibris' Archimedes also wrote a book on balances which has been lost [43].

For a symmetric balance in equilibrium (Fig. 4.25) the following relationship holds

$$W_l \frac{l}{2} = W_c \frac{l}{2} \tag{4.16}$$

where W_l is the weight force of the load and W_c the weight force of the counter-weight and l the length of the beam. Because

$$W = m g_n \tag{4.17}$$

where m is test mass and g_n the gravitational acceleration we have likewise

$$m_l \frac{l}{2} = m_c \frac{l}{2} \tag{4.18}$$

When the balance is in equilibrium in a horizontal position of the beam, we find

$$W_l = W_c \quad \text{or} \quad m_l = m_c \tag{4.19}$$

The comparison principle makes the double armed balance to be a very sensitive instrument because in this way a ground mass is compensated and small differences can be resolved within a limited measuring range.

4.3.2 Asymmetrical Counterweight Balances

Naga, an indigenous tribe in Assam, India, uses a club as unequal-armed lever bal-ance (Fig. 4.26). Haeberle [44] suggests that this could be the original type of scales. However, any roots of such a design cannot be traced back and so there is no evi-dence for that assumption. More likely, after some thousand year's use of the sym-metric balance, practical needs, as well as trial-and-error, led to the development of technologies leading to balances with unequal arms. Balances with unequal arms

Fig. 4.26 Single-pan lever
scales of the Naga, a
primitive tribe, Assam, India.
© Haeberle

were in use in Hellenistic countries as early as the fourth and fifth centuries BC, before Archimedes and other thinkers of that era gave a mathematical demonstration
of its theoretical foundations [45].

For asymmetric scales where the central pivot is not located in the centre more
general equations must be applied. Furthermore, weighing in air is influenced by
buoyancy which depends on the volume and, hence, on the density of the bodies
to be weighed. In general, load and weight consist of materials of different density.
The general balance equation reads

$$m_l \left(1 - \frac{\rho_a}{\rho_l} \right) l_l g_n = m_c \left(1 - \frac{\rho_a}{\rho_c} \right) l_c g_n \tag{4.20}$$

where ρ_a is the density of air ρ_l density of the load material and ρ_c density of the
weight and l_l and l_c length of the part of the beam at the load side and of the counterweight side, respectively. The gravitational acceleration g_n varies with geographic
situation but is constant locally. For less accurate purposes in daily life we may
neglect buoyancy corrections and use the equation

$$W_l l_l = W_c l_c \tag{4.21}$$

We learn that with a balance with unequal arms and operated in equilibrium it is
advantageous to weigh heavy loads with small counterweights. Then the counterweight must be multiplied by the quotient of both armlengths.

With regard to the construction similar considerations as for symmetrical balances are valid: The beam should be rigid but light.

4.3.2.1 Besemer

About 400 BC the Bismar, Besmer or Besemer appeared in the Eastern Mediterranean area (Figs. 4.27) [46]. It consisted of a rod of wood with a large weight fixed

at one end. At the other end was a hook for the goods to be weighed. The user held a rope or metal loop as the beam's variable pivot that was slid along the rod until it balanced. The weight was read from a series of notches or nail heads hammered in to the underside of the rod. It was not very accurate but traders liked it, probably for the wrong reasons and it became very popular throughout Europe. Aristotle condemned it as an instrument of deceit, but nonetheless it spread across the world under different names. The Normans called it the Auncel, in Russia it was the Bezmen, in India and the Far East it was the Dhari. Banned twice in England in the space of a hundred years, its use did not decline until it was condemned for public weighing by Henry II. However, it is still used today in the Baltic and Eastern European countries (sometimes wrongly called Scandinavian steelyard) and in the Far East. Figure 4.27c show modern Finnish fish balances.

4.3.2.2 Steelyard

In wide use, established in ancient times and surviving to modern times, was the so-called Roman balance, the steelyard. It consists of a straight beam which is suspended from a defined pivot very close to one end of the beam. Two arms of the beam flank the pivot. The arm from which the object to be weighed is hung, is short. The other arm is longer, is graduated and incorporates a counterweight which can be moved along the arm until the beam is balanced in horizontal position about the pivot. The weight of the load is indicated by the position of the counterweight and can then be calculated by multiplying the known value of the counterweight by the ratio of the distances from the beam's fulcrum using Eq. (4.10) or read directly from the graduation marks.

Originally known as the statera, the English word steelyard comes from the German 'Stalhof', the name of the London base of the Hanseatic League in the 14th-century, who used the instrument extensively in their businesses [46]. The steelyard was already in use among Greek craftsmen of the fifth and fourth centuries BC. From such instruments around 200 BC improved types were developed within the Roman Empire, generally made from bronze (Figs. 4.28, 4.29). Such steelyards may have up to three suspension hooks for weighing objects of varying weights. Heavy objects would have used the hook nearest the end and light objects would have been weighed on the central hook. The sliding counterweight is often designed artistically. Scales of this type with wooden beam and only one hook had been in use in China.

Steelyards of different sizes have been used to weigh loads ranging from ounces to tons. A small steelyard could be some ten centimetres in length and thus conveniently used as a portable device which merchants and traders could use to weigh small ounce-sized items of merchandise. In other cases a steelyard could be a meter long and used to weigh sacks of flour and other commodities. Such a balance is shown on a gravestone found in Trier (Germany) (Fig. 3.20). Even larger steelyards were three stories tall and used to weigh fully laden horse-drawn carts. On the other hand small steelyards had been constructed and are still available as pocket letter scales.

Fig. 4.27 (a) Besemer,
principle. (b) Middle Ages
besemer. (c) Modern Finnish
fish besemer

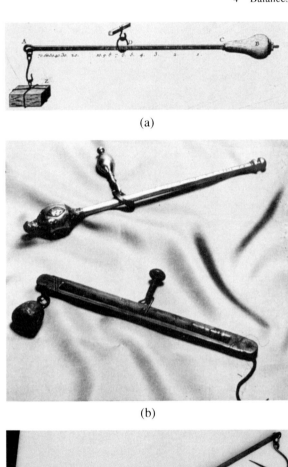

(a)

(b)

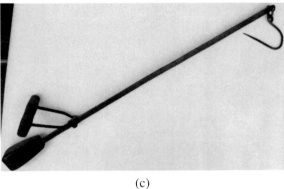

(c)

The invention of the steelyard is also claimed by the Chinese under the name gan-
chang [47]. As early as 200 BC, China began to make a scale of this type big enough
to weigh several hundred pounds (Fig. 4.30a). The arm or beam measured about
1.5 metres long, graduated with the weight units jin (~500 g) and liang (~30 g). The
hook, hanging from one end of the arm, was used to lift up the object to be weighed.
Hanging from the other part of the arm was the free moving weight, attached on

Fig. 4.28 Roman steelyard and weight from Caernarfon. Copper alloy. © National Museums & Galleries of Wales

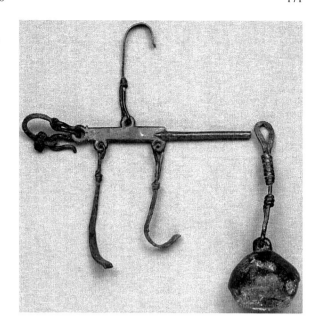

a looped string. On the arm was fixed one, two or three lifting cords, placed much closer to the hook than to the other end. Anything to be weighed should be picked up by the hook, while the weigher lifted up the whole steelyard, holding one of the cords. He then slid the weight left or right until he found a perfect horizontal balance of the beam. He then read the weight from the graduation mark on which the weight-string rested.

This kind of steelyard is still in widespread use at market gatherings in China. They may be made in varying sizes working by the same principle, with the large ones to weigh food grain in bulk, pigs or sheep or their carcasses, and medium-sized ones for smaller transactions. There is also a miniature steelyard only about one third of a metre long, graduated with liang (~30 g) and qian (~3 g). Used to weigh medicinal herbs and silver or gold, it first appeared about 1000 years ago. In Fig. 4.30b a Chinese steelyard scale (do'tchin) is shown from ~1840 consisting of an ivory scale rod with a brass pan and weight, all stored in a varnished, wooden, paddle shaped case [48]. The ivory scale rod has black dots marking off the weight scale. The brass pan is joined to the rod with four strings and the detached oval weight is also attached to string. It was used for weighing gold in Australian mines.

Steelyards of different type are still in use (Figs. 4.31) but will be replaced more and more by electronic scales.

4.3.2.3 Deflection Balance with Suspended Pan

Deflection or inclination scales are usually called single arm balances. Indeed they are equipped with two arms: one for supporting the sample pan and the other side

Fig. 4.29 Steelyard, bronze
2nd century AD, Rome,
Muzeum Książąt
Czartoryskich, Kraków,
Poland

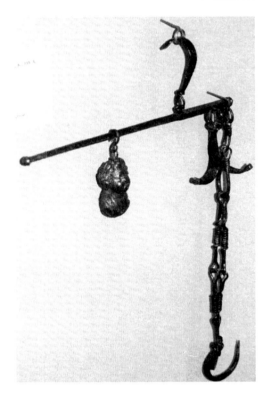

equipped with a counterweight. However, this counterweight part can be short and can be bent with respect to the load arm (Fig. 4.32). When loaded with an article the load arm sinks, the bearing of the sample pan describes a circular path whereby the distance to a vertical line through the beam suspension is reduced. On the other hand the counterweight ascends on a circular path and its distance to that vertical line increases. As soon as

$$m_l l_l \sin \alpha_l = m_c l_c \sin \alpha_c \qquad (4.22)$$

the balance achieves equilibrium and the deflection can be indicated by means of a pointer fastened at the beam against a fixed scale. Here m_l is the mass of the article to be weighed and m_c the mass of the counterweight, l_l is the length of the load arm and l_c the length of the counterweight arm and α_l and α_c are the angles of the beam parts against a vertical line through the beam suspension.

The earliest sketches of such deflection scales are from Leonardo da Vinci (1452–1519) designed as gravitational hygrometers (Fig. 4.33) [49]. In the early 18th century a special Nuremberg deflection coin balance [50] and a hydrostatic balance of Johann Adam Cass were known [51] In 1758 Johann Heinrich Lambert (1728–1777) developed a deflection balance and described a large number of modifications [52–54] (Fig. 4.34). About ten years later Philipp Matthäus Hahn (1739–1790), a Swabian priest produced the 'comfortable house scales [55–57] (Fig. 4.35).

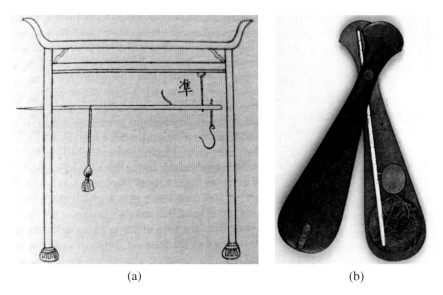

(a)   (b)

Fig. 4.30 (a) Large Chinese steelyard. (b) Chinese steelyard, ~1840–1860. Beam ~220 mm long. © Powerhouse Museum, Sydney, NSW, Australia

In such a system the movable counterweight can be regarded as a force sensor. When a small additional mass m_φ is deposited on the balance, the beam turns by angle φ. Also other force sensors measure just the force of the sample due to gravity, e.g. the spring in the spring balance. Because this is a direct measurement and not a comparison, in principle the relative sensitivity or resolution is low compared with the equal armed balance.

4.3.3 Symmetrical Top Pan Balances

The balance with suspended pan has the disadvantages of having strings in the way of the user and placing of the sample and weights on a rocking scale needs some skill. Platform balances, with scale on top allowing loading from above, are much more convenient. In special cases it is possible to refrain from using pivots for the test pan and attach instead a scale or platform firmly to the beam. Then, however, the arm length of the lever given by the distance of the centre of the sample is undefined and such a balance can only be used for samples of similar shape which are placed on fixed points on that platform. Examples are so-called rocker scales; simple spoon balances for liquids, powders and other pourable materials and weight checkers for eggs, coins and letters, etc. (Fig. 4.36).

It is reported that an equal-armed platform scale appeared in China earlier than the steelyard with a sliding weight [47]. A scale of the former description with a

Fig. 4.31 (a) Antique Greek
steelyard. Museum Olympia.
(b) Complicated steelyard.
(c) 19th century steelyard

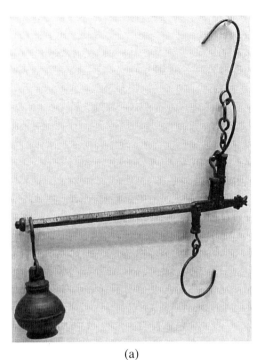

(a)

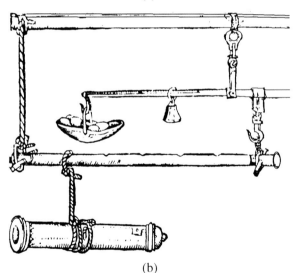

(b)

complete set of weights was discovered lately from a tomb near Changsha, Hunan
Province, which dates back to the Period of the Warring States (475–221 BC). It is
in size similar to those in use today and its component parts are found to be in good
proportions.

Fig. 4.31 (Continued)

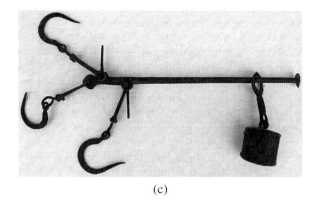

(c)

Fig. 4.32 Schematic diagram of a deflection or inclination balance

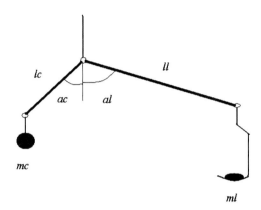

A pan can be placed above the beam by means of counterweights which are fastened to a rod at the bottom of that scale and hung below the beam (Fig. 4.37). It is essential that the leverage of that pendulum mass is greater than the scale + load, to preserve stability. That counterweight detracts the maximum load of the balance and of course the pan can swing. Such constructions are applied, e.g. in thermogravimetry where the oven should be placed on top.

4.3.4 Symmetrical Top Pan Balances with a Parallelogram Beam

The basic element of such a counterweight balance is an arrangement of four rods connected with bearings at the each end to form a parallelogram (Fig. 4.38). Using two of the parallel rods as balance beams on central pivots, weight forces exerted at the ends are attracted in the direction of the parallelogram edges. This method is applied for almost all platform scales. Though more difficult to manufacture than balances with suspended scales and therefore more expensive, these instruments were dominating as market and household scales until the dawn of the 20th century.

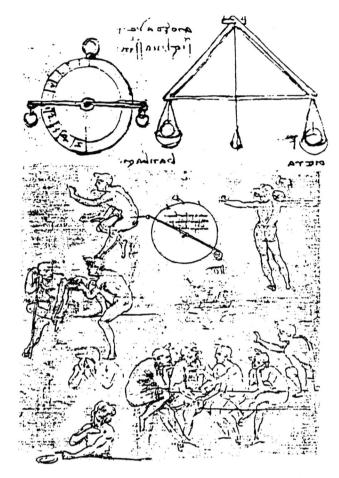

Fig. 4.33 Sketches of deflection scales are from Leonardo da Vinci (1452–1519) designed as gravitational hygrometers

4.3.4.1 The Roberval Balance

A mechanical top-loading platform balance was first was presented in 1669 to the French Academy of Sciences by the French mathematician Gilles Personne de Roberval (1602–1675) [58]. The principle was demonstrated to illustrate an enigmatic point in mechanics. The idea was not used, apparently, until such scales were made in the early 19th century. In these scales, two identical horizontal beams are arranged one above the other as a parallelogram with pivots at the edges (Fig. 4.39). By two central pivots both beams are supported by the balance stand so that the lateral sides of the parallelogram are always vertical rods on top of which horizontal plates are fastened suitable for placing the weighing article and calibrated weights, respectively. These masses can exert only downwards vectors because no horizontal movement is possible, as in conventional balances, on the lateral pivots of the beam.

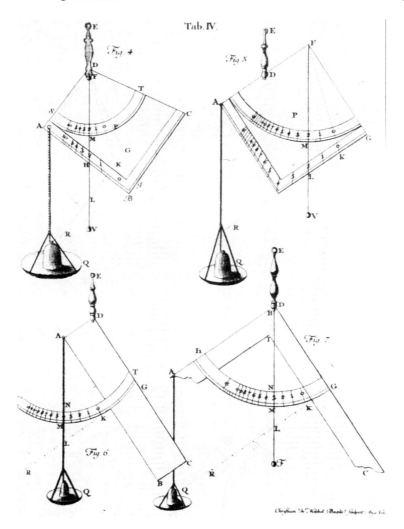

Fig. 4.34 Johann Heinrich Lambert's derivations of the theory of the deflection balance, 1758

The masses do not need to be centred on the plates. An arrow on the lower horizontal beam (and perpendicular to it) and a mark on the vertical column may be added to aid in levelling the scale.

The Roberval Balance is a lever scale; the length of the lever is limited by the pivots of the lateral vertical rods of the parallelogram. Any lateral attachments to those rods do not contribute to the length of the lever and any masses placed on it exert forces only in the vertical direction of its pivots. The arc of motion of a mass attached to the vertical rod is independent of its position. Wherever the weight is placed on the pan, it moves the same vertical distance. This is the 'Roberval Principle' common to all of these kinds of balances. Such an arrangement is in

Fig. 4.35 Philipp Matthäus
Hahn (1739–1790). The
'Bequeme Hauswaage'
(convenient household scales)

Fig. 4.36 Rocker scales.
Weighing on a spoon

indifferent equilibrium as is a balance beam with a pivot exactly centred. To obtain
a balance with stable equilibrium it is necessary to add a mass below the lower
beam's centre. This additional mass will draw back (in equilibrium) the beams into
the horizontal position when deflected.

Roberval scales had been widely used in commerce and household. As post office
and letter balances they were often used to check weights; that means that it was
observed whether a mass given by a fixed counterweight was surpassed.

Fig. 4.37 Thermogravimetric top pan balance with
counterweight for the load
pan. The beam is a quartz
tube with plug connection.
1 suspension hook, *2* moving
coil, *3* ferrite core, *4* taut band
mounting of the beam,
5 sample cup,
6 thermocouple, *7* sample
spoon, *8* counterweight,
9 stationary coil, *10* inductive
sensor, *11* gold band as
movable current supply,
12 plug connection.
© Linseis, Selb

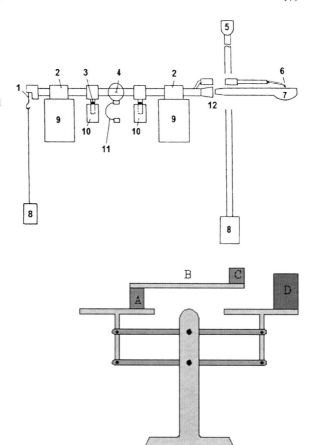

Fig. 4.38 Arrangement of
four rods connected with
bearings at the each end to
form a parallelogram (system
Roberval). If the weights
$A + B + C = D$, then the
balance is in equilibrium,
disregarding the situation of
the weights

4.3.4.2 The Béranger and the Pfanzeder Balance

The Roberval Balance is arguably less accurate and more difficult to adjust. In 1849
the French scale engineer Joseph Béranger presented an improved equal-arm top-
pan balance. In this design the pans are supported at multiple spots, the side forces
and friction were reduced by using small secondary beams to replace the force
spindles (Fig. 4.40). Other improvements of such a balance were made by Georg
Pfanzeder (Fig. 4.41). Such balances were more accurate and could be officially
calibrated. In England, the weigh bridge was developed with an 'equalizer bar' to
enable heavy duty weighing without having to use a large weight scale.

4.3.4.3 Kent's Torsion Balance

In 1887 the United States Torsion Balance Company, of New York, presented Kent's
torsion balance [59]. This is a top-loading Roberval system with two platforms
applied for a construction in which all bearings are replaced by flexible strips

Fig. 4.39 Roberval kitchen balance. © Heimatmuseum Seulberg

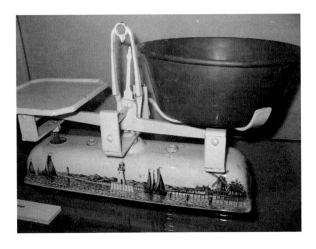

Fig. 4.40 Béranger balance

(Figs. 4.42). Whereas the main part of the mass of the weighing article is compensated by counterweights, the rest causes a small deflection of the balance and a pointer indicates on a scale of mass units.

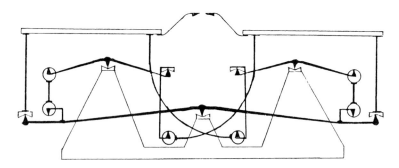

Fig. 4.41 Pfanzeder balance

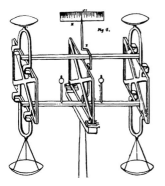 

Fig. 4.42 Kent's Torsion balance

4.3.5 Asymmetrical Top Pan Balances with a Parallelogram Beam

The parallelogram principle can also be applied for asymmetric counterweight scale constructions with fixed or sliding counterweight; a great variety of such counter-tops dominated the market until the middle of the 20th century. In such types of mechanical balances either an inclination device or a steelyard device of a sliding weight is combined with the Roberval principle holding the sample pan in an upward direction. Certified balances of that type, though more difficult to manufacture than balances with suspended scales and therefore more expensive, dominated in commerce. There followed the semi-automatic scales with one or two platforms, then came automatic scales, before gradually being replaced by 'electronic' scales with strain gauge sensors or with electromagnetic restoring systems. The mass of such balances is indicated on a display and the values can be processed by a computer, e.g. to calculate the price of the article and to print it.

4.3.5.1 Platform Scales

A platform scale has a system of levers below the platform using the Roberval principle, but essentially modified with unequal lever beams. This allows for equilibration of heavy objects with minor counterweights. The mass is transferred independently when the load is placed on the platform. The arrangement also prevents damage to the bearings caused by lateral forces. Many different weighing methods have been applied: loose weights on classical beam scales, steelyard, spring, pendulum and hydrostatic forces. The first true platform scale is attributed to John Wyatt of Birmingham, England ∼1741. Improvements were made by Thaddeus Fairbanks (1776–1886), USA ∼1830 (Fig. 4.43) [60, 61]. A modern type is shown in Fig. 4.44.

Typical platform deflection scales as used in household and industry are based on one movable weight, or two, arranged symmetrically delivering the balancing force. A smaller additional weight can to be used for more precise weighing (Fig. 4.45). This weight is attached slidable and is moved along the beam to the equilibrium

Fig. 4.43 Fairbanks platform scales. © Wikipedia

position. Platform deflection scales with one or two platforms for industry and commerce had been equipped with large gauge dials (circular, 'V' or an upside-down 'V') (Fig. 4.46). These indicators could be used not only to show the weight of the object but also the corresponding price. In addition some of these types were equipped with an internal weight support to allow changing the weight range. The beam then had a chart of weight graduations.

4.3.5.2 Decimal Scales

The decimal scale is a compound lever scale with a lever ratio of 10:1 for use with the metric system of weights (Fig. 4.47). The weighing object is on a platform on top of the weighing mechanism based on the Roberval principle. The counterweight side has a suspended pan and in addition often a sliding weight. The decimal scale was devised by the Benedictine Friedrich Alois Quintenz, born in 1774 in Gegenbach, Germany, and working at Strasbourg, and first patented on 9 Feb. 1822. He died shortly after his patent was granted on 17 April 1822. Jean-Babtiste Schwilgué made some useful corrections to this system which helped its wide distribution. Other authors assume the basic invention by the clockmaker Schwilgué and modifications by Quintez [62]. J.F.H. Rollé, J.B. Schwilgue, W. Schwarzenbach, all in the Alsace

Fig. 4.44 Commercial
platform scales with
indicator, Duthion & Cie.
Pont l'Évêque, France,
calvados distillery Père
Magloire, platform balance

Fig. 4.45 Household
platform balance (kitchen
scale). Rocking-horse
steelyard, kitchen scale with
sliding weight, Krups Comet,
Heimatmuseum Seulberg

region, improved and manufactured the instruments. There had been also scales of
that type fabricated with a lever ratio of 100:1. Also platform scales as described
above often have a transformation of forces, e.g. Fairbanks scales.

Fig. 4.46 Bizerba merchants
top loader inclination balance
equipped with large gauge
dials (circular, 'V' or an
upside-down 'V')

Fig. 4.47 (**a**) Decimal scale.
© Heimatmuseum Seulberg,
Germany. (**b**) Quintez
decimal scale, Brückenwaage,
platform scale

(a)

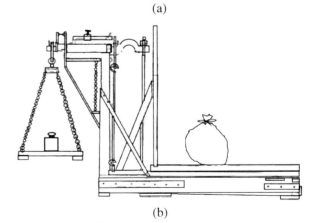

(b)

Fig. 4.48 Principle of inclination balances, (**a**) fixed chart and pointer at the beam, (**b**) fixed pointer and chart at the beam

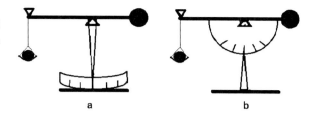

a   b

Fig. 4.49 Sauter pendulum scale with total circle as chart. © Phillipp-Mathäus-Hahn-Museum, Onstmettingen

4.3.5.3 Platform Deflection Scales

Simple inclination top pan balances, also referred to as pendulum or quadrant scales are still widely used as letter scales (see special chapter below). There are pendulum scales in which the chart is fixed to the frame, and the indicator or pointer moves with the mechanism when loaded (Fig. 4.48). Also, the reverse exists: the chart moves with the mechanism along a frame-fixed pointer. With the bilateral pendulum scales, the pointers as well as the charts, move when loaded. Some industrial pendulum scale designs have a full circle as chart (Fig. 4.49). Most of the pendulum scales have a chart with the form of a part of a circle; the size is almost a quarter (Fig. 4.50). Therefore these pendulums are also called quadrant scales (Figs. 4.35). The counterweight of some pendulum scale models is rotatable to a stable second position. Here the scales have two separate weighing ranges (Fig. 4.51a).

Fig. 4.50 Principle of the deflection or quadrant balance, Johann Heinrich Lambert

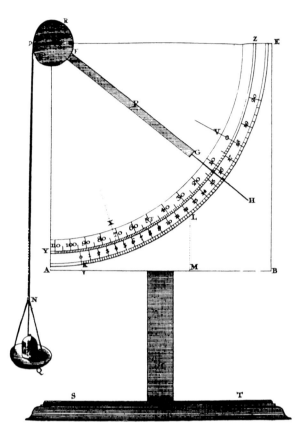

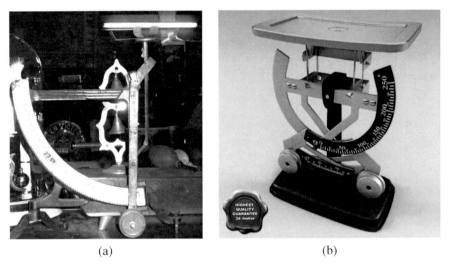

(a)   (b)

Fig. 4.51 (a) Pendulum scale with rotatable counterweight. (b) Double pendulum letter scale, Maul

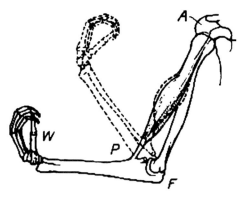

Fig. 4.52 Forearm as a lever balance. © Mahidol Physics Education Centre

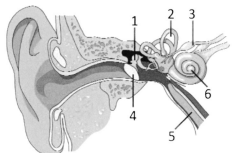

Fig. 4.53 Ear, cochlea. *1* ossicles (bones of the middle ear), *2* labyrinth, *3* auditory nerve, *4* ear drum, *5* Eustachian tube, *6* cochlea

4.4 Bioforce Balances

We have no sense for mass but for the weight force. The earliest and most commonly used method to assess the mass of an article is lifting by hand with the bending arm and compensating its weight force by means of the muscles of the upper arm. Using the stretched arm for that purpose we have a lever device actuated by dorsal muscles according to Fig. 4.52. Most neighbouring bones in our skeleton have joints which convert them into levers. The motion of the lower arm relative to the upper arm, when it moves in the elbow is a rotation [63]. The axis of rotation is the axis of the joint, while the bones, which can rotate about the axis of the joint, are levers. The motions of the bones are caused by the muscles. These are stretched between two independent points, A and P. As they contract, they turn the lower arm towards the upper arm. It is a one-arm lever; the points of attack of the load W and the force P lie on the same side of the axis through F. The bioforce counteracting the gravitational force is a complex one, including elastic and electrical forces as well as forces due to chemical reactions and osmosis. The chemical force based transducer provides electrical signals conducted through nerves to the brain.

All complex living bodies depend on gravitational sensors. The effect is not displayed in figures but used to control the development of the body and its actions. Besides other environmental influences, the growth of plants and also the development of an embryo are controlled by gravity. Only weak influences of the gravitational field had been observed in some recent biological experiments in zero gravity.

However, modifications of the body observed by astronauts after long residence under zero gravity conditions can be explained by absence of gravity.

The bodies of living beings obviously possess various gravity sensors which until now are only partially detected. The most important gravity sensor of human beings is situated in the ear (Fig. 4.53). This organ detects the position of the body with respect to gravity and the motion of the body. Just above the cochlea are two interconnecting chambers filled with endolymph, the sacculus and utriculus. On their inner surface are patches of hair cells to which are attached thousands of tiny spheres of calcium carbonate ($CaCO_3$). Gravity pulls these downward. As the head is oriented in different directions, these ear stones or otoliths shift their position. The action potentials initiated in the hair cells are sent back to the brain. Some people also suffer severe dizziness because otoliths have become dislodged from their utriculus (e.g. following a blow to the head) and settled in a semicircular canal.

Motion of the body is detected in the three semicircular canals at the top of each inner ear, each one oriented in a different plane. There is a small chamber at one end of each canal containing hair cells. Whenever the head is moved, the fluid within the canals lags in its motion so that there is relative motion between the walls and the endolymph. This stimulates the hair cells to send impulses back to the brain.

4.5 Elastic Force Balances

Elastic forces can be used to counter the weight force of an article under investigation. These forces had been investigated by Robert Hooke (1635–1703). He presented his work on elasticity in 1678 [64, 65] and mentioned that he was the first scientist who succeeded in establishing the complete theory and that he had realised the basic law in about 1660. Indeed in 1676 he noticed the law of elastic forces in an anagram 'ceiiinosssttuu' which means "Ut tensio, sic vis" (As the extension, so the force) [66]. Mathematically, Hooke's law of elasticity (spring equation) states that

$$F = -kx \qquad (4.23)$$

Where x is the displacement of the end of the spring from its rest position, F is the restoring force exerted by the material; and k is the force constant (or spring constant). There is a negative sign on the right hand side of the equation because the restoring force always acts in the opposite direction of the displacement. Hooke's Law is an approximation that states that the extension of a spring is in direct proportion with the weight force (in newtons) of the load added to it as long as this load does not exceed the elastic limit of the material. Because in the gravitational field the weight force is directly proportional to the mass, the scaling for everyday use is in mass units, grams or kilograms. Materials for which Hooke's law is a useful approximation are known as linear-elastic or 'Hookean' materials. These include hardened metals like steel, glass and quartz. In nanotechnology carbon nanotubes had been used; however, in this case it is not the deflection but the resonance shift of oscillations that are observed (see Sect. 4.10).

The elastic element can be in the form of strings, wires, tubes, spirals or coiled springs, bands or various shaped bodies. These may be bent, twisted, stretched or compressed. Because the counteracting elastic force is not well defined and in general cannot be calibrated as exactly as the gravitational force the relative sensitivity is poor in comparison with that of a counterweight balance. However spring balances are cheap, robust and easy to handle. They are ideal with regard to vacuum technology because they are simple in construction, take up little space, the sections have a small surface area, and they are bakeable. Their surfaces can be easily protected for use in corrosive environments. Spring scales are also used as accelerometers and to measure momentum, but its main uses are industrial, especially related to weighing heavy loads such as trucks, storage silos, and material carried on a conveyor belt. A special type, load cells equipped with strain gauges, have been developed to very sensitive devices and are widely applied today.

4.5.1 Spring Balances

Robert Hooke not only established the theory of elasticity, in addition he realised "new sorts of philosophical-scales, of great use in experimental philosophy" (Fig. 4.54) and he conducted experiments. He investigated whether the gravitational force decreases with altitude of the balance location. He failed in this, but suspected that altitude differences in his environment were not large enough to find a measurable variation in the weight force. Furthermore he used a helical balance for density determinations.

At that time various spring devices had been described also by others without mentioning the work of Hooke. So it is not clear whether he was the first who invented spring balances [67]. In his book published 1698 Christoff Weigel reports on a spring balance [68]. Johannes Dolaeus presents figures of helical spring balances (Fig. 4.55) [69] and likewise did J. Ozanam [70] (Fig. 4.56). The scaling of these scales was obviously not linear in contrast to Hooke's law.

A spring balance consists of a spring fixed at one end with a hook to attach an object at the other (Figs. 4.57). It works by Hooke's Law, which states that the force needed to extend a spring is proportional to the distance that spring is extended from its rest position within its elastic limits. Therefore the scale markings on the spring scale are equally spaced. A pointer attached to the spring indicates force or mass on a graduated scale. The indication mostly is observed by eye; for scientific balances detection is eventually by optical means, microscope or telescope.

As spring scales are gravitational balances they should be calibrated for the accurate measurement of mass in the location in which they are used. Its limited measuring range is certainly a disadvantage, as a force measurement is carried out instead of a comparison of masses. Most of them are not licensed for official use or trade due to the approximate nature of the theory used to mark the scale and the gravitational correction is thus not so relevant. Also, the spring in the scale can permanently stretch with repeated use. Spring scales come in different sizes. There are

Fig. 4.54 Different types of spring balances according to Robert Hooke 1678. Figure *1* helical spring, Fig. *2* spiral spring, Fig. *3* longitudinal thread

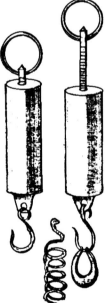

Fig. 4.55 Helical spring balances of the report of Johannes Dolaeus 1689

three kinds of spring balances: rod, spiral and helical spring balances. Mostly bending of the element is applied. In addition, Hooke considered the elastic lengthening of wires.

Fig. 4.56 Spring balances of
the report J. Ozanam 1696

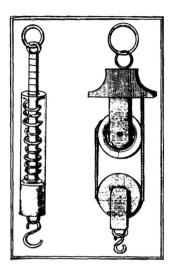

(a)

(b)

Fig. 4.57 (**a**) Soehnle kitchen spring balance. © Matthias Müller. (**b**) Alpha kitchen spring balance. © Heimatmuseum Seulberg, Germany

The first spring balance in Britain was made around 1770 by Richard Salter of West Bromwich. He and his nephews John & George founded the firm of George Salter & Co., still notable makers of scales and balances, who in 1838 patented the spring balance. They also applied the same spring balance principle to steam locomotive safety valves, replacing the earlier deadweight valves.

Most often spring balances are equipped with a helical coil spring (Fig. 4.58). For heavier loads different spring types are used. In the robust Mancur scale a flat

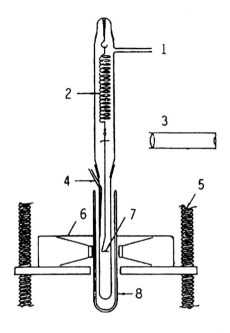

Fig. 4.58 Quartz spring balance with helical coil shaped spring for the measurement of magnetic susceptibility (Faraday balance). *1* quartz tube, *2* quartz helical coil, *3* telescope, *4* thermo-element, *5* screw for moving the magnet, *6* permanent magnet, *7* sample pan, *8* dewar

band ring is applied (Fig. 4.59). This scale is suited for rough weighing only and was especially used in the farming industry. The sector balance also utilises a flat band spring. Besides this, scales with elliptical, bow, and torsion springs do exist. The more complex spring balances have a lever mechanism between the load, spring and the indicator tag. These are applied in cheap mechanical top pan balances, e.g. bathroom scales.

Spring type balances also were used as letter scales. With a restricted load range the required sensitivity can be obtained. The mass or postage is directly read off the chart. There are table or desk models and hand held models. The simple spring balances make use of an extension or a compression coiled spring which is placed in line with the load to be determined. Because of the resemblance, the English called certain compression spring balances 'candlesticks' (Fig. 4.60). The vast majority of candlestick postal scales were manufactured by Robert Walter Winfield and the brothers Joseph and Edmund Ratcliff. Winfield worked in the Birmingham area from 1829 to 1860 and had a factory on the Baskerville Estate [71].

Jolly's spring balance (Fig. 4.61) is designed for the rapid determination of the specific gravity of solids, for sorting of minerals etc. The two pans enable to weigh an object first in the upper pan in air, and then transfer it into the lower pan suspended in water in a beaker. The specific gravity is determined by dividing the extension of the spring by the object in air by the difference between the extensions in air and water. The instrument was invented by the German physicist Johann Philipp Gustav von Jolly (1809–1884).

Fig. 4.59 Mancur spring
balances with band shaped
spring; canner balance, sack
weigher centner balance

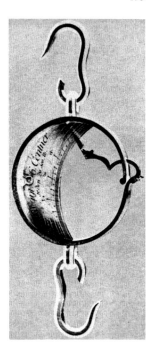

4.5.1.1 Rod Spring Balances

An elastic rod is clamped in an approximately horizontal direction at one end, and
the other end is loaded with the mass m to be determined [9] (Fig. 4.62). The lowering y may be observed with a microscope. For a circular cross section with a
radius r, the following equation is valid:

$$y = \frac{4l^3 g}{\pi E r^4} \left\{ \frac{m}{3} + \frac{m_f}{8} \right\}$$ (4.24)

For the sensitivity, lowering per change in mass, we then obtain:

$$\frac{dy}{dm} = \frac{4l^3 g}{3\pi E r^4}$$ (4.25)

Here, l is the length of the rod, E its Young's modulus, m_f that of the self-extension
of the relevant mass, and g the gravitational acceleration.

A microbalance constructed on this principle with a glass fibre 100 mm long and
with a radius of 50 μm was described by Salvioni in 1901 (Fig. 4.63) [72]. For reading the mass it had a scale and a microscope with an eyepiece micrometer. Giesen
[73] added a vacuum-tight casing to this kind of balance system in order to make
gas density, humidity and sorption measurements possible. In later models of this
balance, quartz fibres were always used. In order to weigh plutonium Cunningham
constructed a simple device consisting of a quartz fibre, about 12 cm long and 1/10
of a mm in diameter, suspended at one end, with a weighing pan hung on the other

Fig. 4.60 Letter spring
balance known as the
'candlestick' type, by
R.W. Winfield, Birmingham,
England. © Matthias Hass
Germany

end. Then the depression of that end of the fibre with the pan containing the sample would relate to the weight of the sample that was weighed. Cunningham then measured the depression of the quartz fibre with a microscope. He invented this balance independently, although he found out later that an Italian named Salvioni had invented it earlier, and so it became known as the Salvioni balance. Cunningham [74] calculated the sensitivity taking into account the different lengths of pointer and lever arm and an additional flector. With $y/l > 1/5$, $y(m)$ becomes distinctly non-linear. A self-extension $y_s = l/10$, is an advantage because a further tenth of the fibre length then remains as a useful extension. In this case, the radius of the fibre with a density ρ is:

$$r_{opt} = \sqrt{\frac{8}{30\pi\rho}\frac{dy}{dm}} \tag{4.26}$$

and the length is:

$$l_{opt} = \sqrt[3]{3\pi r_2 E \frac{dy}{dm}\frac{1}{4g}} \tag{4.27}$$

The sinking resulting from the mass of the balance pan is added to the self-extension of the fibre. By twisting the clamping, the fibre is so adjusted that it halves the area between its course during self-extension and that at maximum capacity.

With the scanning probe microscopy (SPM) by means of a small leaf spring forces down to about 1 pN can be detected.

Fig. 4.61   Jolly's spring
balance

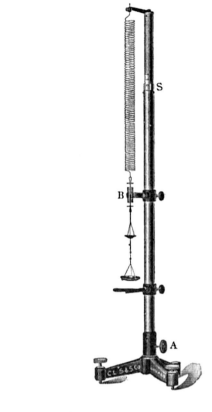

Fig. 4.62   Diagram showing
the principle of the rod spring
balance

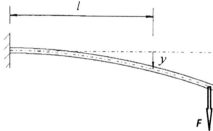

4.5.1.2 Helical Spring Balances

Common types of spring balances consist of a helical spring fixed to a suspension
ring at the upper end and a sample hook at the lower end. It is enclosed in a casing
with flat front with an equally spaced scale (Fig. 4.64). At the lower end of the spring
coil is an indicator tag which can move through a slot in the front. More sensitive
spring balances have two tubular casing sections which can be telescoped into each
other, and surround a helical coil spring.

In science the helical spring balance has a wide application due to the fact that
the springs with a sensitivity of up to 100 mm/mg are relatively easy to manufacture,

Fig. 4.63 Quartz fiber
balance of Salvioni

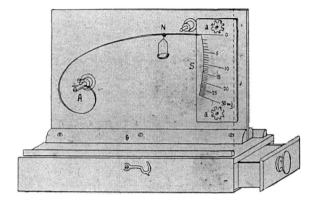

Fig. 4.64 Emich helical
spring balance. (Type of
Joly's spring balance)
0.5 mg/mm

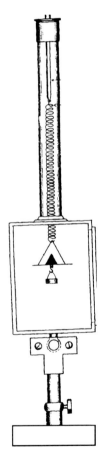

are of high quality and are generally available in great variety, and the construction
of these balances is simple. As a result of its excellent chemical resistance, very
small expansion coefficient and extremely good elastic properties, quartz glass is

particularly popular as the spring material. It can be used in contact with highly reactive gases and vapours over wide temperature ranges. However, other materials are also suitable, such as the mechanically more resistant tungsten.

Kirk and Schaffer [75, 76] give a detailed account of the manufacture of precise helical springs made of quartz glass. The quartz fibres are wound around a cylinder rotating with a constant angular velocity by means of a guiding device moving uniformly on-axis. A revolving burner focused on the fibre softens it as it is being wound round the cylinder. In another method [77] in one working cycle the fibre is drawn horizontally off a quartz rod with a burner at each end, one focused upwards and the other downward. The fibre becomes rigid in the process of being wound up on a cylindrical mandrel which rotates at a peripheral velocity of approximately 100 mm/s, the quartz rod and burners undergoing a uniform transversal displacement.

The inter-relationship between the extension y and the load m is being sought. An axially loaded helical spring balance with N windings of a radius r corresponds unwound to a rod with a length $l = 2\pi N r$ which is subjected to a torsional moment mgr. The twisting Θ of the rod follows from the torque:

$$M_\varphi = rmg \tag{4.28}$$

and from the rod's polar moment of inertia J_p with a circular section of the radius r_s:

$$J_p = \pi r_s^4 / 2 \tag{4.29}$$

with the modulus of torsion G_T, this gives:

$$\Theta_\varphi = \frac{M_\varphi}{J_p} = \frac{2rmg}{\pi G_T r_s^4} l_0 = \frac{4r^2 N mg}{\pi G_T r_s^4} \tag{4.30}$$

The following is valid:

$$\Theta_\varphi = \frac{y}{r} \tag{4.31}$$

and therefore:

$$\frac{y}{m} = \frac{4r^3 N g}{\pi G_T r_s^4} \tag{4.32}$$

Here, the simplified assumption is that:

$$\frac{r}{r_s} \gg 1 \quad \text{and} \quad y \ll N r$$

and that the initial gradient of the spiral is practically zero. The factor G_T / E also enters into a more accurate analysis:

$$y = \frac{4m r^3 N g}{G_T r_s^4} \left\{ \frac{\cos \varphi_s}{1 + \frac{3 \cos^4 \varphi}{16(\frac{r^2}{r_s^2} - 1)}} + \frac{2 G_T}{E} \sin \varphi_s \, \mathrm{tg}\, \varphi_s \right\} \tag{4.33}$$

where φ is the angle of slope and E is Young's modulus for $r/r_s = 3$, $\varphi_s = 5°$ and $G_T/E = 0.35$ a correction of 2 % results. The self-extension of a helical spring is:

$$y_s = \frac{2r^3 N^2 m_f g}{G_T r_s^4} \tag{4.34}$$

in which m_f is the total mass of the uniform spring. Provided that the load m is essentially greater than the dead weight, for the period of the fundamental oscillation we then have:

$$\tau_\varphi = 2\pi \sqrt{\frac{4m N r^3}{G_T r_s^4}} \tag{4.35}$$

The maximum permissible capacity is decided by properties of the spring's material and the degree of linearity required. The extension should not exceed a total of 1 radius/winding:

$$y_{total} = y_s + \frac{4m_{max} r^3}{G_T r_s^4} \approx Nr \tag{4.36}$$

The maximum capacity can be determined from this. Helical spring balances can also be automated, in which case the damping, which is slight in vacuum, can be improved [78].

4.5.2 Torsion Balances

As long as they are not twisted beyond their elastic limit, torsion springs obey an angular form of Hooke's law:

$$\tau = k_t \Theta \tag{4.37}$$

where τ is the torque exerted by the spring in newton-meters, and Θ is the angle of twist from its equilibrium position in radians. k_t is a constant with units of newton meters/radian, variously called the spring's torsion coefficient, torsion elastic modulus, rate, or just spring constant, equal to the torque required to twist the spring through an angle of 1 radian. It is analogous to the spring constant of a linear spring.

There are two types of torsion balances

- the torsion pendulum in which a horizontal bar is suspended on a vertical torsion fibre
- a balance beam which is suspended by means of two stretched torsion fibres or taut bands attached lateral at the beam's fulcrum.

A torsion system can be operated in a different way. The mass attraction in the gravitational field of the Earth or—in case of the pendulum—of the reference body causes twisting of the wire until the equilibrium position is attained and the deflection is measured. The other possibility is stimulating of oscillations and observation of frequency changes due to mass variations. The latter method is discussed in a separate chapter on oscillators.

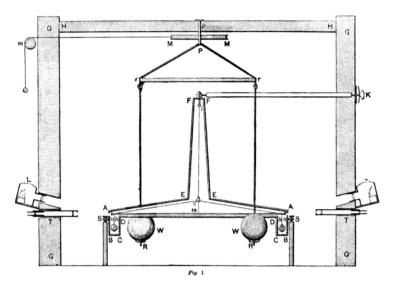

Fig. 4.65 The torsion pendulum, torsion balance or Michell/Cavendish apparatus

4.5.2.1 The Torsion Pendulum

The torsion pendulum or torsion balance is "one of the jewels of physics, a marvel of simplicity and beauty" [79]. It consists of a bar horizontally suspended from its middle by a thin fibre loaded at its ends by two weights. The fibre acts as a very weak torsion spring. Its mechanical characteristics can be pre-determined by means of calculations described in Sect. 4.2.1. If a force is applied at right angles to the ends of the bar, the bar will rotate, twisting the fibre, until it reaches equilibrium between the force applied and the restoring elastic force. Then the magnitude of the force is proportional to the angle of the bar. The sensitivity of the instrument comes from the weak spring constant of the fibre, so a very weak force causes a large rotation of the bar. It can be used to determine the fundamental gravitational constant, the Earth's mass, electric and magnetic forces and the elastic properties of the fibre material. The deflection is observed by means of a needle attached to the bar or a luminous pointer

In about 1760 the Reverend John Michell (1724–1793) invented the apparatus (Fig. 4.65). Most probably he never used it. The apparatus was obtained by Henry Cavendish (1731–1810), and in 1798 he used it to measure the gravitational force between two masses to calculate the density of the Earth, leading later to a value for the gravitational constant (Eq. 1.5) [80, 81]. In his report to the Royal Society he emphasised Michell's work. His experiment is often referred to as having "weighed the earth" because from the knowledge of R, the earth's radius, and the gravitational field strength g at the surface of the earth, it is possible to determine the mass M of the earth according to (Eq. 1.4).

To measure the unknown force, the spring constant of the torsion fibre must first be known. This is difficult to measure directly because of the small magnitude of the force. Cavendish overcame this by a method widely used since then—measuring the resonant vibration period of the balance. If the free balance is twisted and released, it will oscillate slowly clockwise and counterclockwise as a harmonic oscillator, at a frequency that depends on the moment of inertia of the beam and the elasticity of the fibre. Since the inertia I of the beam can be found from its mass, the spring constant k_s can be calculated. If the damping is low, this can be obtained by measuring the natural resonance frequency of the balance, so:

$$k_s = (2\pi f_n)^2 I \qquad (4.38)$$

where k_s [N m rad^{-1}] is coefficient of torsion spring, f_n [Hz] undamped (or natural) resonance frequency, I [kg m^2] moment of inertia.

In 1784 Charles-Augustin de Coulomb (1736–1806) described investigations with a torsion pendulum (Fig. 4.66) [82, 83]: "How to construct and use an electric balance (torsion balance) based on the property of the metal wires of having a reaction torsion force proportional to the torsion angle." In Coulomb's experiment, the torsion balance was an insulating rod with a metal-coated ball attached to one end, suspended by a silk thread. The ball was charged with a known charge of static electricity, and a second charged ball of the same polarity was brought near it. The two charged balls repelled each other, twisting the fibre through a certain angle, which could be read from a scale on the instrument. By knowing how much force it took to twist the fibre through a given angle, Coulomb was able to calculate the force between the balls. Determining the force for different charges and different separations between the balls, he showed that it followed Coulomb's law:

$$F = \frac{1}{4\pi\varepsilon_0} \frac{Q_1 Q_2}{r^2} \qquad (4.39)$$

where $\varepsilon_0 = 8.859 \times 10^{-12}$ C V^{-1} m^{-1} the electric field constant, Q the electric charge and r the distance.

Coulomb first developed the theory of torsion fibres and the torsion balance in his 1785 memoirs, *Recherches theoriques et experimentales sur la force de torsion et sur l'elasticite des fils de metal & c.* This led to its use in other scientific instruments, such as galvanometers, and the Nichols radiometer which measured the radiation pressure of light. In 1906 Loránd Eötvös (Baron Roland von Eötvös) (1848–1919) modified the Michell-Cavendish apparatus. He again determined the gravitational constant and investigated the gravitational gradient on Earth's surface. On account of this of the equivalence of gravitational and inertial mass (the so-called weak equivalence principle) such experiments are called Eötvös experiments.

Eötvös was the first to make the torsion balance suitable especially for the solution of practical geological problems by changing the design of the torsion balance already known. He suspended one of the loading masses below the horizontal part of the beam (Fig. 4.67). This improvement enabled him not only to measure horizontal directing forces, as was possible with the first type, but also the alteration of gravity in vertical direction, the so-called 'gradient of gravity.' This Eötvös horizontal variometer measures both horizontal components of the Earth gravitational acceleration g_n: the gravitational force in direction to Earth's centre and the effect

Fig. 4.66 Coulomb's torsion balance

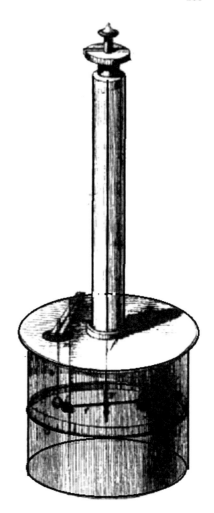

of the rotation of Earth. A deflection of such a balance with two weights in different levels is produced by irregularities in the subsoil, since the directions of attractions of such masses are different at the two weights, thus causing a small horizontal component. Naturally there occur no differences in those directions of attractions over the centre of heavy masses in the subsoil, but they reach their highest amount over the edges, meaning that the greatest deflection is produced where the gradient of gravity is highest.

The described balance system is enclosed in a triple casing in order to protect it against air currents and temperature changes. Generally two balances by the side of each other are used in the same instrument. The deflections of the two beams are observed either visually in three azimuths, or the instrument is moved automatically in

Fig. 4.67 First Eötvös
torsion balance
1898 designed for geodetical
surveying. *1* three-leaf wall
for temperature equalisation,
2 mirror, *3* sample, *4* vertical
adjustment, *5* turning
adjustment, *6* suspension
wire, *7* platinum beam,
8 telescope, *9* axle,
10 suspension wire,
11 thermometer

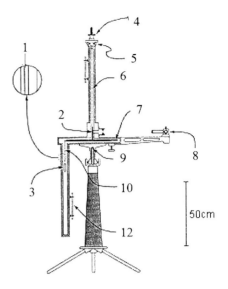

these three positions, taking records photographically after the balances have come to rest. He used successfully the instrument in petroleum prospecting.

Today torsion balances are still used in physics where the system may be operated in high vacuum and at cryogenic temperatures (see Chap. 1).

4.5.2.2 The Torsion Balance

Torsion balances consist of a beam the fulcrum of which is stretched laterally by torsion fibres or taut bands. This technique was developed first for vacuum microbalances with maximum load of milligrams, first shown in the year 1837 by Wilhelm Weber [84]. Where the deflection of the beam is observed and calibrated in mass units, we have a real torsion balance. In many cases the deflection is compensated to zero and the compensating force is measured. In this case the torsion wire acts as a bearing and it is not a torsion balance. In 1887 a two-pan top-loader kilogram balance was presented with flexible string bearings (see Sect. 4.3.4). Kent's torsion balance is actually a hybrid because most of the article's mass is compensated by weights and only small deviations are measured by deflection.

For sensitive laboratory balances, beam balances with knife edges and rider weights were used likewise in the sub-gram range. For vacuum microbalances such devices proved to be unfavourable. In 1914 Petterson constructed a balance entirely of quartz and presented a detailed discussion of the theory of the instrument [85, 86]. He made the balance completely symmetrical in every detail except for the difference between sample and counterweight; the suspension consisting of a pair of quartz fibres. A magnetic arrestment was employed. Deflections had been compensated by means of an 'air-weight,' a buoyancy method developed by Steele and Grant [87, 88]: by the upward displacement on a bulb being varied by varying

Fig. 4.68 Gulbransen &
Andrew balance with quartz
beam. *1* vacuum and gas
connection, *2* counterweight,
3 tungsten wire, *4* quartz
beam, *5* earthing device,
6 pyrex tube, 22 mm ∅,
7 quartz tube, *8* sample

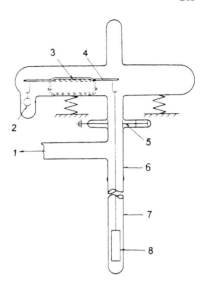

the gas pressure within the enclosing case, or by a 'magnetic weight', involving the
use of a measured electric current. The range of weighings depended on the size of
beam: with one 5 cm in length the maximum load is between 100 and 200 mg, and
the setting can be made to within 10^{-7} mg. Also in this the torsion wires acted only
as a bearing and Petterson's balances cannot be regarded as a torsion balance.

In 1942 Gulbransen constructed a series of vacuum balances equipped with a
quartz beam [89] (Fig. 4.68). For advanced types tungsten support wires with silver
chloride seals to the quartz were used for samples with a mass below 0.1 g. The
deflection of 10^{-6} g per 10 μm of the beam's end had been observed by means of a
conventional micrometer microscope [15].

In the 20th century a series of torsion balances for chemical and pharmaceutical
laboratories had been produced (Fig. 4.69). Whereas such balances need to be calibrated mechanically, the dimensions of the torsion wire or band can be calculated in
order to obtain the desired load capacity and sensitivity. On account of the limited
load range today such balances have been replaced by electro-dynamical types.

Special torsion balances (tensiometer) are used for the investigation of the surface
tension of liquids and molecular films on liquids (film balance) (see Chap. 5).

4.5.2.3 Molecular Torsion Balance

Fischer, Schweizer and Diederich used the elastic behaviour of organic molecules
to measure the enthalpy of reactions [90]. An indole-extended molecular torsion
balance has the geometry for measuring a truly orthogonal noncovalent interaction
between a C–F bond dipole and an amide carbonyl group (Fig. 4.70). Employing a
double-mutant cycle approach, negative interaction free enthalpies were determined.
Thus orthogonal dipolar interactions can be a new tool for stabilising protein-ligand
complexes and assembling supramolecular architectures.

Fig. 4.69 Laboratory torsion balance. W. Gambichler, Basel. Virtuelles Museum. © Virtuelles Museum der Wissenschaft. http://www. amuseum.de/physik/alwami/ alwamikilo.htm

24.5 cm

4.5.3 The Load Cell

The load cell is a flexible metal body, the shape of which is adapted to the special task of weighing (Fig. 4.71). On the surface of that body, in the region where the maximum of deformation occurs, a sensor is attached or connected by a lever system. Load cells mostly are applied for industrial purposes, and the type of sensor used depends on the desired measuring range. The following sensors are usable:

- Strain gauge
- Piezoelectric crystal
- Hydraulic gauge
- Oscillating string.

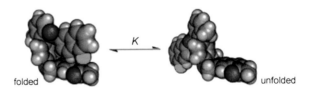

Fig. 4.70 An indole-extended molecular torsion balance has the geometry for measuring a truly orthogonal noncovalent interaction between a C–F bond dipole and an amide carbonyl group (see picture, *green* F, *red* O, *blue* N). © Wiley-Interscience

Fig. 4.71 Different types of strain gauge load cells. © Siemens

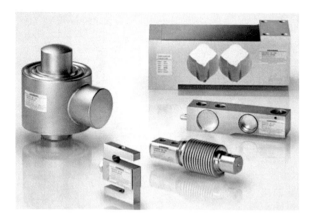

4.5.3.1 Strain Gauge

The strain gauge (DMS = Dehnungsmessstreifen) is a resistance wire folded on the surface of a flexible metal body [91, 92]. It is attached, usually by gluing (Figs. 4.72a, b) but today also by sputtering (Fig. 5.34). By deformation (extension or compression) its resistance varies. The basic resistance is in the range 0.1 to 1 kΩ and the variation of the resistance is in the range of mΩ. Variations of mV are generated and measured on a Wheatstone bridge or a modified arrangement (Fig. 4.73). The principle was described in 1856 by Lord Kelvin. The device was invented independently by Edward E. Simmons Jr. (1911–2004) [93] and Arthur Claude Ruge (1905–2000) [94]. Ruge developed the strain gauge principle whereas Simmons invented the force transducer.

The resistance of the strain gauge is a function of applied strain.

$$\frac{\Delta R}{R} = S\varepsilon \tag{4.40}$$

where R is the resistance, ε is the strain, and S is the strain sensitivity factor of the gauge material. Amongst types of strain gauges, an electric resistance wire strain gauge has the advantages of lower cost and of being an established product. Thus it is the most commonly used type of device. Other types of strain gauges are acoustic, capacitive, inductive, mechanical, optical, piezo-resistive, and semi-conductive. Consider a wire strain gauge, as illustrated above. The wire is composed of a uniform conductor of electrical resistivity ρ with length l and cross-section area A. Its resistance R is a function of the geometry given by

$$R = \rho \frac{l}{A} \tag{4.41}$$

The resistance of the unstressed strain gauge is

$$R = \rho \frac{l}{A} = \rho \frac{4l}{\pi D^2} \tag{4.42}$$

where l is the length of the resistance wire, A the cross section area, D the diameter and ρ the resistivity. The resistance change rate is a combination effect of changes in

Fig. 4.72 (**a**) Load cell with
strain gauges applied to the
top and the lower surface.
© Mettler. (**b**) Load cell with
strain gauge glued to the top.
© Sartorius. (**c**) Balance
module with strain gauge
load cell for 50–500 kg.
© Hottinger Baldwin
Messtechnik GmbH.
(**d**) Balance module with
strain gauge load cell for
20–200 t. © Hottinger
Baldwin Messtechnik GmbH

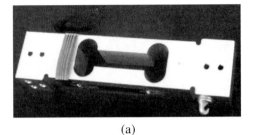

(a)

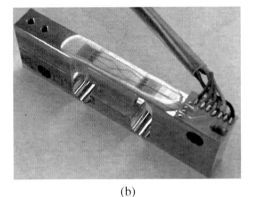

(b)

(c)

length, cross-section area, and resistivity. The resistance of the stressed strain gauge
is

$$R + \Delta R = (\rho + \Delta\rho)\frac{4(l + \Delta l)}{\pi(D - \Delta D)^2} \tag{4.43}$$

Conversion and series expansion according to Taylor results in the resistance variation:

$$\frac{\Delta R}{R} = k\frac{\Delta L}{L_0} = k\varepsilon \tag{4.44}$$

Fig. 4.72   (Continued)

(d)

Fig. 4.73 Strain gauge, circuit diagram. The folded resistance wire is compared in a Wheatstone bridge with three known resistors R_2, R_3, R_4; U is a voltage supply

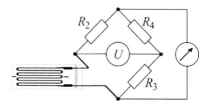

where k for metals is about 2: the relative extension ε is given by:

$$\varepsilon = \frac{F}{AE} \qquad (4.45)$$

where E is he modulus in flexure of the material. The strain gauge can be made of metal or of semiconductor materials.

4.6   The Gyro Balance

A gyroscope consists of a disc rotating on a spin axis at constant angular velocity ω (Fig. 4.74a). This spin axis is suspended on gimbals (cardanic mounting), freely movable on two axes perpendicular to the spin axis. That free suspension can also be done in a cage (Fig. 4.74b). Such a device may be regarded as a closed system in which conservation of the angular momentum is maintained. The system remains in stable position. If, however, an external force is applied e.g. at the axis designated

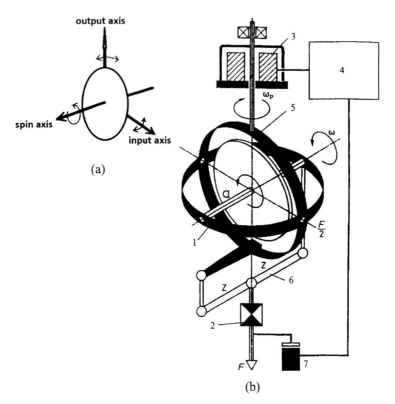

(a)

(b)

Fig. 4.74 (a) Principle of the Gyro scale. (b) *1* inner gyroscope casing, *2* shoulder bearing, *3* servo-motor, *4* control circuit, *5* outer gyroscope casing, *6* rotary lever, *7* proximity switch. © Kochsiek/Gläser: Comprehensive Mass Metrology, Wiley-VCH

as input axis, a movement of the output axis takes place. That axis starts precession turns, the frequency of which is proportional to force exerted [95, 96].

4.7 Buoyancy Balances

On the one hand buoyancy is a troublesome source of error in weighing. On the other hand the application of buoyancy as a counteracting force allows for very sensitive measurements. Because this is indeed a density determination by buoyancy, the volume of the body must be known in order to calculate the mass. Hydrostatic measurements are mostly applied for density determinations and are discussed in that special chapter. Archimedes density determination as well as hydrometers, areometers, spirit gauges, lactometers fit in this category.

The depth of a floating body can be used as a measure of its load. This method is applied to determine the loading of ships by measuring the draught. The draught graduation usually is determined experimentally. A hydrostatic balance was con-

Fig. 4.75 Hydraulic mass
comparator. Hydrostatic
comparator balance BTB
Braunschweig *1* connection
2 six floaters in a tank of
liquid, *3* pan, *4* weight.
© PTB, Braunschweig

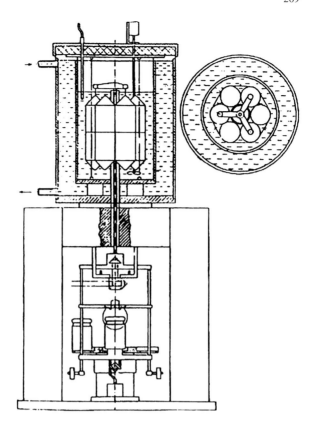

structed by the 'Physikalisch Technische Bundesanstalt', Braunschweig, Germany as a mass comparator. It consists of six floating bodies in a tank filled with a liquid, usually water (Fig. 4.75) [96]. The hydraulic principle has been applied also for letter balances.

In 1926 A. Stock and G. Ritter reported on the determination of gas densities using a microbalance (Fig. 4.99) [97]. They measured the buoyancy as a function of gas pressure, restoring the balance in its zero position at each run. That balance equipped also with electromagnetic compensation is described in a subsequent chapter.

4.8 Hydraulic Balances

The weight force can be compensated by lifting a liquid column as a variable counterweight. The height of the column can be graduated in mass units. Furthermore the pressure increase in closed vessel filled with a liquid or a gas can be measured by means of manometer and transferred into mass units.

Weighing of heavy objects can be made by means of a hydraulic press. For example, the Eiffel Tower in Paris has a mass of 10 000 tons and the mass could be measured by means of hydraulic presses in the foundations which had been planned for the correction of the upright position of the tower [98].

4.9 Balances with Electric Force Compensation

The following section of this chapter is concerned with balances in which the gravitational force on the weighing article is compensated by electrostatic forces. According to Coulomb's law only weak forces can be exerted if not very high voltages are applied. Therefore, electrostatic counteraction in weighing is used for special tasks only, in particular for the determination of electric units and electric parameters of materials. However, capacitive sensors are widely applied to control the deflection of balances.

4.9.1 Electrostatic Force Elements

Electrostatic force elements make it possible to compensate very small forces and moments with voltages that are easy to measure. Applications of dual-plate capacitors with a quadratic steady-state characteristic and tri-plate capacitors with a linear steady-state characteristic are examples of this [99]. If A is the area, d the distance between the electrodes, U the adjacent voltage, ε the relative permittivity of vacuum (DK) of the surrounding medium and ε_0 the DK of the vacuum, for the attractive force we obtain:

$$F = \frac{\varepsilon \varepsilon_0 U^2 A}{2d^2} \qquad (4.46)$$

With three plates with the distance d on the potential $-U/2$, $+U_m$ and $+U/2$, the following is yielded for the force on the median electrode:

$$F = \varepsilon \varepsilon_0 U U_m \frac{A}{d^2} \qquad (4.47)$$

The configuration exhibits a negative recovery capacity, allowed by the compensation of elastic and quasi-elastic forces or moments. The dependence of the force on the DK of the gas, and the influence of the electrode's contact potentials are a disadvantage. Cylinder capacitors also make suitable electrostatic force elements. A stationary cylinder capacitor pulling a glass tube attached to the beam into its gap is also used as a force element [100].

4.9.2 The Electrometer

In the 18th century the effects of static electricity produced by friction of glass tubes and similar bodies had been investigated thoroughly. Faced with the surface of other

Fig. 4.76 Gold leaf
electroscope, principle. *1* gold
leaves, *2* glass balloon,
3 insulating bar loaded by
rubbing

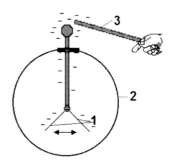

materials, attractive or repulsive forces were observed and it was attempted to measure the strength of such effects. Already in 1743 Christian August Hausen, professor of mathematics in Leipzig and member of the 'Königlich Preußische Akademie
der Wissenschaften' at Berlin, reported in Latin that he equilibrated by means of
weights recipients consisting of copper and glass, which were fastened at levers
and attracted by a rubbed tube [101, 102]. A little later, in 1746, an anonymous
author wrote a letter about weighing the "Strength of the Electrical Effluvia" in order to "measure electrical powers and to compare the quantities and strength of the
electrical virtue". For that purpose he used a balance equipped with balls [103].
These early attempts obviously were not successful, and later on the measurements
were executed by other methods. A simple arrangement is the gold leaf electroscope
(Fig. 4.76). Accurate measurements of the electric forces had been performed first
in the early 1780's by Charles Augustin Coulomb (1736–1806) using a torsion device, described in Sect. 4.5.2 of Chap. 4 (Fig. 4.66). Alessandro Volta (1745–1827)
showed that a force field develops in the surrounding of a body with electrostatic
charges; for demonstration he used diverging straw [104]. In 1781 Volta also constructed a sensitive electrometer for measuring electric voltage. Such instruments
already operated quite efficiently. Results agreed with those of electrometers operating on other principles [105].

4.9.2.1 The Absolute Electrometer of William Thomson

The 'absolute electrometer# of William Thomson (1824–1907), presented in the
year 1855 and reported in 1860 [106], was a very important advance in measuring
electrical potentials. This instrument consisted of two circular electrodes; one of
them was suspended on a balance and was insulated from it. The second plate facing
the first one was movably arranged and connected to ground. The first plate was
surrounded by a protection ring in order to keep the lines of force parallel resulting
in a homogeneous field. The original quadrant electrometer uses a light aluminium
sector suspended inside a drum cut into four segments. The segments are insulated
and connected diagonally in pairs. The charged aluminium sector is attracted to one
pair of segments and repelled from the other. The deflection is observed by a beam of
light reflected from a small mirror attached to the sector. Figure 4.77 shows a slightly

Fig. 4.77 The absolute electrometer of W. Thomson (Lord Kelvin), ca. 1860

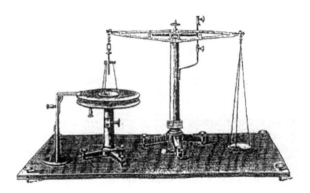

different form of this electrometer, using four flat plates rather than closed segments. The plates can be connected externally in the conventional diagonal way. or in a different order for specific applications. Thomson's publication was not illustrated and the figure is drawn from a subsequent paper [107].

4.9.2.2 The Absolute Electrodynamometer of Helmholtz

Hermann Helmholtz (1821–1894) too used the advantages of the electrodynamic balance in some of his experiments (Fig. 4.78). He determined for example the electrodynamic equivalent of the electric current, which corresponded to one gram of a counter weight on his balance [108], and also the magnetic moments of magnets [107]. The electrodynamometer of Helmholtz (1863) was depicted and described in detail by K. Kahle in a further improved design [109]. Kahle also used such instruments for the precise measurement of the voltage of the Clarc element [110].

4.9.2.3 Modern Electrometers

Gast has developed a self-compensating electrometer based on a three-plate capacitor which, in reverse, enables radiometric forces and Knudsen forces to be measured [111]. Typically a modern electrometer is a highly sensitive electronic voltmeter whose input impedance is so high that the current flowing into it can be considered, for most practical purposes, to be zero. The actual value of input resistance for modern electronic electrometers is around 10^{14} Ω. Figure 4.78b shows a diagrams of Gast electromagnetic balances used to measure electrostatic forces. Among other applications, electrometers are used in nuclear physics experiments as they are able to measure the tiny charges left in matter by the passage of ionising radiation.

Fig. 4.78 (**a**) The absolute electrodynamometer of Helmholtz (1863). (**b**) Gast electromagnetic balance used to measure electrostatic forces. *1* balance beam, *2* taut band, *3* sample, *4* counterweight, *5* moving coil, *6* stator coil, *7* permanent magnet, *8* movable capacitor plate, *9* fixed capacitor plates, *10* dc supply, *11* switch

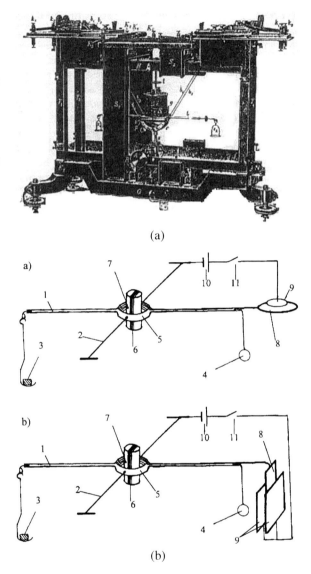

(a)

(b)

4.9.3 Voltage Balances

A voltage balance is usually an equal armed beam balance, equipped at one side with an electrode. Typically this movable electrode is suspended vertically and moves parallel to a fixed electrode (Fig. 4.79). In this arrangement the force of only a little

Fig. 4.79 Principle of a
voltage balance. *1* movable
electrode, *2* high-voltage
electrode, *3* auxiliary
electrode, *4* column of the
sliding unit, *5* carriage for the
sliding unit, *6* drive,
7 interferometer for
measuring the displacement
path, *8* position sensor,
9 flexible bearing of the
balance, *10* balance beam,
11 counterweight, *12* mass
standard. © Kochsiek/Gläser:
Comprehensive Mass
Metrology, Wiley-VCH 2000

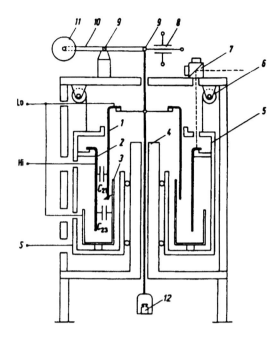

mass up to 10 g can be compensated by the electrostatic force. The voltage U is given as follows:

$$U = \sqrt{2mg\left(1 - \frac{\rho_g}{\rho_m}\right)\frac{\partial s}{\partial C}} \qquad (4.48)$$

where m is the counterweight mass, g the local gravitational constant, ρ_g and ρ_m are the density of the protecting gas (e.g. nitrogen) and the density of the counterweight, respectively.

A Kelvin electrometer with horizontal electrode has been used at the 'Laboratoire Centrale des Industries Electrique (LCIE)', Fontenay-aux-Roses, France to compensate a mass of 5 g electrostatically.

A liquid electrometer had been realised at the Commonwealth Scientific and Industrial Research Organisation (CSIRO), Australia (Fig. 4.80). It consists of a mercury bath the surface of which is used as one electrode opposed to a fixed metal electrode. Voltage is applied at two distances d_1, d_2 which raises the mercury surface by the height difference z_1, z_2 respectively. The difference in voltage is then

$$\Delta U = \sqrt{\frac{2\rho g}{\varepsilon}}(d_2\sqrt{z_2} - d_1\sqrt{z_2}) \qquad (4.49)$$

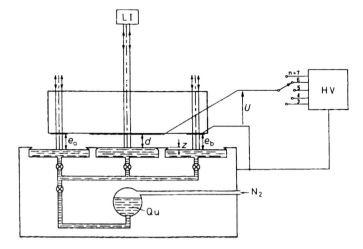

Fig. 4.80 Diagram of a liquid electrometer. D distance between electrodes, e_a, e_b distance between reference electrodes, z change in distance due to applied voltage U. The distance between electrodes is controlled by laser interferometer. *HV* high voltage source, *QU* mercury reservoir, *LI* laser white light interferometer

Fig. 4.81 Electromagnetic balance V. Crémieu 1901

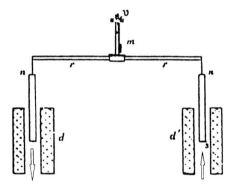

4.9.4 Electrostatic Balances

A notable sensitive balance was presented by V. Crémieu in the year 1901, which could be used as a galvanometer, an electrodynamometer and an absolute electrometer (Fig. 4.81) [112]. This balance was not supported by knife-edges, but it was suspended on ribbons, as already shown in the year 1837 by Wilhelm Weber [84].

An electrostatic field can be applied especially in high vacuum. In gases difficulties are encountered because of the different dielectric properties [30]. On account of the high tensions needed, the gas between the electrodes is ionised producing a current or a flashover. An electrostatically operated vacuum microbalance was described by Mayer et al. [22] (Fig. 4.82). The balance was equipped with a capacitive indicator and the deflection was compensated electrostatically as well.

Fig. 4.82 Electrostatically
operated vacuum microgram
balance of Mayer,
Niedermayer, Schroen,
Stünkel and Göhre.
1 solenoid, 2 Cu cylinder,
3 magnet, 4 mirror, 5 torsion
wire, 6 knife edge,
6 calibration pan,
7 suspension fibre

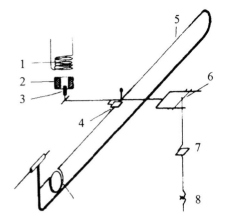

4.10 Balances with Electromagnetic Force Compensation

Today many balances work on an electrical, mostly digital, display. Often the balance itself is hidden in a black box and the notation 'electronic' does not reveal its principle. The following chapter deals with balances with electromagnetic force compensation [4, 113–115]. This principle is proved to be the most favourable for correct weighing in an extended measuring range. It can be realised in two ways:

- Electromagnetic force compensation: a magnetic core, usually a permanent magnet connected to the balance lever (beam) is influenced by a current in a fixed coil (Figs. 4.19/4.83a) (or the reverse).
- Electrodynamic force compensation: the current in a fixed coil induces a current in moving coil connected to the balance lever (beam) (Figs. 4.20a/4.83b). The fixed coil may be wound on a fixed permanent magnet in order to enforce its inductivity.

Often no clear distinction is made between these two terms and they are commonly used as synonyms. Furthermore in self-compensating balances both principles are applied simultaneously.

Because weight force and electric counterforce are compared by means of a mechanical balance such electric balances indeed are electromechanical devices. In order to extend the measuring range such balances sometimes are equipped in addition with a counterweight scale or with a mechanical feeder for (automatic) support of counterweights. Such devices may be regarded as hybrid balances.

In 1820 Hans Christian Ørsted (1772–1851) discovered the deviation of a magnetic needle in the vicinity of a conductor carrying an electric current (Fig. 4.84). Shortly after André-Marie Ampère (1775–1830) formulated the basic laws to describe electric currents (Fig. 4.85). Instruments were constructed to measure the intensity of the current, based on the deflection of a magnetic pointer by an electric current passing a nearby situated conductor or a surrounding coil. The results—together with the those of electrometers—provided the basis for the absolute system of electric units introduced by Wilhelm Weber [116].

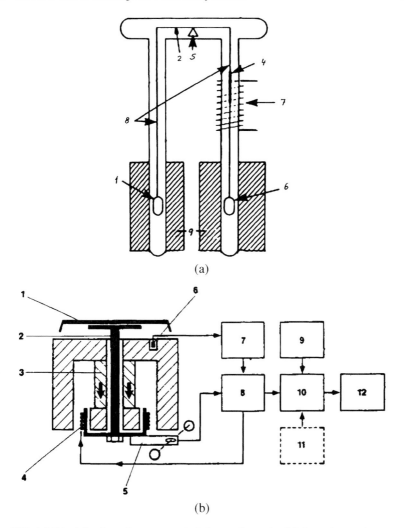

(a)

(b)

Fig. 4.83 (**a**) Principle of an electromagnetic balance. *1* sample, *2* balance beam, *4* permanent magnet, *5* knife-edge, *6* counterweight, *7* field coil, *8* suspension wire, *9* thermostat. (**b**) Principle of a top pan balance with electromagnetic force compensation: *1* scale pan; *2* carrier of the pan; *3* magnet; *4* coil; *5* deflection sensor (photo cell); *6* temperature sensor; *7* source of constant current; *8* controller/compensator; *9* tare device; *10* microprocessor; *11* unit of function; *12* digital display. © Mettler Greifensee

4.10.1 Current Balances

Early galvanoscopes had been relatively rough instruments. The deflection effect of the current could be amplified by increasing the number of windings. Johann S. Christoph Schweigger (1799–1857) and Johann Christian Poggendorff (1799–1877) constructed almost independently of each other, the first multipliers [117].

Fig. 4.84 Deflection of a magnetic needle by the direct electric current according to H.Ch. Oersted (1820)

Fig. 4.85 Arrangement to demonstrate the mutual action of two currents acc. to A.-M. Ampère (1820)

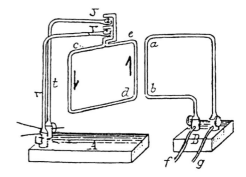

The principle of the electrodynamometer by Wilhelm Eduard Weber (1804–1891) designed in 1846 is based on the reciprocal action of two electric currents [116]. One of the two coils was fixed stationary, the other one suspended by bifilar strings (Fig. 4.86). The variation of the position of the movable coil with respect to the fixed one was measured as a function of the two currents. The force on the coil was found to be proportional to the square of the current.

4.10.1.1 The Current Balance of Becquerel

The first precise current balance (Ampère balance) was built in 1837 by Antoine César Becquerel (1788–1878). He used a balance of equal arm length, with resolution of fractions of a milligram (Fig. 4.87a). The pans of this balance were fixed closely below the beam. Two ferromagnetic bars were fastened to the pans, both with the north pole downwards. The coils adjustable in height were arranged below the pans carrying the magnets. The coils consisted of glass tubes spooled with

Fig. 4.86 The
electrodynamometer of
Wilhelm Eduard Weber
designed in 1846 is based on
the reciprocal action of two
electric currents. One of the
two coils is fixed stationary,
the other one suspended by
bifilar strings

copper wires, insulated by means of glass silk. Every magnet was adjusted in such
a way that the beam deflected always to the same side with maximum amplitude.
Then the two coils were connected to each other, in order to cumulate the effects of
both magnets. The balance was equilibrated by weights. The intensity of the current
was found to be proportional to the counterbalancing mass. The current balance of
Becquerel was built by the mechanic Barthelmy Urbain Bianchi in Paris. The instrument is shown today in the Musée National des Techniques in Paris (Fig. 4.87b)
[118]. Becquerel's electromagnetic balance can be regarded as the ancestor of the
modern electronic laboratory balances.

4.10.1.2 The Current Balance of Lenz and Jacobi

H.F. Emil Lenz and M.H. Jacobi changed the arrangement of the coils in Becquerel's
balance in order to improve its efficiency [119]. The arrangement of the magnetic
bars and the direction of the current of Becquerel's balance had been chosen in such
a way, that one of the magnets was attracted whereas the other one was repelled.
This resulted in oscillations of the balance and in difficulties to reproduce the equilibrium position. When the balance was swinging, the measurements could not be
reproduced correctly. Lenz and Jacobi arranged the coils in such a way that the repulsive forces were at both sides of the balance, in order to drive it in the same
direction (Fig. 4.88) [120]. Herewith stable equilibrium could be achieved.

Fig. 4.87 (**a**) Current
balance of C.A. Becquerel,
1837. (**b**) Current-Weigher of
C.A. Becquerel, 1837. The
coils consisted of glass tubes
spooled with copper wires,
insulated by means of glass
silk. © Musée des
Techniques, Paris

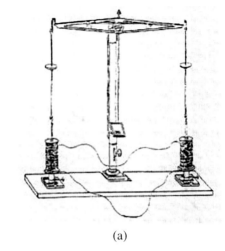

(a)

(b)

4.10.2 Watt Balances

The force F_z exerted on a coil through which the current I_g flows in an inhomogeneous magnetic field can only be calculated if the flux density, the field distribution and the dimensions of the coil are known:

$$F_z = -I_g \frac{\partial \phi}{\partial z} \tag{4.50}$$

where Φ is the magnetic flux and z is the vertical coordinate. The missing parameters can be obtained by a second experiment in the same configuration by measuring the voltage U_m induced in the coil when it is moved in a speed v_z in the magnetic field:

$$U_m = v_z \frac{\partial \phi}{\partial z} \tag{4.51}$$

Fig. 4.88 The varied current
balance of Lenz and Jacobi,
1839

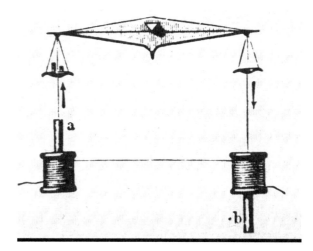

Combination of both equations and regarding buoyancy leads to

$$U_m I_g = v_z mg \left(1 - \frac{\rho_g}{\rho_m} \right) \qquad (4.52)$$

where m is the counterweight mass, g the local gravity, ρ_g and ρ_m are the density of
the protecting gas (e.g. nitrogen) and the density of the counterweight, respectively.
Corresponding experiments had been made at the British NPL with a kilogram prototype. At the NIST, USA, a superconducting magnet was used. The Swiss OFMET
constructed a watt balance for a 100 g mass.

4.10.3 Electromagnetic Balances

The central component of the electromagnetic balance is the movable coil with an
electric current flowing through it, faced by a fixed permanent magnet. The coil is
connected to the load pan. In the vicinity of the conductor through which the current
flows, an electromagnetic force field is formed, the magnitude of which is proportional to the intensity of the current. The force produced by this field interacts with
magnetic or electromagnetic forces and is related to the loaded mass. If an additional mass is loaded on the pan, the gravitational force tries to change the position
of the coil relative to the magnet. But, if an additional direct current of an equivalent
intensity is sent through the coil, the force of the magnetic field surrounding the coil
is increased. The weight force of the added mass is then compensated. The coil re-
turns to its original position, or it can even remain in place, if the electric current is
adjusted quickly enough. The change of the additional current flowing through the
coil is directly proportional to the weight force of the mass considered. If the local
acceleration of gravity is taken into consideration, the measurement of the intensity
of the current will lead to the mass to be determined. If the balance is calibrated by

known masses, which are equivalent to known weights, the result corresponds to the mass to be determined, where the effect of buoyancy of the air has also to be taken into account.

Already in the nineteenth century attempts were made to weigh electrically. First instruments were constructed to measure the fundamental units, then scientific micro and precision balances were developed. At the turn of the 19th century balances for continuous registration of chemical processes had been constructed followed by registration balances for industry and commerce. At that time a large variety of balances for mass determination based on the principle of electrodynamic force compensation had been designed. Taking advantage of the progress in electronics in the 1950s the balances were improved. The basic principle of electromagnetic or electrodynamic force compensation, however, remained unchanged. Today electronic instruments supersede mechanical balances in all fields of application.

4.10.3.1 Cazin Balance

In 1863 Achilles Cazin combined the principle of the electrodynamometer with a sensitive balance [121, 122]. His electrodynamic balance for precision measurement of current was similar to Thomson's electrometer (Fig. 4.89). It was equipped with rectangular coils arranged horizontally and parallel to each other. The attraction between the two coils was equilibrated on the opposing side of the balance by weights. Cazin could confirm Weber's electrodynamic fundamental laws with high precision. These laws included the relation of the parameters of the electric current to those of the absolute mechanical measuring system. In addition Cazin used his balance to examine the mass of weights and changed the principle of the 'current weighing' into an electromagnetic method of mass determination.

4.10.3.2 Ångstrøm's Balances

In 1884 F.C.G. Müller designed a balance galvanometer, in which the influence of the electric forces was compensated by rider weights at the graduated beam of an equal-armed balance (Fig. 4.90) [123]. With regard to the beam A similar apparatus was constructed by Knut Ångström (1857–1910) a few years later, however with unequal arms, for the determination of the strength of magnetic fields (Fig. 4.91a) [124].

Later on, in 1895 Knut Ångström described an electromagnetic microbalance, which allowed weighing of a mass of 1 g with an accuracy of at least 10 micrograms (Fig. 4.91b) [125]. A feature of this balance was the finely shaped balance beam made of aluminium. The balance beam was suspended by cocoon filaments. The pans were fixed on the two ends of the balance beam in the same manner. Ångström reports that such a suspension had been already indicated by Weber [84]. One of the hangdown pans had a two-tier pan system so that substitution weighing could be executed in the range from 1 to 1000 mg [126]. On the opposite pan a little

Fig. 4.89 The electrodynamic balance of A. Cazin, 1864

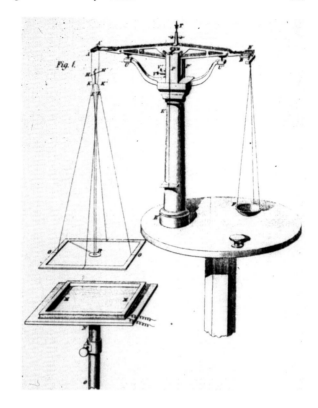

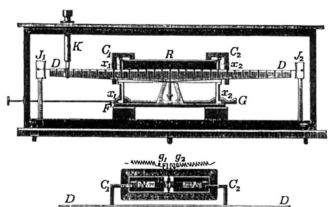

Fig. 4.90 Balance galvanometer with a rider weight, of F.C.G. Müller, 1884

permanent magnet was attached and arranged to move within an induction coil. After substituting the mass components in the milligram range, Ångström performed the determination below the milligram by reading inclinations. For that purpose he employed the known method of Poggendorff using the mirror S, which was fixed to

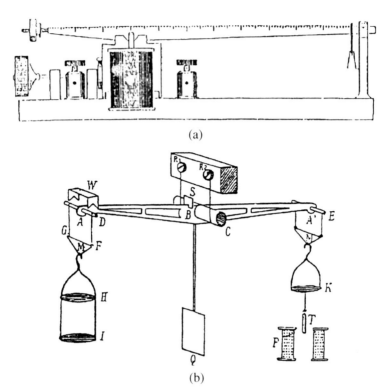

Fig. 4.91 (**a**) Electromagnetic Microbalance of Ångstrøm 1889 for the determination of the strength of magnetic fields. (**b**) Electromagnetic microbalance balance of K. Ångstrøm, 1895

the beam [127]. In order to damp the oscillations of the balance, Ångström sent a weak current through the coil P. He found out that restoring of the beam by means of a current was more sensitive than measuring the inclination. Weak currents, measured by means of a mirror galvanometer, were nearly proportional to the mass changes [128].

4.10.3.3 The Kruspe Balance

Three years later Hans Kruspe equipped a precision balance with an electrodynamic compensation system. The balance principle of electrodynamometer was patented in the year 1898 [129]. The load pan and the electrodynamic system were on the same side of the balance. On the other side a movable counter-weight was installed, in order to enable the balance to be tared in the zero position (Fig. 4.92). The coil f, which was fixed at the beam b and thus also at the pan d, was freely movable within the second coil. The two coils could be connected by means of the compensation vessel system measured by the galvanometer m.

Fig. 4.92 Precision balance
according to the principle of
the electrodynamometer, of
H. Kruspe, 1898

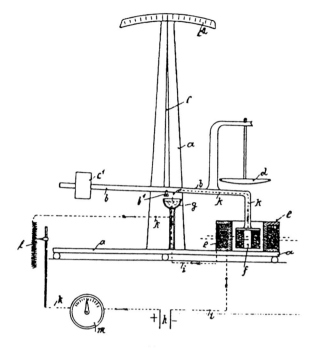

When the pan was loaded with a mass to be determined, the arm of the balance
with the coil f went down. Then the current was switched on and regulated by the
rheostat so that the pointer of the balance turned to zero again. According to the deflection of the galvanometer the mass could be calculated, after the system had been
suitably calibrated. Allegedly the balance was designed for operation in vacuum by
avoiding mechanical weight changes.

4.10.3.4   Oertling Balances

In 1908 W.E. Ayrton et al. [130] presented a new current weigher of Ludwig
Oertling in London (Figs. 4.93) for the determination of the absolute electromotive force of the Normal Weston Cadmium Cell. The authors also gave a review of
a number of publications about the absolute measurement of the electric current,
in particular by weighing [130]. Oertling's construction was a long beam precision
balance with a load capacity of 50 kg.

4.10.3.5   The Waltenhofen Apparatus

In the year 1870 Adalbert Carl von Waltenhofen had already carried out measurements, in order to acquire knowledge about the magnetic behaviour of iron tubes
(Fig. 4.94) [131, 132]. He established that metallic materials are only magnetisable
to a definitive limit of saturation.

Fig. 4.93 (**a**) Electrodynamic
balance for the measurement
of the absolute electromotive
force of the Weston Cadmium
element, of L. Oertling, 1908.
(**b**) Balance of Oertling,
principle of the windings of
the movable and of the fixed
coils

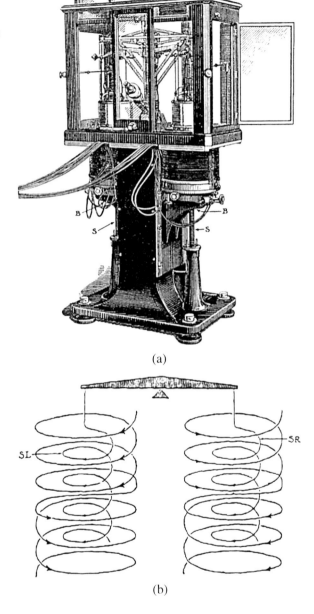

(a)

(b)

4.10.3.6 The Du Bois Balance

H.G. du Bois showed in 1892 that by using a balance it is possible to measure the
permeability of iron, in order to determine the number of lines of force (Fig. 4.95)
[133].

Fig. 4.94 Electromagnetic balance for measuring the magnetic behaviour of iron tubes, of A.C. von Waltenhofen, 1870

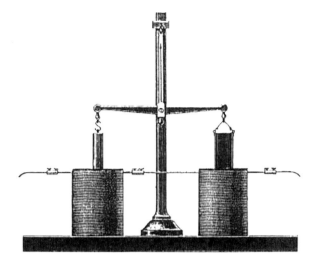

Fig. 4.95 Electrodynamic balance for measuring the permeability of iron, of H.G. du Bois, 1892

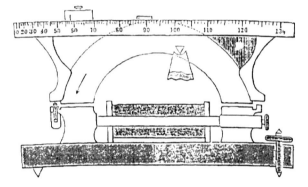

4.10.3.7 The Crémieu Balance

A notable sensitive balance was presented by V. Crémieu in the year 1901, which could be used as a galvanometer, an electrodynamometer and an absolute electrometer [112]. This balance was not supported by knife-edges, but it was suspended on ribbons, as already shown in the year 1837 by Wilhelm Weber (Fig. 4.86) [84].

4.10.3.8 Janet and Pellat Balances

Two publications in the year 1908, one of P. Janet (Fig. 4.96) [134] and the other one of H. Pellat (Fig. 4.97) [135], had been concerned with the determination of the electromotive force of the Weston Normal Element, also by using precision balances for measuring the electric effects. The uncertainty of the result was estimated to be 4×10^{-4}, based especially upon the insufficient knowledge of the gravitational acceleration at the place of observation. The error of the weighings was clearly smaller.

Fig. 4.96 Electromagnetic
balance for the determination
of the electromotive force of
the Weston normal element,
of P. Janet, 1908

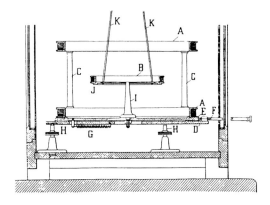

Fig. 4.97 Electromagnetic
balance for the determination
of the electromotive force of
the Weston Normal Element
of H. Pellat, 1908

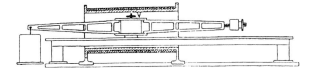

4.10.3.9 The Pascal Balance

Furthermore the paper of P. Pascal from the year 1910 should be mentioned, in
which the measurement of the magnetic susceptibility was described using an electromagnetic balance [136]. The peculiar feature of this balance was that its total load
always remained constant, because the weighing procedure took place according to
the substitution principle (Fig. 5.101a) [126, 137]. Measurement of the magnetic
susceptibility is discussed in the next Chap. 5.

4.10.3.10 The Thermo-Analytical Electromagnetic Microbalance of Urbain

According to a short report from 1912, G. Urbain designed a microbalance using
the principle of electromagnetic force compensation [138]. The beam of Urbain's
balance had a length of 80 mm and consisted of a fine glass rod 0.4 mm in thickness. The load capacity was claimed to be about 100 mg, and weighings could be
executed down to 10 micrograms and even less. The microbalance was designed
for the determination of the efflorescence rate of crystals at elevated temperature.
For that purpose the load pan was surrounded by an electric heater and the sample space could be evacuated. Just a few results obtained with this arrangement
are published without an illustration. Urbain may be considered the first one, who
described a thermobalance with electromagnetic force compensation. Like other investigators before him who made thermoanalytical investigations, he did not use the
specification 'thermobalance'. Thus usually the priority for the first thermobalance
is attributed to K. Honda [139].

Fig. 4.98 Electromagnetic microbalance of F. Emich, 1916

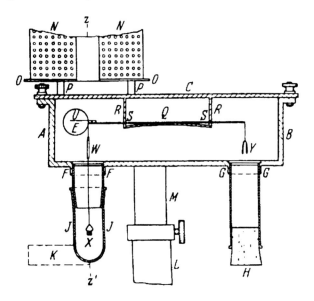

4.10.3.11 The Electromagnetic Microbalance of Emich

In 1916 Friedrich Emich (1860–1940) constructed a microbalance in order to shift the detection range down to 15 micrograms, whereby the load capacity was restricted to 0.2 milligrams [140, 141]. For this purpose he reduced the proportions of all components consequently (Fig. 4.98). According to Emich's instructions the accuracy of the weighings mainly depended on the observation of the current intensity indicated at the galvanometer. At full scale deflection he obtained a resolution of 1:104.

Emich arranged his balance as a 'zero instrument', in order to be able to work always under the same conditions. A small magnet was fixed at the load side of the balance beam within the force field of a large electromagnet arranged outside the balance casing. At the other end of the beam a small counterweight was fastened. In this way he performed weighings according to the substitution techniques. The system was arranged in a glass casing which could be evacuated. Emich's microbalance was the prototype for many subsequent constructions.

4.10.3.12 The Electromagnetic Microbalance of Wiesenberger

E. Wiesenberger, a pupil of Emich, modified this type of microbalance later and made it still more sensitive [142]. The very tiny shaped balance beam had a diameter of 1 mm in its middle region and it was reduced at the ends to 0.5 mm. According to the publication in the year 1932 it was possible to obtain correct results when analysing samples of a few micrograms in mass.

Fig. 4.99 Electromagnetic microbalance for the measurement of gas densities, of A. Stock and G. Ritter, 1926

4.10.3.13 The Electromagnetic and Buoyancy Balance of Stock and Ritter

In 1926 A. Stock and G. Ritter published a detailed report on the determination of gas densities using a microbalance which was supported by tip-shaped pivots and counterforced by a balloon shaped glass [97]. They measured the buoyancy as a function of gas pressure, restoring the balance in its zero position at each run. Later a permanent magnet was fused to the beam and an electromagnet was installed opposite to it. The system was calibrated by measuring the intensity of current as a function of changing pressure. With the use of the electromagnet it was possible to attain a much better accuracy in comparison to the initial procedure (Fig. 4.99).

4.10.3.14 Improved Microbalances with Electromagnetic Force Compensation

In the following period an intensification of the use of electromagnetic and electro-dynamic force compensation is to be observed. These methods proved to be very favourable, especially for the determination of extremely small quantities of substances which were difficult to be determined by purely mechanical weighing methods. The arrangements of the following investigators should be mentioned:

E. Wedekind made investigations with a mass-produced microbalance for magneto-chemical measurements in 1928 [143].

James W. McBain and H.G. Tanner built a very sensitive microbalance for weighing adsorbed layers in 1929 [144].

E. Lehrer and E. Kuss designed an improved gas density microbalance for weighing adsorbed layers in 1932 [145].

In 1951 Frank C. Edwards and Robert R. Baldwin arranged a microbalance in a closed system for analytical purposes like reactions of a solid with a gaseous compound (Fig. 4.100) [146].

4.10.4 Recording Balances with Electromagnetic Force Compensation

The use of recording balances was established early for the continuous observation of chemical processes which were accompanied by mass decrease. In these registration balances the driving force for the time axis was the spring of clockwork. As examples the recording balances of Paul Stückrath in 1881 and of the Paul Bunge Workshop 1910 should be mentioned [114].

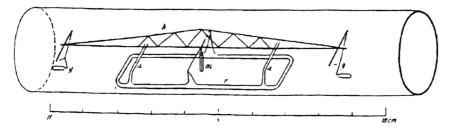

Fig. 4.100 Electromagnetic microbalance for analytical use in close systems, of Edwards and Baldwin, 1951

4.10.4.1 The Odén-Keen Balance

In 1916 a recording balance was described by Sven Odén, designed for the examination of deep-sea deposits. Sedimentation was measured automatically by means of a collecting disc suspended from the balance beam [147]. The beam deflection was compensated using a coil/magnet system at the other side of the beam (Fig. 4.101a).

A few years later Odén's balance was improved and supplemented for the automatic recording of chemical reactions as a function of time [148]. In the Odén-Keen balance the deflection was controlled electromagnetically. A cylindrical magnet attached underneath the one pan plunged into a fixed current coil (Fig. 4.101b). The electromagnetic field was adjusted varying the current by means of a graduated rheostat. For this purpose the sliding contact of the rheostat was coupled to two little adjusting motors which could move it upwards or downwards. Contacts were also installed at both sides of the pointer, which were connected to the two motors also. As a consequence of a small deflection of the pointer one of the contacts was closed and the corresponding motor activated. The position of the sliding contact was recorded automatically on a paper roll. When the sliding contact reached its end position caused by the weight decrease an electrical circuit was closed. Then a horizontally fixed disc was turned until a spherical weight was released which rolled through a channel onto the corresponding pan of the balance (Fig. 4.101c). By this means the loss of weight resulting from the chemical reaction was compensated. The arrangement of the pans had to be exchanged, when the chemical reaction was accompanied by an increase of weight.

In the Odén-Keen balance the principle of electromagnetic force compensation was realised for smaller loads. Likewise, electrical control components were used comprehensively and thus providing a basis for subsequent developments.

Later on electronic components were used for recording balances of any size as well as for highly sensitive microbalances which were meanwhile developed to a high state of art (Fig. 4.102) [21, 113–115]. Based on such balances and its electronic equipment at the beginning of the 1950s a development of the weighing techniques started, resulting to an unpredictably high level.

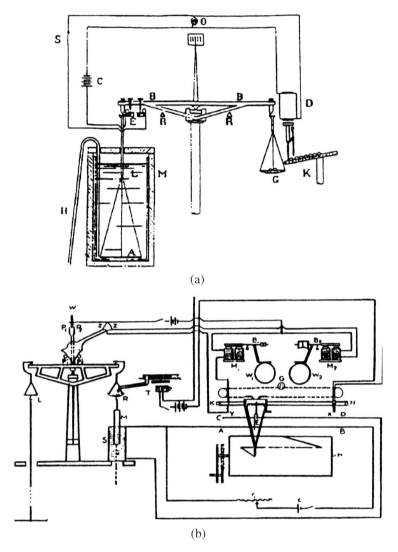

Fig. 4.101 (a) Electromagnetic registration balance of S. Odén, 1916. (b) Automatic electromagnetic registration balance of Odén and Keen, 1924. (c) Odén-Keen-Balance, Fig. 29, principle of the automatic weight feeding apparatus

4.10.5 Electromechanical Force Elements

Automatic compensation can be carried out mechanically by the deposition or displacement of masses, or winding a chain and twisting a taut band or torsion wire. Display or registration then follows electrically or mechanically from the positions of the control unit [111].

Fig. 4.101 (Continued)

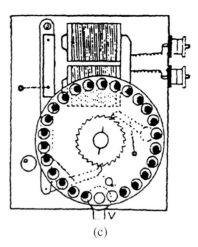

(c)

Fig. 4.102 Automatic electrodynamic registrating Sartorius microbalance of Th. Gast, 1944. *1* suspension wire for sample and counterweight, *2* quartz tube beam, *3* quartz ring with moving coil inside, *4* permanent sinter magnet, *5* stator coil, *6* torsion wire as beam suspension, current supply

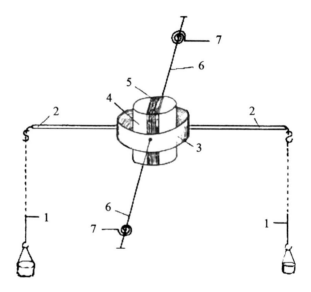

Proportionally acting electromechanical force elements work electrodynamically or electromagnetically. In the first of these, a fixed coil exerts forces or moments upon magnets attached to the balance beam [125]. They are bakeable and do not require movable supply lines. Their sensitivity to external magnetic fields is a disadvantage, making astatisation or screening necessary.

Electrodynamic force elements can be divided into rotating coil systems, plunger coil systems and swivelling coil systems (Figs. 4.103/4.104). Rotating coil systems dispense with large angles of deflection in favour of high flux density. The constructional connection to the beam is made in various ways. The coil is entirely in strained suspension. Thanks to the very limited deflection and the orthogonality of the coil field and the permanent field, the steady-state characteristic is largely linear.

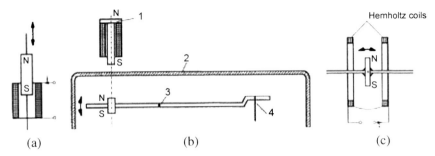

Fig. 4.103 Electromagnetic force compensation with bar magnet in the external field of an electromagnet. (**a**) Vertically arranged bar magnet. (**b**) Electromagnetic force compensation with bar magnet. *1* soft iron core, *2* casing wall, *3* axle, *4* end joint. (**c**) Bar magnet rotating in Helmholtz coils

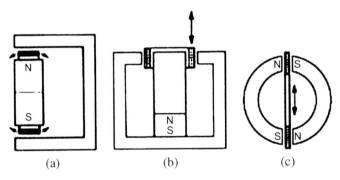

Fig. 4.104 Electromagnetic force compensation with coil. (**a**) Rotating coil. (**b**) Axially rotating coil (plunger coil). (**c**) Swinging (swivelling coil)

If l is the length of the coil in the magnetic field, d its mean diameter, N the number of windings, and B the flux density in the gap, the compensation current I gives rise to a torque:

$$M = I N B l d \tag{4.53}$$

Strong external magnetic fields generate deviations in sensitivity. In the plunger coil system, the cylindrical coil is in a concentric magnetic system with a core magnet and soft magnetic casing. If N is the number of windings in the field, B the flux density and d the mean coil diameter, for the force F at current I holds

$$F = I N B \pi d \tag{4.54}$$

The eccentricity of the coil in the gap results in a small second-order error which is usually negligible. A noticeable non-linearity is due to the reaction of the current-carrying coil to the flux density in the gap. In the first approximation the change in flux density is proportional to the current, the proportionality factor being K_b:

$$\Delta B = K_b I \tag{4.55}$$

Combination with preceding equations leads to:

$$F = B\left(1 + \frac{K_b}{B}I^2\right)\pi d \tag{4.56}$$

An auxiliary winding lying in series with the plunger coil around the magnet neutralises the disturbing flux, thereby linearising the steady-state characteristic.

In the swivelling coil system, the winding sections of a flat oil that are in the gap between two U-magnets generate forces in the same direction perpendicular to the magnetic flux and current. In total:

$$F = 2INBl \tag{4.57}$$

is valid, where N is the number of windings and l the length of the gap. If the swivelling coil is attached to one end of the balance beam, the effective point of impact of the compensating force is unaffected by the casing. A displacement of the beam's axis of rotation, e.g. an offset of the knife-edges during de-taring causes arm-length errors which can be cancelled by a symmetrical arrangement of two swivelling coil systems. The development of magnetic materials with a high specific energy product has made a considerable contribution to the improvement of force elements [149].

4.10.6 The Servo-Loop of a Self-Compensating Microbalance

In Figs. 4.105, the balance beam, sensor, amplifier and force element all form a closed action circuit, the beam representing the controlled member. The input quantity is the difference between the weight force, to be regarded as a disturbance variable, and the compensating force, which can be considered a regulated quantity produced by the force element. The regulating current, which is proportional to the compensating force by a factor as constant as possible, is displayed. The resonance frequency in the open loop results from the moment of inertia of the beam, including the pans, object and counter weight, from the directing force of the taut bands and from the distance between the centre of gravity and the pivot. It therefore depends upon the load on the balance. However, in the loop, the beam exhibits a much higher resonance frequency because a far greater electric directing force K_{el} is added to the mechanical directing force K_T. In the differential equation an additional term K_{el} appears and the period of oscillation is now:

$$\tau_\varphi = 2\pi \sqrt{\frac{J + \frac{m_l l_B^2}{2}}{2m_l h_0 + m_b h_b g + K_T + K_{el}}} \tag{4.58}$$

If the signal for a deflection φ is delayed on its way through the sensor, amplifier and force element, e.g. in the first order by the time constant τ_k, for the dragging moment the following is valid:

$$M_{el} + \tau_k \dot{M}_{el} = K_{el}\varphi \tag{4.59}$$

Fig. 4.105 Block diagram of
an electronic microbalance.
1 balance beam, position,
3 sensor, *4* control unit,
5 force element

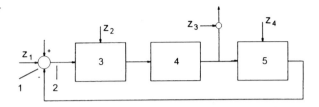

Fig. 4.106 Damping circuit
of a self-compensating beam
balance

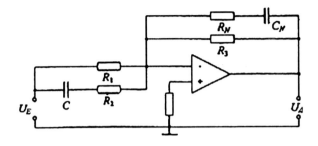

and in approximation:

$$M_{el} = K_{el}(\varphi - \tau_k \dot{\varphi}) \tag{4.60}$$

A negative damping term now appears in the differential equation, i.e. the system is
oscillatorily unstable. To remedy this, a positive damping term at least as great must
be introduced into the differential equation. The air cushioning effective at normal
pressure is cancelled, and eddy-current damping is not without risk due to possible
ferromagnetic impurities of the damper vane. The sensor signal is therefore conducted via a derivative element which adds the first derivation with a factor $\tau^* > \tau k$
to the antiderivative $\varphi(t)$.

According to Fig. 4.106, a RC element $R_1 C_1$ is connected in front of the invert-
ing opening of an operational amplifier [150]. The resistor R_3 lies in the reverse
feedback path. According to Kirchoff's first law, for $R_2 \ll 3$:

$$\frac{1}{R_1}(U_E + \omega R_1 C_1 \dot{U}_E) + \frac{U_a}{R_3} = 0 \tag{4.61}$$

$$U_a = -\frac{R_3}{R_1}(U_E + \omega \tau^* \dot{U}_E) \tag{4.62}$$

is valid. The amplification above a cut-off frequency is kept constant by the additional resistor, otherwise there would be strong noise. The correcting variable is
limited due to energy. Consequently, the required aperiodic limiting case of the
damping can be reached only within a certain range of deflection, and this decreases
with increasing loop amplification. The conversion constant of the chain beam—
sensor—amplifier is designated as K_s:

$$K_s = \frac{I}{\Delta mgl_b - M} \tag{4.63}$$

Fig. 4.107 PID control with operational amplifier of a self-compensating beam balance

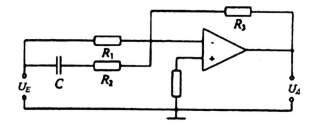

where K_m is the constant of the force element. For the output current we have:

$$I = K_s(\Delta mgl - K_m I) = \Delta mgl \frac{K_s}{1 + K_s K_m} \approx \frac{\Delta mgl}{K_m} \qquad (4.64)$$

i.e. the conversion constant with its drifts resulting from time and temperature loses with $K_5 K_m \gg$ its influence on the indicating device. On the other hand, high amplification endangers the stability and in disturbances may lead to multiple ballistic effects. As a remedial measure, the amplification factor in the proportioning band is set in such a way that in an appropriate deflection range the correcting variable does not reach its limits. As shown in Fig. 4.107 the amplifier also receives a differentiating reverse feedback. Its output voltage controls a current source. The equation for the regulating current is now:

$$I = k_d \dot{\varphi} + k_p \varphi + k_p \int \varphi d\tau \qquad (4.65)$$

As a consequence of the integral, I changes until the beam is fully restored to the zero position. The resistor R_3 limits the increase of amplification with decreasing frequency which, if the resolution of the servo-component is poor, may lead to low-frequency instabilities.

For the integration a motor can also be used which directly controls the electrical correcting variable via a potentiometer or adjusts a mechanical compensation element, by twisting a taut band, for example [111], whose movement is directly recorded. Very fast compensation can be achieved with a state controller [151]. The control can be carried out by computer which also executes other tasks such as corrections and data processing.

4.10.7 Examples of Self-compensating Balances

Vieweg and Gast's balance has a transformerless force element consisting of a rotating coil mounted on a taut-band suspension in a receptacle which supports the balance beam, and a pair of Helmholtz coils fed with undulating direct current outside the receptacle (Fig. 4.108) [21].

When the field coil is excited by the undulating direct current the deflection of the beam induces an alternating voltage in the rotating coil, the amplitude of which is a measure of the angle and the phase of which depends on its polarity sign. After

Fig. 4.108 Diagram of a
balance of Vieweg and Gast
with Helmholtz coils

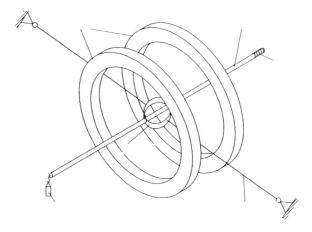

Fig. 4.109 Measurement of
permeation through a plastic
diaphragm (balance turned at
an angle of 90°). *1* counter
balance, *2* rotating coil,
3 field coil, *4* P205, *5* taut
band, *6* zero point setting,
7 seals, *8* plastic diaphragm,
9 H₂O

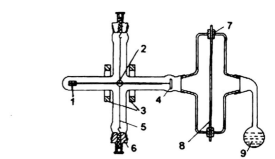

amplification, phase-dependent rectification, and smoothing, there is a direct voltage
which, after re-amplification is conducted back into the rotating coil as compensation, flowing through the indicating device on its way. Adapting the rotating coil to
resonance increases the sensitivity of angle detection and eliminates the electrodynamic recovery capacity in the HF field. The homogeneous stator field neutralises
translation movements of the beam that are due to vibrations or changes in load.
Because of the high carrier frequency, the disadvantage of the compensation current is small, enabling a high loop amplification to be applied. Figure 4.109 shows
the basic configuration selected for permeation measurements, Fig. 4.105 the block
diagram of a circuit build with electron tubes, and Fig. 4.110 the photograph of a
permeation diagram. From this, the permeation constant can be calculated:

$$K_P = \frac{d\,\Delta m}{Ap\Delta\tau}\ \left[\mathrm{g\,cm^{-1}\,h^{-1}\,Pa^{-1}}\right] \tag{4.66}$$

For measurements of the loss factor according to Lertes, the weighing system was
also used with a vertical axis of rotation [152].

   The further development of this balance and its various applications has been
the subject of a number of works. Sartorius & Co. adopted the principle and with
the co-operation of T. Gast developed the construction and technical details of the
circuitry to the type manufactured recently (Figs. 4.111/4.112). In the scope of the

Fig. 4.110 Photograph of a permeation diagram, Trolitul/H_2O, 25°C, sample area $A = 20$ cm^2, sample density $d = 1.15$ mm

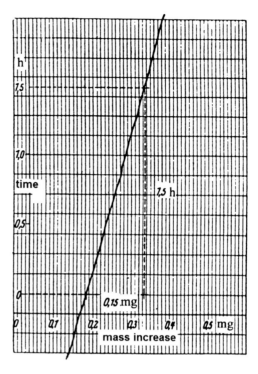

Fig. 4.111 Gast vacuum microbalance as produced by Sartorius, Göttingen. Maximum load 2 g, resolution 0.1 μg

development of the dust balance undertaken by E. Gast [111], miniaturisation, automisation and adaptation to severe operating conditions was carried out. Instead of Helmholtz coils, a transversely magnetised cylindrical ceramic internal magnet was used which carried the winding for the generation of the high frequency field for the angle detection and which was separated from the vacuum by a glass envelope. Translation movements of the beam still had no influence on the display, which also applies to the later model with an ALNICO internal magnet and frame-shaped HF field coils acting on the face of the rotating coil. Constructional and technical improvements made it possible to achieve a useful resolution of 0.1 μg in the finest measurement range of 2 mg at a maximum measurement range of 200 mg. In order

Fig. 4.112 Arrangement of
the Gast Vacuum
microbalance as produced by
Sartorius, Göttingen. Two
quartz tubes are arranged in
form of a ladder. The beam is
operated to zero position by
two electromagnets.
Maximum load 50 g,
resolution 0.1 µg

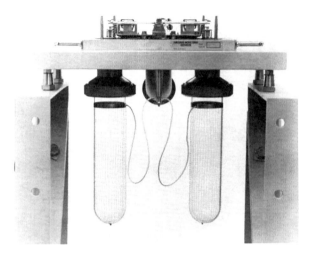

to increase the carrying force, after preparatory work by Gast, Sartorius brought out
a microbalance with a ladder-shaped beam. The capping pieces of quartz tubing are
connected by metal staves. The beam is mounted in a taut-band suspension with approximately 10 N tensile force and supports symmetrically arranged plunger coils
which engage with the permanent magnet systems, and at the same time serve as
position sensors in the high frequency field of stator coils. The pans are held by a
pair of suspension bands on each side. With a carrying force of 25 g, the sensitivity
is 1 Skt/µg. The largest measuring range is 2 g, and the finest 20 mg.

4.11 A Balance with Light Pressure Compensation

Zinnow and Dybwad [29] describe a nanogram balance with which samples of a few
hundred milligram in mass can be weighed. It is a symmetric balance with counterweight. A sensitivity adjustment is made directly via the tensile force of the taut
bands. Weight changes are compensated by radiation pressure. Figure 4.113 shows
a diagram of this balance. The beam is a glass tube 250 mg in weight and 18 cm
long with 10 µm thick tungsten suspension wires joined by casting. It is connected
to the tungsten wires by a W-shaped spring, the middle of which is raised or dropped
with the beam when the elastic force changes, thus changing the distance between
the centre of gravity and the pivot. A pencil of rays from a halogen lamp falls onto
a mirror attached to a torsion balance of an intensity-measuring instrument, and
is then thrown onto another mirror fixed to the left-hand end of the balance beam,
which then experiences a vertical, upwards force with the reflection. The rays' angle
of incidence is the same on both mirrors. The compensation force from the torsion
of the suspension wires of the torsion balance and its arm-length is therefore cal-
culable. The balance beam maintains constant oscillation amplitude by means of
light flashes. At crossover, and the difference in the time intervals of successive

Fig. 4.113 Zinnow and
Dybwad torsion balance with
beam recovery by light ray.
1 force-compensating light
ray, *2* and *3* light ray for
detecting beam deflection

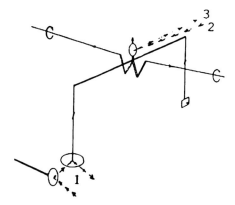

crossovers is used as a measure of deviation from equilibrium. This difference controls the intensity of the halogen lamp via a servo-circuit. Thus the servo-loop which
maintains the balance's equilibrium is closed.

4.12 Momentum Balances

According to the Theory of Relativity gravitational mass is indistinguishable from
inertial mass. In momentum or impulse balances the inert mass is measured almost
independently from the gravitational field. Instead of gravity an acceleration field is
established between the weighing article and the mass sensor. Either the weighing
article is accelerated with respect to a sensor in rest or the sensor is moved on the
surface of the weighing article. In industry usually the momentum weighing method
is applied for continuous mass determination in flowing media, e.g. in a production
line. In general more environmental influences have to be taken into account in
determination of the inert mass than of the gravitational mass. In addition the results
have a larger uncertainty for following reasons. It is difficult to determine the exact
orbit of the moved conglomerate and the position of its centre of gravity. Whereas
the reference mass in gravitational measurements is the Earth, in the measurement of
the inert mass it is only the mass of the sensor. The smallest uncertainties achieved
with such methods are about 1 %. Comparison with mass standards is impossible;
so the methods are unsuitable for absolute mass determination.

   In contrast the determination by mass spectrometry of the inert mass of atomic
particles is more accurate because the gravitational force of a particle is very small.

   Mass determination of astronomical objects is made by observation of the relative
motion under the influence of a neighbouring object and application of Kepler's
Laws. Here the reference masses are large; however the distances are large as well.
Furthermore, the uncertainty of the value of the gravitational constant G is 1×10^{-4}.
That means that the accuracy of all formula including G is limited to the per mille
range. Many calculations in geological, meteorological, astronomical as well as of
space operations are burdened with a basic uncertainty of per mille.

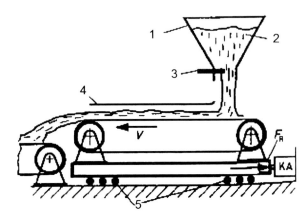

Fig. 4.114 Momentum belt weigher. *1* funnel, *2* bulk material, *3* sliding valve, *4* guard plate, *5* horizontal guiding devices. The bulk material is accelerated by the momentum of the belt to a velocity *v*. This results in a action of horizontal force F_R on the belt which is measured by the force transducer *KA*. © Kochsiek/Gläser

A special application of the measurement of the inert mass is weighing in an alternating acceleration field as produced by oscillating bodies. That method produces highly accurate results because oscillating acceleration fields much stronger than the gravitational field can be generated.

The following part of the chapter is concerned to the mechanical impulse weighers. The last three applications are described in special chapters.

4.12.1 The Momentum Belt Weigher

In the momentum belt weigher the material to be weighed is poured onto a moving belt and accelerated there from an initial velocity v_0 (mostly $= 0$) to the belt velocity v_1 (Fig. 4.114). The force F applied for that acceleration in the time interval $t_1 - t_0$ is dependent on the mass of the moved material

$$m(v_1 - v_0) = \int \mathrm{d}t \tag{4.67}$$

The integral represents a difference in momentum. A continuous stream of material accelerated in this way is given then by

$$\frac{\mathrm{d}m}{\mathrm{d}t} = \frac{F(t)}{v_1 - v_0} \tag{4.68}$$

4.12.2 The Deflector Plate

In deflector plate flow measurement the bulk material is poured continuously onto a plate and deflected there (Fig. 4.115). The force on the plate generated by the change of momentum is measured. In the calculation of the mass flux the angle between introduced flow and the deflector plate as well as of the angle between

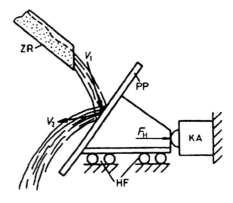

Fig. 4.115 Deflector plate
flow measurement.
PP deflector plate, *ZR* feed-in
channel, *KA* force transducer,
F_H force, *HF* horizontal
guide. v_1, v_2 mean velocity of
particles. © Kochsiek/Gläser

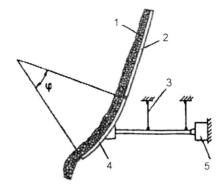

Fig. 4.116 Measuring chute.
1 bulk material, *2* guide
chute, *3* plate spring parallel
guide, *4* reversing chute,
5 load cell. The bulk material
is redirected at an angle ϕ
onto the chute. This results in
a horizontal force being
exerted on the load cell
proportional to the centrifugal
force, and this is therefore the
mass. © Kochsiek/Gläser

deflector plate and reflected flow is to be considered. Some part of the deflection is
inelastic and waste heat is produced besides the mechanical energy to be measured.
Thus for each material a factor must be determined for the evaluation of the mass to
be determined.

4.12.3 The Measuring Chute

On a measuring chute the bulk material is guided along a curved chute and the force
exerted is measured by a load cell (Fig. 4.116). This horizontal force is proportional
to the centrifugal force, and therefore to the flowing mass. Beside dependence on the
geometry of the device the results are dependent and need to be calibrated. Furthermore results are dependent on the bulk material which is to be taken into account by
a factor.

Fig. 4.117 Momentum liquid
flow measurement. The tube
section is uncoupled from
main tubing by expansion
bellows B. The force
$F = F_1 + F_2$ is measured by
the force transducer KA. V is
a thermal velocity detector.
PF is a parallel guide.
© Kochsiek/Gläser

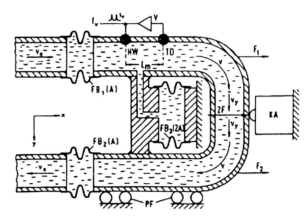

4.12.4 Momentum Liquid Flow Weigher

The mass flux of a liquid can be determined by measuring the force exerted on a
U-tube (Fig. 4.117). The tube section is uncoupled from main tubing by expansion
bellows. The force generated by the 180° redirection of the flow is measured by a
force transducer. The measuring device usually is supplemented by a thermal flow
meter, temperature and pressure sensors.

4.12.5 Coriolis Force Instruments

In a rotating reference system besides centrifugal forces, Coriolis forces also occur.
Velocities of a moved particle are influenced and forces become dependent on the
distance of the particle from the centre. In a rotating system, bulk materials or liquids
are guided in a radial motion, thereby producing a torque, a change in the angular
momentum. The mass conveyed can be calculated by an integration of the torque
over the time.

   Instead of a circular motion, Coriolis forces can also be generated by oscillations
on a bent path, e.g. by a vibrating tube (Fig. 4.118). By redirection a flowing liquid
twists the tube by the Coriolis force and the tube starts to oscillate. Coriolis force
measurements may also be applied in belt weigher systems.

4.13 Oscillator Balances

All balances are oscillating systems. The oscillations may be used to determine the
equilibrium position as described in a preceding chapter. In a subsequent chapter a
method is described to stabilise the location of suspended particles to be weighed.
In the present chapter we describe methods to measure the mass or mass variations
directly by observing the frequency shift of oscillations.

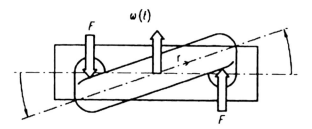

Fig. 4.118 Vibrating tube. The tube is lying perpendicular to the focal plane. A liquid whose density or mass is to be determined flows into the left arm of the tube, is led into the curve and flows out of the right arm. The tube oscillates with a frequency ωt in the direction indicated by the arrow. The tube is twisted by the Coriolis force F in the way shown. © Kochsiek/Gläser

Mass and force determination based on mechanical oscillators is a fairly new measuring technique though the influence of tension on a string was known since string instruments existed [153, 154]. Experimental and theoretical studies of elastic waves were already established 200 years ago [155]. However, in the 19th century mechanical beam balances achieved the extraordinary relative sensitivity (sensitivity/maximum load) of 10^{-9}. So there was no incentive to be concerned with other principles for mass determination. Indeed, with oscillator balances a relative sensitivity of only about 10^{-6} can be attained, however today extremely high resolutions down to the zeptogram region could be realised.

An oscillatory mass sensor creates its own accelerating field which can be by far stronger than the terrestrial gravitational field [153, 156–158]. In a crystal balance the sample is excited to oscillations of high frequency together with the oscillator. Using high frequencies, the method is restricted to measuring mass changes of samples which are firmly connected to the sensor. This may be deposited films or adsorbed layers. Direct results are obtained in vacuum or in a gaseous atmosphere using Sauerbrey's equation (see below), but also results of measurements in liquid medium can be evaluated [159]. However, in a sample of loose powder it is not the mass but the impedance of the mechanical wave that is measured.

Mechanical oscillators provide lower frequencies and the connection of the sample with the oscillator's surface is less critical. Such devices can be used e.g. for dust measurements. Furthermore, oscillating strings and similar items can be applied like a conventional sensor and connected via levers to the sample receptacle. Piezoelectric, electrodynamic, electrostatic and magnetostrictive transformers can be used to excite the oscillation, and displacement, velocity and acceleration sensors used for detection. Sensors with binary output signals are also suitable. To generate self-starting autonomous oscillations, the sensor is coupled back to the exciter, a phase-locked reaction coupling being frequently used.

An elastic body can be stimulated to oscillations of different shape and mode as indicated in Figs. 4.119. In general mass and force sensors are operated with continuous harmonic oscillations, though their modelling often may be simplified, also discrete spring-mass-oscillators are applied. The sensitivity of the mass determination is restricted by the interval during which the resonance frequency is subject to

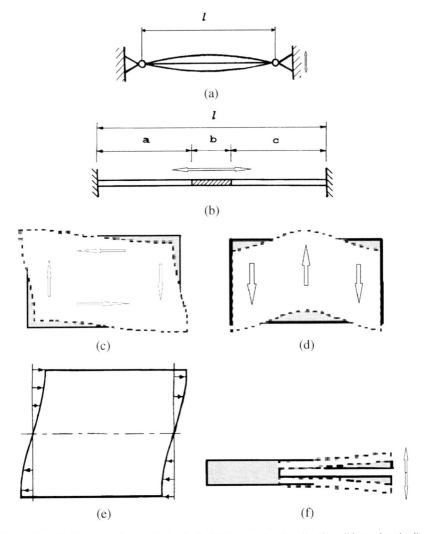

Fig. 4.119 (a) Transversally oscillating belt. (b) Longitudinally vibrating ribbon—longitudinal mode. (c) Flexural resonator—flexural shear mode. (d) Flexural resonator—flexural shear mode. (e) Thickness shear vibration of a quartz crystal (SC-cut). (f) Tuning fork

stochastic fluctuations. Mechanical oscillators react sensitively to ambient parasitic indications such as temperature, pressure, humidity and interfering fields. Systematic deviations related to these are either compensated or corrected. Ageing can also be a source of error.

Summarising:

- Oscillatory mass sensors create their own accelerating field which can be million times stronger than the terrestrial gravitational field; thus its sensitivity is remarkably high.

- Being independent of the Earth's gravity, the location of the balance can be freely chosen and is variable. Also mass measurements of moving and vibrating samples are possible.
- The sample must be connected strongly enough to the surface of the sensor so that it can take over its oscillations.

4.13.1 Crystal Oscillators

With the single crystal balance mass changes can be determined of a sample which is deposited firmly at the surface of the crystal. The effect of added mass on crystal frequency has been known since the early days of radio when frequency adjustment was accomplished by a pencil mark on the controlling quartz crystal. In 1957 Sauerbrey [160, 161] was the first to investigate theoretically and experimentally the quartz crystal's suitability for mass determination and invented the quartz crystal microbalance (QCM). Probably independently, Warner and Stockbridge developed a QCM based on long years of research on application of quartz crystals [162]. At the same time Wade and Slutsky [163] reported on adsorption measurements of water vapour and hexane at a vibrating quartz crystal [163]. A period of lively activity and enthusiastic research and development was initiated, in the course of which the QCM underwent considerable improvements and became widely used. Already by 1971 King Jr. had been asked to present a survey [164] and in 1984 the applications were reviewed in a book [165].

Recently quartz-homeotypic gallium orthophosphate $GaPO_4$ has been recommended as crystal material [166]. It is suitable for operation up to 970 °C. Plano-convex resonators of Y-cuts with 10 mm diameter have been used as sensor for the crystal microbalance (CM) whereby high temperature stability was achieved by means of compensation.

The local mass sensitivity of the CM depends on the local intensity of the inertial field developed on the crystal surface during crystal vibration. The field intensity is measured by the acceleration developed at a certain point. The maximum intensity of the inertial field, in the centre of the quartz resonator, is a million times higher than the intensity of the gravitational field on the Earth [157]. Experimental results reveal that the product between the minimum detectable mass and the intensity of the field acting on that mass is a constant for both CM and beam balances, explaining thus, why CM is more sensitive than conventional analytical microbalances. The apparent effect of a liquid's viscosity on the frequency response of a quartz crystal resonator in contact with the liquid is, in fact, the result of the field intensity dependency of the mass sensitivity; thus this makes it clear that CM is really a mass sensor.

4.13.1.1 The Principle

Determination of mass by means of mechanical oscillators is based on inertia. The coupling of an additional mass changes the resonance frequency, and this can be registered quickly and with high precision. The connection between frequency and mass may be calculated by means of an analytical equation which expresses the basic physical law. As a rule, it is preferable to calibrate the system, whereby due to the relatively small changes, an empirical linear interrelationship between the changes in mass and frequency can be assumed. For a thickness shear quartz oscillator the interrelationship between the mass m_s deposited on the crystal with a mass mq and the change in frequency Δf can be expressed in simplified form by the equation:

$$\frac{m_s}{m_q} = \frac{-\Delta f}{f_q} \tag{4.69}$$

where f_q is the frequency of the unloaded crystal. This relationship given by Sauerbrey is applicable to mass loads that constitute layers that are thin in comparison with the thickness of the crystal. For thick layers, different geometries of the oscillating body, other materials, modified equations have been derived.

The shape and mode of motion of the crystal are of crucial importance for its function. A mode of function extremely well suited to mass determination is the thickness shear mode of a thin lamina, the upper and lower surfaces of which have antiparallel movement and remains undistorted (Fig. 4.11e). In the case of crystal oscillator certain orientations of the surface to the crystal's axis (AT cuts, BT cuts) result in these modes of motion and enable the temperature coefficients of the resonance frequency to be optimised.

A single CM employs a sensor that consists of a thin crystalline disc sandwiched between two electrodes. Small oscillations in the sensor at resonance frequencies are created by applying an AC voltage across the electrodes. Piezoelectric crystals combine long-term stable behaviour with little damping, good constancy of the elastic properties as well as low disturbing sorption. A CM is distinguished by its simple design, small size and low cost. The measuring equipment is robust, resistant to aggressive surroundings, shock-proof and not sensitive to ground vibrations. Its output, the frequency, is easily and quickly registered and can be directly processed digitally. Because the balance creates its own acceleration field it is independent of gravity. For mass determination with a CM, the following fundamental conditions must be fulfilled [167]:

- The mass must consist either of a homogeneous film that evenly covers the oscillating region of the single crystal or of evenly distributed fine particles.
- The mass must completely adhere rigidly to the crystal surface.

The resonance behaviour of a piezoelectric single crystal is described by an electrical equivalent circuit diagram consisting of the parallel connection of an acceptor circuit with a capacitor (Fig. 4.120) [168]. This system has a parallel and a serial resonance. Only an operation with serial resonance corresponding to the mechanical resonance frequency can be used in weighing.

As a rule, a single crystal resonator produces only one output quantity, and that is the change in resonance frequency. If several measurement quantities are to be registered several crystals are to be used. Another method makes use of the different sensitivities of resonance frequencies of different modes in one crystal. This method

Fig. 4.120 Equivalent circuit of a single crystal resonator

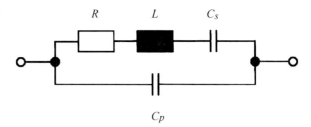

Fig. 4.121 Block diagram of a quartz crystal microbalance (QCM). *1* quartz sensor, *2* resonance thermometer, *3* reference quartz, *4* oscillator sensor, *5* oscillator reference, *6* mixer stage, *7* low-frequency output, *8* temperature

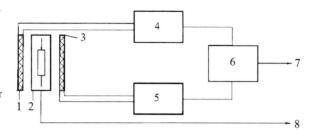

is applied, for instance, in simultaneous mass and temperature determinations. In order to eliminate temperature or other disturbing influences comparison with the signal of a reference crystal is recommended (Fig. 4.121).

4.13.1.2 Applications

The following examples give some idea of the variety of both sensor development and applications of the single crystal balance which mostly refer to a quartz crystal microbalance.

Oscillators are ideal from the point of view of vacuum technology because of their small surface and their bakeability, and they are not affected by buoyancy. For quite some time now the quartz crystal microbalance has served to control and regulate thin-film deposition processes, as in vacuum metalising, sputtering and ion implantation. Thickness and mass of deposited layers can be monitored *in situ*. The QCM is a reliable instrument for the automation of these processes, a condition for reproducible results and consistently high quality. Furthermore QCMs are used in the quantitative determination of surface contamination.

In gravity-free space, oscillation systems which generate their own acceleration field can be used for mass determination. Otherwise indirect methods must be applied. One of the typical tasks is the monitoring of the technical functions of objects, surface contamination and evaporation in space (Fig. 4.122), and, for comparison, in simulation chambers.

Another area of application is testing of new materials with regard to evaporating components. A device for determining outgassing rates has been described by Glassford [169]. It contains an effusion cell whose molecular mass flow escapes into the space around the quartz balance through an opening equipped with a sealing cap.

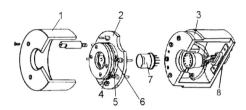

Fig. 4.122 Berkeley quartz crystal microbalance designed for contamination records in space. Diagram of Berkeley quartz crystal microbalance. *1* cover, *2* plug-in crystal/heater assay, *3* connector assay, *4* sensing crystal, *5* heater/temperature monitor, *6* hybrid circuit oscillator mixer, reference crystal, *7* connector

4.13.1.3 Chemical and Biological Sensing

Quartz crystal microbalance, also known as thickness shear mode or piezoelectric quartz crystals, and surface acoustic wave devices (SAW) are the main devices that have been employed as transducer elements in chemical and biological sensing [170, 171]. In chemical or biological sensing, a layer is added to the device surface that can recognise and bind the analyte. Binding transfers the analyte from the medium being analysed to the device surface where it alters some property of the acoustic wave. A wide range of selective layers including bioreceptors and polymer films can be employed for sensor applications [172]. The addition of the film alters the resonant frequency of the sensor.

In both QCM and SAW devices the sensitivity is dependent on the square of the resonant frequency. SAW devices operate at a higher frequency and therefore have higher predicted sensitivities although this is not necessarily realised. SAW devices have the advantage that they can be miniaturised with precise and reproducible characteristics using photo-lithographic techniques. In addition, lithographic fabrication capability permits a complex circuit to be present on the substrate surface. The major advantage of the QCM is that they have higher mechanical strength and therefore have higher stability. More recently there has been increasing interest in miniaturisation of QCMs using microfabrication techniques.

Acoustic wave devices have been most widely used for gas or vapour phase sensing. Owing to wide applicability there has been increasing interest in the use of these devices for the liquid phase. SAW devices have been little used in the liquid phase since liquid-phase operation is precluded in devices which have surface normal particle displacements.

4.13.1.4 Vapour Sensing

Vapour sensing requires a chemical layer to collect and concentrate vapour molecules from the gas phase to the device surface [173]. If the chemical layer

is rigid, then the frequency change of a piezoelectric quartz crystal is described by
the Sauerbrey relation [160].

$$\Delta f = -2.3 \cdot 10^6 f_o^2 \frac{\Delta M_s}{A} \qquad (4.70)$$

where Δf is the change in frequency of the quartz crystal in Hz, f_o is the fundamental frequency of the quartz crystal in MHz, ΔM_s is the mass of material deposited or sorbed onto the crystal in g and A is the area coated in cm^2.

The factors which influence sensitivity of the sensor include (i) strength with which the chemical layer sorbs the vapour (sensitivity increasing with stronger bonding) (ii) thickness of polymer film and (iii) dielectric effects. When the chemical layer is conducting then interaction of analyte and chemical layer will, in the case of SAW devices, lead to changes in propagation of the Rayleigh wave and the associated electric potential wave. An enhanced response will thus be observed due to changes in both mass and dielectric effects.

The selectivity and reversibility of acoustic wave devices is entirely dependent on the chemical coating. If the analyte and chemical interface have very high bonding strength then the device will be highly selective but has poor reversibility. Conversely if the bonding strength of the analyte material interface is very low then the selectivity will be very low but the reversibility will be very good. Selectivity and reversibility are thus mutually exclusive properties. Two major approaches have been proposed to overcome this problem. In the first approach, a sampling device, such as a denuder tube, is introduced prior to the detection system. The walls of the denuder tube act as a perfect sink for the analyte. The sample passes through the sampling device with the analyte becoming bound to the walls of the tube. Analysis is then performed by thermal desorption of the analyte in a reference stream which is directed to the detector. Selectivity is thus primarily carried out by the denuder tube and the detector is only required to be partially selective and therefore has good reversibility [174].

The second method of overcoming problems of selectivity and reversibility involves the use of an array of coated sensors with each sensor element only having partial selectivity for the analyte. The sensors are reversible since they are only partial selective. The specificity is obtained from the pattern of responses that act as a fingerprint for the analyte. Pattern recognition can be performed using a variety of standard statistical and soft computing based methods. In general, data from the sensor array can be analysed using supervised or unsupervised methods. Unsupervised learning methods such as principal component analysis (PCA) are used in exploratory data analysis; they attempt to identify a gas mixture without prior information. PCA is a commonly used multivariate technique that acts unsupervised, it finds an alternative set of axes (principal components) about which a data set may be represented. The axes are orthogonal to one another and are designed to provide the best possible view of variability in the independent variables of a multivariate data set. If the principal component scores are plotted they may reveal natural clustering in the data and the samples. PCA provides an insight into how effective the pattern recognition system will be at classifying the data. Plotting the load data enables the factors (original data columns) to be compared with one another, if two

Fig. 4.123 Quartz crystal microbalance sensor with hygroscopic covering used for moisture measurements.© Mitchell GmbH

factors show little separation then it is likely that the measurements are correlated and are not truly independent.

Supervised learning techniques classify a vapour sample by developing a mathematical model relating training data [173, 174]. A neural network is an example of a supervised method that is able to solve non-linear problems, it is dynamic and self-adapting. Neural networks are based on the cognitive processes of the human brain and are efficient in comparing unknown samples with a number of known references. A neural network is a collection of units that are connected in some pattern. A unit is a simple processor which has a rule for combining the inputs and an activation function that takes the combined input to calculate an output. A weight is specified by (i) unit the weights connect from (ii) unit the weight connects to and (iii) number that denotes weight value. A negative weight value will inhibit activity of the connected to unit and a positive value will strengthen connection to unit. The pattern of connectivity refers to which units connect, direction of connection and the connection weights. The task that a network is required to perform is coded in the connection weights i.e. the connection weights represent the memory of the network. Fuzzy set theory is capable of dealing with vague and uncertain data and has been used for pattern recognition [173]. Fuzzy sets differ from classical sets in that they allow for an object to be a partial member of a set. A fuzzy set is fully defined by its membership function.

4.13.1.5 Moisture Measurement

QCM technology for moisture measurement is based on monitoring the frequency modulation of a hygroscopic-coated quartz crystal with specific sensitivity to water vapour (Fig. 4.123). Bulk adsorption of water vapour onto the coated crystal causes an increase in effective mass, which reduces the resonant frequency of the crystal, in direct proportion to the water vapour pressure. This sorption process should be fully reversible with no long-term drift effect, giving a highly reliable and repeatable measurement. Moisture analysers have been developed to provide highly reliable, fast and accurate measurement of moisture content (ppb and ppm by volume) in a wide variety of industrial pure gases. Modern instruments incorporate automatic calibration using an internal wet reference source.

4.13.1.6 Liquid Sensing

For measurements in liquids the sensor can be pre-coated with a huge range of thin films from metals, polymers, ceramics to biomolecules, bacterial and living cells [172, 175]. The addition of the film alters the resonant frequency of the sensor. This frequency change is directly proportional to the film's mass, if the film is rigid. For soft or viscous films, the Sauerbrey relation is not valid since the outer region of the film does not follow the sensor's shear motion.

Kanazawa and Gordon have described the resonant frequency change in liquid [176] as:

$$\Delta f = -f_0^{3/2} \left(\frac{\rho_L \eta}{\pi \mu \rho_q} \right)^{1/2} \tag{4.71}$$

where ρ_L, ρ_q are respectively the density of liquid and quartz crystal, μ is the shear modulus and η is the viscosity of the liquid. The resonant resistance of the quartz crystal reflects its mechanical resistance and is given by:

$$R = \frac{(2\pi f \rho_L \eta)^{1/2} A}{k^2} \tag{4.72}$$

where k is the electro-mechanical coupling factor. In the case of liquid operation, the measurement of dissipation (lost energy per oscillation divided by the total energy stored in the system) is crucial. Simultaneously, the change of the resonance frequency and the dissipation factor for up to four different resonance frequencies e.g. basic frequency plus 3rd, 5th and 7th harmonic overtone are measured. Dissipation factor measurement of the oscillation allows accurate thickness estimations and viscoelastic properties of floppy films such as polymer multilayers, cell and bacteria adhered to functionalised surfaces, elongated proteins or polymers with no specific structure adsorbed to solid surfaces. It is possible to observe structural changes e.g. phase changes, creation of networks, absorption of water.

In contact with a liquid the resonance frequency is influenced by the fluid. For small sized immersed colloidal particles ($\varnothing \sim 10$ nm) a change of the resonance frequency with concentration is observed. Large particles ($\varnothing$ 400 or 1000 nm) have no influence on the resonance frequency which is equal to that of the pure liquid [177]. A double arrangement of quartz crystals is able to detect density and viscosity of fluid [178, 179].

Immunosensors based on QCM have been developed [180, 181]. An AT cut quartz coated with gold electrodes positioned in a special flow through system detects a mass accumulation caused by immobilisation of synthetic peptides. With this system antibodies against HIV have been detected. Advantage is in situ measurement avoiding complicated equipment and procedures.

Measurements in liquids were improved by design and experiment of horizontally polarised shear wave devices. The shear wave is only propagating along the surface and not coupled with compression wave [182]. Hence cross sensitivities were remarkably reduced. A wide variety of bioactive components, either affinity or catalytic, can be employed including enzyme/enzyme substrate, antigen/antibody (more selective than enzymes), DNA/complementary strand and DNA/RNA. Immobilisation of the bioactive component can be carried out by physical adsorption, covalent bonding and polymer entrapment. Physical adsorption is easy but because the bonding is weak the bioactive component can be easily washed away. In the case of covalent bonding the bioactive component is more strongly held and the direction of the bioactive component can be more easily controlled. Polymer entrapment is a simple method of immobilisation but thick films will adversely affect the kinetics of mass transfer of analytes.

4.13.2 Mechanical Oscillators

Various types of mechanical oscillators have been developed with regard to the special task of mass monitoring. The type of devices varies in size though mostly micromechanical sensors are used in practice. The frequency range of operation is decisive for its application.

4.13.2.1 The Vibrating String

To increase the frequency of a note produced by stimulation of a string in a musical instrument we increase the string tension. The increase in tension can be carried out by means of a weight suspended at the end of the string and thus a similar device can be used for mass determination. Load cells equipped with vibrating string sensors are applied in commercial scales as well as in belt weighers.

Independence of the gravitational field allows for vibration immune weighing e.g. of material on transportation belt. Achieving high gravimetric feeding performance on the process line requires discriminating between weight data and the contaminating effects of inertial forces induced from ambient vibration and shock. Dynamic digital filtering algorithm continuously identifies and extracts spurious inertial components from the weight measurement, even in severe process environments.

Likewise, by increase of the mass suspended on a rope or chain the frequency of a standing wave increases. Because for loads of several kilograms these frequencies are low it can easily be observed and a sensitivity of about 1 percent may be attained. This method is applied, e.g. when hoisting sunken ships.

4.13.2.2 The Vibrating Ribbon

Considerations that an oscillating substrate of lower quality than quartz should be suitable for many applications, if the ratio of precipitated to substrate mass is large enough, leading to development of longitudinally and transversally vibrating ribbon, as it provides measurement and documentation feasibility, e.g. of the mass rate versus time. Pertinent experiences comprise experiments in vacuo [183], dust

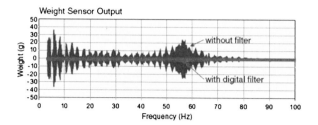

Fig. 4.124 Effect of vibration across the frequency spectrum found in a plant environment. Digital filtering of the output signal of a vibrating string load cell results in a constant mass indication of the sample. False weight readings caused by plant vibrations are eliminated. © K-Tron International, Pitman, NJ, USA

measurements (Fig. 4.124) [184, 185] and the measurement of solid particles in water [8] with the transversally oscillating ribbon and dust concentration, viscosity and elasticity measurements with the longitudinally vibrating ribbon [186–188]. The strong dependence of the eigenfrequency on temperature and mechanical exertions required among other things a sophisticated thermostat. Therefore considerations to improve the measurement method were made by modelling the object patch more strictly and getting the results by calculating the quotient of sensitive to non-sensitive eigenvalues. Further quantities may be detected by measuring more than two eigenfrequencies [189].

Bahner and Gast reported on measurement of solid particles in water by means of a transversally oscillating ribbon [187]. After loading in water the ribbon was dried and then the mass change measured. Dust concentration measurements in liquids may be performed *in situ* by means of a longitudinally vibrating ribbon [188, 190]. In order to measure viscosity of liquids and elasticity of layers the longitudinally vibrating ribbon should be placed between walls in short distance.

4.13.2.3 The Vibrating Body

Little tuning fork shaped bodies stimulated to oscillations were applied as sensors for laboratory balances and others (Fig. 4.119f).

As cheap resonators thin-walled hollow metal bodies of different shape oscillating at low frequencies are used to measure dust concentrations. For routine particle deposition a tapered element oscillating microbalance (TEOM) was developed. TEOM consists of a thin-walled tapered element, its broad end fixed to a solid base (Fig. 4.125b). The free end executes autonomous oscillations, the frequency of which depends on the mass and the stiffness of the element and on the additional mass connected to the free end. The oscillations are excited electrostatically, and detection is optoelectronic. The instrument is calibrated by connecting it to a reference mass. From the differences between the oscillation frequencies with and without the reference mass, a constant is determined giving a factor in Sauerbrey's equation (56.1) [191]. The measur-

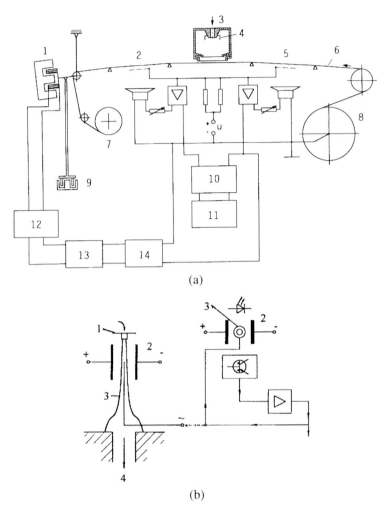

(a)

(b)

Fig. 4.125 (a) Differential system for dust measurements by means of a vibrating band. *1* moving coil for adjustment of the band tension, *2* measurement with dust, *3* contaminated gas inlet, *4* high voltage charge electrode, *5* measurement without dust, *6* elastomeric band, *7* take-up roller, *8* band store, *9* magnetic sensor for adjustment of the winding, *10* mixing stage, *11* frequency meter, *12* dc amplifier, *13* RC element, *14* discriminator. (**b**) Tapered element oscillating microbalance. *1* filter, *2* electrodes, *3* conical shaped tube, *4* to pump. The contaminated gas is pumped through a filter at top of the conical shaped tube and the mass of dust deposited is determined by shift of frequency of the tube in the field of two electrodes

ing range is between ten to several hundred mg. Areas of application cover the continuous measurement of particle emissions of diesel engines [192] and monitoring of mass changes by deposited dust particles in atmosphere control.

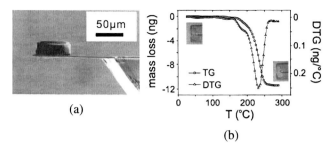

(a)   (b)

Fig. 4.126 (a) Scanning electron micrograph of a zeolite single crystal on a micromechanical cantilever. © Rüdiger Berger. (b) Heating of the sample (TG) resulted in a mass loss of 12 ng. Furthermore the differential mass loss is (DTG) is plotted. © Rüdiger Berger

4.13.2.4 Nano-oscillators

Scanning probe microscopy (SPM) typically covers atomic force microscopy (AFM), scanning tunneling microscopy (STM), near-field scanning optical microscopy (NSOM, or SNOM) and related technologies. Since its invention in early 1980s, SPM is now regarded as one of the major driving forces for the rapid development of nanoscience and nanotechnology, and the tool of choice in many areas of research [193]. SPM devices have been successfully used in gravimetric thermal analysis (thermogravimetry TG, differential thermogravimetry DTG) (Fig. 4.126) [194]. For the temperature depending frequency variation in micromechanical TG the following equation holds [195]

$$\frac{1}{f}\frac{df}{dT} = \frac{1}{2K}\frac{dK}{dT} - \frac{1}{2m}\frac{dm}{dT} \tag{4.73}$$

where f is the frequency, T the temperature, m the sample mass, and K an instrumental constant. Tracking the resonance frequency f as a function of T permits determination of changes of the total mass.

Nanotube-based physical sensors can measure pressure, flow rate, temperature, or the mass of an attached particle. Moreover, carbon nanotubes are strong fibres, which can undergo severe deformations without showing any damage in their atomic structure. Nanoscale mechanical devices in combination with oscillating elements can be used for mass determination down to the mass of an atom. By use of carbon nanotubes, of diameter up to 75 nm and length of a few micrometers, which were resonantly excited in the flexural mode by an alternating voltage of a few MHz, the mass of attached spheroidal carbon particles has been determined to 22 ± 4 fg [196] (Fig. 3.44). Observing the frequency shift of oscillating carbon nanotubes or of silica nanorods, masses or mass changes in the attogram and zeptogram range have been observed recently [197–199]. A nanomechanical resonator has been used to measure quantum effects of thermal Brownian motion [200, 201]. Today it is possible to weigh clusters of a few atoms. Aligned ZnO nanowires and nanobelts can be used for converting mechanical energy into electrical signals and vice versa.

Fig. 4.127 Single graphite
sheet, planar honeycomb
lattice structure of SP^2
bonded carbon atoms.
© Mazdak Taghioskoui

Fig. 4.128 Graphene foil
resonator. © Arend van der
Zande

Such devices can be used for force measurement in microscopic regions [202]. Even
the adsorption of few molecules could be observed [199, 203].

Recently electrical-readout nanomechanical resonators made from graphene
have been described [204, 205] (Fig. 4.127/4.128). The devices, which consist of
vibrating sheets of graphene suspended over micron-sized trenches, could be used
as highly sensitive, robust, mass detectors. Graphene sheets are sheets of carbon
that are just one atom thick. As well as having remarkable electronic properties,
graphene is extremely stiff and strong. This means that the material can be made
into bridge-like resonators that vibrate at very high frequencies. Because such a resonator has an extremely small mass, its resonant frequency changes each time a
molecule is adsorbed onto its surface. Although graphene shares these advantages
with carbon nanotubes, which have also been used to make highly sensitive mass
detectors, it has the added bonus of being a 2D sheet that can be carved into any
shape. This gives a better control over the properties of the finished resonator.

Graphene based resonators that can be optically or electrically actuated have been
fabricated [206] by exfoliating single or multilayer sheets of graphene over prepared
trenches of SiO_2 substrates. The suspended graphene sheets can be described as
micron-scales cantilevers clamped to the surface by van der Waals forces. These
cantilevers are set into vibrating motions through the application of a time-varying

Fig. 4.129 Graphene tube
resonator. © Adrian Bachtold

radio frequency voltage or by focusing a diode laser onto the surface. The resulting
MHz-range vibrations have a peak around 65 MHz that depends on the geometry
of the device. The frequency can also be adjusted with a DC voltage applied to the
gate, which introduces tension to the sheet. When an object is placed on the device,
the frequency changes—and the change is detected with the electrodes, and used
to calculate the mass of the molecule. The vibrations are observed by optical interferometry. Unlike Si, graphene shows no degradation in Young's modulus with
decreasing thickness. The bonds holding the carbon atoms together in graphene are
similar to the bonds that hold together the carbon atoms in diamond. What sets
grapheme apart from diamond, however, is that it stays strong and stiff even as a
single layer of atoms. Mass sensing using a cantilever system relies on detecting
shifts in the frequency of vibrations in response to mass. By using a single layer
of graphene atoms as a cantilever, the ultimate limit of two-dimensional nanoelectromechanical systems, a sensor could detect ultrasmall masses. Graphene is also
robust enough to resist long-term wear, improving the lifetime and cost-efficiency
of such balances.

Recently a mass sensor was made capable of weighing a single proton—which
has a mass of 1.7 yoctograms (1.7×10^{-24} g) [207–209]. It consists of a single
suspended carbon nanotube (Fig. 4.129) 150 nm in length and 2 nm in diameter that
resonates at the respective frequency in GHz range. When a tiny particle sticks to
the tube, this resonant frequency drops and the shift in resonant frequency can be
used to calculate the mass of the particle. Weighings are started by passing a large
electric current through the tube to clean its surface. That nanobalance was used
first to detect single naphthalene molecules and small numbers of xenon atoms in
ultrahigh vacuum and at temperatures around 4 K. It is likely that chemical reactions
can be followed by this technique.

4.13.3 Rotational Pendulum

The rotational pendulum consists of a body horizontally suspended by a thin fibre
above its centre of gravity. The fibre acts as a weak torsion spring. If a force is
applied at right angles to that body, it will rotate, twisting the fibre, until it reaches
equilibrium between the force applied and the restoring elastic force and starts to
swing harmonically. Alternatively such a body, sometimes in form of a wheel, is
rotatable supported on a pin bearing and equipped with a spiral spring.

Fig. 4.130 Diagram of a rotary pendulum for adsorption measurements on a powder sample. *1* torsion wire, *2* stabiliser, *3* disc, *4* grove with sample, *5* mirror to reflect a laser beam

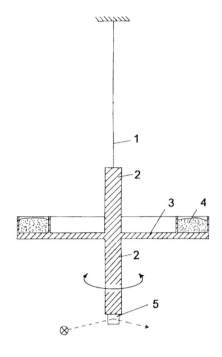

4.13.3.1 Keller's Adsorption Pendulum

A very troublesome effect in traditional weighing under gravity with changing gas pressure, for example, in adsorption measurements, is buoyancy. In particular the density of the sample is often unknown, or it changes during the measurement as a result of temperature effects or reactions with the environment. Buoyancy can be avoided if the mass is determined by means of an inertia measurement. Keller used the slow oscillations of a horizontally arranged rotary pendulum (Fig. 4.130) to measure adsorption of gases at dispersed solids [210]. The pendulum is equipped with a ring-shaped groove for the sample.

4.13.3.2 The Watch Balance

A mechanical watch is mostly equipped with two springs: a spiral band spring as energy support and a clock generator based on a spring (echapment). The clock generator is an oscillating element, usually in the form of a wheel, which is rotatable, supported and equipped with a spiral spring (Fig. 4.131). In old wooden clocks instead of a spring a twisted cord is used (Fig. 2.17a).

This part of a mechanical watch is called balance, pendulum wheel or wheel fly. It is a harmonic oscillator whose resonance frequency sets the rate of the watch. The resonance frequency is regulated, first coarsely by adjusting the momentum with weight screws set radially into the rim of the wheel, and then more finely by adjusting with a regulating lever that changes the length of the balance spring. This pendu-

Fig. 4.131 Echapment of the clock including an oscillating element, usually in form of a wheel, which is rotatable supported and equipped with a spiral spring (wheel fly). *1* escape wheel, *2* wheel fly, *3* spring. © Wikipedia

lum wheel was described first by Christiaan Huygens (1629–1695) [211]. However Robert Hooke (1635–1703) claimed that he invented it 17 years before [66].

4.13.3.3 The Horizontal Pendulum

In 1830 Lorenz Hengler invented the horizontal pendulum and in 1869 it was re-invented by M.F. Zöllner [212–214] which has since been associated with his name (Fig. 4.132). It consisted of a beam with two weights, suspended on a thread and drawn by a second thread so that the beam was a little bit inclined. A mirror on the pendulum was used to reflect a light beam to a scale, where the motion of the pendulum, as magnified by the optical lever, was directly observed. The instrument was installed in the cellar of the university in Leipzig; Zöllner could detect significant movement of the pendulum due to the filling up of the auditorium on the second floor of the building.

The pendulum was built in order to observe changes in the direction of gravity due to tidal forces, Earth's rotation, influences of Sun and Moon. Already Zöllner suggested that it might also be valuable as a seismometer. For geophysical investigations the instrument was improved by E. von Rebeur-Paschwitz [215] and R. Ehlert [216]. There were several different suspensions used in these later

Fig. 4.132 Zöllner's
horizontal pendulum. *1* stand,
2 pendulum bar, *3* suspension,
4 taut wire, *5* counterweight,
6 weight, *7* mirror

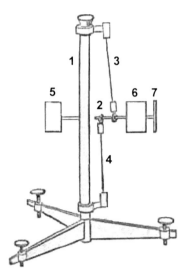

horizontal-pendulum instruments. For the oscillation of the pendulum the following equation holds:

$$t^2 mgs \sin\varphi = \pi^2 K \tag{4.74}$$

where t is the time for one oscillation, m the mass, g gravitational acceleration, s distance of the centre of gravity to the beam, K the moment of inertia, and ϕ the angle of the suspension to the vertical.

4.14 Radiometric Belt Weigher

The intensity of radiation, coming from a radiation source, is measured by a scintillation counter during passage of the material on the conveying system [96] (Fig. 4.133). From the resulting measuring signal the mass flow rate is calculated. That signal depends on type of the material, on the conveyor system and the product speed. The relative uncertainty is about 5 %. Advantageous is the contactless measurement of the moved material.

4.15 Combination of Balance Types

The features of balances working on the different basic principles are very different. In particular, some of such principles allow for measurements in a wide range whereas others are highly sensitive but only in a restricted area. Furthermore, the operation of the balance and its response time or the time for attaining equilibrium, respectively, is different. Thus the combination of weighing systems may be favourable.

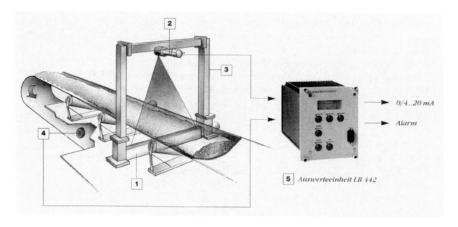

Fig. 4.133 Radiometric belt weigher (density determination). *1* capsulated and shielded radiation source mounted below the conveyor, *2* scintillation detector, *3* measuring and shielding frame, *4* tachometer, *5* processor module with digital display

For low requirements of accuracy beam balances are equipped with sliding weight to obtain equilibrium, or with a chain of variable length. The weighing range may be extended by means of a rotating weight or by electromagnetic force compensation.

With a gravitational balance a wide measuring range is covered with high sensitivity, but the weighing procedure is rather complicated and slow. Inclination balances are fast but measuring range and relative sensitivity are low. Advantages are given by the combination of weighing with an equal-armed beam balance almost in equilibrium, but the last digit of the result then calculated from a remaining inclination. That inclination may be compensated by a small spring and the force applied is used to calculate last digits. Also electromagnetic compensation may be used for that purpose.

Oscillating systems can be combined with the observation of bend of the elastic sensor in the force field. Further examples of the combination of weighing principles are applied when measuring different forces simultaneously in multicomponent systems (see Sect. 5.14)

References

1. E. Robens, A. Dąbrowski, The measuring range of balances. ICTAC News **37**(2), 80–83 (2004)
2. H.R. Jenemann, Zehntausend Jahre Waage? Teil 1. Mass Gewicht, Beih. Z. Metrol. **21**, 470–487 (1992)
3. H.R. Jenemann, Zehntausend Jahre Waage? Teil 2. Mass Gewicht, Beih. Z. Metrol. **22**, 509 (1992)

4. H.R. Jenemann, The early history of balances based on electromagnetic and elektrodynamic force compensation, in *Microbalance Techniques*, ed. by J.U. Keller, E. Robens (Multi-Science, Brentwood, 1994), pp. 25–53

5. M. Kochsiek (ed.), *Handbuch des Wägens*, 2nd edn. (Vieweg, Braunschweig, 1985)

6. M. Kochsiek, M. Gläser (eds.), *Comprehensive Mass Metrology* (Wiley/VCH, Berlin, 2000)

7. DIN, *DIN 8120, Teil 3: Begriffe im Waagenbau*. Berlin Beuth (1981)

8. H.R. Jenemann, Die Entwicklung der Präzisionswaage, in *Handbuch des Wägens*, ed. by M. Kochsiek (Vieweg, Braunschweig, 1989), pp. 745–779

9. T. Gast, T. Brokate, E. Robens, Vacuum weighing, in *Comprehensive Mass Metrology*, ed. by M. Kochsiek, M. Gläser (Wiley/VCH, Weinheim, 2000), pp. 296–399

10. H.R. Jenemann, A.M. Basedow, E. Robens, Die Entwicklung der Makro-Vakuumwaage. PTB-Berichte, vol. TWD-38. Braunschweig, Bremerhaven (1992). Wirschaftsverlag NW. 70

11. A.W. Czanderna, J.M. Honig, Anal. Chem. **29**, 1206–1210 (1937)

12. J.A. Poulis, W. Dekker, P.J. Meeusen, The use of pivot bearings in sensitive balances, in *Vacuum Microbalance Techniques*, ed. by K.H. Behrndt (Plenum, New York, 1965), pp. 49–58

13. E. Hering, R. Martin, M. Stohrer, *Physik für Ingenieure*, 3rd edn. (VDI Verlag, Düsseldorf, 1989)

14. W. Espe, *Werkstoffkunde der Hochvakuumtechnik*, vol. 1 (VEB Deutscher Verlag der Wissenschaften, 1960)

15. E.A. Gulbransen, K.F. Andrew, An enclosed physical chemistry laboratory: the vacuum microbalance, in *Vacuum Microbalance Techniques*, ed. by M.J. Katz (Plenum, New York, 1961), pp. 1–21

16. T. Gast, Elektrische Mikrowägung mit Hilfe trägerfrequenter Regelkreise. ETZ, Elektrotech. Z., Ausg. A **87**, 9–13 (1966) (Sonderheft zum 70. Geburtstag von R. Vieweg)

17. E.G. Woschni, *Meßgrößenverarbeitung* (Verlag Chemie, Weinheim, 1969)

18. T. Gast, Gesichtspunkte für die Gestaltung einer elektrischen Mikrowaage zur Messung von Oberflächenspannungen. Wägen + Dos. **2**, 59–62 (1983)

19. G. Luce, *Balkenlose Schwebewaage mit negativer magnetischer Federkonstante nach dem Prinzip der unterlagerten Stromregelung* (Technische Universität, Berlin, 1991)

20. D. Kisch, Konstruktion einer hochstabilen elektronischen Mikrowaage für Diffusionsmessungen. Z. Phys. Chem. **77**, 176–184 (1972)

21. R. Vieweg, T. Gast, Registrierende Mikrowaage für Diffusionsmessungen an Kunststoff-Membranen. Kunststoffe **34**, 117–119 (1944)

22. H. Mayer et al., On some modifications of a torsion microbalance for use in ultrahigh vacuum, in *Vacuum Microbalance Techniques*, ed. by K.H. Behrndt (Plenum, New York, 1963), pp. 75–84

23. T. Gast, K.P. Gebauer, Measuring of the density of gases with the aid of free suspension. Thermochim. Acta **51**, 1–6 (1981)

24. H. Tischner, in *Entwicklungstendenzen und Neukonstruktionen im Analysenwaagenbau*. Dechema-Monographien, vol. 27 (Dechema, Frankfurt am Main, 1956)

25. H. Büchel, Beitrag zur Lagerung von Hebelfeinwaagen. Universität Stuttgart (1972)

26. J. Wang, Design of gas bearing systems for precision applications. Proefschrift. Eindhoven, Technical University (1993)

27. P. Holster, Gaslagers met uitwendige drukbron. T.b.v. praktikum 4N023 (Lagers), werkeenheid Aandrijf- en Tribotechniek, vakgroep WOC. Eindhoven Technische Universiteit (1992)

28. W. Seifert, *Ein Beitrag zur exakten Wägung unter dem Einfluß wechselnder Beschleunigung* (Technische Universität, Berlin, 1975)

29. K.P. Zinnow, J.P. Dybwad, Pressure of light used as restoring force on an ultramicrobalance, in *Vacuum Microbalance Techniques*, ed. by A.W. Czanderna (Plenum, New York, 1971), pp. 147–153

30. A.W. Czanderna, S.P. Wolsky, *Microweighing in Vacuum and Controlled Environments* (Elsevier, Amsterdam, 1980)

31. J. Thölden, Haligraphia, das ist Gründliche und eigendliche Beschreibung aller Saltz Mineralien (1603)
32. H.R. Jenemann (ed.), *Zur Geschichte der Dichtebestimmung von Flüssigkeiten insbesondere des Traubenmostes in Oechsle-Graden*, Schriften zur Weingeschichte, vol. 98 (Gesellschaft für Geschichte des Weins e.V, Wiesbaden, 1990)
33. J.M. Honig, Use of pivotal microbalance design for determination of mass changes, in *Vacuum Microbalance Techniques*, ed. by M.J. Katz (Plenum, New York, 1961), pp. 55–68
34. T. Gast, E. Robens, Free suspension systems in mass determination, in *Microbalance Techniques*, ed. by J.U. Keller, E. Robens (Multiscience, Brentwood, 1994), pp. 65–72
35. T. Gast, Microweighing in vacuo with a magnetic suspension balance, in *Vacuum Microbalance Techniques*, ed. by K.H. Behrndt (Plenum, New York, 1963), pp. 45–54
36. T. Gast, W. Pahlke, Magentic coupling for a microbalance. J. Therm. Anal. **37**, 1933–1941 (1991)
37. H.R. Jenemann, Über die Aufhänge- und Arretierungsvorrichtung der ägyptischen Waage der Pharaonenzeit. Ber. Wiss.gesch. **11**, 67–82 (1988)
38. H.R. Jenemann, The development of the determination of mass, in *Comprehensive Mass Metrology*, ed. by M. Kochsiek, M. Gläser (Wiley/VCH, Berlin, 2000), pp. 119–163
39. Aristoteles, a.t., *Questiones mechanicae*. Kleine Schriften zur Physik und Metaphysik., ed. by P. Gohlke (Paderborn, 1957)
40. T.N. Winter, The mechanical problems in the corpus of Aristotle, in *Classics and Religious Studies, Faculty Publications, Classics and Religious Studies Department* (University of Nebraska, Lincoln, 2007)
41. T. Heath, *The Works of Archimedes* (Dover, Mineola)
42. Archimedes, *Archimedes Werke* (Darmstadt, 1983)
43. A.G. Drachmann, Fragments from Archimedes in Heron's Mechanics. Centaurus **8**, 91–146 (1963)
44. K.E. Haeberle, *10 000 Jahre Waage* (Bizerba, Balingen, 1966)
45. M.J. Schiefsky (ed.), Even without math, ancients engineered sophisticated machines. http://www.fas.harvard.edu/home/news_and_events/releases/math_10012007.html. Archimedes Projec., Harvard University, Cambridge, MA (2009)
46. Avery-Weigh-Tronix, The History of Weighing. http://www.wtxweb.com/. Avery Weigh-Tronix (2008)
47. China-Window, Chinese Steelyard—Gancheng. http://www.china-window.com/china_culture/china_culture_essentials/chinese-steelyard-ganchen.shtml. China Window (2008)
48. S. Thompson, Steelyard made in China. http://www.powerhousemuseum.com/. Powerhouse Museum Sidney EMu collection information system and research files (2007)
49. L. da Vinci, Codex atlanticus—Saggio del Codice atlantico, ed. by Aretin. Vol. fol. 249 verso-a + fol. 8 verso-b Milano (1872)
50. J. Leupold, Theatrum staticum – das ist: Schauplatz der Gewichtskunst. Theatrum staticum universale, Leipzig (1726)
51. H. Michel, P.A. Kirchvogel, Messen über Zeit und Raum. Stuttgart (1965)
52. J.H. Lambert, Theoria Staterarum. Acta Helvetica, Physico-, Mathematico, Anatomico-, Botanico-, Medica **3**, 13–22 (1758)
53. H.R. Jenemann, Early History of the Inclination Balance. Equilibrium, Quarterly Magazine of ISASC (1983), pp. 571–578; 602–610. International Society of Antique Scale Collectors
54. H.R. Jenemann, Zur frühen Geschichte der Neigungswaage. Mass Gewicht **11**, 210–215, see also 248–253 (1980)
55. A. Munz et al., Waagen und Wiegen – Die Geschichte des Waagenbaus in und um Onstmettingen im Zollernalbkreis & Der Arbeitskreis "Waagen und Gewichte". Förderverein Philipp-Matthäus-Hahn-Museum e.V, Albstadt (2006)
56. H.R. Jenemann, Der Mechaniker-Pfarrer Philipp Mathäus Hahn und die Ausbreitung der Feinmechanik in Südwestdeutschland. Z. Württemb. Landesgesch. **46**, 117–161 (1987)
57. H.R. Jenemann, Die wägetechnischen Arbeiten von Philipp Matthäus Hahn, in *Philipp Matthäus Hahn*, 1739–1790, ed. by C. Väterlein (Stuttgart, 1989)

58. L. Auger, *Un Savant Méconnu, Gilles Personne de Roberval* (Blanchard, Paris, 1962)
59. W. Kent, Kent's torsion balance. Sci. Am. **24**, 601, Supplement (1887)
60. M.A. Crawforth, *Handbook of Old Weighing Instruments* (International Society of Antique Scale Collectors, Chicago, 1984)
61. Wikipedia, Thaddeus Fairbanks. http://en.wikipedia.org/wiki/Thaddeus_Fairbanks (2010)
62. Wikipedia, Dezimalwaage. http://de.wikipedia.org/wiki/Dezimalwaage (2011)
63. R. Radok, General Theory of Motion and Force. http://mpec.sc.mahidol.ac.th/radok/physmath/PHYSICS/B5.htm. Mahidol Physics. Education, Centre, Bangkok, Thailand (2009)
64. R. Hooke, De Potentia Restitutiva or of Spring, explaining the Power of Springing Bodies. Lectiones Cutleriana or a collection of Lectures. Physical, Mechanical, Geographical, & Astronautical. Early Science in Oxford, vol. VIII (Gunther, London, 1678/1679)
65. H.R. Jenemann, Robert Hooke und die frühe Geschichte der Federwaage. Ber. Wiss.gesch. **8**, 121–130 (1985)
66. R. Hooke, A description of helioscopes and some other instruments. Tract VI of: R. Hooke: Lectiones Cutleriana or a Collection of Lecturs Physical, Mechanical, Geographical, & Astronomical. Early Science in Oxford: The Cutler Lectures of Robert Hooke, vol. VIII (Gunther, Oxford & London, 1676)
67. W.A. Benton, The early history of the spring balance. Trans. Newcom. Soc. **22**, 65–78 (1941/1942)
68. C. Weigel, *Gemein-Nützliche Hauptstände*. Regensburg, 1698
69. J. Dolaeus, De libella nova. Miscellanus curiosa sive Ephemeridum medico-physicarum Germanicarum Academiae Imperialis Leopoldina 2. Decuria, 8, pp. 295–296 (1689)
70. J. Ozanam, *Recréations mathematiques et physiques*. Paris (1696)
71. M. Hass, Scales and Weights. A collection of historical scales and weights from different periods of the past 3000 years (2010). http://www.s-a-w.net/
72. E. Salvioni, *Misura di Masse Compresa Fra G 10 $^{-1}eg10^{-6}$* (University of Messina, Messina, 1901)
73. J. Giesen, Ann. Phys. **10**(4), 830 (1903)
74. B.B. Cunningham, Microchemical methods used in nuclear research. Nucleonics **5**(5), 62–85 (1949)
75. P.L. Kirk, F.L. Schaffer, Rev. Sci. Instrum. **19**, 785 (1948)
76. P.L. Kirk, F.L. Schaffer, Rev. Sci. Instr., 250 (1951)
77. W.A. Robertson, in *Ultra Micro Weight Determination in Controlled Environments*, ed. by S.P. Wolsky, E.J. Zdanuk (Interscience, New York, 1969), pp. 86–87
78. C. Moreau, Recording MacBain balance, in *Vacuum Microbalance Techniques* (Plenum, New York, 1965), pp. 21–33
79. L. Hodges, The Michell-Cavendish experiment. http://www.public.iastate.edu/~lhodges/Michell.htm (1998)
80. H. Cavendish, Experiments to determine the density of the Earth. Philos. Trans. R. Soc. Lond. (Part II) **88**, 469–526 (1798)
81. H. Cavendish, Experiments to Determine the Density of the Earth, in Scientific Memoirs, vol. 9: The Laws of Gravitation, ed. by A.S. MacKenzie (American Book Co., 1798/1900), pp. 59–105
82. C.-A.d. Coulomb, Recherches théoriques et expérimentales sur la force de torsion et sur l'élasticité des fils de metal. Histoire de l'Académie Royale des Sciences, 1784, pp. 229–269
83. C.-A.d. Coulomb, Premier Mémoire sur l'Electricité et le Magnétisme; Construction et usage d'une Balance électrique, fondée sur la propriété qu'ont les Fils de métal, d'avoir une force de réaction proportionelle à l'usage de Torsion. Histoire de l'Académie Royale des Sciences, avec les Mémoires de Mathématique & de Physique (Paris) (Mém), 1785, pp. 569–577
84. W. Weber, Über drei neue Methoden der Konstruktion von Waagen, in *Wilhelm Weber's Werke* (Berlin, 1892), pp. 489–496

85. H. Pettersson, A new micro-balance and its use. Diss. Stockholm, in *Göteborg's Vet. of Vitterh. Samhalle's Handlinger* (Göteborg, 1914)
86. H. Pettersson, Experiments with a new micro-balance. Proc. Phys. Soc. Lond. **32**, 209–221 (1919)
87. D. Steele, K. Grant, Proc. R. Soc. Lond. Ser. A, Math. Phys. Sci. **86**, 270 (1912)
88. D. Steele, K. Grant, Proc. R. Soc. Lond. Ser. A, Math. Phys. Sci. **82**, 580 (1909)
89. E.A. Gulbransen, Trans. Am. Electrochem. Soc. **81**, 327–339 (1942)
90. F.R. Fischer, W.B. Schweizer, F. Diederich, Molecular torsion balances: evidence for favorable orthogonal dipolar interactions between organic fluorine and amide groups. Angew. Chem., Int. Ed. Engl. **46**(43), 8270–8273 (2007)
91. Wikipedia, Dehnungsmessstreifen. http://de.wikipedia.org/wiki/Dehnungsmessstreifen (2007)
92. B. Meißner, C.U. Volkmann, Prüfung von Dehnungsmeßstreifen-Wägezellen. PTB-Bericht, vol. Me-30. Braunschweig, PTB (1981)
93. E.E. Simmons, Jr., Method of making strain gauges, U.S.P. Office, Editor. USA, 1948
94. A.C. Ruge, Electrical load weighing apparatus. U.S.P. Office, Editor. The Baldwin Locomotive Works, USA, 1949
95. H. Hinderer, Die Kreiselwaage – ein neuartiges Wägeprinzip. Wägen + Dos. **5**, 102–104 (1974)
96. M. Gläser, Methods of mass determination, in *Comprehensive Mass Metrology*, ed. by M. Kochsiek, M. Gläser (Wiley/VCH, Berlin, 2000), pp. 184–231
97. A. Stock, G. Ritter, Gasdichtebestimmungen mit der Schwebewage. Z. Phys. Chem. **119**, 333–367 (1926)
98. E. Schneider, Daten zu Eiffelturm. http://www.baufachinformation.de/denkmalpflege.jsp?md=1988017121187 (2010)
99. T. Gast, Neue Anwendungen der selbsttätigen Kompensation. AEÜ, Arch. Elektron. Übertrag.tech. **1**, 114–121 (1947)
100. R.J. Kolenkow, P.W. Zitzewitz, A microbalance for magnetic susceptibility measurements, in *Vacuum Microbalance Techniques*, ed. by P.M. Waters (Plenum, New York, 1965), pp. 195–208
101. C.A. Hausen, Novi profectus in historia electricitatis (Leipzig, 1743)
102. D. Gralath, *Geschichte der Elektrizität. Part VI.* Versuche und Abhandlungen der Naturforschenden Gesellschaft zu Dantzig (1747), pp. 175–304
103. Anonymus, Letter to mr. John Ellicot, F.R.S.: about weighing the strength of electrical effluvia. Philos. Trans. R. Soc. Lond. A **44**, 96–99 (1746)
104. E. Gerland, F. Traumüller, Geschichte der physikalischen Experimentierkunst (Leipzig, 1899)
105. F. Rosenberger, Die Geschichte der Physik, II (Braunschweig, 1884)
106. W. Thomson, Measurement of the electrostatic force produced by a Daniell's battery. Proc. R. Soc. Lond. **10**, 319–326 (1860)
107. J. Frick, *Physikalische Technik*, 7th edn., vol. II/1 (Braunschweig, 1907)
108. H. Helmholtz, Über eine electrodynamische Wage. Wiedemanns Ann. Phys. Chem. **250**(14), 52–54 (1881)
109. K. Kahle, Das Helmholtz'sche absolute Elektrodynamometer. Z. Instrum.kd. **17**, 97–109 (1897)
110. K. Kahle, Das Helmholtz'sche absolute Elektrodynamometer und eine Anwendung desselben zur Messung der Spannung des Clark-Elementes. Wiedemanns Ann. Phys. Chem. **295**(59), 532–574 (1896)
111. T. Gast, Wirkungsweise und Anwendungsergebnisse der registrierenden Staubwaage. Chem. Ing. Tech. **24**, 505–508 (1952)
112. V. Crémieu, Sur une balance très sensible pouvant servir de galvanomètre et d'électromètre absolu. C. R. Hebd. Séances Acad. Sci. **132**, 1267–1270 (1901)
113. H.R. Jenemann, Über die Grundlagen und die geschichtliche entwicklung elektromechanischer wägesysteme. teil I - III. CLB, Chem. Labor Betr. **36**, 393–396 (1985), see

also 500–504, 629–632

114. H.R. Jenemann, Über die Grundlagen und die geschichtliche Entwicklung elektro-mechanischer Wägesysteme. Teil IV–VI. CLB, Chem. Labor Betr. **37**, 169–172 (1986), see also 344–345, 631–633

115. H.R. Jenemann, Über die Grundlagen und die geschichtliche Entwicklung elektro-mechanischer Wägesysteme. Teil VII. CLB, Chem. Labor Betr. **38**, 240–246 (1987)

116. W. Weber, Elektrodynamische Maassbestimmungen. Poggendorffs Ann. Phys. Chem. **149**(73), 193–240 (1848)

117. E. Hoppe, Geschichte der Elektrizität (Leipzig, 1884)

118. H.R. Jenemann, La balance électro-magnétique de Becquerel. Le Système métrique. Bull. Soc. Métr. Fr. (1987), pp. 321–327

119. O. Frölich, Die Entwickelung der elektrischen Messung (Braunschweig, 1905)

120. E. Lenz, M. Jacobi, Ueber die Gesetze der Elektromagnete. Poggendorffs Ann. Phys. **123**(47), 225–270 (1839)

121. A. Cazin, Mémoire sur l'Evaluation en Unités de poids des Actions életrodynamiques. Ann. Chim. Phys. 4. série **1**, 257–276 (1864)

122. A. Cazin, Beschreibung der electrodynamischen Wage. (Ed. Ph.) 1, in *Repertorium für Physikalische Technik, für Mathematische und Astronomische Instrumentenkunde 1*, ed. by P. Carl (1866), pp. 42–46

123. F.C.G. Müller, Neue galvanische Apparate für den Unterricht, sowie für den technischen Gebrauch. Z. Instrum.kd. **4**, 119–125 (1884)

124. K. Ångström (ed.), *Eine Wage zur Bestimmung der Stärke magnetischer Felder*, Repertorium der Physik, ed. by P. Carl, vol. 25 (1889), pp. 383–387

125. K. Ångström, Två metronomiska hjälpapparater. Oefversigt af Kongl. Vetenskap-Akademiens Foerhandlingar **52**, 643–655 (1895)

126. H.R. Jenemann, Zur Geschichte der Substitutionswägung und der Substitutionswaage. Technikgeschichte **49**, 89–131 (1982)

127. J.C. Poggendorff, Ein Vorschlag zum Messen der magnetischen Abweichung. Poggendorffs Ann. Phys. **83**(7), 121–130 (1826)

128. G. Gorbach, Die Mikrowaage. Mikrochemie **20**(2/3), 236–254 (1936)

129. H. Kruspe, Präzisionswaage nach Art des Elektrodynamometers, in *Deutsches Reichspatent* (1898)

130. W.E. Ayrton, T. Mather, F.E. Smith, A new current weigher and a determination of the electromotive force of the normal Weston cadmium cell. Philos. Trans. R. Soc. Lond. Ser. A, Math. Phys. Sci. **207**, 463–541 (1908)

131. A.C. Waltenhofen, Über einen einfachen Apparat zur Nachweisung des magnetischen Verhaltens eiserner Röhren. Sitzungsber. Akad. Wiss. Wien, Mat.-Nat. Klasse **62**, 438–440 (1870)

132. A.C. Waltenhofen, Über einen einfachen Apparat zur Nachweisung des magnetischen Verhaltens eiserner Röhren. Repertorium für Experimentalchemie, für Physikalische Technik, Mathematische und Astronomische Instrumentenkunde, vol. 6 (1870), pp. 119–125

133. H.E.J.G. du Bois, Eine magnetische Waage und deren Gebrauch. Z. Instrum.kd. **12**, 404–408 (1892)

134. P. Janet, F. Laporte, R. Jouaust, Détermination par un Electrodynamomètre absolu de la Force Electromotrice des Elémens au Cadmium. Bull. Soc. Int. électr., 2. série **8**, 459–522 (1908)

135. H. Pellat, Nouvel Electrodynamomètre absolu et détermination de la Force Electromotrice de l'élément du type Weston. Bull. Soc. Int. électr., 2. série **8**, 573–633 (1908)

136. P. Pascal, Mesure des susceptibilités magnétiques des corps solides. C. R. Hebd. Séances Acad. Sci. **150**, 1054–1056 (1910), see also 1514

137. B. Weber, *Skript zur Vorlesung Molekulare Magnete* (2009)

138. G. Urbain, C. Boulanger, Sur une balance-laboratoire à compensation électromagnétique à l'étude des systèmes qui dégagent des gaz avec une vitesse sensible. Compt. Rend. **154**, 347–349 (1912)

139. K. Honda, On a thermobalance. Sci. Rep. Tohoku Univ., Ser. 1 **4**, 97–103 (1915)

140. F. Emich, Einrichtung und Gebrauch der zu chemischen Zwecken verwendbaren Mikrowaagen, in *Handbuch der biochemischen Arbeitsmethoden*, ed. by E. Abderhalden (Berlin/Wien, 1919), pp. 55–147

141. F. Emich, Einrichtung und Gebrauch der zu chemischen Zwecken verwendbaren Mikrowaagen, in *Handbuch der biologischen Arbeitsmethoden*, ed. by E. Abderhalden (Berlin/Wien, 1921), pp. 183–269

142. E. Wiesenberger, Die Anwendung der elektromagnetischen Mikrowaage bei der Ausführung von Rückstandsbestimmungen und Elektrolysen nach dem Gammaverfahren. Mikrochemie **10**, 10–26 (1932)

143. E. Wedekind, Über eine magnetische Mikrowaage. Z. Elektrochem. Angew. Phys. Chem. **41**, 358–363 (1928)

144. J.W. McBain, H.G. Tanner, A robust microbalance of high sensitivity, suitable for weighing sorbed films. Proc. R. Soc. Lond. Ser. A, Math. Phys. Sci. **125**, 579–586 (1929)

145. E. Lehrer, E. Kuss, Eine verbesserte Gasdichtewaage mit elektromagnetischer Messeinrichtung. Z. Phys. Chem. **163**, 73–81 (1932)

146. F.C. Edwards, R.R. Baldwin, Magnetically controlled quartz fiber microbalance. Anal. Chem. **23**, 357–361 (1951)

147. S. Odén, On the size of the particles in deep-sea deposite. Proc. R. Soc. Edinb. **36**, 219–235 (1916)

148. J.R.N. Coutts et al., An automatic and continuous recording balance (The Odén-Keen-balance). Proc. R. Soc. Lond. Ser. A, Math. Phys. Sci. **106**, 33–51 (1924)

149. T. Gast, Permanent magnets in measuring techniques, in *Proceedings of the 8th International Workshop on Earth Magnets and their Application* (Dayton, 1985)

150. U. Tietze, C. Schenck, *Halbleiter-Schaltungstechnik* (Springer, Berlin, 1976)

151. R. Hönl, Ein Stauchscheibenmeßfühler mit Kraftkompensation als selbstkalibrierendes Meßsystem. Dissertation, vol. d 83 (Technische Universität, Berln, 1987)

152. T. Gast, E. Alpers, Ponderometrische Bestimmung dielektrischer Größen. Z. Angew. Phys. **1**, 228–232 (1948)

153. T. Gast et al., Survey on mass determination with oscillating systems. Part I: Fundamentals and history. J. Therm. Anal. Calorim. **71**(1), 19–23 (2003)

154. T. Brokate et al., Survey on mass determination with oscillating systems. Part II: Instruments and weighing of matter from gaseous environment. J. Therm. Anal. Calorim. **71**(1), 25–29 (2003)

155. E.H. Weber, E.W. Weber, *Wellenlehre, auf Experimente gegründet* (Leipzig, 1825)

156. V.M. Mecea, Fundamentals of mass measurement. J. Therm. Anal. Calorim. **86**(1), 9–16 (2006)

157. V.M. Mecea, Is quartz crystal microbalance really a mass sensor? Sens. Actuators A, Phys. **128**(2), 270–277 (2006)

158. V.M. Mecea, J.O. Carlsson, R.V. Bucur, Extension of the quartz-crystal-microbalance technique. Sens. Actuators **53**, 371–378 (1996)

159. K. Doblhofer, K.G. Weil, Application of the quartz microbalance in electrochemistry. Bunsenmagazin **9**(5), 162–172 (2007)

160. G. Sauerbrey, Verwendung von Schwingquarzen zur Wägung dünner Schichten und zur Mikrowägung. Z. Phys. **155**, 206–222 (1959)

161. G. Sauerbrey, Wägung dünner Schichten mit Schwingquarzen. Phys. Verh. **8**, 193 (1957)

162. A.W. Warner, C.D. Stockbridge, Mass and thermal measurements with resonating crystalline quartz, in *Vacuum Microbalance Techniques*, ed. by R.F. Walker (Plenum, New York, 1962), pp. 71–92

163. W.H. Wade, L.J. Slutsky, Adsorption on quartz single crystals, in *Vacuum Microbalance Techniques*, ed. by R.F. Walker (Plenum, New York, 1962), pp. 115–128

164. W.H. King Jr., Applications of the quartz crystal resonator, in *Vacuum Microbalance Techniques*, ed. by A.W. Czanderna (Plenum, New York, 1971), pp. 183–200

165. C. Lu, A.W. Czanderna, *Applications of Piezoelectric Quartz Crystal Microbalances* (Elsevier, Amsterdam, 1984)

166. H. Thanner et al., GAPO$_4$ high temperature crystal microbalance demonstration up to 720 °C. J. Therm. Anal. Calorim. **71**(1), 53–59 (2003)

167. H.K. Pulker, J.P. Decostered, Applications of quartz crystal microbalances for thin film deposition process control, in *Applications of Piezoelectric Quartz Crystal Microbalances*, ed. by C. Lu, A.W. Czanderna (Elsevier, Amsterdam, 1984), pp. 63–123

168. W.G. Cady, *Piezoelectricity* (McGraw Hill, New York, 1946)

169. A.P.M. Glassford, Application of the quartz crystal microbalance to space system contamination studies, in *Applications of Piezoelectric Quartz Crystal Microbalances*, ed. by C. Lu, A.W. Czanderna (Elsevier, Amsterdam, 1984), pp. 281–305

170. Z. Ali, K. Pavey, E. Robens, Survey on mass determination with oscillating systems. Part III: Acoustic wave mass sensors for chemical and biological sensing. J. Therm. Anal. Calorim. **71**(1), 31–35 (2003)

171. Z. Ali, Recent advances of quartz crystal microbalances in chemical and biological sensing. J. Therm. Anal. Calorim. (2002)

172. Q-sense, Q-sense AB, Stena Center 18, SE-41292 Göteborg

173. Z. Ali et al., Gas-sensing system using an array of coated quartz crystal microbalances with a fuzzy inference system. J. Therm. Anal. Calorim. **55**(2), 371–381 (1999)

174. Z. Ali et al., Denuder tube preconcentration and detection of gaseous ammonia using a coated quartz piezoelectric crystal. Analyst **117**, 899–903 (1992)

175. T.P. Burg et al., Weighing of biomolecules, single cells and single nanoparticles in fluid. Nature **446**(4), 1066–1069 (2007)

176. K.K. Kanazawa, J.G. Gordon, Anal. Chim. Acta **175**, 99 (1985)

177. J. Bell, T. Köhler, D. Woermann, Change of the resonance frequency of a quartz crystal microbalance in contact with an aequeous dispersion of solid particles. Ber. Bunsenges. Phys. Chem. **101**(6), 879–883 (1997)

178. H. Nowotny, E. Benes, General one-dimensional treatment of the layered piezoelectric resonator with two electrodes. J. Acoust. Soc. Am. **82**(August), 513–521 (1987)

179. H. Nowotny, E. Benes, M. Schmid, Layered piezoelectric resonators with an arbitrary number of electrodes (one dimensional treatment). J. Acoust. Soc. Am. **90**(September), 1238–1245 (1991)

180. Kösslinger et al., A quartz crystal biosensor for measurement in liquids. Biosens. Biolectron. **7**, 397–404 (1992)

181. Woias et al., Biosensors for HIV-immunodiagnosis, in *In Vivo Chemical Sensors: Recent Developments*, ed. by T. Alcock (Cranfield, 1993)

182. Drobe et al., Acoustic sensors based on surface localized HPSWs for measurements in liquids. Sens. Actuators A, Phys. **37–38**, 141–148 (1993)

183. D. Büker, T. Gast, Kontinuierliche gravimetrische Staubmessung durch mechanische Resonanz. Chem. Ing. Tech. **39**(16), 963–966 (1967)

184. T. Gast, Microweighing in vacuo with the aid of vibrations of a thin band, in *Vacuum Microbalance Techniques*, ed. by C.H. Massen, H.J. van Beckum (Plenum, New York, 1970), pp. 105–107

185. T. Gast, Pramanik, Verfahrenstechnik **7**, 4 (1973)

186. H. Bahner, Th. Gast, Massebestimmung disperser Stoffe mit Hilfe transversal schwingender Filterbänder. Tech. Mess. **50**(1), 3–13 (1983)

187. H. Bahner, T. Gast, GIT, Z. Labortechnik **25**(9) (1981)

188. K.U. Kramm, *Bestimmung von Massen, Viskositäten und E-Moduln mit dem longitudinal schwingenden Band* (Technische Universität, Berlin, 1985)

189. T. Gast, T. Brokate, Progress in mass determination with the aid of a vibrating ribbon. Measurement **17**(3), 141–149 (1996)

190. T. Gast, The longtudinally oscillating ribbon as a sensor for mass changes in controlled atmospheres. Thermochim. Acta **112**(1), 67 (1987)

191. H. Patashnick, E.G. Rupprecht, Continuous measurement using the tapered element oscillating microbalance. J. Air Waste Manage. Assoc. **41**(8), 1079–1083 (1991)

192. C.C.J. French, Advanced techniques for engine research and design. J. Automob. Eng. **203**(D3), 169–183 (1989)
193. H.P. Lang, M. Hegner, C. Gerber, Cantilever array sensors. Mater. Today **4**, 30–36 (2005)
194. R. Berger, J. Gutmann, R. Schäfer, Scanning probe methods: from microscopy to sensing. Bunsenmagazin **2**, 42–53 (2011)
195. R. Berger et al., Chem. Phys. Lett. **294**, 363–369 (1998)
196. P. Poncheral et al., Electrostatic deflections and electromechanical resonances of carbon nanotubes. Science **283**, 1513–1516 (1999)
197. B. Ilic et al., Attogram detection using nanoelectromechanical oscillators. J. Appl. Phys. **95**(7), 3694–3703 (2004)
198. S. Gupta, G. Morell, B.R. Weiner, Electron field-emission mechanism in nanostructured carbon films: a quest. J. Appl. Phys. **95**(12), 8314–8320 (2004)
199. J. Wood, Mass detection finds new resonance. Mater. Today **4**, 20 (2004)
200. M.D. LaHaye et al., Approaching the quantum limit of a nanomechanical resonator. Science **304**(5667), 74–77 (2004)
201. C. Sealy, Probing the quantum world with uncertainty. Mater. Today **6**, 9 (2004)
202. Z.L. Wang, The new field of nanopiezotronics. Mater. Today **10**(5), 20–28 (2007)
203. K. Jensen, K. Kim, A. Zettl, An atomic-resolution nanomechanical mass sensor. Nat. Nanotechnol. **3**(7), 533–537 (2008)
204. C. Chen et al., Performance of monolayer graphene nanomechanical resonators with electrical readout. Nat. Nanotechnol. **9** (2009)
205. M. Taghioskoui, Trends in graphene research. Mater. Today **12**(10), 34–37 (2009)
206. J.S. Bunch et al., Electromechanical resonators from graphene sheets. Science **315**, 490 (2007)
207. J. Chaste et al., High-frequency nanotube mechanical resonators. Appl. Phys. Lett. **99**, 213502 (2011)
208. A. Bachtold, Graphene ElectroMechanical Resonators (2012)
209. B. Dumé, The yoctogram weighs in. Physicsworld.com (2012)
210. J.U. Keller, Theory of measurement of gas-adsorption equilibria by rotational oscillations. Adsorption **1**, 283–290 (1995)
211. C. Huygens, Une nouvelle invention d'horloges très-justes & portatives. J. Savans, 68–70 (1675)
212. M.F. Zöllner, Zur Geschichte des Horizontalpendels. Kgl. sächs. Gesellsch. der Wissensch. zu Leipzig, math.-phys. Klasse, 1872(11)
213. M.F. Zöllner, Ueber einen neuen Apparat zur Messung anziehender und abstoßender Kräfte. Kgl. sächs. Gesellsch. der Wissensch. zu Leipzig, math.-phys. Klasse, 1869(11)
214. M.F. Zöllner, Ueber einen neuen Apparat zur Messung anziehender und abstoßender Kräfte. Kgl. sächs. Gesellsch. der Wissensch. zu Leipzig, math.-phys. Klasse, 1871(6)
215. E. von Rebeur-Paschwitz, Das Horizontalpendel. Nova acta der Kais. Leop. Carol. Akad. **60**(1) (1894)
216. R. Ehlert, Horizontalpendelbeobachtungen, in *Beiträge zur Geophysik*, ed. by E. Gerland (Stuttgart, 1895), p. 131

Chapter 5
Balances for Special Applications

Most balances are used for determining the mass with moderate accuracy of an article in air at ambient temperature under static conditions of state. However, in science and technology weighing has been made in various gases and liquids at various temperatures and pressures. In addition mass changes are observed at conditions varying with time. Often a high resolution is demanded. For those special tasks balances had been adapted. In this chapter special modifications of balances are classified with regard to particular applications.

5.1 Laboratory Balances

A predominant task of chemical and physical laboratories is analysis [1, 2]. "Weighing lies at the heart of all quantitative analysis … and the design of balances is, therefore, an important element in the history of analytical chemistry…" [3]. An analytical balance is a typical laboratory balance as used for weighing of any sample. The highest possible readability combined with very high absolute sensitivity is the important criterion [4, 5]. Until recently, mechanical analytical balances had been arranged for use as macro-analytical balances with readability down to 0.1 mg per scale division and a load capacity of 200 g, sometimes up to several kilograms. Semimicro-, micro-, and ultramicro-analytical balances had readabilities down to 10 μg, 1 μg or 0.1 μg respectively, and high relative resolution [6]. So-called precision balances showed a restricted resolution, but their load capacity was up to several tons. Nowadays mechanical analytical balances are superseded by electronic or hybrid types with high resolution in an extended measuring range and with high load capacity. Therefore the classification mentioned above has become obsolete.

In the development of analytical techniques a trend is to use samples as small as possible. For such work chemical microbalances had been required. The term 'microbalance' is used for microchemical balances with a discrimination below the milligram range [7]. The term 'microbalance' is somewhat indefinite. In general, it is used for small balances of any kind which exhibit a high sensitivity. We may,

E. Robens et al., *Balances*, DOI 10.1007/978-3-642-36447-1_5,
© Springer-Verlag Berlin Heidelberg 2014

Fig. 5.1 Oldest balance
beam, Egypt, made of
limestone, 85 mm in length.
(The ropes are added for
demonstration). Upper Egypt,
pre-dynastic Negade period.
© Petrie Museum. London

however, distinguish between two types: the microchemical balance as a standard
balance in the chemical laboratory and a variety of different types which are applied
as vacuum or thermobalances. The microchemical balances with symmetric beam
were developed on the basis of assay balances which were already in use in the
Middle Ages. Also such microchemical balances today are superseded by electronic
balances.

Already in ancient times it was clear that small masses should be weighed with
small balances. The oldest relict, a balance beam consisting of limestone was only
about 85 mm in size (Figs. 3.39/5.1). Figure 5.2 shows a relief on a Hittitian tomb of
a man weighing probably precious metal. This is not a coin balance as assumed by
some researchers, because at that time, about 2000 BC, coins were not yet invented.

Scientific study of chemistry began in Egypt and Greece at the beginning of the
first millennium AD and was developed subsequently in Arabian countries and most
probably sensitive assay balances then existed. Until now, we have only the description of the 'Balance of Wisdom' of Abd ar-Rahmān al-Chāzinï or Abu al-Fath Abd
al-Rahman Mansour al-Khāzini [8, 9] from 1122. The balance was designed in particular for density determinations by means of hydrostatic weighing. Chāzinï came
from Greece and he was conversant with Archimedes' work (Figs. 3.42/5.3). Figure 5.4 shows the original diagram and Fig. 3.42 the reconstruction [10]. It may
be hoped that additional descriptions of instruments are found in old books from
Arabian libraries which are not yet examined thoroughly.

Via Northern Africa and Spain the knowledge in ancient science came to Europe
where the Arabian alchemy was continued and developed. An early representation
of an assay balance we find on a painting from 13th century of a pharmaceutical
workshop for the preparation of drugs (Fig. 5.5a). Several prints in books show
pharmacies of the middle ages (Fig. 5.5). On a painting of a university lesson of
about 1430 the balance is already installed in a case to protect the instrument from
environmental disturbances [11]. In Georg Agricola's book from 1555 about mining
and metallurgical work [12] we find three analytical balances (Fig. 5.6). The most

Fig. 5.2 Hittitic weigher
Bas-relief ~2000 B.C.
Kahramanmaraş, Eastern
Turkey. © Musée du Louvre,
Paris

Fig. 5.3 Principle of
Archimedes density balance
with Arabic caption

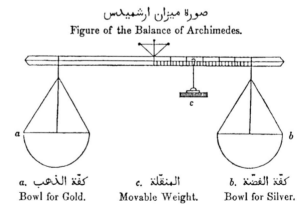

صورة ميزان ارشميدس

Figure of the Balance of Archimedes.

a. كفّة الذهب   *c.* المنقّلة   *b.* كفّة الفضّة

Bowl for Gold.   Movable Weight.   Bowl for Silver.

sensitive one is installed in a case and from the depicted weights we learn, that the discrimination was about 1/2 mg.

In about 1750, Joseph Black (1728–1799) developed the analytical balance based on a light-weight beam balanced on a wedge-shaped fulcrum [13]. Each arm carried a pan on which sample or weights were placed. It far exceeded the accuracy of any

Fig. 5.4 Balance of Wisdom
of Al Chazini ∼ 1100 density
balance (Arabic)

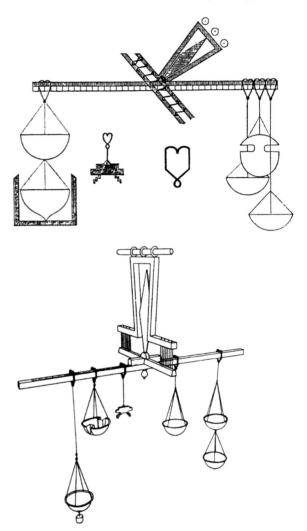

other balance of the time and became an important scientific instrument in most
chemistry laboratories.

In the middle of the 18th century Leonhard Euler deduced a first theory of the
balance from which the design of the beam could be optimised. With increasing demands of analytical chemistry at the beginning of the century the discrimination of
mechanical balances with symmetric beam was improved for weighing down to the
microgram range. This progress was supported from many manufacturers especially
in Europe and as the most important we may mention Paul Bunge. A Bunge assay
balance from 1866 with typical little beam is depicted in Fig. 5.7.

On account of the environmental influences there was no stimulus to improve
the discrimination and the sensitivity of mechanical balances. Further developments
resulted, therefore, in the improvement of the convenience of handling and the eval-

le defendre tenue. Et la culur muera. et metez prez grec en
la grandur que deuant der auum que sert table en pulbre

(a)

(b)

Fig. 5.5 (**a**) Medieval drugstore. Preparation of drugs with balance 13th century. (**b**) Medieval drugstore. (**c**) Laurntius Phriesen/Othonem Brunfels: Spiegel der artzney (Refelction of medicine). Depiction of a medieval drugstore. (**d**) Medieval sickbay and drugstore and; print

uation of results. The beam was equipped with rider weights for the compensation of small deviations and with air damping (Fig. 5.8). In the 1950s analytical balances were designed according to the substitution principle with unequal arm and fixed counterweight for only one suspending balance pan [14] (Figs. 5.9/5.10). The mass of the sample and mass changes were substituted by saddle weights (Fig. 5.11). The electric output signal was displayed, evaluated and recorded in digital form (Fig. 5.12). Then, electrodynamic compensation was introduced. Today, the whole measuring range is compensated using a plunger coil in a pot magnet

(c)

Fig. 5.5 (Continued)

(Figs. 5.13/4.83b). Furthermore, the arrangement with suspended balance pan was replaced by balances with pan on top. Typical are the two triangular levers guiding in parallel and the compensating plunger coil. Today, in few examples, other principles of weighing are also applied in analytical balances: the load cell with strain gauge and oscillatory sensors.

At the end of the 19th century some researcher perceived that it was impossible to modify common balance types in such a way that measurement of very low masses

(d)

Fig. 5.5   (Continued)

and mass changes in environments diverging widely from ambient temperature and pressure could be performed. A variety of unusual instruments were designed, the first one of Warburg and Ihmori in 1886/7: a beam balance pivoted on a razor blade (Fig. 5.14). In 1901 Salvioni [15, 16] designed a simple instrument with a thin glass fibre, clamped at one end and the sample suspended from the free end (Fig. 4.63). The deflection was observed by means of a microscope with ocular micrometer. This was the first balance designated 'microbalance'. Later on a quartz rod balance of Nernst was built in series (Fig. 5.15). Though suspended in a taut wire it worked as an inclination balance. The very important step was the application of electro-magnetic observation and compensation of the mass first realised for such balances by Ångstrøm 1895 (Fig. 4.91).

Oscillatory systems expanded the measuring range down to the molecular mass region. In particular crystal oscillators which are made of almost inert materials are suitable for chemical analysis. Meanwhile we learned to evaluate the complex signals obtained in liquid environment which usually is present in chemical reactions. Several 'femtobalances' small enough to weigh viruses and other nanoscale particles is one application for newly-discovered electronic and micromechanical properties of carbon nanotubes and grapheme foils. Here an electrical voltage is used to induce electrostatic deflection and vibrational resonance. This ability to selectively deflect or induce resonance in such micromechanical devices opens new potential applications for the tiny structures, which may be even smaller than the finest features on modern microcircuits.

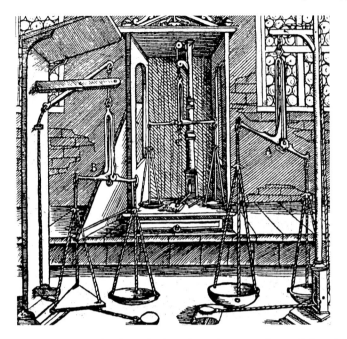

Fig. 5.6 Three assay balances for determination of metal content of ores of Georgius Agricola (1555), the most sensitive one in a case

5.2 Balances for Vacuum and Controlled Atmosphere

Following the work concerning air pressure of Evangelista Toricelli (1608–1647), Otto von Guerike (1502–1686), Robert Boyle (1626–1691), Edme Mariotte (1620–1684) and Daniel Bernoulli (1700–1782) it has become clear that air is weighable matter and that it should produce buoyancy like a liquid. At the end of the 18th century it was realised that sensitive weighings should be corrected with regard to this effect if the density of sample and counterweight differed. In highly accurate weighings, as is the case in metrological work and in gravimetric analytical experiments, adsorption effects and buoyancy should be kept to a minimum. There are a number of reasons for the use of special vacuum balances:

- Minimisation of undesirable influences on the weighing result, particularly that of buoyancy,
- Elimination of atmospheric gases to avoid unwanted reactions with the sample,
- Avoidance of reaction products escaping from the weighing chamber into the atmosphere,
- Determination of mass in production processes carried out at low pressure or in a controlled atmosphere, and measurements in space.

In principle each balance type can be used in vacuum. However it should be possible to operate and to observe the balance from outside of the vacuum vessel.

Fig. 5.7 Paul Bunge's
original short-beam assay
balance, Hamburg, Germany
1866

Fig. 5.8 Laboratory balance
with rider weight and air
damping, Sauter, Ebingen,
Germany

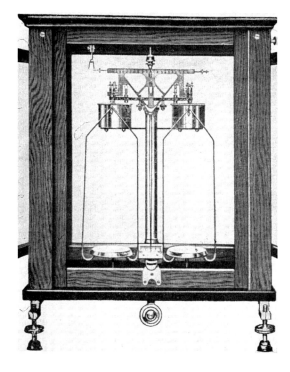

Fig. 5.9 Laboratory
substitution balance with
suspended scale, © Mettler,
Greifensee, Switzerland

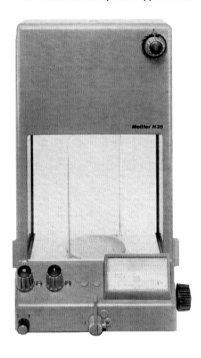

Fig. 5.10 Laboratory
substitution balance with
suspended scale, Sauter
Monopan, Ebingen, Germany

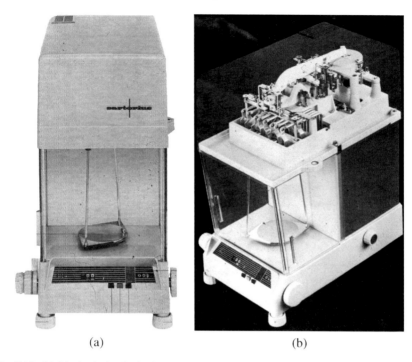

(a)   (b)

Fig. 5.11 (**a**) Mechanical substitution balance with suspended scale. Sartorius, Göttingen, Germany. (**b**) Without cover

5.2.1 Tasks and Methods of Weighing in Vacuum and Controlled Atmosphere

In all gravitational weighing buoyancy is an annoying influence. For highly accurate mass determination as required in metrological work in addition adsorption effects must be avoided. In gravimetric observations of chemical reactions, the influence of the atmosphere must be eliminated and harmful reaction products should not come into contact with sensitive parts of the balance or be released into the atmosphere. For many processes air must be excluded and the gas phase must be controlled. Technical processes occur often at high pressure of a surrounding gas. For such purposes, so-called vacuum balances are used; some of these are suitable also for high gas pressures. These instruments should fulfil the requirements of vacuum technology. Their components should not release volatiles and should not contain any cavities or materials in which gas could accumulate. Outgassing of its surface at elevated temperature should be possible. Mass determination in vacuum or a controlled atmosphere can be done in different ways:

- The weighing of a test object is in a closed vacuum container on a conventional balance (Fig. 5.16a). The mass of the container must be ascertained beforehand in a separate weighing; continuous measurements are not possible. Weighing of

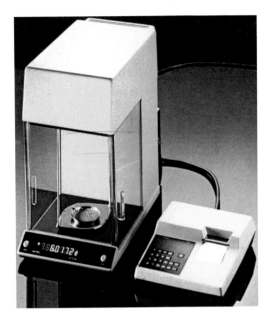

Fig. 5.12 Laboratory balance with electric registration and calculation unit. © Sartorius, Göttingen, Germany

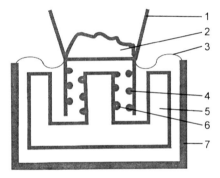

Fig. 5.13 Diagram of an electrodynamic load cell. *1* sample pan, *2* sample, *3* metal bellow *4* moving coil, *5* permanent magnet, *6* stator coil, *7* casing

gases in steel cylinders in order to determine their mass and density fall into this category. Stas [17] used two almost identical gas cylinders on an analytical balance to determine the molecular mass of hygroscopic substances. One cylinder was for the substance to be weighed and the other for the counterweight. When the cylinders were evacuated simultaneously, buoyancy did not affect the weighing.

- The entire balance with the sample and the counterweight or the counteracting device is placed under vacuum (Fig. 5.16b). For this, special vacuum balances are necessary which will be dealt with in this chapter.
- The balance remains in the atmosphere, and the sample in a separate receptacle is coupled in suspension to the balance by means of controlled magnets (Fig. 5.16c). A separate section is devoted to suspension balances.
- Adsorption measurements at high pressures had been made in receptacles situated on the pans of a symmetrical beam balance and connected to vacuum pump

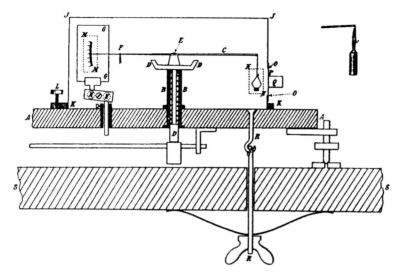

Fig. 5.14 Vacuum microbalance with razor blade knife edge of Warburg and Ihmori

and gas supply by a springy helical-shaped tube (Fig. 5.16d) [18]. The adsorption at the sample surface was compared with a smooth dummy at the counterweight side. These methods will not be discussed in detail here, as conventional balances are used.

Several types of vacuum balances had been developed: metrological comparator balances for control of the 1 kg mass prototypes, scientific laboratory balances for samples in the gram to kilogram region and scientific balances for samples with mass of a few grams or below, so-called vacuum-microbalances. In chemical and physical analysis and in technical fields, processes as a function of time are to be observed, where changes of mass are programmed under conditions different from those of the ambient atmosphere, such as pressure, temperature and the composition of gases. Balances greatly differing in construction have been developed for the purposes mentioned above. Electromagnetic compensation beam balances for loads in a range from 1 g to 100 g are widely used. For measurements with corrosive gases, spring balances and suspension balances are used. Furthermore industrial load cells and load cells for scientific investigations had been developed. Recent developments include nanomechanical oscillating systems for samples of a mass of milligrams or below extending the measuring range down to molecular mass dimensions.

- For measurements in gravity-free space, mass must be determined either indirectly, or weighing systems must be used that generate their own acceleration field by use of oscillation systems. Actually many types of weight cells and force transducers e.g. flexural bodies equipped with a strain gauge, piezoquartz, vibrating strings and moving coils are suitable for use in vacuum. Such transducers equipped with levers can weigh loads of any size and transfer the electrical signal

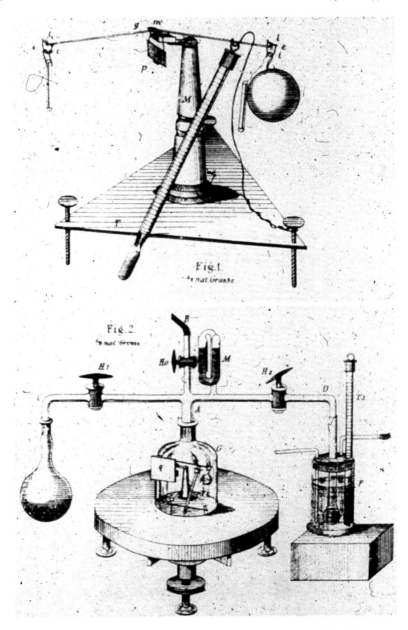

Fig. 5.15 Quartz-rod inclination balance of Walther Hermann Nernst and Julius Donau

to outside the vacuum vessel. In this way mass flow of large amounts of material can be controlled in vacuum or in any, even hazardous, atmosphere. Today in many cases the balance has become only a part of minor importance integrated in a complex apparatus.

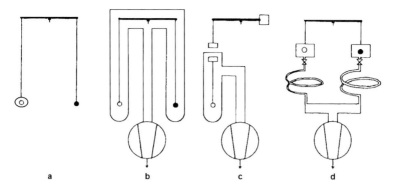

Fig. 5.16 (**a**) Weighing of a test object in an evacuated container. (**b**) Comparison weighing with test object and counterweight on a vacuum balance in a common chamber. (**c**) Weighing a test object with a suspension balance, sample and suspension magnet being in separate containers. (**d**) The container holding sample and counterweight are connected to the pump by a springy helical-shaped tube

5.2.2 Metrological Comparator Balances

The need for weighing in vacuum first became apparent in metrology when the international mass standard was established [19]. Most probably Borda who was the chairman of the relevant commission of the Académie Royale des Sciences was the first who requested in 1791 that the weighings of the mass prototype should be related to vacuum. Before the French Revolution the Royal Commission had declared that this should correspond to 1 cm^3 of water at 0 °C (later 1 dm^3 at 4 °C) in vacuum. However, liquid water in an open vessel cannot be handled in vacuum. As the technique of weighing water in vacuum could not be mastered, hydrostatic weighings were carried out to determine the volume and the mass of the prototype, which was a copper cylinder. The masses of the weights and the prototype, which were all made of copper, were compared in air assuming uniform density of the material. Later standard masses consisted of various materials: platinum, brass, and mercury-filled glass vessels. However, it was found that each of those materials had not a consistent density. The results of the first measurements were not reduced to vacuum conditions and differing air pressure was not regarded in their initial measurements [20]. Later, air pressure-dependent buoyancy was eliminated by complicated calculations, but the influence of ambient atmosphere on the surface of the prototype was not clearly understood. So a desire to weigh directly in a vacuum was stimulated.

For this reason, in 1861 Louis-Joseph Delieul and his son Jean-Adrien constructed a vacuum balance (Fig. 5.17) which was used by Henri-Victor Regnault to control the mass standards [21, 22]. The basis was the famous Deleuil balance which had a resolution of 5 μg at a maximum load of 3 kg. Such a balance was incorporated into an iron case which was equipped with a glass window to allow for the observation of the pointer by means of a telescope to avoid any temperature rise. Arresting of the balance beam could be made by means of a lever operated from

Fig. 5.17 The first vacuum balance: metrological kilogram balance, Deleuil/Regnault, 1860/61. The Deleuil balance was encased in iron housing with windows, and a telescope was used for observation in order to avoid influences of temperature. This apparatus still exists and today is kept at the Bureau Inernational des Poids et Mesures (BIPM) in Sèvres (France)

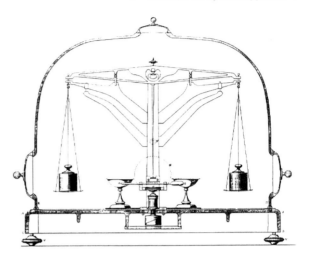

outside via a stuffing box. Operation of weights or transposition weighing, which was necessary for comparator balances, however, could be performed only after equilibrating the case to atmospheric pressure and removing of flanges. The Delieul balance in its iron case still exists and is exhibited in the Musée des Techniques du Conservatoire des Arts et des Métiers at Paris (France).

The basic construction of kilogram comparator balances of maximum sensitivity was established in the last century; they are mechanical beam balances. The exchange of load required opening of the balance case. Eddying air could enter and the balance was warmed up by heat radiation from the operator. At the request of Friedrich Arzberger the mechanic Paul Böhme designed a balance in 1876 in which transposition of load and weight could be carried out by mechanical remote control within the closed case (Fig. 5.18a). Figure 5.18b shows the cross-shaped slitted turntables for an adequately shaped sample and counterweight. In addition the balance was equipped with a manipulator for rider weights. This balance was improved by Rueprecht (Fig. 5.19) and delivered to the Bureau International de Poids et Mesures (BIPM) at Paris as well as to other national measurement offices.

To meet the difficulties of handling H.W. Miller ordered a balance from Ludwig Oertling in London (UK) in 1871 which could be fully operated in vacuo (Fig. 5.20). Transposition weighing was performed using two carts on rails which transported weights and load including the pans to the opposite side. Simultaneous change of the pans, however, provided a new source of uncertainties. Another drawback was that only nearly equal loads could be compared. This balance is preserved in the Science Museum of London.

In 1872 the German imperial board of weights and measures ordered a balance from Paul Bunge which should realise all metrological demands. The balance of which no picture is kept was finished in 1876. An improved model was delivered to the BIPM at Sèvres (Fig. 5.21). All manipulations were operated by rods 4 m in length. The specific sensitivity of that master piece of mechanical techniques was about 2.5×10^{-9}, a value hardly improved by novel constructions. The Bunge trans-

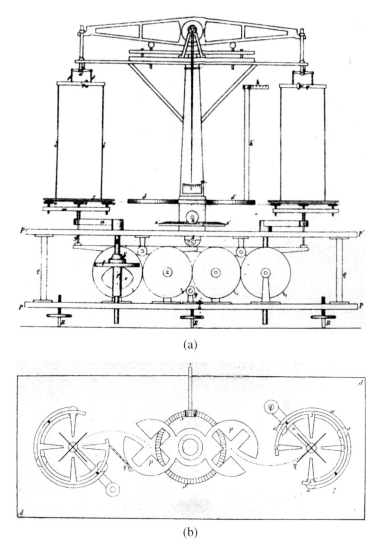

(a)

(b)

Fig. 5.18 Equal armed metrological balance of Arzberger and Böhme 1876 for transposition weighings

position balance was used for more than 80 years between 1879 and 1951 and is now in the museum of the BIPM. This type of balance was offered from the Hamburg Bunge workshop until 1939 but it is unknown whether any more were manufactured. A balance somewhat improved in details was ordered by the German board from Bunge´s competitor Stückrath in Berlin in 1879 (Fig. 5.22). Subsequently, Stückrath made one more vacuum balance, but none has survived.

In Wien (Vienna, Austria) Nemetz also built a comparator vacuum balance on the basis of plans of Kruspér of Budapest (Fig. 5.23). Another one (Fig. 5.24) he made

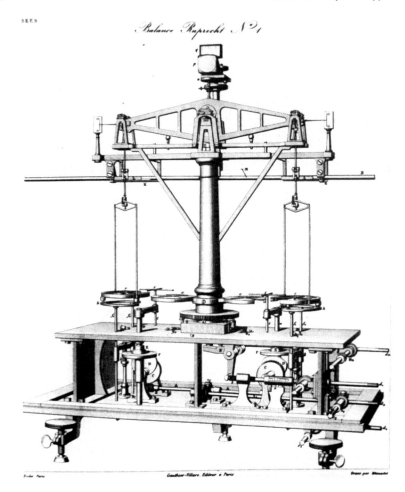

Fig. 5.19 Equal armed metrological balance of Rueprecht with transposition mechanism

for Mendeleev in St. Petersburg who had been using already for 20 years before a vacuum balance of Oertling, London [23]. Another manufacturer of vacuum balances in this time was Stollnreuther at München (Germany) who worked on design plans of Jolly, details however are unknown.

The history of metrological weighing actually carried out in vacuum is limited to one relatively brief episode, brought about by the unresolved problem of sorption on the surface of the balance and the weights. Heating of the standards seemed to be unsuitable, so in a modest vacuum the adsorbed layer was only partly removed to an undefined extent. In a text book from 1911 [24] we find this comment: "It follows that mass prototypes must be protected from changes in the environs, but under no circumstances should they ever be misused for application in vacuum". Until recently in comparator test apparatus, vacuum containers were used only to protect balances.

Fig. 5.20 Metrological
vacuum balance of L.
Oertling, London 1870

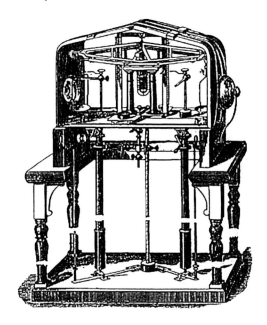

Fig. 5.21 Vacuum
comparator balance of Paul
Bunge 1879 with exchange
mechanism operated via rods
4 m in length, Bureau
Inernational des Poids et
Mesures (BIPM) in Sèvres
(France)

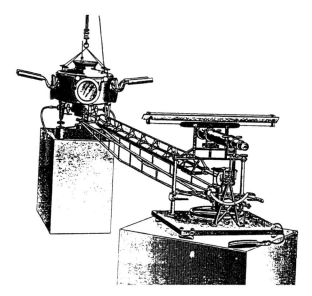

With oil-free high-vacuum pumps, sealing materials that are almost non-gassing, surfaces with better finishes [25, 26] and well-tested cleaning methods [27], a vacuum of better than 10^{-3} Pa can be generated today in a comparator balance recipient. In the 1980's, experiments were therefore started again to weigh mass standards in vacuum [28]. In NRLM, the Japanese state metrological laboratories, a symmetrical kilogram prototype was constructed on the three knife-edge principles

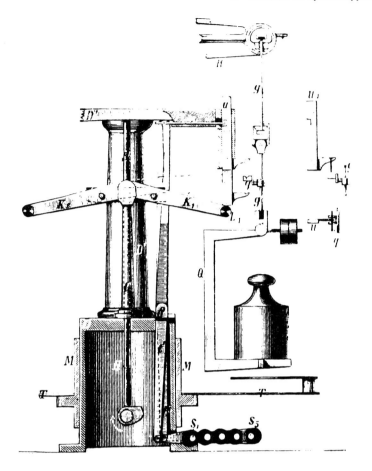

Fig. 5.22 Transposition mechanism of Stückrath/Krummach metrological vacuum balance

Fig. 5.23 Metrological
vacuum balance of
Nemetz/Kruspér, Budapest
'Heube'

Fig. 5.24 Kruspér/Nemetz
metrological vacuum balance
for Mendeleev

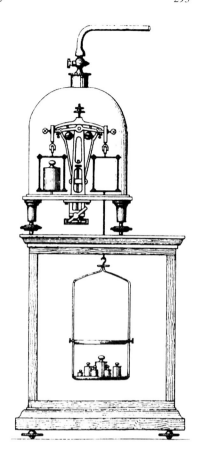

(Fig. 5.25). The balance is operated according to the substitution principle, whereby
two fixed counter weights each weighing 500 g serve as substitution masses [29, 30].
The lifting and exchange mechanism is worked hydraulically. In the weighing process the mass standard lies on the right-hand side on the cross-shaped support P
connected to the balance's rod assembly. For lifting, the corresponding slit rotary
plate W is raised, and simultaneously two counterbalance masses each of 500 g to
the left and right of the rotary plate are lowered onto the rod assembly. In this way
the balance is always under full load, and no arrestment is necessary. The cross-
shaped rotatory plate makes it possible for four mass standards to be compared
successively with the substitution mass. Differences in mass are balanced by rider
weights. The influences of buoyancy and adsorption are compared by comparison
weighing between weights of the same mass but with different surfaces and volumes
(Fig. 5.26). It has been possible to lower the buoyancy correction error to 10 μg by
using more accurate measuring instruments for pressure, temperature and air humidity and by applying the new international gas equation. The error due to adsorption
proved to be smaller. Using weights of equal mass but different surface it could be
demonstrated that at conditions of a high vacuum of 10^{-2} Pa errors by adsorption

Fig. 5.25 Comparator balance: symmetrical metrological kilogram prototype balance of the Japanese laboratory of metrology (NRLM)

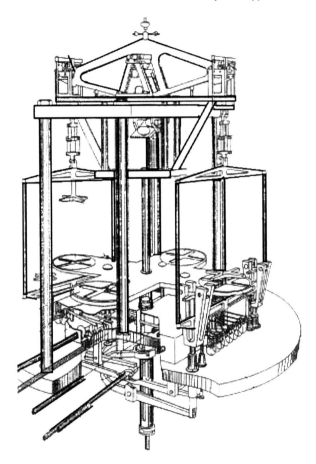

Fig. 5.26 Weights for testing the influences of sorption and buoyancy on weighing

	1 : 1	1,95 : 1
	216,516 cm³	0,001 cm³

are below the uncertainty of measurement caused by errors of the balance which amount to about 1 µg [30].

Similar experiments in air density determinations have been reported by the German 'Physikalish-Technishe Bundesansalt' [31]. The modified comparator balance from Mettler that was used is a hybrid construction and enclosed in a metal bell jar which can be evacuated by means of a turbo molecular pump. On a bottom-pan substitution balance with an unsymmetrical beam with two knife-edges and a fixed counter weight, mass differences of up to 150 mg have been balanced by electromagnetic force compensation and larger differences by electrically operated substitution weights. There is an electromechanical exchange mechanism for four mass standards. No arrestment is necessary; when the balance is relieved of weight, the beam meets a buffer.

Today a so called 'supercomparator' is able to weigh standard prototypes of 1 kg mass with an uncertainty below 50 ng in vacuum and below 100 ng at ambient conditions [32]. These fully automatic mass comparators are designed for use in national and international primary standard laboratories. They comprise automatic load alternation for weights up to 50 kg. Repeatability is not affected by the user, direct comparison of weight groups, four weight positions accommodate 5 g to 20 kg.

At present, the most accurate balance is an equal-arm beam balance for loads up to 2,5 kg of the Bureau International des Poids et Mésures at Sèvres, France (Figs. 3.4a/b, 5.27) [33]. This balance has flexural strips made from a copper-beryllium alloy instead of bearings and is equipped with an automatic sample changer. The monolithic beam and most other components are made of an aluminium alloy. An electromagnetic servo-control maintains the beam at constant position and provides the imbalance reading. The instrument is arranged in a vacuum vessel. The standard deviation of the balance in vacuum is 9 ng and in air 50 ng. Indeed a relative sensitivity of 3×10^{-12} could be achieved.

Although it has been demonstrated that today, comparisons of mass in vacuum are more accurate than those in air, the metrological application still remains doubtful, as ultimately commercial weights are compared with the mass standard in air.

5.2.3 Chemical Vacuum Macrobalances

Besides mass comparators, vacuum macrobalances, comparable to the 'analytical balances' described above, have been designed for the determination of atomic mass and other chemical needs [34]. In chemistry, vacuum balances are used to exclude reactions with components of the air, to eliminate humidity and to remove volatile components from the sample rather than to diminish the effect of buoyancy, because the required specific sensitivity of 10^{-6} to 10^{-7} is much lower than in metrology. Furthermore, in many cases, the work using a vacuum balance is made in controlled protective or reactive gas rather than in vacuum. Typical tasks of chemical vacuum balances are investigations of hygroscopic samples and the determination of their molecular masses. The observation of chemical and physical reactions in a defined environment requires continuous recording of mass and of additional parameters like temperature, pressure and composition of the surrounding gas as well as of species evolved from the sample. Typical applications of that kind are investigations of sorption and of catalytic processes and thermogravimetric analysis.

In 1865 the Belgian Jean Servais Stas [17] published a paper on the determination of the molecular mass of silver halogenides and similar compounds. He described a vacuum method using a conventional laboratory balance which was operated in the atmosphere. Sample and counterweight were surrounded by evacuated glass vessels of equal mass and volume according to Fig. 5.15a. In this way both

Fig. 5.27 (**a**)/(**b**) Photos and
(**c**) schematic diagram of the
metrological flexural strip
comparator balance with
electromagnetic
compensation of the Bureau
International des Poids et
Mesures at Sèvres, France.
(**c**) B beam, $L_1 = L_2$ lever
arm length, C centre of the
flexural strip, E_1, E_2 flexural
strip suspensions, G_1–G_4
cross bearings, P_1, P_2 weight
pans, M_T tare mass, M_1, M_2
test weight and standard, m_c
auxiliary sensitive weight, T
exchange mechanism, S
buffer, D_1 magnetic damping
for longitudinal pan
movements, D_2 magnetic
damping for transversal pan
movements, O optical
position detector for the
beam, U electromagnetic
compensation mechanism.
© BIPM

(a)

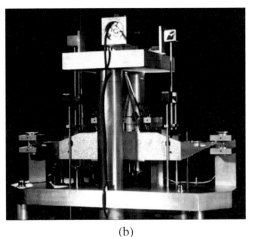

(b)

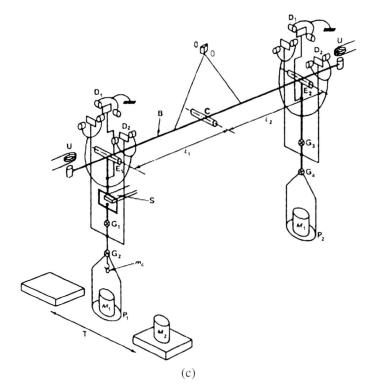

(c)

Fig. 5.27   (Continued)

the atmosphere was withdrawn from the sample and buoyancy on account of differing density avoided. In 1959 this method was re-invented by de Vreese and applied immediately in 1991 by Serpinsky et al. [18].

Crookes and later on Blount used vacuum balances of Oertling of London (Fig. 5.28). Weight changes could be compensated by means of rider weights moved from outside, but weight and sample could be operated only at atmospheric pressure. The Oertling balances were equipped with a short beam as first applied by Bunge. All these pure mechanical balances and their cases were ordered on account of a special research project and especially designed for that purpose. Increasing demand of vacuum macrobalances in the 20th/21st century was satisfied by only a few serial products.

Electrically operated accessories could be utilised for vacuum balances in different respects: to operate remote control, e.g. for arresting the beam, to place riders and bigger weights, to operate the sample, to observe the deflection of the beam or to compensate its deflection. Often force compensation was combined with mechanical weight feeders to give a hybrid balance.

At about 1950 Sartorius, Göttingen, built a sensitive balance 'Vakua' (Fig. 5.29) with a maximum load up to 200 g under a glass vessel. The weight change mecha-

Fig. 5.28 Laboratory vacuum balance Casella of L. Oertling, London

Fig. 5.29 Sartorius 'Vakua'
('Golden calf')

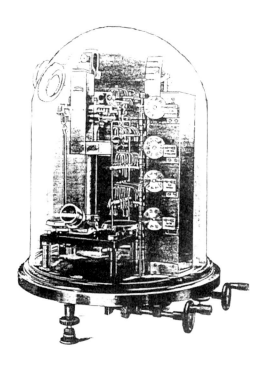

nism was operated by means of solenoids. All parts of the balance were gold coated giving rise to the nick name 'golden calf'.

A series of vacuum balances from micro to macro range was offered in the 1970s by Ainsworth of Denver, Colorado, USA (Fig. 5.30).

Fig. 5.30 Ainsworth gravimetric apparatus with vacuum balance. © Ainsworth

5.2.4 Vacuum Microbalances and Thermo Microbalances

Obviously, all the work done with 10 g to 1 kg samples in vacuum macrobalances and large thermobalances could be done better with small samples using microbalances. By the end of the 19th century the trend of developments approached in that direction. The discrimination of such balances went down to the nanogram range and the maximum capacity was between milligrams and some hundred grams. Typical vacuum microbalances are often used in thermogravimetric apparatus as thermobalances. In thermogravimetry the sample is heated in general in air or protecting gas at atmospheric pressure. Expensive vacuum equipment was avoided; also the operation of such an apparatus was much simpler. However, large disturbances by convective gas may occur. In order to minimise this effect the sample was often arranged laterally or above the balance beam.

5.2.4.1 Spring Balances

On account of their simplicity spring balances are particularly appropriate for use in a vacuum apparatus (Fig. 5.31). The few materials necessary for their construction enable measurements under very clean conditions. A quartz spring is suitable also for investigations in corrosive atmosphere. The extension of the spring is observed optically or by means of an inductive sensor. Unlike beam balances a direct measurement of the force of the load in the gravitational field is performed and not a difference measurement with reference to a counterweight. Buoyancy cannot be corrected. Thus, spring balances can never reach as low values of relative sensitivity as beam balances.

Fig. 5.31 Diagram of a
vacuum spring balance with
heat shielding

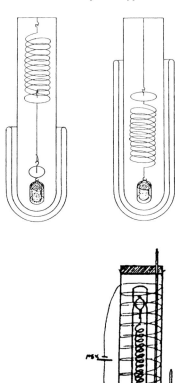

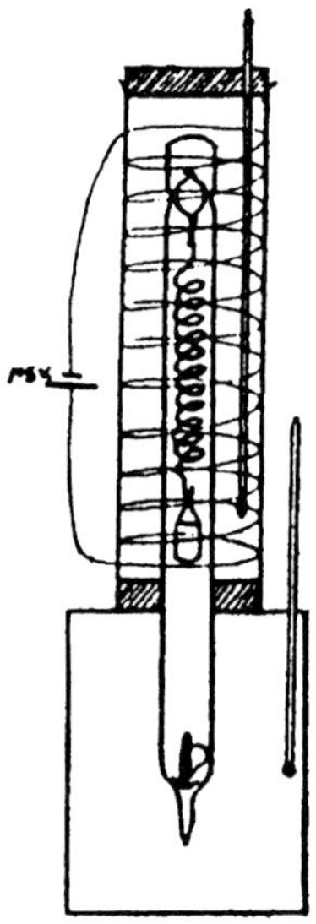

Fig. 5.32 McBain-Bakr
vacuum spring balance for
sorption measurements

The Salvioni rod spring balance (Fig. 4.63) has been already mentioned. Emich [35] was the first to describe a helical spring vacuum balance in 1915 (Fig. 4.64). The name of the McBain-Bakr balance was based on comprehensive sorption experiments which the two authors carried out with quartz helical spring balances [36, 37] (Fig. 5.32). In about 1965 AMINCO of Silver Spring, Maryland, USA, constructed a series of thermogravimetric apparatus equipped with a helical spring balance with load capacities ranging from 5 mg to 230 g and a sensitivity of ±0.5 % (Fig. 5.33). Changes in mass of 5 mg could be observed; with a spring loaded up to 10 g, it was 0.1 mg. The spring was fixed in a glass tube, and its change in length was transferred to a coil outside the tube by a core of soft iron attached to the spring. As far as we know, today no spring balances for use in a vacuum are manufactured. Only springs of quartz or metal are offered for that purpose, but the observation equipment has

Fig. 5.33 Scheme of a
thermogravimetric apparatus
with spring balance. *1* spring,
2 position transducer,
3 weight pan, *4* damper,
5 vacuum connection,
6 furnace, *7* crucible,
8 thermocouple. © AMINCO,
Silver Spring, MD, USA

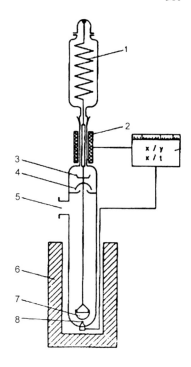

Fig. 5.34 Load cell with
sputtered strain gauge for
vacuum applications.
© Sartorius, Göttingen,
Germany

to be added. On principle, spring balances can be operated by means of electronic
control in the same way as is practised with beam balances.

For technical measurements and control in vacuum and pressure controlled environment load cells may be used. Advantageously the strain gauge is sputtered on
the flexible metal body (Fig. 5.34). Such an instrument was suggested for adsorption
experiments in the low pressure atmosphere on Mars (Fig. 5.13) [38].

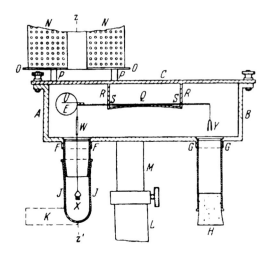

Fig. 5.35 Electromagnetic compensating vacuum microbalance of Emich

5.2.4.2 Compensating Beam Balances

On beam balances, mass comparisons are carried out in a gravitational field, enabling the difference in the object's change in mass to be measured against a constant comparison mass. This principle is of particular advantage for vacuum microbalances. Many of the reactions investigated in vacuum occur on the surface, whereas the greater part of the object (bulk material) does not react. The mass of this non-reacting part of the object together with the mass of the device holding the object can be compensated before the actual measurement, and only the change in mass is sensitively measured. With commercially available vacuum microbalances, a sensitivity of 10^{-8} referred to the maximum capacity is achieved [39]. Typical examples of this kind of balance are shown in the illustrations that follow.

With symmetrical beam balances the thermal expansion of the object is compensated to the largest possible extent by using counter weights of the same density and ensuring that load and counter weight are at the same temperature. With an unsymmetrical arrangement, the influences must be taken into account by calculations. If the balance's beam is short enough, sample and counterweight may be arranged in the same thermostat or calorimeter. Influences can be further determined and compensated by using a second balance loaded with a dummy.

In 1916 Emich designed a vacuum microbalance on the basis of the already known principle of electromagnetic compensation described by Ångstrøm. The balance was arranged in a vacuum vessel (Fig. 5.35) and may be regarded as the prototype of vacuum microbalances with electromagnetic force compensation [40, 41] though later balances of that type had been constructed independently. Today's vacuum balances are special type beam balances, the beam usually suspended by a horizontally stretched taut band. In all the commercially manufactured instruments the beam deflection is observed by an inductive or photoelectric sensor and is restored electrodynamically. Three subassemblies may be distinguished: the balance itself, the controller and the output unit. Whereas the mechanical design, the compensating magnet/coil system and the deflection sensors did not change in principle, improvements were made to the controllers as a result of the developments in electronics. The most impressive changes concern the data output. Digital records allow online data processing, comprising correction and transformation of results and display in real time. In the text following we will nevertheless present examples for the balance system, because its design and finish are decisive for the suitability in various applications rather than convenience in the output.

Figure 4.112 show Sartorius models on the basis of Gast balances. The production expired at the end of the last millennium. The beam is a quartz tube, for higher loads two parallel tubes form a ladder. The surface of the beam is covered with silver to allow an electrical connection to the sample pans. For loads up to two grams two spherical sapphire cups are pivoted on diamond pins as bearings at each end. For loads of some ten grams the sample pans are suspended via two short metal bands at each side.

A Cahn balance for loads up to 50 g is equipped with a metal tube beam and a coil system as a position sensor. The 1 g Cahn balance (Fig. 5.36) is equipped with a metal wire beam. At the load side the beam has an additional suspension bearing near the pivot in order to allow weighings of several grams load at reduced sensitivity. The deflection is detected by means of a photocell. The beam of the Perkin Elmer balance (Fig. 5.37) is a rhombus-shaped wire. A two-dimensional wire framework constitutes the beam of VTI-SA sorption balance of TA Instruments (Fig. 5.38). Maximum load is 1.5 g, the dynamic range 150 mg and the sensitivity 0.1 μg. The balance can be operated in a temperature range 5–60 °C. Although not available commercially, we recall the Czanderna balance, which might have had the best relation of detection limit to maximum load of all vacuum microbalances. The beam consists of a three-dimensional framework of quartz wires (Fig. 4.4). The bearings consist of quartz cups placed on tungsten tips.

The triangular-shaped massive metal beam of Setaram is punched (Fig. 5.39). It is only 6 cm in length and allows for arranging the sample and a reference sample in a common thermostatted tube. The magnet/coil system is rather large. The balance of CI Electronics is designed as a flat triangular metal band framework (Fig. 5.40). A combination of linear and circular flat metal framework constitutes the beam of the Q5000 SA balance of TA Instruments (Fig. 5.41). Maximum load is 1 g and sensitivity < 0.1 μg. The balance can be operated in a temperature range between 5–85 °C.

A thermobalance with quartz tube beam and plug connection is produced by Linseis (Fig. 4.37). This enables the extension of the beam and an optional top-position of the sample. Of course, in this way the usable sensitivity is restricted. An unsymmetrical thermobalance with two measuring units is offered by TA Instruments (Fig. 5.42). The object and the comparison material in a parallel arrangement can be heated in the laterally arranged oven, and operation is made easier by the one-sided access.

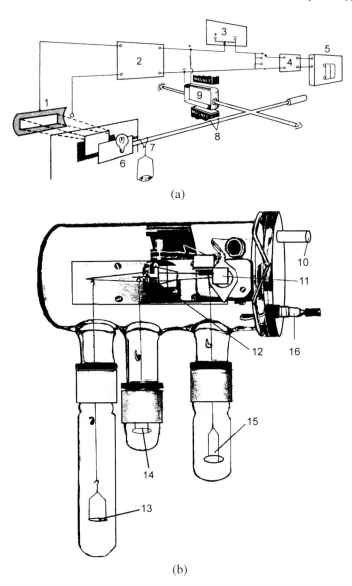

(a)

(b)

Fig. 5.36 Vacuum microbalance with metal wire beam from Thermo Scientific Cahn, Cerritos, CA, USA. © Thermo Scientific Cahn. *1* photo cell, *2* amplifier, *3* calibrator, *4* filter, *5* recorder, *6* lamp, *7* coil, *8* suspension for 20 mg sample, *9* suspension for 100 mg sample, *10* vacuum line, *11* photo cell, *12* moving coil, *13*, *14* sample pans, *15* counterweight pan, *16* electric connection

5.2.4.3 Oscillators

Resonator systems for the measurement of mass changes are realised as oscillating strings, bands and in the form of quartz resonators, which achieve an extremely high

Fig. 5.37 Vacuum balance with wire beam of Perkin Elmer Corporation, Norwalk, Connecticut, USA.
1 suspension, *2* beam, *3* seal, *4* force transducer, *5* position sensor, *6* base plate. © Perkin Elmer

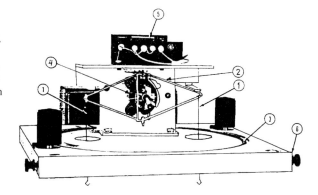

Fig. 5.38 VTI-SA type balance of TA Instruments, New Castle, DE, USA. © TA Instruments

Fig. 5.39 Vacuum microbalance with a 6 cm punched metal beam from SETARAM, Caluire, France. © SETARAM

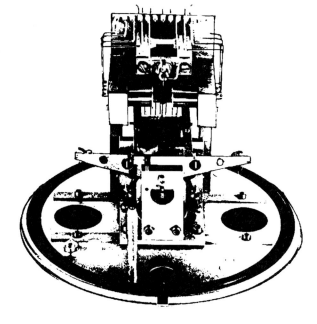

Fig. 5.40 (**a**) Microbalance with metal framework beam of CI Electronics in a glass vacuum recipient. © CI Electronics. (**b**) Diagram of the microbalance of CI Electronics. © CI Electronics

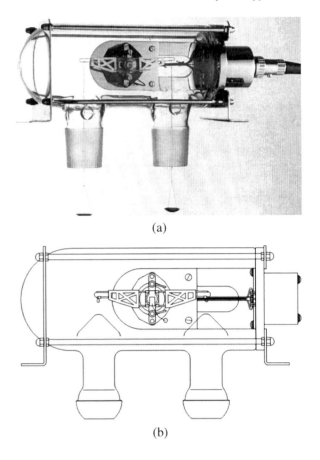

(a)

(b)

Fig. 5.41 Q5000 SA type balance of TA Instruments, New Castle, DE, USA. © TA Instruments

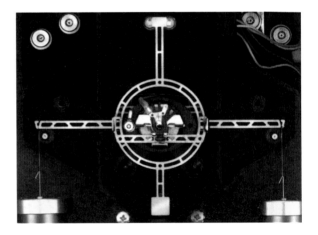

Fig. 5.42 Thermobalance
with two parallel mass
sensors from TA Instruments,
Newcastle, DE. *1* photocell,
2 force transducer, *3* sample
and comparison material,
4 oven

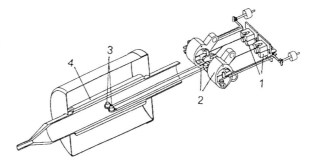

sensitivity but restricted measuring range. The measuring system is very small and
clean and can even be heated. So it is very suitable in vacuo. Quartz resonators are
used to control vacuum metalising and other evaporation and sputtering processes
and to record contamination. Such sensors are widely applied as layer thickness
monitors in apparatus for vapour deposition. Because resonator type balances need
no gravitational field they can be applied in space (Fig. 4.125). The mass determination is restricted to samples which are strongly connected to the surface of the sensor.
Recently devices developed from atomic force microscopes were applied to weigh
clusters of few molecules. For that purpose the cantilever was excited to oscillations
electrostatically and the frequency change was measured. Furthermore adsorption
measurements have been carried out using oscillating graphene monolayers.

5.2.5 Vacuum Equipment for Balances

The different tasks of weighing in vacuum require different qualities of the vacuum
and this should be considered in the choice of the pump. The balance needs to be
suitable for vacuum and the tubes of the apparatus should be designed in such a
way, that the gas flow is not restricted. For that purpose simple rules should be
observed, which, however, seem to be unknown for many balance manufacturers
and experimentalists [42]. In the following section we discuss vacuum conditions
on the investigation of a clean solid surface.

5.2.5.1 Theory

The basic equation describing the pressure in a vacuum chamber is

$$p S_{eff} = -V \frac{dp}{dt} + q \tag{5.1}$$

where p is the pressure, t the time, S_{eff} the effective pumping speed and q the
influx of gas. To obtain a low ultimate pressure, it is necessary to have a high pumping speed and a low influx. It might appear from Eq. (5.1) that the volume of the

chamber V affects only the pumpdown speed. However, since desorption of gas contributes to the influx, size of the walls and the geometry of the chamber will influence the ultimate pressure as well. The ultimate pressure is reached at the steady state between gas influx and removal, i.e. $dp/dt = 0$. In this case Eq. (5.1) reduces to

$$p = \frac{q}{S_{eff}} \tag{5.2}$$

S_{eff} is to be determined near to the sample, because "vacuum is a very viscous liquid" and the flow of which is hindered remarkably by the tubular system, including valves and other obstacles. This is expressed by

$$\frac{1}{S_{eff}} = \frac{1}{L} + \frac{1}{S_p} \tag{5.3}$$

where S_p is the speed of the pump and L the total conductance of the tubular system which results from the conductance of the parts L_i by

$$\frac{1}{L} = \sum \frac{1}{L_i} \tag{5.4}$$

The conductance of a circular aperture in the molecular flow region is expressed as

$$L_a = 0.918 \times 10^5 d_a^2 \tag{5.5}$$

where L_a is the conductance in $1\,s^{-1}$ for air at 20 °C and the diameter d_a in m. The molecular flow region is that region where the mean free path of a gas molecule $\bar{l}$ is much larger than the diameter of the tubing d_t, characterised by the Knudsen number

$$Kn = \frac{\bar{l}}{d_t} \tag{5.6}$$

For typical microbalance arrangements this corresponds to a pressure about 0.1 Pa. For negligible outgassing tubes the conductance can be calculated using

$$L_t = \frac{12.12 d_t^3}{l_t + \frac{4 d_t}{3}} \tag{5.7}$$

The conductance of commercially produced parts in general is indicated. For outgassing parts, like bellows, the pressure drop due to gas influx must be considered.

5.2.5.2 The Ultimate Vacuum

Demands on vacuum should correspond to the requirements, because the costs for vacuum apparatus and difficulties of its operation increase exponentially from rough vacuum generated by means of a water jet pump, to fine vacuum produced by a rotary vane pump, to high and ultrahigh vacuum.

If physical or chemical reactions with the sample surface should be observed, the surface needs to be clean at the beginning of the investigation. A clean surface can be

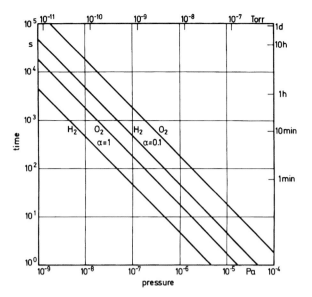

Fig. 5.43 Time required for the coverage of a smooth surface with a 10 percent contaminating layer in dependence of the pressure and the sticking probability α

generated by vapour deposition or by sputtering in ultrahigh vacuum (UHV) below 10^{-5} Pa. Also cleaving of bulk material in UHV may be applied; however, the new surface may be covered with contaminants occluded in the bulk material. A widely used cleaning technique is baking under vacuum. Then, for thermogravimetric or sorption measurements a high vacuum better than 10^{-3} Pa may be sufficient.

After preparation of the sample some time is required to adjust the measuring conditions, e.g., the temperature. Within this period of time the sample surface may be contaminated once again. The diagram (Fig. 5.43) shows the time period in which a smooth surface will be covered with a 10 percent sorbate layer depending on the partial pressure of the sorptive gas and for different sticking probabilities, α. The diagram demonstrates that after degassing, the measurements should be initiated without delay, beginning in a very good vacuum free from condensable components of the pump fluid, of monomers from the sealing material as well as from active molecules like oxygen.

If a dry pump is not used, the apparatus must be kept free of condensable constituents by means of a cold trap or catalysts, respectively. When cooling the sample, the walls of the cooled balance tubes (Fig. 5.44) act as a cold trap and, therefore, the oven for sample degassing should be exchanged by the cryostat as fast as possible. Fortunately, with the balance contamination can be observed.

Another value for the required ultimate vacuum is given by thermo-molecular flow. By temperature gradients a gas flow is generated which affects reactive forces on the hangdown wires and the pans in the vicinity of temperature gradients. For typical arrangements maximum of that so-called Knudsen forces is observed at $Kn = 1$. To avoid disturbances the ultimate pressure should be below 10^{-2} Pa.

It may be mentioned that in such arrangements in the pressure region above 10 Pa disturbances by Knudsen forces vanish, whereas up to 100 Pa disturbances due to

Fig. 5.44 (a) Diagram of a gravimetric sorption apparatus. *1* sample, *2* Dewar vessel with liquid nitrogen, *3* vacuum aggregate, *4* pressure gauge, *5* microbalance, *6* recipient with sample tube, usually glass. The wall of the cooled balance tube acts as a cold trap. (**b**) Resulting adsorption isotherm. m_a mass adsorbed. p/p_s relative pressure = gas pressure p: saturation pressure p_s

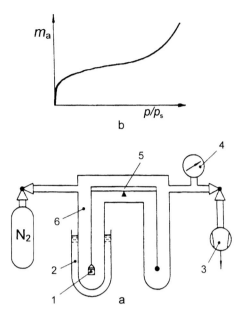

convection are still negligible. The region between 10 and 100 Pa, therefore, is very suitable for thermogravimetric investigations.

5.2.5.3 Choice of the Pump

Any pump which produces the required ultimate vacuum is suitable for its combination with a balance. The choice of vacuum pump depends on the intended application. Investigating humid samples the vapour pressure of the moisture is limiting for the attainable vacuum and a mechanical forepump of rotary vane or diaphragm type or even a water jet pump may be sufficient. If corrosive gases are involved, the use of an oil or mercury diffusion pump with condensation trap and series-connected fore pump is indicated. It is possible to make the entire apparatus including the diffusion pump from glass or quartz and to use low-gassing synthetic sealed (e.g. polytetrafluorethylene, Teflon, Hostaflon, etc.) glass or quartz valves. A jet pump may serve as a forepump. High vacuum diffusion pumps work with oil and a cold trap is required to protect the receptacle from oil vapours. For universal high vacuum investigations of surfaces, vacuum aggregates with an oil free pump should be preferably used combined with oil-free mechanical pump. In ion getter pumps high magnetic and electrical fields are applied which may disturb the balance. Modern turbomolecular pumps run so quietly that they can be flange-mounted directly on the balance case, with only very small loss of suction.

The transmission of vibrations from the backing pump can be prevented by means of a plastic tube. By means of baffles and cold traps the balance vessel must be protected from vapours backstreaming from jet pumps. For all types of pumps after turn off, dry air or a protecting gas should be fed into the apparatus from above the pump, so that no pump oil or contaminants from traps are swept into the balance chamber.

From Eq. (5.3) it follows that the speed of the pump need not to be much larger than the total conductance of the tubular system. By experience we found, however, that a speed of about $100 \, 1\text{s}^{-1}$ is required to obtain always a good vacuum in an appropriate time.

5.2.5.4 Pressure Measurement and Control

There is a large range of commercial pressure gauges available, suitable for measurements from high-vacuum to high pressure. Instruments including mercury have passed out of use. It should be noted that ionisation gauges emit ions and steam jets. In order to avoid dynamic effects and contamination, the flange of the measuring head should never be directed straight at the balance beam or pan. The measuring head must, however, be as close as possible to the balance, as substantial drops in pressure may occur in vacuum. A measurement on the pump connecting piece may simulate a vacuum that is better by powers of ten. Furthermore, in high vacuum, pressure differences occur between cooled and heated parts of the apparatus (Knudsen effect). If two vessels with different temperatures T_1, T_2 are connected by a tube with a diameter d_r, a pressure drop p_1 to p_2 occurs in the same direction. In the limiting case, if the free travelling path of gas molecules $\lambda \gg d_r$, the ratio of the pressures reaches the maximum value [43].

$$\frac{p_1}{p_2} = \sqrt{\frac{T_1}{T_2}} \qquad (5.8)$$

This means that the vacuum on a cooled specimen hanging in a narrow connecting piece is somewhat better than the vacuum in the balance. In cooling with liquid nitrogen the effect is reinforced by the pumping action as a result of condensation on the tube wall.

For the fine vacuum range up to moderate pressure, a diaphragm and quartz helical gauge with electrical output which can be used to control and record the pressure and to automate the measurement are recommended, (it should be noted that the force-measuring methods used with these instruments correspond to those used with vacuum microbalances). If the kind of gas used is not changed during a measurement, the buoyancy can be used to measure the pressure.

The gas pressure for the range above 10^{-2} Pa in the balance chamber can be regulated by means of metering and solenoid valves controlled by the pressure gauge. A large variety of gas pressure and vacuum gauges with electrical output is available for that task. An example of a gravimetric apparatus is shown in Fig. 5.45. In this special case a second balance using buoyancy and Knudsen forces is the pressure sensor [44, 45]. If the pressure gauge indicates a deviation from the set point gas is either introduced by opening of solenoid valves of different size (depending on pressure) or pumped off by slow operation of a motor-controlled metering valve.

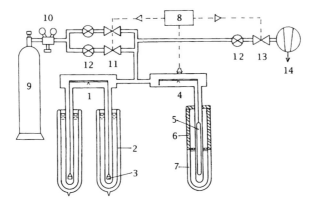

Fig. 5.45 Gravimetric sorption apparatus with gravimetric pressure measurement and control in the pressure range 10^{-2} Pa to 10^{6} Pa. *1, 4* compensating vacuum microbalances, *2* cryostat or heater, *3* sample pan, *4* buoyancy/Knudsen pressure gauge, *5* void glass balloon, *6* thermostat, *7* cryostat, *8* pressure control, *9* (N_2) gas cylinder, *10* reducing valve, *11* solenoid valves, *12* needle valves, *13* motor metering valve, *14* vacuum pump (turbo molecular pump + rotary vane pump

For high and ultrahigh vacuum a dynamic procedure may be applied [46]. A steady gas stream is introduced into the balance chamber and steadily pumped off by the pump.

Vapour pressure can be controlled by thermoregulation of the liquid in the storage tank. To avoid the vapour unintentionally condensing in parts of the apparatus, these should be thermoregulated to a higher temperature.

5.2.5.5 Choice of the Balance

For good vacuum the balance should be both easy to clean and able to withstand heat. For sorption and thermogravimetric measurements high resolving beam balances are most suitable. The relative resolution (discrimination threshold related to maximum capacity) of commercial instruments is of the order of 10^{-8} and may be improved to 10^{-10}. Disadvantageously the sensor cannot be heated to temperatures above 100 °C. Furthermore, electronic parts and coils may be affected by organic or corrosive vapours [47]. In order to attain ultra high vacuum, it is necessary to place the transducer outside the vacuum system [48]. This is not realised in any commercial beam balance. On the contrary, ancillary equipment like sample changers are to be tolerated which may additionally worsen the vacuum.

The vacuum chamber of the magnetic suspension balance (Figs. 4.22/4.24) contains only the suspension magnet, sample pan and sample. It can be heated up to 200 °C. The magnet may be protected against aggressive gases by a quartz envelope. In some suspension balances, however, so many parts are built in, that the advantage of separation of sensor and sample is called into question.

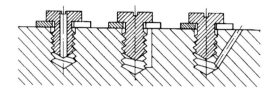

Fig. 5.46 Several possibilities to vent screw-holes. (*a*) central drill hole, (*b*) lateral groove-pin, (*c*) lateral drill hole. In addition the washer should provided with a sparing

With regard to its low surface and its heatability, spring balances of quartz, tungsten or special alloys are most suitable for vacuum. Regarding its low relative resolution of 10^{-5} its application is limited.

Quartz resonators and other oscillating sensors are used for special tasks like thickness monitoring and contamination measurements. For such devices the sample needs to be connected strongly to the sensor surface. Strain gauges and other force sensors exhibit a low relative resolution comparable to spring type balances.

5.2.5.6 Design of the Apparatus

The conditions in high vacuum in which gas transport takes place in the form of a molecular flow are factors deciding the design of the pump and apparatus. The gas transport is restricted by the flow resistance of the pipework and the components between the high-vacuum pump and the sample under investigation [49]. In general, the diameter of the narrowest part is decisive. An aperture of 1 cm^2 has a conductance of 11.6 l/s of air at room temperature. The pump's suction flange which is generally adjusted to the suction capacity by the manufacturer is an indication. This flange and the respective size of the pump must not be much larger than the narrowest part of the pipework. The connection tube should be short; long bellows should preferably be avoided. In most cases standard component parts are used whose conductance is stated by the manufacturer. The reciprocal values of the conductance of the component parts arranged in series are cumulative and so their diameters should be chosen wide enough but comparable in size. Parts built in should not have any hollow side-space with small entrance. In such a void liquids like oil or water may be trapped which evaporate slowly and worsen the ultimate vacuum. Screw holes and similar parts need to be vented by means of drill holes as shown in Fig. 5.46. In addition, washers should be provided with a groove.

Typical microgravimetric apparatus with beam balances [50] are shown in Figs. 3.9, 5.48–5.51. In an apparatus for water vapour measurements described by Willems [51] (Fig. 5.48) the speed of the 80 $1 s^{-1}$ oil diffusion pump is dropped to about 1 $1 s^{-1}$, caused by long tubes and stainless steel bellows. The pressure sensor is arranged at the vacuum chamber not far from the sample.

In Fig. 5.48 120 $1 s^{-1}$ turbo-molecular pump is hung in free suspension on 10 cm long soft expansion stainless steel bellows. The connection can be made with a

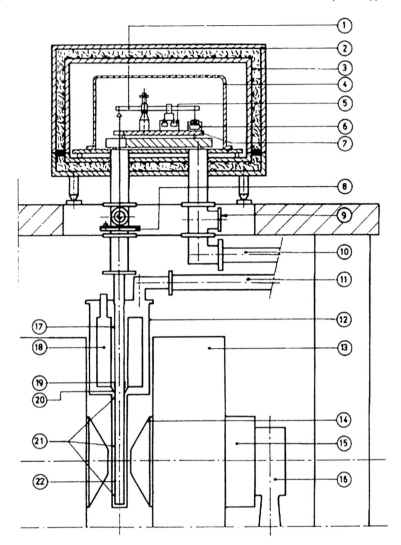

Fig. 5.47 Cross-sectional diagram of the balance system for measurements of magnetic susceptibility of solids. *1* balance beam, *2* thermal shield of the balance, *3* copper coils for stabilising the temperature of the balance to within o.e °C, *4* iron vacuum chamber, *5* loudspeaker magnet with coil, *6* microformer M with core k, *7* base plate, *8* alignment, system for the Dewar, *9* vacuum-sealed electrodes, *10* connection to vacuum fore pump, *11* connection to high vacuum pump, *12* metal dewar, *13* coils of the magnet, *14* pole shoes, *15* return yoke, *16* magnet support, *17* thin-walled monel tube which thermally isolates the sample tube, *18* cryogenic liquid reference container, *19* electric heating element, *20* heat leak:thin copper ribbon, *21* copper-constantan thermoelements, *22* copper sample tube. © A. van den Bosch

longer tube corresponding in diameter to the connecting piece, as viscous or Knudsen flow occurs here. The pumping speed is limited mainly by the long balance tubes

Fig. 5.48 Microbalances
with vacuum pump stand.
1 microbalance, *2* Penning
vacuum gauge, *3* sample pan,
4 stand (hollow pillar),
5 footing of stand (solid steel
beam), *6* stainless steel
expansion bellows, *7* Pirani
vacuum gauge,
8 turbo-molecular pump,
9 plastic tube, *10* rotary vane
pump, *11* shock absorbers

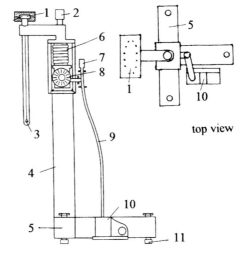

top view

Fig. 5.49 Arrangement of
six vacuum microbalances
under a bell jar. *1* bell jar,
2 microbalances, *3* flow
protection shield, *4* solenoid
valves, *5* high vacuum pump,
6 stand, *7* flexible tube,
8 rotary vane pump

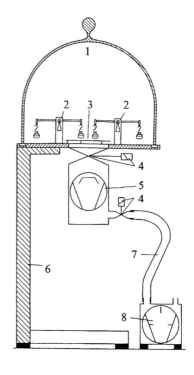

to about $8\,\mathrm{l\,s^{-1}}$. The valves should be chosen with great care for quick and jerk-free operation. Electro-valves are practical, allowing the apparatus to be automated. The backing pump of rotary vane type is placed separately at the ground and connected via a flexible tube. A Penning gauge arranged just above the suction valve provides only a rough impression of the vacuum in the manifold.

Fig. 5.50 Principle of the
magnetic suspension balance
according to Gast.
1 connection to balance,
2 soft iron casing, *3* carrying
bar magnet, *4* force coil,
5 sensor coil, *6* dielectric
window, *7* silver plate (eddy
current sensor), *8* suspended
magnet, *9* sample pan.
© Theodor Gast

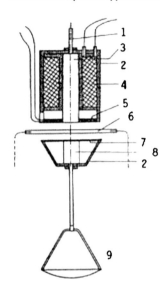

In Fig. 5.49 an apparatus with several balances is depicted. The pump with main suction valve is clamped directly to the vacuum chamber. The vibrations of a 500 l s^{-1} turbo-molecular pump did not disturb the weighing. A parallel small suction line including a metering valve is not shown. It is required for slow pre-pumping in order to avoid scattering of powdery samples.

Moderate heating considerably shortens the pumping process and improves the final vacuum. The vacuum chamber should contain only few organic materials (insulation, electronic components, and glue) and the material necessary must be suitable for vacuum. Large areas of contact should be avoided in the construction or modelled so that there are no hollow spaces where access is difficult.

5.3 Magnetic Suspension Balances

In gravimetric measuring techniques the term 'suspension balance' designates a scale in which the sample hangs directly below the mass sensor, usually of spring type or any load cell. The magnetic suspension indeed is a coupling device between a balance (usually an electronic laboratory balance) and the sample (Fig. 5.50). In free suspension no mechanical connection between the sample and the measuring set-up exists. Until now, attempts to use the coupling device directly for mass determination did not result in suitable technical solutions.

For numerous applications it is unfavourable to locate a mass sensor in the measuring chamber itself (e.g. at high temperatures, high pressures, aggressive atmospheres). In many of these cases magnetic suspension balances could be a suitable solution. By means of magnetic suspension balances it is possible to accurately measure the weight changes of a sample located in a closed measuring cell made of metal or glass. In these cases, the balance (a commercial analytical balance or

Fig. 5.51 Distance control of the magnet of the suspension system from outside of the balance. *1* carrying magnet, *2* suspended magnet, *3* Hall sensors, *4* force coil, *5* current regulator, *6* amplifier, *7* current/voltage transformer. © Theodor Gast

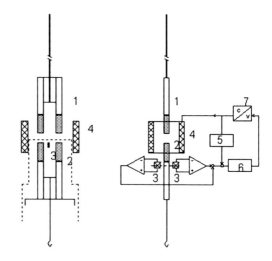

microbalance) is located outside the measuring cell at ambient atmosphere and the sample is weighed contactless of the mass sensor. The key component of the magnetic suspension balance is the magnetic suspension coupling which consists of an electromagnet, a suspension magnet, a position sensor, and a control system. The electromagnet is attached to the underfloor weighing hook of the balance and maintains a freely suspended state of the suspension magnet. To achieve the freely suspended state of this permanent magnet, its position is detected by a position sensor and controlled via a direct analogue control circuit. The position sensor may be a photo cell, an electromagnetic device or a Hall sensor (Figs. 5.52/5.53). Hermetically sheltered from the atmosphere and the measuring system, the sample can be held under extremely clean conditions e.g. in ultrahigh vacuum, treated with corrosive gases under variable temperature and pressure, whilst the balance system is outside. The suspension balance is used advantageously if during the thermal treatment a corrosive gas is set free or if the investigation includes the treatment of the sample with a gas which attacks the instrument. As a vacuum suspension balance, the sample, usually in a sample basin, is suspended at a permanent magnet and enclosed in a receptacle. This magnet is held floating by means of an electromagnet (coil with soft iron core) from outside which is suspended on a recording balance.

Free magnetic suspension was considered in 1947 by Clark as the basis of an analytical balance [52] and in 1950 by Beams [53] as an aid to weighing. In 1959 Gast [54] proposed suspending the balance pan and specimen in an enclosed space with a controlled atmosphere and transferring its weight force magnetically through the wall of the container to an analytical balance without loss of information.

The Gast magnetic suspension balance consists of a conventional electronic balance to which the sample is connected by magnetic force at a short distance below or above the measuring system. There are several ways of establishing and controlling suspension [55]. Most practical application involved controlled ferromagnetic attraction; an upper bar magnet suspended from the beam of an electromagnetic balance carries a lower magnet with a pan. A nonmagnetic, preferably dielectric, wall

Fig. 5.52 Principle of a
suspension vacuum balance
with photo cell position
sensor. Diagram of magnetic
coupling. *1* balance beam,
2 carrying magnet,
3 suspended magnet, *4* pan,
5 recipient, *6* light source,
7 condenser lens,
8 photodiode, *9* control unit.
© Theodor Gast

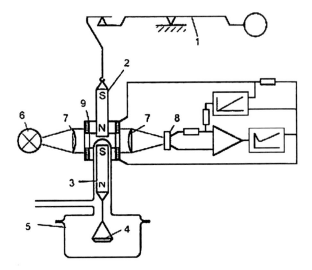

Fig. 5.53 Magnetic
suspension balances. *Left*: top
loader, *right*: suspended pan.
© Theodor Gast

between the attracting poles separates the reaction chamber from the balance. The
distance of the magnets is measured by an inductive sensor while a superimposed
control loop keeps the controlling power at a minimum.

In the development of electromagnetic balances, progress in sensor technology
and permanent magnetic materials, electrical components and circuitry have extended the range of electromagnetic compensation from a small fraction of the measurement range to the full capacity of the balance used. The balance beam could be
replaced by a system of parallel levers, which guide the balance pan in a vertical
path of deflection. In prototypes any mechanical guidance could be omitted and replaced by artificially stabilised magnetic attraction or repulsion which can also be
combined. With magnets of rare earth-cobalt and neodymium-iron-boron [56] the
maximum energy produced has been increased tenfold compared with ALNICO.
Due to the small diameter of the magnets, an enclosure for the lower magnet can
be made which withstands pressures up to 5 kbar without distance control from the
outside being obstructed. The precision of the distance measurement could be considerably improved by the application of a phase locked loop. With the aid of an im-

Fig. 5.54 Principle of a top loader suspension balance. *1* self-compensating balance and suspension control, *2* balance pan with carrier magnet, *3* electromagnet, *4* receptacle, *5* suspended magnet with sample pan, *6* position sensor. © Theodor Gast

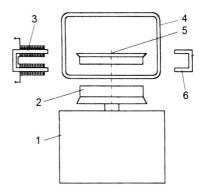

proved eddy current sensor, a frequency signal is obtained, which allows weighing without external balance with an uncertainty of 1×10^{-5}. Distance control is also possible by extracting the velocity signal, which is induced in the control winding by vertical motion of the magnet, from the superimposed regulating voltage [57]. Further improvement allows the coupling device to be used in addition to transmit data for other parameters, like the sample temperature [58]. These principles have been applied to balances with the sample below the measuring system as well as for top loaders (Figs. 5.54–5.56). Recent developments concerned the arrangement of the distance control from outside the balance [59, 60]. Thus, the dead load of the balance could be minimised in such a way, that the full capacity of a microgram balance is available for the sample. Instead of using an electromagnet in the suspension, the compensating force is generated by a stator coil which influences the (virtual) mutual attraction of four permanent magnets by superimposing an external field. Using a Hall probe as a distance sensor, a window of poor conducting material between the suspension magnets can be avoided. Thus, the sample space can be enclosed in a high pressure autoclave (Fig. 5.56). The control mechanism for this type of suspension is depicted in Fig. 5.56. By equipment with a suspension system the applicability of a balance is remarkably extended.

Problems arise if additional values from the sample, e.g. its temperature are to be measured. With free suspension of the sample holder, wireless transmission of measured values is desirable. Capacitive, inductive and optoelectronic devices can be applied for this purpose [58]. Figure 5.57 shows a diagram of an inductive transmission at the Gast magnetic suspension coupling.

5.3.1 The Sartorius Suspension Balance

Sartorius. Göttingen, first produced a commercial model of a suspension balance (Fig. 5.58) developed by Theodor Gast [61]. It was based on a single arm type electromagnetically compensating laboratory balance enclosed in a thermostatted vessel. At the suspension rod a bar magnet is fastened. Below a glass tube which could be closed with a ground-in connection, the electromagnet is placed within the

Fig. 5.55 Diagram of a
top-loader suspension balance
for vacuum and high
pressure. *A* pan, *B* force coils,
C metal ing, *D* sensor coils,
E force coil, *J* cooling spiral,
K oven. © Theodor Gast

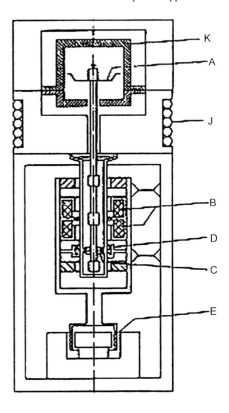

Fig. 5.56 Block diagram of
the regulated magnetic
suspension system.
1 oscillator, *2* discriminator,
3 PD controller, *4* integrator,
5 controlled current source,
6 load suspension. © Theodor
Gast

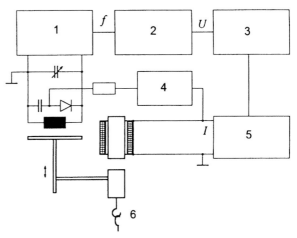

tube with the sample pan suspended below. The distance of the magnets is measured
by an inductive sensor while a superimposed control loop keeps the controlling
power at a minimum. Besides that model for vacuum and ambient pressure another
model was equipped with high pressure autoclave (Figs. 5.55b/5.59). On account of

Fig. 5.57 Transmission of
measuring values in Gast's
suspension balance.
1 oscillator, receiver,
3 holding electromagnet,
4 signal receiving coil,
5 signal transmitting coil,
6 sensor coil, *7* ferrite plate,
8 RF-receiving coil,
9 magnet, *10* multivibrator,
11 low-pass filter, *12* to
counter. © Theodor Gast

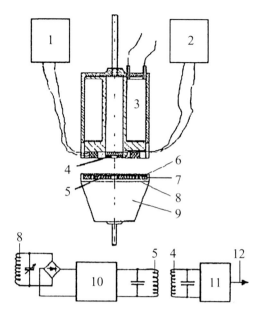

only little demand Sartorius stopped the production of suspension balances in the
1990s.

5.3.2 The Rubotherm Suspension Balance

The Rubotherm suspension coupling is shown in Fig. 5.60 [62–64]. It consists of an
electromagnet, hanging at the underfloor weighing hook of a conventional electronic
laboratory balance located outside the measuring cell, and a suspension permanent
magnet inside the measuring cell to which the sample is connected. The electromagnet voltage is modulated by a controlling unit in such a way that the suspension
magnet and the connected sample achieve a constant vertical position in the measuring cell. In this position the magnet and the sample are freely suspended and their
mass is transmitted to the microbalance through the wall of the (glass or metal) pressure vessel. The sample is linked to the permanent magnet via a load coupling and
decoupling device. By means of the magnetic suspension coupling, weight changes
of the sample are transmitted without contact to the balance. Only by changing the
control principle decisively has it been successful to develop a new type of magnetic
suspension balances that could be used for a great variety of different applications.
To avoid any influence of a zero-point drift of the balance the sample can be de-
coupled from the suspension magnet by lowering it to a zero-point position (tare
position) about 5 mm below the measuring position shown in Fig. 4.2b. Since only
the permanent magnet is weighed in this position a simple and fast zero-point check
is possible at all times. All movements of the suspension magnet and the sample

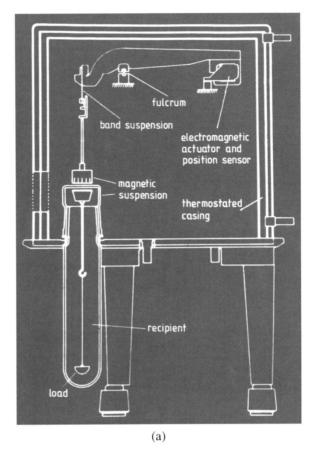

(a)

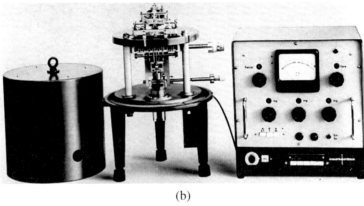

(b)

Fig. 5.58 (**a**) Sartorius suspension balance. © Sartorius, Göttingen, Germany. (**b**) Sartorius suspension balance. © Sartorius, Göttingen, Germany

Fig. 5.59 High pressure suspension balance.
1 electrodynamic balance,
2 carrying magnet with control coil, *3* suspension magnet, *4* bell oven, *5* sample pan, *6* autoclave. © Sartorius, Göttingen, Germany

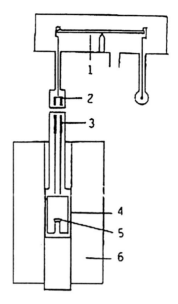

Fig. 5.60 Rubotherm suspension coupling on elecctrodynamic laboratory balance. *1* microbalabce, *2* electromagnet, *3* pressure vessel wall, *4* permanent magnet, *5* sample, *6* reactor pressure vessel, *7* measuring gas. © Rubotherm, Bochum, Germany

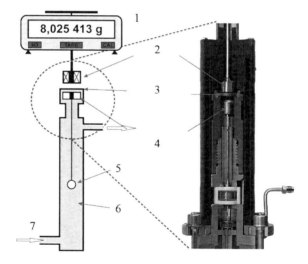

vessel are electronically controlled so that any vibration is avoided. The measuring accuracy at the balance is not adversely affected by the magnetic suspension coupling. The balance can be operated with a load range of 10 g or 100 g with resolution of 1 μg or 10 μg, respectively.

Magnetic suspension balances are especially applied for measurement of the density and the viscosity of fluids, surface tension, thermogravimetry, gas adsorption. The universal sorption measuring apparatus (Fig. 5.61) allows for sorption measurements to be carried out with almost any pure or mixed gas or liquid in the pressure

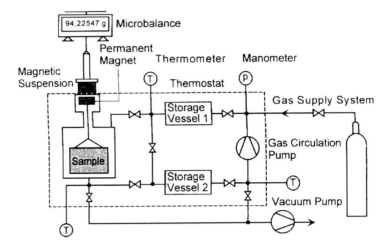

Fig. 5.61 Rubotherm Universal sorption measuring apparatus with magnetic suspension balance. © Rubotherm, Bochum, Germany

range from ultra high vacuum to 100 MPa (10^3 bar) at temperatures between 77 K and 700 K.

5.3.3 *The Linseis Suspension Balance*

Linseis developed a magnetic top loader suspension balance (Fig. 5.62). In this balance the sample is in a separate chamber made of stainless steel or quartz glass. The coupling to the thermal balance L81 is done magnetically. The electromagnetic compensating thermobalance is equipped with the permanent magnet. The separated sample container is situated above the beam of the thermobalance and it is also equipped with a permanent magnet which is repelled by the magnet below. Thus there is no mechanical connection between sample holder and thermal balance mechanism. The sample holder is held in a vertical position by an electromagnet. Due to the magnetic coupling all weight changes of the sample will be transferred without any mechanical connection through the balance. To compensate the drift of the balance the magnets are adjusted and controlled in their position. Due to this operation it is easy to do a zero adjustment for the measurement. All mechanical changes to the measuring system are compensated electronically.

5.4 Thermogravimetry

Mostly balances are used for single weighing of a sample mass. Thermogravimetry (TG) is the observation of changes in mass with time as a function of temperature

Fig. 5.62 Linseis magnetic top loader suspension balance. *1* counterweight, balance beam, *2* coil for radial magnetic suspension, *3* thermocouple, *4* crucible, *5* protection tube, *6* gas inlet, *7* gas outlet, *8* radial magnetic suspension, *9* stainless steel tube, *10* vertical magnetic suspension. © Linseis, Selb, Germany

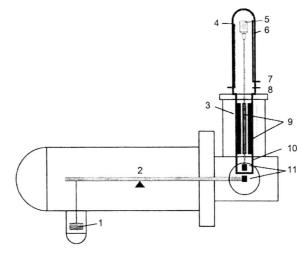

[65]. Thermogravimetry has become a widely used analytical technique in science and industry and so special thermobalances for such routine work have been developed. As shown in the preceding chapter many types of vacuum balances and suspension balances are simultaneously suitable for observations of weighing in a wide temperature range and can be designated as thermobalances. The presumably first record of a gravimetric observation of a thermo-chemical process was by Vitruvius in 27 BC by observing the loss of weight during lime-burning. He wrote [66, 67]:

> "Whatever the weight of the limestone thrown into the kiln, it equals not the weight of that what comes out of it. But when weight and the amount was the same, one can observe that it lost about one third of its weight."

The history of thermogravimetry and thermobalances is described by C. Duval [68], S. Iwata [69, 70], C.J. Keattch [71–73], and J. Šesták [74]. The invention of thermogravimetry, however, must be antedated, as discovered by C. Eyraud [40, 73, 75, 76]. In 1833 Talabot equipped a laboratory at Lyon with thermobalances for quality control of Chinese silk (Fig. 5.63) and that may be regarded as the beginning of thermogravimetry. The 'conditioning apparatus' of Talabot-Persoz-Rogeat were widely used for moisture determination of any material, e.g. in Hungary for determining the moisture of wool. In 1844 August Oertling exhibited a conditioning apparatus for silk at Berlin. The first scientific thermogravimetric assembly seems to have been described by Walther Hermann Nernst (1864–1941) and E.H. Riesenfeld in 1903: a quartz beam balance (improved by Julius Donau (*1877) (Fig. 5.15)) with electric oven. In 1915 K. Honda was the very first to use the term 'thermobalance' for his instrument [77, 78] (Fig. 5.64). Further developments on thermogravimetry were made in particular in France by M. Guichard [79], P. Dubois [80], P. Chevenard [81] and C. Duval [68] and also in Japan [69].

The first commercial vacuum balance with electromagnetic force compensation was a conventional milligram balance, presented in 1953 by Charles Eyraud

Fig. 5.63 Désiccateur Talabot, Lyon 1833. Conditioning apparatus at Lyon for the measurement of the humidity of silk

Fig. 5.64 Honda was the very first using 1915 the expression 'thermobalance' for his instrument

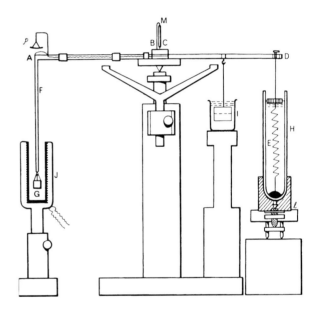

(†2009) and Ivan Eyraud [82, 83] and produced by the company A.R.A.M./D.A.M./ SETARAM of Lyon, France (Fig. 5.65). The deflection of the beam was detected by means of a photocell and compensated using a permanent magnet suspended at one side of the symmetric beam and forced by an electromagnetic field produced in coil surrounding the balance tube. For an improved model, the maximum load of the beam was up to 200 g and the range of electrical compensation up to 20 mg. In addition, up to 800 mg could be placed by mechanical manipulators.

Fig. 5.65 (**a**) The thermogravimetric apparatus of Charles and Ivan Eyraud was the first commercial apparatus equipped with an electromagnetic balance. Charles Eyraud with the first prototype. (**b**) The commercial setup of the Eyraud apparatus. (**c**) Diagram of the balance Eyraud

(a)

(b)

Fig. 5.65 (Continued)

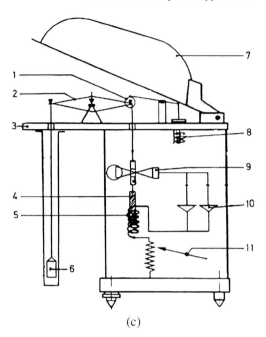

(c)

In 1953 W.A. de Kaiser developed the differential thermogravimetric method [84]. Important developments were made in Budapest (Hungary) by the brothers J. and F. Paulik [85–87], by G. Liptay [88] and others.

In contrast to real thermoanalytical methods in which amounts of heat are measured, in thermogravimetry the amount of material taken up from the surrounding gas phase or conveyed to it is determined. Indeed, thermogravimetry and calorimetry complement each other because there are reactions with heat evolution without evaporation of a volatile component and also reactions with evaporation and only small thermal effect. Unlike other chemical weighing processes in thermogravimetry the substance is not only weighed in its initial and end states, but the course of the reaction is followed continuously. According to the objective of the test, various working methods have evolved: isobaric, isothermal, and isobaric/isothermal [89, 90].

Thermogravimetric analysis (TGA) sets out to detect and determine concentrations of certain components in a compound. This involves usually measuring the evaporation of the components at linear increasing temperature T under isobaric conditions. The sample mass m_s usually stored at ambient atmospheric conditions is heated continuously and the thermogram is plotted:

$$(m_s)_{p=const} = m_s(T) \tag{5.9}$$

The decrease in mass dependent on temperature runs a typical course which can be used as a 'fingerprint' to identify the material and its composition. Typical curves are collected in 'atlases' [88]. The method is widely used in the analysis of ceramic substances and synthetics [91]. For that purpose special apparatus have been de-

Fig. 5.66 Oven configurations for thermobalances

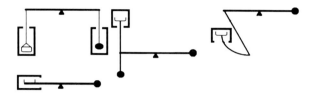

veloped. To facilitate its operation and to avoid disturbances by thermal air flow top-loaders or side-loaders are preferred with corresponding arrangements of the oven. Typical examples are depicted in Fig. 5.66. For routine measurements some thermogravimetric apparatus are equipped with automatic sample changer.

As the partial pressure of evaporating components in the ambient atmosphere is generally low, vacuum is not necessary for most measurement tasks. Thermogravimetric devices with milligram balances were first developed from conditioning appliances used to determine the moisture content of textile fibres. In more recent times however, some thermogravimetric apparatus were supplied with vacuum pumps and their thermobalances placed into vacuum vessels. Mass decrease is a quicker process in vacuum than in a gas atmosphere, as the diffusion on the surface and in the pores is not inhibited. Figure 5.67 is a diagram of a thermogravimetric apparatus [85]. Differential thermogravimetry (DTG) is a special form of TG. It

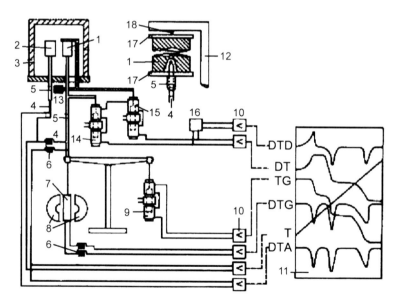

Fig. 5.67 Diagram of a thermogravimetric apparatus: Derivatograph according to Paulik, for simultaneous TD, DTD, TG, DTG and DTA. *1* sample, *2* reference sample, *3* oven, *4* thermocouple, *5* quartz tube, *6* flexible conductor, *7* coil, *8* magnet, *9* differential transformer, *10* recorder, *11* chart, *12* quartz rod, *13* counter weight, *14/15* counter-connected differential transformers, *16* derivative transformer, *17* quartz disk, *18* quartz wedge. On the right hand side typical records of thermogravimetric analysis. T temperature, TG thermogravimetry, DTG differential thermogravimetry, DTA differential thermo-analysis, TD dilatometry, DTD differential dilatometry

consists of two essentially very different methods, one being a comparison of mass change with the respective standard substance and the other a differentiation of the weight curve. Only the latter method is widely used.

To determine reaction kinetic data it is usually necessary to carry out isothermal measurements. The temperature dependence of the velocity constants of outgassing and ingassing makes it possible in many cases to characterise the reaction mechanism. The method is applied in dry metal corrosion, in adsorption as the fractional part of heterogeneous catalysis, and in desorption processes such as the drying of lacquer, synthetics and foodstuff.

Coats and Redfern [92] developed a method for the determination of kinetic data from non-isothermal thermogravimetry. Although the method is based on mathematical approximations and assumptions, it has been used to determine activation energies of reactions [93, 94].

The combination of the isothermal and the isobaric method in the quasi-isothermal method allows for the determination of thermodynamic data and the pore width distribution of porous materials. This method has been developed by F. Paulik [87] and P. Staszczuk [95, 96]. Only with extremely fast-running reactions can measurements also be carried out at continually changing temperatures. Sorption enthalpies can be calculated from the temperature dependence of adsorption and absorption isotherms. The measuring method can be used to investigate catalytic processes and to determine phase diagrams.

For practical purposes thermogravimetry is combined with differential thermal analysis (DTA), as the same temperature program is run through with both methods. DTA can be carried out on the balance pan with a restriction on the sample mass or with a comparison sample next to the balance pan. The balance may also be used in dilatometry [85, 86]. Also a combination with calorimetry is sensible. It is an advantage if the gases released in thermogravimetry are analysed in gas chromatographs or mass spectrometers. However, since for optimal use of the different methods different sample size also is advantageous, separate apparatus used in parallel are favoured.

As mentioned, in the 1960s, AMINCO manufactured a series of thermogravimetric apparatus with helical spring balances (Fig. 5.33). A thermogravimetric apparatus with a radiation oven and optical registration for investigating kinetic reactions with corrosive gases has been described by Gérard [97] (Fig. 5.68). This balance's sensitivity is 10 scale divisions/mg.

There exist a great variety of commercially available thermogravimetric apparatus, some of them designed for special applications. As examples we mention the Netzsch Tarsus thermo-microbalance which has a measuring range 2 g and a resolution 0.1 μg. (Fig. 5.69) and the thermogravimetric analyser Q5000 SA of TA Instruments (Fig. 5.70).

Using micromechanical cantilevers as applied in micro-mechatronic-systems (MEMS) it is possible to investigate samples with nanogram mass and mass changes in the pictogram range [98] (Fig. 4.126). Observing the resonance frequency of the elastic lever the mass curve in dependence of temperature can be registered.

Fig. 5.68 Gérard's thermogravimetric apparatus with helical spring balance. *1* quartz spring, *2* thermal insulation, *3* optical reading device, *4* vacuum pump with condensation trap, *5* infrared heater, *6* sample, *7* gas inlet, *8* polyethylene bag with gloves, *9* sample

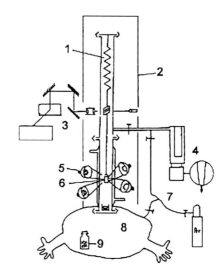

Fig. 5.69 Netzsch TG 209 F3 Tarsus - Thermo-Microbalance. Temperature range: 20–1000 °C, Heating and cooling rates: 0.001–100 K/min, Cooling time: 20–25 min in the temperature range 1000–100 °C, measuring range: 2000 mg, Resolution: 0,1 μg. *1* furnace cooling, *2* radiation shields, *3* thermostat control, *4* pressure release, *5* protectin gas, *6* gas outlet, *7* sample, *8* purge gass, *9* sample holder lift, *10* weighing chamber *11* microbalance. © Netzsch Instruments, Selb, Germany

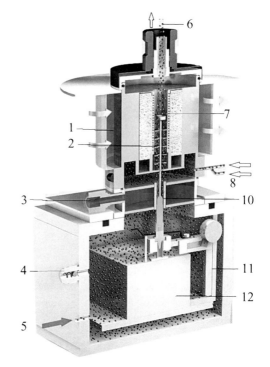

5.5 Sorptometry

For the understanding of gas sorption processes with a solid material the amount of the reacting gas adsorbed dependent on the variables of state, temperature and

Fig. 5.70
Thermogravimetric analyser
Q5000 SA. *1* reference
chamber, *2* humidity sensor,
3 humidity chamber,
4 humidifyer, *5* dry gas input,
6 microbalance, *7* sample
chamber. © TA Instruments,
New Castle, DE, USA

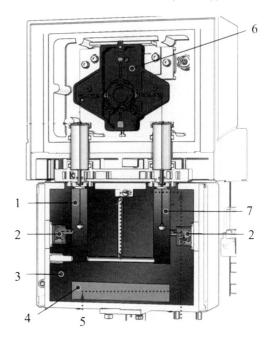

pressure, should be known. For that purpose usually sorption isotherms are measured. In an adsorption process, a gas (adsorptive) is concentrated on the surface
of the solid matter sample with mass m_s, and the adsorbed or condensed amount
m_a is measured as a function of the adsorptive pressure at a constant temperature
[99–102]. The amount of adsorbate is plotted against the relative pressure p/p_0 to
give the adsorption isotherm

$$(m_a)_{T=const} = m_a\left(\frac{p}{p_0}\right) \tag{5.10}$$

where p_0 = saturation pressure of the adsorptive gas at the measuring temperature.
Adsorption and desorption isotherms with non-reacting test gases like nitrogen and
noble gases allow for the determination of the surface structure of solids (specific
surface area, pore size distribution, surface fractality) [103]. A variety of measuring methods including gravimetric methods have been developed for that purpose
and standardised. Special methods are applied for the determination of humidity of
materials and of the humidity.

5.5.1 Measurement of the Sorption Isotherm

For the measurement of sorption isotherms in particular two concurring methods
have been established: volumetry and gravimetry. With the gravimetric method the
amount of adsorbed mass is measured directly with a microbalance, independent

Fig. 5.71
Thermogravimetric check of
the degassing of a sample. τ
time, m sample mass, T_1
temperature too low, T_2
optimal temperature, T_3
temperature too high:
chemical decomposition of
the sample, *1* balance, *2* oven,
3 sample, *4* counterweight,
5 vacuum pump

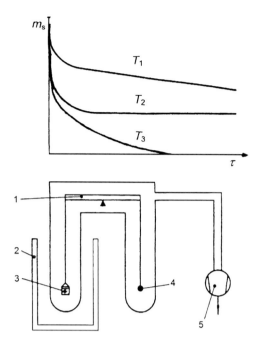

of the gas pressure measurement. With the volumetric method the mass adsorbed is calculated from gas pressure, temperature and a calibrated vessel. Nevertheless, the indirect volumetric method is preferred because the apparatus is much simpler and cheaper and the sophisticated operation of a microbalance is avoided. The most serious drawback of gravimetry is buoyancy whereas in volumetry that is the error of the dead volume: the difference between sample volume and container volume. We have shown that both errors are similar, because both are based on the density of the sample [104–106]. Though both methods may achieve the same sensitivity the gravimetric method reaches a far better resolution of the fine structure of the measured isotherm and clarifies much better effects and disturbances.

An additional advantage of gravimetric sorption measurement is that the preparation of the sample is also done on the balance and can be observed. This thermogravimetric test allows finding of optimum degassing conditions (maximum temperature, degassing period, vacuum) at which the sample is not damaged (Fig. 5.71). Furthermore, the dry mass of the sample m_s is determined to which the results should be referred.

Sorption phenomena in general are small effects occurring at the surface of a relative large mass of the object under investigation. For its gravimetric observation sensitive balances for continuous observation are required which nevertheless should bear relatively large loads. For the automatic measurement of sorption isotherms a variety of commercial instruments are offered, the first was by Sandstede and Robens [107], produced for some time by Sartorius (Fig. 5.72). The apparatus measures isotherms with nitrogen or other gases at stepwise changed pressure to allow for attaining equilibrium at each step. Gravimetric technique can be improved if the

Fig. 5.72 Gravimat of
Sandstede and Robens,
produced by Sartorius,
Göttingen. Apparatus for the
automatic measurement of
adsorption isotherms. On the
left a Gast vacuum
microbalance sticks out.
© SMS, UK

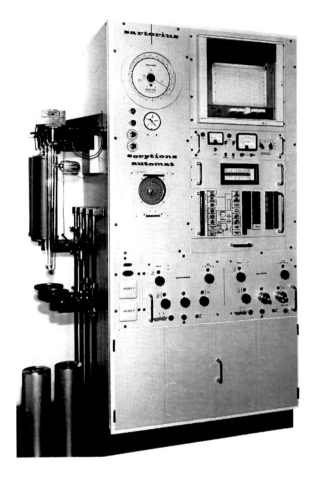

high resolution balance and the sample under investigation are physically separated:
the sample then can be exposed to the measuring atmosphere while the balance is
located under ambient conditions. This is achieved by using a magnetic suspension
coupling, which transmits the force acting on the sample due to gravity through the
wall of a pressure and temperature resistant vessel to the high resolution balance
located outside the vessel (Figs. 5.51–5.60). Of course, that solution is expensive.

Prior to measuring the adsorption isotherm, the buoyancy of the balance and the
sample must be determined in the measuring gas at room temperature. Then the
sample is thermostated at the measuring temperature. As the sample has no contact
with the thermostat it must be ensured that it achieves the measuring temperature
by means of a suitable long oven configuration and by shielding. The measurement
should be started in a vacuum of better than 10^{-3} Pa in order to avoid a zero-point
error from thermal gas flow. Because adsorption and desorption are slow processes
mostly the adsorptive gas pressure is changed stepwise and held until an equilibrium
mass is indicated.

Fig. 5.73 Record of automatic gravimetric measurement of a nitrogen sorption isotherm at 77 K using the Sartorius Gravimat

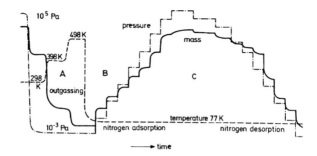

Fig. 5.74 Nitrogen sorption isotherm and regions for its evaluation using the methods of Henry or Dubinin/Radushkewitsch; Brunauer, Emmett and Teller (BET); Kiselev or Barrett, Joyner and Halenda (BJH)

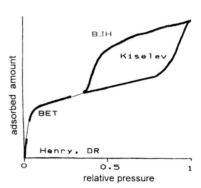

Standard methods for surface structure analysis are based on the measurement of adsorption isotherms with nitrogen at 77 K or with krypton or argon at 77 K or 90 K (Fig. 5.73). For that purpose the sample is cooled with liquid nitrogen or oxygen, respectively. Also water vapour at ambient temperature for hydrophobic surfaces and some organic vapours may be used. Such adsorptive gases are bound weakly at the sample surface by van der Waals forces and form several adsorbate layers dependent on gas pressure. The representation of the isotherms and its evaluation is standardised [108].

From the lower part of the isotherm the specific surface area can be calculated (Fig. 5.74), preferably using the standardised two-parameter equation given by Brunauer, Emmett and Teller (BET) [109]. That equation allows for the calculation of the capacity of the sample for a dense monolayer of the test gas, usually nitrogen [110]. More precise results may be obtained by application of thermodynamic theory [111–115].

From the beginning of the isotherm some information on pores with width below 2 nm (micropores) can be obtained [100, 116].

Isotherms on porous materials with pore width between 2 and 200 nm usually exhibit a hysteresis between the adsorption and desorption branch at elevated relative pressures. At increasing pressure the adsorbate layer grows on the surface and in the pores until at a value that is characteristic for every pore width, the pore is completely filled. In the desorption process the condensate forms a meniscus on the pore opening corresponding to the pressure of the sorptive. The pore size distribution can be calculated from those isotherms using preferably the standardised method of Barrett, Joyner and Halenda (BJH) [117, 118] which is based on an equation given by Thomson (Lord Kelvin) [119].

Many porous materials are fractal in bounded size ranges, their surfaces showing the symmetry property of self-similarity. To obtain the surface fractal dimension, according to Avnir and Pfeifer [120–122] the specific surface is determined dependent on the molecule diameter or the space required, by measuring several isotherms with molecules of various sizes. The thermodynamic method developed by Neimark [123–126] is more practicable. In this, the radius of curvature of the meniscus of a liquid condensed in pores is used as a scale. This means that only one isotherm has to be measured and evaluated in the range of the hysteresis loop. With nitrogen, structures between about 2 and 100 nm are then registered, and the limits of the fractal range ascertained [127].

5.5.2 Hygrometry

Of wide public interest but likewise of technical importance is weather forecasting and for that a knowledge of atmospheric humidity is essential [128]. In hygrometry the amount of free water is determined as a function of temperature. For that purpose water sorption isotherms from the atmosphere are recorded. The aim of the first gravimetric adsorption measurements was weather forecasting by determination of atmospheric humidity [129]. The very first to describe such an instrument was the German cardinal Nicolaus Cusanus (Nikolaus Cryfftz) (1401–1463). In his book [130, 131], published in 1450, 'Idiota de Staticis Experimentis' that means 'The layman about experiments with a balance' (Fig. 7.17), he let an ignorant person, probably his mechanic, suppose:

"If anyone hangs at one side of a big balance dry wool and loads the other side with stones until equilibrium is established, at a place and in air of moderate temperature he could observe that with increasing humidity the weight of the wool increases and with increasing dryness of the air it decreases. By these differences it is possible to weigh the air and likely one might perform weather forecasting."

About 20 years later the Italian architect and painter Leo Battista Alberti (1404–1472) described a similar device [132, 133]:

"We know that a sponge becomes wet from atmospheric humidity and by this fact we make a balance with which we weigh the weight of the air and the dryness of the winds".

From Leonardo da Vinci (1452–1519) we have three designs of inclination balances [134, 135] loaded with a sponge or with cotton. One of these drawings (Fig. 4.33) is found on a slip also showing sketches of the Last Supper, which was painted during 1495–97. Perhaps he intended to choose between fresco and oil with regard to the humidity of the room. With reflected face he added to the sketches:

"To recognize the quality and density of the air and to forecast rain"
and
"Means to detect, when the weather will break-up".

Fig. 5.75 The paper disc hygrometer includes a stack of paper discs, a balance and a scale Inscribed on it is "Adams Maker London Coventry Inventor". © Museo Galileo, Firenze, Italy

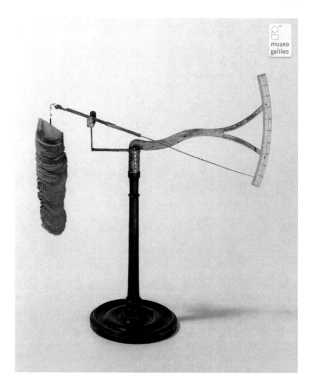

More than two hundred years later, in the 17th and 18th century the first accurate hygrometers were invented. Santorio Santorio (1561–1636) used cord stretched by a weight that hangs against a graduated index. Later, Fernando II de' Medici, Francesco Folli and Vincenczio Viviani made similar and improved instruments. The traditional folk art device known as a 'weather house' works on this principle with a sensor being catgut or hair.

A large variety of hygrometers, based on very different principles were tested, some of them based on weighing. The paper disc hygrometer includes a stack of paper discs suspended at a balance [136] (Fig. 5.75). Inscribed on such an instrument exhibited in the Museo Galileo, Florence, Italy is "Adams Maker London Coventry Inventor" indicating that it was made by one of the two Adams', father (1704–1786) or son (1750–1795), who were instrument makers and opticians in London in the 18th century.

Today the gravimetric method is applied as a primary standard for humidity measurements [137]. The design employs an automated, continuous-flow gas-collection and water-collection system for determination of the mixing ratio. The gas-collection system monitors the rate of gas flow by interferometrically measuring the displacement of pistons in two, precision-bore glass tubes. The mass of the water is determined by weighing the water-collection tubes before and after passage of the total amount of gas through the hygrometer. The NIST gravimetric hygrometer has an uncertainty of 0.1 % mass fraction.

Fig. 5.76 Infrared drying of
candies at - - - - 90 °C,
— · — · — 95 °C,
· · · · · 100 °C, -------- 110 °C,
and water content measured
using the Karl Fischer
method —-, © Isengard

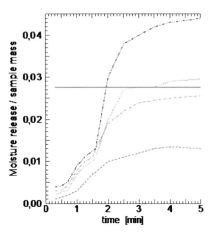

A suitable mass sensor is the quartz crystal balance whose surface is covered by
a hygroscopic film or by a microporous substance [138, 139]. As the mass of the
crystal changes due to adsorption of water vapour, the frequency of the oscillator
changes. Ultrathin Linde Type A (LTA-type) molecular sieves grown on a quartz
crystal microbalance have high sensitivity, good reversibility and long life at low
humidity ranges [140].

5.5.3 *The Moisture of Materials*

Moisture can be bound in very different ways and with different strengths in solid
or liquid materials. Furthermore, water near a solid surface or within a material can
form agglomerates of different structure [128]. The common term humidity also
comprises liquid contents other than water, e.g. solvents [141]. It is not easy to
define the moisture content and the dry state of a material.

The peculiar characteristic of the water molecule and its anomalous features facilitates its identification and allows the application of very different methods to
determine water content and humidity depending on the ambient conditions. The
gravimetric measurement of moisture content, water sorption isotherms or isobars,
are widely applied. Here always the sum of volatile substances within the material is
measured. On the other hand the different strength of binding is revealed as can be
seen in the example of isothermal drying curves of candies [142, 143] (Fig. 5.76).

With the conditioning instruments of Talabot, described in the section 'thermogravimetry' the moisture content of textiles was measured. Soon afterwards such
instruments were used to investigate the metabolism of plants (Fig. 5.77). In 1877
Hannay [144] described a gravimetric method to determine dehydration isotherms
of salts. He conveyed water vapour, which was released from a heated sample into
a cylinder that was filled with an adsorbent or absorbent, by means of a carrier gas.
The cylinder was weighed periodically. Warming up samples in a heating chamber

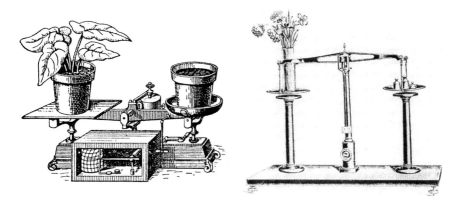

Fig. 5.77 Balances for the investigation of the metabolism of plants

and sequential measurements of the decreasing sample mass in order to determine the desorption isotherm is still widely used [145, 146].

Oven-drying is a widely used method. Here the sample is dried at a constant temperature. Humidity is removed by circulating air. The sample is weighed after reaching mass constancy. Such measurements give reliable 'true' results of the moisture content only in case of well known drying characteristics of the material. Weighing at intermediate times and deriving a kinetic curve may provide more information. Many microbalances described earlier were used first for moisture measurement. Today the measurement of materials humidity is facilitated by so called drying or moisture balances, which are laboratory scales equipped with an infrared or microwave heater (Fig. 5.78) [147].

An apparatus of gravimetric water sorption consists of a microbalance and a thermostat. Water vapour pressure is adjusted and varied by means of a carrier gas flow loaded with water vapour (Fig. 5.79a). Alternatively, using a vacuum balance, either portions of water vapour are added or the pressure is adjusted by means of a thermostated water reservoir (Fig. 5.79b). An apparatus designed especially for the continuous measurement of water vapour isotherms at ambient pressure is offered by Surface Measurement Systems (SMS) (Fig. 5.80).

The standard contact porometry was developed by Volfkovich, Bagotzky, Sosenkin and Shkolnikov [148]. Here, the sample is brought into contact with a porous standard sample with defined water content (Fig. 5.81). In thermodynamic equilibrium, the liquid into the whole pore system has an identical chemical potential. The humidity of the sample can be derived from the equilibrium water mass in the standard. The standard sample is weighed before and after the contact with the sample under investigation. A set of standards allows the determination of an isotherm even near saturation pressure.

Fig. 5.78 Drying or moisture
balance Sartorius, with
halogen infrared heater

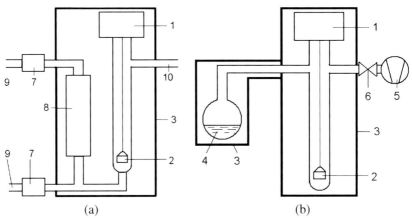

(a)   (b)

Fig. 5.79 (a) Water vapour pressure is adjusted and varied by means of a carrier gas flow loaded
with water vapour. (b) Alternatively, using a vacuum balance, either portions of water vapour are
added or pressure is adjusted by means of a thermostated water reservoir

5.5.4 Reaction Kinetics

Reaction kinetics may be measured isothermally or isobarically by means of a vacuum microbalance. Starting from vacuum the pressure is rapidly increased. As technical reactions often subside very quickly, the balance's response time must be short
[149]. It must also be taken into account that temperature changes of the sample
caused by the heat of reaction falsify the kinetics at the start of the course. If nec-

Fig. 5.80 Gravimetric water sorption measurement apparatus of Surface Measurement Systems, UK. *1* dry gas flow controller, *2* thermostat, *3* humidifyer, *4* microbalance, *5* sample, *6* counterweight, *7* temperature and pressure sensors

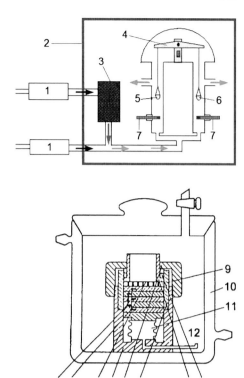

Fig. 5.81 Conditioning apparatus for standard contact porometry. *1* sample, *2* standard material of defined humidity, *3* perforated plate, *4* screwed cup, *5* gasket, *6* sample tube, *7* gas outlet, *8* centring shoulder, *9* gasket, *10* thermostat, *11* bellows, *12* vacuum/pressure connection. © Volfkovich, Bagotzky, Sosenkin and Shkolnikov

essary the temperature and pressure can be reduced to slow down the reaction in order to be able to measure it. As an example of this in Figs. 5.82/5.50 an apparatus containing six microbalances is depicted, designed for the measurement of metal corrosion in oxygen in the presence of nuclear radiation [150]. Additional irradiation with beta rays up to 10^6 rad/h did not cause any change in the uptake of oxygen; the course of the reaction, and the measured values are exactly the same as for the unirradiated sample (Fig. 5.83).

5.6 Surface Tension

Pliny the Elder (23–79) reported on calming sea waves by means of oil. In 1757 Benjamin Franklin described an experiment on a pond at Clapham, UK [151]:

> "I fetched out a cruet of oil and dropped a little of it on the water. I saw it spread itself with surprising swiftness upon the surface… Though not more than a teaspoonful, produced an instant calm over a space several yards square which spread amazingly and extended itself gradually till it reached the lee side, making all that quarter of the pond, perhaps half an acre, as smooth as a looking glass."

Reason for this phenomenon is the surface tension of water and generation of a monomolecular film at the surface which influences the surface tension. Surface

Fig. 5.82 (**a**) Apparatus with 6 Gast microbalances for corrosion measurements in presence of nuclear radiation. *1* thermostated stainless steel vessel, *2* electronic microbalances, *3* reaction tube, *4* diffusion pump, *5* butterfly valve, *6* lead shield against nuclear radiation, *7* turbomolecular pump, rotary vacuum pump. © Battelle-Institut eV, Frankfurt am Main, Germany, (**b**) Diagram of the apparatus with 6 Gast microbalances. *1* thermostated stainless steel vessel, *2* electronic microbalances, *3* reaction tube, *4* samples, *5* β-source, *6* diffusion barrier, *7* pressure gauges, *8* combined exhaust for reaction and protecting gas, *9* heat shields, *10* turbomolecular pump, *11* rotary vane vacuum pump, *12* radiation-protecting stainless steel block, *13* baffle, *14* oil-diffusion pump, *15* butterfly valve. © Battelle-Institut eV, Frankfurt am Main, Germany

(a)

tension is a property of the surface of a liquid or a solid caused by the cohesion of molecules whose forces are not completely saturated at a surface. It has the dimension of force per unit length = energy per unit area. Surface tension causes resistance to forces exerted to the surface so that some objects (e.g. water striders) can float on the surface of water, even though they are denser than water. Surface tension of a liquid is an important property characterising wetting and cleaning processes. It can be measured by means of a balance.

Agnes Pockels (1862–1935) investigated the surface tension of monolayers of hydrophobic and amphiphilic substances and the influence of impurities in her kitchen [152]. To measure the tension she developed the Pockels trough, precursor to the Langmuir film balance (Fig. 5.84) or Langmuir-Blodgett trough [153]. This simple device was a trough made from a tin pan with tin inserts for determining the size of the surface and a little beam balance with a 6 mm $\varnothing$ disk on one end to measure the force required to pull the disc from the surface. Using this device she described the general behaviour of surface tension with varying surface concentrations of oil.

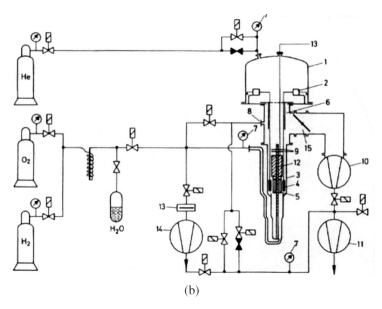

(b)

Fig. 5.82 (Continued)

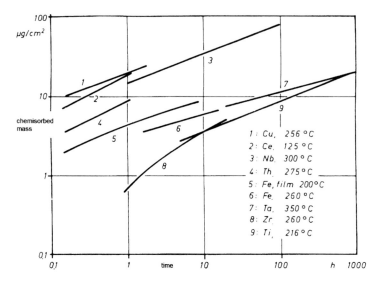

Fig. 5.83 Kinetics of the corrosion of metal foils in dependence of irradiation. © Battelle-Institut eV, Frankfurt am Main, Germany

Several modifications of the Pockels/Langmuir film balance have been designed and its application extended for the investigation of interfacial properties. The interfacial tensiometer according to Lecomte du Noüy (1883–1972) [154, 155] is based on a single-armed torsion balance. At the end of the beam of that balance a ring-

Fig. 5.84 Pockels-Langmuir
film balance

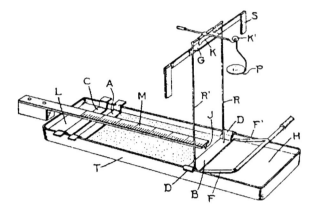

shaped platinum-iridium wire is applied (Fig. 5.85). The balance is situated on a
stand which can be moved vertically by means of a screw. It can be lowered so that
the platinum-iridium ring is immersed into the liquid under investigation. Then by
means of a revolving dial the torsion wire is twisted and the ring is pulled out of the
liquid but held by a lamella which ruptures finally. The force at the point of rupture
is measured as the value of the surface tension.

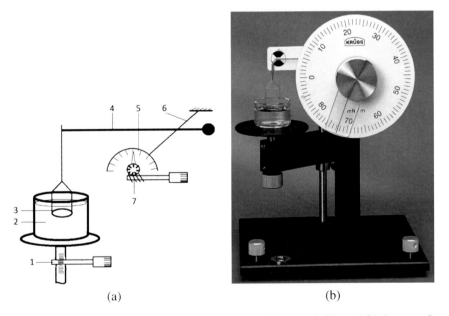

(a)   (b)

Fig. 5.85 (a) Principle of the Tensiometer according to Lecomte de Noüy. *1* friction gear for
adjustment of the liquid level, *2* measuring liquid, *3* platinum-iridium ring, *4* balance beam, *5* dial
(surface tension), *6* torsion wire (balance beam suspension), *7* worm pinion for twisting the torsion
wire. (**b**) Hand operated tensiometer. © Krüss GmbH, Hamburg, Germany

Fig. 5.86 Schematic diagram
of a scanning tunnelling
microscope (STM).
1 piezo-quartz crystal for
height control of the sensor
point, *2* sensor point, *3* way
of the sensor point,
4 molecular surface of the
sample, *5* sensor point control

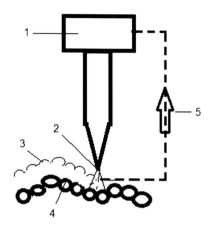

5.7 Mass of Single Particles

It is possible to weigh a single particle by means of electromagnetic balances down
to about 100 ng. Described below are cantilever methods which can be used derived from techniques applied in the scanning tunnelling microscope (STM) [156].
Other methods are based on free suspension of charged particles in an electric or
electromagnetic field. Furthermore buoyancy in a liquid or gas can be applied.

5.7.1 Cantilever Methods

The scanning tunnelling microscope according to Gerd Karl Binnig (*1947) and
Heinrich Rohrer (*1933) consists of a tip made of tungsten, platinum-iridium or
gold, with sharp end, connected via a microlever to a piezoelectric position transmitter (Fig. 5.86). When that tip is brought very near (0.4–0.7 nm) to a surface to
be examined, a voltage difference applied between the two can allow electrons to
tunnel through the vacuum between them (quantum tunnelling effect). If the tip is
moved across the sample in the x–y plane, the changes in surface height and density of states cause changes in the tunnelling current between surface and tip. These
changes are mapped in images. The change in current with respect to position can
be measured itself, or the height of the tip corresponding to a constant current can
be measured [157].

Binnig, Calvin, F. Quate (*1923) and Christoph Gerber invented the atomic force
microscope (AFM) in 1986. Besides imaging the AFM allows manipulating matter at the nanoscale. Electric potentials and forces and can also be scanned using
conducting cantilevers. It is possible by contacting the tip at the surface and from
the resulting rush of current to take single or groups of atoms out of the surface
[158–161]. The mass of such samples can be determined either via changes in the
tunnelling current, by means of the piezoelectric sensor or by exciting the lever to
oscillations and observing the frequency shift. Atomic gripper systems have been

developed which allow in similar way mass determination of molecules and groups of molecules. The sensitivity is down to the zeptogram range. A nanomechanical resonator has even been used to measure quantum effects of the thermal Brownian motion [162, 163].

Especially sensible are devices which apply oscillating systems (Fig. 4.12a). Such a system has been described based on a multiwalled carbon nanotube which is excited electrically to such oscillations and deflections. When an additional mass is attached the tube is bent and a shift in the resonance frequency can be related to the mass increase [158]. In this way the mass of a 22 femtogram graphite particle attached to the end of a resonating nanotube could be determined (Fig. 3.44). Manipulation and observation was made by means of a transmission electron microscope.

A micromechanical cantilever sensor consists of two major components: the cantilever that is used as transducer element and a coating [164]. The type of coating gives the sensitivity of the sensor to different stimuli. It has been possible to measure the resistance change of metallic line coatings due to bending similar to load cells. Arrays of micromechanical cantilever sensors can be integrated on a single chip in a small package of only a few mm^3 in volume. Such an array of several micromechanical cantilevers was made as a polymer brush.

5.7.2 Elementary Charge

The mass of a charged particle can be determined by observing its motion in an electric field. Benjamin Franklin suggested in 1747 for the first time, evaluating his experiments about electricity, that electricity may be quantised and may consist of something similar to atoms [165, 166]. His presumption was confirmed in 1833 by Michael Faraday's experiments on electrolysis. Discharging one and the same amount of electricity in different electrolytes produced always chemically equivalent amounts of dissociation products. Because the exchanged amounts of charge during electrolysis were of the same magnitude, it could not be concluded from these that Faraday laws were beyond doubt, because during electrolysis the charge of an ion is determined by an averaging process over many hypothetical charge units at the same time.

1897 J.J. Thomson proved the existence of electrons by his famous cathode ray tube experiment at the Cavendish Laboratory at University of Cambridge. But he determined here only the specific charge e/m_e of an electron (with e = elementary charge and m_e = mass of the electron).

5.7.3 Millikan's Droplet Experiment

Only a completely new type of experiment, which was realised by H.A. Wilson in 1903 delivered success. He observed the motion of ionised water droplets under the

Fig. 5.87 Millikans's experiment for the determination of e/m using oil droplets, sinking drop/rising drop. F_G gravitational force, F_E Coulomb electric force, F_R Stokes' friction force

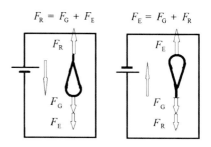

influence of a homogeneous electric field. Wilson was able to quote a value for the elementary charge, but these values were widely scattering; the explanation for this poor result is that the water droplets were easily evaporating and, therefore, the mass of a water droplet was not constant during measurement.

In the so-called Millikan experiment charged particles are suspended in a constant electric field to determine the relation e/m of charge to mass of a suspended particle. In 1909 Ehrenhaft and Millikan could weigh particles of 10^{-15} g ($= 1$ femtogram) mass [167]. R.A. Millikan improved Wilson's experiment by using charged oil droplets instead, which were not so easily evaporating and made longer observations with constant mass possible [168]. He observed its movement in a constant electric field to determine the relation e/m of charge to mass. Between two electrodes on an oil droplet of mass m_{oil} with quantity of electric charge q and suspended in the air the following forces are acting (Fig. 5.87):

Gravitational force $\quad F_G = m_{oil} \cdot g$ $\qquad\qquad\qquad\qquad$ (5.11)

Buoyancy $\quad F_b = m_{air} \cdot g$ $\qquad\qquad\qquad\qquad\qquad$ (5.12)

Coulomb force $\quad F_C = q \cdot E$ $\qquad\qquad\qquad\qquad\qquad$ (5.13)

Stokes' force of friction for a sphere $\quad F_{Stokes} = 6\pi r \eta v$ in case $v \neq 0$

$\qquad\qquad\qquad\qquad\qquad\qquad\qquad\qquad\qquad\qquad\qquad\qquad$ (5.14)

where m_{air} is the mass of air displaced by the oil droplet, g the gravitational acceleration and E electric field strength. For particles of droplet size the viscosity η is determining: it travels with constant velocity v. By observing the movement it is possible to determine the relation e/m and e if m is known. On the other hand m can be determined if e is known.

5.7.4 Straubel's Three-Plate Capacitor

Straubel [169–171] improved this method by superposition of an alternating field (Fig. 5.88). An electrically charged particle with mass in the nano- and picogram range is suspended electrostatically in an inhomogeneous alternating field. He developed a capacitor which consists of three parallel plates, each with a hole in its centre. If an a.c. voltage is applied between the intermediate plate and the ground,

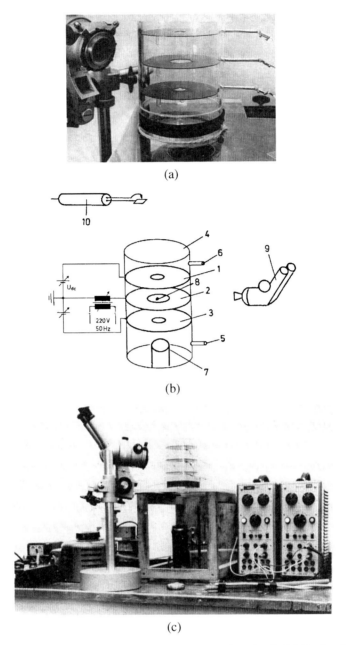

(a)

(b)

(c)

Fig. 5.88 (**a**) Suspension of charged particles in an alternating electric field in Straubel's three plate condensor and circuit for suspension. The mass can be determined by the onset of oscillations of the particle and in addition by compensation of the charge of the particle by an electrical field according to Millikan. (**b**) Arrangement of Straubel's three plate condenser. *1, 3* dc operated Millikan plates, *2* ac operated plate, *4* perspex cylinder, *5, 6* gas supply, *7* lamp, *8* suspended particle, *9* stereo microscope, *10* spatula with corona discharge. (**c**) Straubel's three-plate capacitor arrangement for the measurement of mass changes of electrostatically suspended particles

small particles carrying a sufficiently high charge will be trapped in the inhomogeneous field of the hole. The alternating field keeps the particle in a stable position exactly in the centre of the bore hole somewhat below the horizontal plate according to the weight force. A constant field between the outer electrodes may be applied to counterbalance the mass of the particle (Millikan experiment).

The conditions for a stable position of the particle are given by the following solutions of Mathieu's differential equation:

$$\alpha = \frac{16eU_{dc}}{m\omega^2\pi h R} \tag{5.15}$$

and

$$\beta = \frac{8eU_{ac}}{m\omega^2\pi h R} \tag{5.16}$$

where e is the charge of the particle, U_{dc} the d.c. voltage between the outer electrodes, U_{ac} the a.c. voltage and m the mass of the particle. Ω is the angular frequency of the a.c. voltage, h the distance of the outer electrodes from the intermediate electrode, and R is the radius of the hole in the intermediate electrode. There exists a region in a plot β versus α in which the particle is in a stable position. If the a.c. voltage is raised above

$$\beta_s = \sqrt{0.8245 + 4\gamma^2} \tag{5.17}$$

where γ is a damping factor for the particle in air, the particle will suddenly start to oscillate [172]. By observing the onset of oscillation it is possible to calculate the ratio e/m. In addition, relative weight changes of the particle are obtained from the relation

$$\frac{\Delta m}{m} = \frac{\Delta U_{ac}}{U_{ac}} \tag{5.18}$$

by repeated determination of the start of the onset of oscillation.

Relative mass changes due to sorption or vaporisation are determined from the onset of oscillations caused when the stabilising a.c. voltages is raised. If the charge, which exhibits always discrete values is determined by an independent method, the absolute mass can be determined [173].

The method was applied to investigate the kinetics of sorption and isotherms of suspended particles in air with various vapours at ambient temperature [173, 174]. The device depicted in Fig. 5.88 consisted of three lacquer-coated brass plates 110 mm in diameter, spaced at intervals of 40 mm and attached to a transparent plastic tube. Each plate had a central bore-hole; the hole of the intermediate plate is 30 mm in diameter. The particles were illuminated from below and observed by a lateral stereomicroscope. The particles were charged by friction or in a corona discharge and dusted from above into the capacitor. All particles but one were removed from the bore-hole of the intermediate plate by varying the a.c. voltage or using a charged glass rod. For stabilising the particles, a 50 Hz alternating field with voltages up to 10 kV was applied across the intermediate electrode. Normally, the mass of the particles was not counterbalanced, i.e. the experiment was made in absence of a constant field between the outer electrodes.

Particles of different plastic material, metals, silica gel, activated carbon, dextran and some salts were investigated with size between 2 to 50 μm and mass in the picogramm range. Sorptives were water, acetone and other organics. The minimum mass change observed was 10^{-13} g $= 0.1$ pg. The method can also be used to investigate droplets in sprays [175, 176].

5.7.5 Acoustic Wave Positioning of Particles

Ultrasonic standing waves can be used for positioning of droplets within a surrounding gas (e.g. air or a noble gas) serving as a carrier for the acoustic waves (Fig. 4.23). By low-frequency amplitude modulation the droplet may be stimulated to axial resonance oscillations. The resonance frequency depends on the density and the (electrical) energy required to establish the acoustic field and for positioning of the droplet. By means of a reference droplet of known density whose resonance frequencies are adjusted by tuning of the positioning voltage, the density of the sample can be determined [177].

5.7.6 Ion Traps

By several methods atoms or molecules or aggregates of a few of them can be trapped in electric or magnetic fields or combined systems of and held for some time for investigation [178].

5.7.6.1 The Penning Trap

Frans Michel Penning (1894–1953) invented a cold cathode vacuum gauge. Between the electrodes a discharge is generated by a dc voltage of around 2 kV. A magnetic field arranged crosswise to the electric field forces the electrons in a spiral path which is long enough (meters) so that a collision with gas atoms takes place even at low pressure. The electrons ionise the gas molecules and the pressure-dependent ion current is measured. Hans Georg Dehmelt (∗1922) modified that device using a strong homogeneous axial magnetic field to confine particles radially and a quadrupole electric field to confine the particles axially. This device named Penning trap allows capturing gaseous ions.

Nuclear ground state properties including mass, charge radii, spins and magnetic dipole and electric quadrupole moments can be determined by applying atomic physics techniques such as Penning-trap based mass spectrometry and laser spectroscopy [179]. The mass and its inherent connection with the nuclear binding energy is a fundamental property of a nuclide. Besides, the weak interaction of the

Standard model can be tested. The complex apparatus includes an electron beam ion trap for charge breeding, ion traps for beam preparation, and a high-precision Penning trap system (combination of electric and magnetic fields) for mass measurements and decay studies. Laser spectroscopy of radioactive isotopes and isomers is an efficient and model-independent approach for the determination of nuclear ground and isomeric state properties. A relative mass uncertainty of 10^{-9} can be reached by employing highly-charged ions and a non-destructive Fourier-Transform Ion-Cyclotron-Resonance (FT-ICR) detection technique on single ions stored in an ion cage. The frequency is the measure of the rotation of the charged molecule and depends directly from its mass [180].

In this way the mass could be determined of three Nobelium isotopes with half life between 2.3 s and 1.7 min with accuracy of 10^{-8} [181].

At the European Organisation for Nuclear Research (CERN) antimatter hydrogen could be stored in a Penning trap. It is planned to observe antimatter in the gravitational field in comparison to normal matter, and it is expected to find no differences in its behaviour. The search of antimatter will also be made by means of the Alpha Magnetic Spectrometer (AMS), which is going to be installed at the International Space Station (ISS).

5.7.6.2 The Paul Trap

A quadrupole electric field is used in some mass spectrometers to separate ionised atoms and molecules. Using this principle it is also possible to capture electrically charged particles. There exist several modifications; the most common type is the Paul trap, working on quadrupole field in a three-dimensional arrangement. Wolfgang Paul (1913–1993) used it to capture single ions by combination of a constant dc and radio frequency oscillating ac electric fields. Recently at the Garching Max Planck Institute also electrons could be localised in this way.

5.8 Mass Analyser

It is experimentally much easier and more precise to compare masses of atoms and molecules, that means to determine relative masses than to measure their absolute masses. Masses are compared with a mass spectrometer. Mass spectrometry is an analytical technique that identifies the chemical composition of a compound or sample on the basis of the mass-to-charge ratio of charged particles [182, 183]. It can be used also to determine the mass of a single molecule and to distinguish between isotopes. In mass analysers the sample is evaporated into vacuum, the molecules are fragmented into ions which are charged particles. Those ions are accelerated within an electric field and fed through electric and magnetic fields. The speed of a charged particle may be increased or decreased while passing through the electric field, and its direction may be altered by the magnetic field. Thereby they are separated according to their mass-to-charge ratio.

In 1886, Eugen Goldstein observed rays in gas discharges under low pressure that travelled through the channels in a perforated cathode toward the anode, in the opposite direction to the negatively charged cathode rays. Goldstein called these positively charged anode rays 'Kanalstrahlen' (canal rays). Wilhelm Wien found that strong electric or magnetic fields deflected the canal rays and, in 1899, constructed a device with parallel electric and magnetic fields that separated the positive rays according to their charge-to-mass ratio (Q/m). Wien found that the charge-to-mass ratio depended on the nature of the gas in the discharge tube. J.J. Thomson later improved that device by reducing the pressure to create a mass spectrograph [184].

The following two laws govern the dynamics of accelerated charged particles: Newton's second law of motion:

$$F = ma \tag{5.19}$$

and Lorentz force law

$$F = Q(E + v \times B) \tag{5.20}$$

where F is the force applied to the ion, m is the mass of the ion, a is the acceleration, Q is the ion charge, E is the electric field, and $v \times B$ is the vector cross product of the ion velocity and the magnetic field. Equating the above expressions for the force applied to the ion yields:

$$\frac{m}{Q}a = (E + v \times B) \tag{5.21}$$

This differential equation is the classic equation of motion for charged particles. Together with the particle's initial conditions, it completely determines the particle's motion in space and time in terms of m/Q. When presenting data, it is common to use the dimensionless mass-to-charge ratio m/z, where z is the number of elementary charges e on the ion ($z = Q/e$).

The mass spectrometer has three essential modules: (1) an ion source, which transforms the molecules in a sample into ionised fragments; (2) a mass analyser, which sorts the ions by their masses by applying electric and magnetic fields and (3) a detector, which measures the value of some indicator quantity.

5.8.1 The Ion Source

The sample is evaporated into high vacuum. The ion source ionises the material and the method applied depends on the type of the samples which can be analysed by mass spectrometry. Thermal or electron ionisation and chemical ionisation are used for gases and vapours. In chemical ionisation sources, the analyte is ionised by chemical ion-molecule reactions during collisions in the source. Other types include thermal or electro spraying, creating a plasma, glow discharge, field desorption, electron or atomic bombardment.

Fig. 5.89 Sector mass
spectrometer schematics.
A ion source, *B* magnet,
C detection unit, *D* ratio
output, *1* ionising filament,
2 gas inflow, *3* ion repeller,
4 electron trap, *5* ion
accelerator, *6* beam focusing,
7 bent beam, *8* Faraday
collectors, *9* amplifiers,
10 display

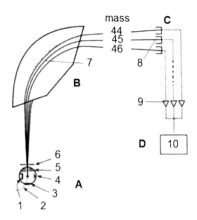

5.8.2 The Analyser

The ions are then sucked-in through the inlet system and accelerated by an electric field into the analyser system in which the movement of the ions is influenced by magnetic and/or electric fields to the detector/analyser. There are many types of mass analysers, using both static or dynamic fields, and magnetic or electric fields, but all operate according to the above differential equation. By tandem arrangement of several stages (MS/MS) the selectivity may be improved.

A sector field mass analyser uses an electric field to accelerate the ion beam and a magnetic field to bend the trajectories of the ions whereby the ions are separated according to their ratio m/Q or m/z, deflecting the more charged and faster-moving, lighter ions more (Fig. 5.89).

The time-of-flight (TOF) analyser uses an electric field to accelerate the ions through the same potential, and then measures the time they take to reach the detector. If the particles all have the same charge, the kinetic energies will be identical, and their velocities will depend only on their masses. Lighter ions will reach the detector first.

Quadrupole mass analysers use oscillating electrical fields to selectively stabilise or destabilise ions passing through a radio frequency quadrupole field [185]. A quadrupole mass analyser acts as a mass selective filter and is closely related to the Quadrupole ion trap, particularly the linear quadrupole ion trap except that it operates without trapping the ions and is for that reason referred to as a transmission quadrupole. A common variation of the quadrupole is the triple quadrupole. With the quadrupole ion trap the ions are trapped and sequentially ejected.

Fourier transform mass spectrometry, or Fourier transform ion cyclotron resonance MS, measures mass by detecting the image current produced by ions cycling in the presence of a magnetic field. Instead of measuring the deflection of ions with a detector such as an electron multiplier, the ions are injected into a Penning trap (a static electric/magnetic ion trap) where they effectively form part of a circuit. Detectors at fixed positions in space measure the electrical signal of ions which pass near them over time, producing a periodic signal. Since the frequency of an ion's cycling is determined by its mass to charge ratio, this can be de-convoluted by performing a Fourier transform on the signal. FTMS has the advantage of high sensitivity (since each ion is 'counted' more than once) and results in much higher resolution and thus precision.

5.9 Density Determination

An important parameter in material science is density. Whereas in classical physics the mass is an invariable property of the material, the volume of a sample depends on temperature and of the degree of aggregation. So, density depends also on state variables like temperature and pressure and in addition type of phase of the material. Sometimes density and volume of a sample are given and then above equation may be used to determine the mass; so that is an indirect method of mass determination.

General definitions are:

- Density ρ is defined as mass m per unit volume V

$$\rho = \frac{m}{V} \tag{5.22}$$

- Relative density (commonly referred to as specific gravity) is the ratio of the density of a material to the density of a reference material, usually water for liquids and air for gases.
- Apparent specific gravity is the ratio of the weight of a volume of the substance to the weight of an equal volume of the reference substance.
- Specific weight (sometimes unit weight) γ is the ratio of weight of a material to its volume. It is connected to density by

$$\gamma = \rho g \tag{5.23}$$

Unlike density, specific weight is not absolute but dependent upon the value of the gravitational acceleration, which varies with location.

Because values of specific gravity and specific weight are nearly similar to density, those notations should be avoided.

There exist several methods to determine the density of a solid, liquid or gaseous sample:

- Measurement of volume and mass
- Comparison with a standard sample
- Buoyancy difference in fluids of different density
- Influence on electromagnetic radiation
- Influence on acoustic waves (seismic waves).

5.9.1 The Measurement of Volume and Mass

Whereas mass is measured by a means of a conventional balance for volume measurement different methods are available and its use depends on the type of sample.

5.9.1.1 Gas Volume Measurement

The volume of a gas is defined by the size of vessel to be filled with that gas. If the receptacle can be extended with flexible wall as a balloon or by means of bellows the volume may be calculated from the geometric properties. For small amounts gas burettes and metering pumps are offered.

A quantity of gas at given temperature can be measured by change of the volume of a receptacle. Such a gasometer was invented by Antoine Laurent de Lavoisier in 1787: a cylinder, closed on top and swimming with the opposite open side in a basin filled with a liquid, usually water [186, 187]. The present day's gasometer type of rounded shape was designed by William Hasledine Pepys in 1802. The weight of the cylinder is partly compensated by counterweights.

Measurement of quantities of gas is an important economic task in industrial and private use e.g. of fuel gas. There exists a variety of measuring methods including two widely applied volumetric methods: wet and dry gas meter. The wet gas meter was invented in 1813 by Samuel Clegg, consisting of a case divided into two parts by a vertically arranged rotating cylinder which is half immersed into a water basin. The cylinder has about four separate chambers with peripherical openings. At the inlet side of the case a chamber is filled and after passage through the sealing liquid emptied. The number of rotations gives the consumed volume.

The dry gas meter works similar to a piston pump. A first patent application was granted in 1820 to John Malam. Improved in 1842 by Defries, industrialised by Siegmar Elster and improved again in 1878 by Emil Haas it is used widely today [188]. It consists of two separate chambers, each of which subdivided by a movable metal sheet with (leather) foil seal. The movement of one sheet controls that of the other in the second chamber and simultaneously connection valves which lead the gas from inlet via a chamber to the outlet. The number of movements is counted as a measure of consumed volume.

5.9.1.2 Gas Density Measurement

In order to determine the mass or the density of a gas it is weighed in a receptacle of known volume. For accurate determination large cylinders and large balances are in use. In industry also weighing of gas at elevated pressure is applied and the density value calculated using any of the following equations. The density of ideal gases like helium or nitrogen can be calculated using the equation of state:

$$\rho_g = \frac{m}{V} = \frac{pM}{RT} \tag{5.24}$$

where m, V and p are mass volume and pressure of the gas, respectively, M the molar mass, R the molar gas constant and T the thermodynamic (absolute) temperature. At moderate pressures the second-order corrections for non-ideality are generally obscured by other experimental influences. In tests with non-ideal gases or in the proximity of the condensation curve, modified gas equations may be used, e.g. a virial equation:

$$\frac{pM}{RT} \cdot \frac{V}{m} = 1 + B(T)p + C(T)p^2 + \cdots \tag{5.25}$$

where $B(T)$, $C(T)$ are virial coefficients. In this case it is recommended that the density of the gas be determined by the buoyancy method using a buoyancy body of known volume.

To describe weighings at different pressure we use

$$m(p) = m_0 - V_s \frac{pM}{RT} \tag{5.26}$$

where $m(p)$ is the value indicated by the balance at pressure p and m_0 the indication in vacuum.

5.9.1.3 Liquid Volume Measurement

Liquid volume may be measured in a graduated beaker or—more accurately—in a measuring flask. The meniscus at the borderline may cause problems of exact reading. A calibrated pyknometer flask is made with a vessel calibrated in volume, usually of glass or quartz with a glass stopper (Fig. 5.90a) and usually equipped with a thermometer (Fig. 5.90b). Such receptacles are weighed empty and filled. Dosage of determined liquid volumes can be made e.g. by means of a burette.

5.9.1.4 Solid Volume Measurement

The volume V_s of a compact solid body of regular shape can be calculated from its geometric parameters. However, if this body contains pores (open or closed) the resulting volume value covers the sum of solid matter and voids (Table 5.1).

For a solid sample with a cleaved or porous or a fractal surface, the borders are unclear and when measuring the surface area the resulting value depends on the size of the ruler applied. If we measure the volume of the sample with a measuring cup using a powdered material, e.g. sand, the resulting volume may differ significantly from the true bulk volume on account of the voids between the particles. Accordingly, in practice we have several different definitions of density.

The true solid volume V_s of an irregular shaped solid body or of roughly broken material can be determined by the measurement of the displacement of any suitable fluid: gas, liquid or fine powder using pyknometers similar to those used for liquids or so-called volumenometers. The measuring fluid should be a wetting liquid which is able to penetrate into the indentations and pores of the sample. Obviously, the result of the displacement measurement depends on the size of the molecules of the measuring medium. A molecule or a particle whose diameter is larger than that of the indentation or of the pore entrance cannot penetrate. The resulting volume V_s of

Fig. 5.90 (a) Pyknometer for liquids and granular materials according to Gay-Lussac. (b) Pyknometer with thermometer. © Assistant

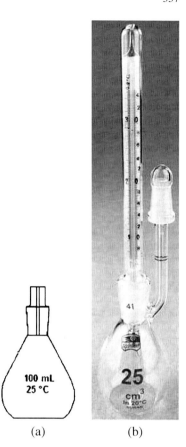

(a)   (b)

the solid part appears to be larger. If closed pores are present which are not accessible for the fluid molecules, the measurement results in a volume and its respective density which includes the sum of the solid part and that of the non-accessible pores.

The apparent volume (or bulk volume) of the sample $V_{apparent}$ defined as the sum of the solid volume V_s and the volume of all closed pores and open pores V_P can be determined by means of a non-wetting liquid or of a fine non-adhering powder which cannot penetrate into pores but into voids between aggregates. Other possibilities are covering a compact sample with a plastic foil or condom or impregnation of the pores, e.g. closing with wax, and measurement with any liquid or gas. Comparison of density determinations using different liquids which can enter pores, or not as a consequence of the surface tension are also used for that purpose. Comparison of the two values yields the porosity of the sample.

For granular materials, powders and dust the bulk volume V_{bulk} is of interest. It depends on mechanical treatment: loosely dumped, compacted by shaking or slightly compressed.

Table 5.1 Volume and density definitions

Notation	Definition	Volume	Density
Bulk volume	Solid + pores + spacings	V_{bulk}	ρ_{bulk}
Packed bulk volume	Solid + pores + spacings	V_{pack}	P_{pack}
Spacings volume		$V_{spacings}$	
Apparent solid volume	Solid + pores	$V_{apparent}$	$\rho_{apparent}$
(True) solid volume	Solid + non-accessible pores	V_s	ρ_s
Pore volume		V_p	

5.9.1.5 The Density of Solids

If we measure the solid volume V_s and sample mass m_s we can calculate the true solid density using its definition

$$\rho_s = \frac{m_s}{V_s} \tag{5.27}$$

or the apparent density:

$$\rho_{apparent} = \frac{m_s}{V_s + V_P}. \tag{5.28}$$

For granular material and powders the apparent density of loosely packed material ρ_{bulk} is of importance:

$$\rho_{bulk} = \frac{m_s}{V_s + V_P + V_{spacing}} \tag{5.29}$$

where $V_{spacing}$ denotes the volume of the interspaces between the particles. ρ_{paxk} allows for the calculation of the volume which is required by a given quantity of granular material or powder. Obviously the following relation exists between the three density parameters:

$$\rho_s > \rho_{apparent} > \rho_{bulk} \tag{5.30}$$

Often the relation apparent to solid density, given in percent, is denoted as relative density:

$$\rho_{relative} = \frac{\rho_{apparent}}{\rho_s} \cdot 100 \, [\%] \tag{5.31}$$

In ceramics this quantity is denoted as packing ratio as percentage of the theoretical density.

On the other hand, we can calculate the solid part of the sample volume by weighing if the density of the material is known. The solid part of porous bricks and other regular bodies $(V_s + V_P)$ may be determined by measuring the outer dimension, than impregnation with a liquid of known density ρ_l and weighing the liquid m_l (or measuring its volume) using

$$V_s = (V_s + V_P) - \frac{m_l}{\rho_l} \tag{5.32}$$

Other methods are based on crystallographic data from X-ray diffraction. From a density measurement of a sample the volume of included closed pores can be derived if the density of the compact material is known.

The porosity P of dispersed material is defined as ratio of bulk to particle density in the following equation:

$$P \text{ (percent)} = \left(1 - \frac{\rho_{bulk}}{\rho_{apparent}} \right) \times 100 \qquad (5.33)$$

5.9.2 Pyknometry

5.9.2.1 Liquid Pyknometry

Pyknometry is a method for measuring and comparing the densities of liquids and solids and determining the porosity of solids.

The density can be determined by comparing the mass of a pyknometer (Fig. 5.90) filled with a wetting liquid and the same pyknometer filled with the solid sample in it. Initially the sample should be ground to powder and then saturated with the liquid in vacuum to avoid air being locked up in the pores.

Widely used is fluid intrusion or extrusion. In the mercury volumenometer the sample is deposited in a calibrated sample vessel and by means of a piston the mercury is slightly pressed into the space between the sample aggregates (Fig. 5.91). Mercury displaced by the sample volume is measured in a graduated glass tube or capillary. Mercury has a high surface tension and cannot penetrate into pores. Comparison with other liquids which are sucked up give an impression of the porosity of the sample. That method has become obsolete because the use of the poisonous mercury is cut down by legislation. An exception is made for mercury porosimeter in which only little mercury is applied. In porosimeters the liquid (mercury, water or organics) is forced to enter the pores by high pressure. High pressure is obtained by means of a hydraulic pressure transformer and exerted on a hydraulic fluid which acts on the liquid column in the capillary (Fig. 5.92). In this way we can measure the pore size distribution: volume introduced in relation to pressure. By means of the Washburn equation we can determine the curve of pore volume as a function of pore width.

5.9.2.2 Saturation and Immersion

Bigger objects, e.g. bricks can be saturated with a wetting liquid, usually water with emulsifier [189]. The wet sample is weighed in air m_w and under liquid m_{wl}. The apparent volume $V_{apparent}$ is given by:

$$V_{apparent} = \frac{m_w - m_{wl}}{\rho_l} \qquad (5.34)$$

Fig. 5.91 Mercury volumenometer. *a* graduated burette, *b* sieve, *c* ground-in connection, *d* sample bulb, *e* mercury reservoir, *f* support on base plate, *g* piston, *h* crank, *i* screw rod, *k* toothed rack

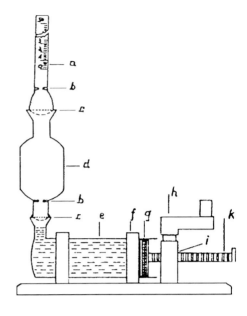

Fig. 5.92 Mercury porosimeter. *1* solid sample, *2* sample flask, *3* mercury, *4* graduated capillary, *5* oil, *6* screw cup, *7* pressure gauge, *8* pressure pump, *9* pressure transformer

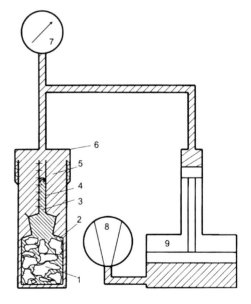

where ρ_l is the density of the liquid.

For the determination of the true (solid) density ρ_s the value of m_w and its inaccuracy is not important because instead the dry mass of the sample m_s is used:

$$\rho_s = \frac{\rho_l}{1 - \frac{m_{wl}}{m_s}} \tag{5.35}$$

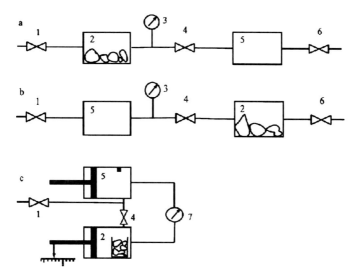

Fig. 5.93 Three procedures of helium pyknometry: (**a**) Filling of the calibrated sample vessel with helium and expansion into a supplementary volume, (**b**) Filling of the supplementary volume and expansion into the sample vessel, (**c**) Comparison of volumes at equal gas pressure. *1* inlet valve, *2* sample vessel, *3* pressure gauge, *4* connection valve, *5* supplementary volume, *6* outlet valve, *7* differential manometer

5.9.2.3 Gas Pyknometry

Gas pyknometry is a simple means to determine the volume V_s and the true density ρ_s of a porous sample. However, closed pores are included in measuring volume. In that case the sample should be milled in order to open up closed pores. Helium (99.99 %) is preferred because it is regarded to be hardly adsorbed at ambient temperature and it can be removed fast by evacuation. The small molecule can enter small pores. Because it behaves fairly as an ideal gas and the measurement is carried out at isothermal conditions Boyle-Mariotte's law can be applied without any corrections. However, careful investigations have revealed that indeed helium is adsorbed by almost all porous solids [105, 190, 191]. Adsorption is very quick: within few seconds a saturation value is obtained. Thus measurement of helium adsorption or an independent control of helium pyknometric measurements is advisable.

There are several modifications applied in commercial gas pyknometers. A volumetric/manometric apparatus (Fig. 5.93a) consists of a sample vessel of volume V_2, an additional volume V_5 and a manometer. The measuring procedure starts with flushing of the apparatus with helium at ambient pressure, closing the valves and adjusting the manometer to zero position. Then the sample vessel V_2 is filled with helium up to pressure p_1. The valve (4) is opened and the decreased pressure p_2 is measured. The value of the sample volume V_s can be calculated using

$$V_s = \frac{p_2 V_2 + p_2 V_5 - p_1 V_2}{p_2 - p_1} \tag{5.36}$$

In another arrangement (Fig. 5.93b) sample vessel and supplementary volume are exchanged.

The differential volumetric method (Fig. 5.93c) compares a cylindrical vessel which contains the sample with an identical reference cylinder, both equipped with pistons with graduated rod bolt. At identical positions of the pistons, gas is introduced. Then the valve which connects both cylinders is closed. The pistons are moved, in the comparison cylinder to a second calibrated position, and in the measuring cylinder so that a differential manometer between the two cylinders indicates no pressure difference. Because in both cylinders always the same pressure is adjusted the sample volume V_s can be calculated from the calibrated volumes of the cylinders in the initial position V_1, the end position of the reference cylinder V_2 and the difference of the positions of reference and sample cylinder ΔV, read at the rod bolt:

$$V_s = \frac{\Delta V \cdot V_1}{V_1 - V_2} \tag{5.37}$$

In this way errors due to non-ideality of the gas are cancelled out.

The accuracy also of such measurements can be substantially improved by comparison with blind measurements using a calibrated dummy with smooth surface of similar volume. To avoid errors by gas desorption in particular porous samples should be dried and degassed in advance, preferably in vacuum. If helium is adsorbed or absorbed by the sample another measuring gas should be used.

5.9.3 Buoyancy Methods

According to Archimedes Principle [192] a body in a fluid is subject of buoyancy, a force which equals the gravitational force acting on the volume of the replaced fluid ($\equiv$ volume of the sample V_s including closed pores) but opposed to it (Fig. 5.94a). That force is proportional to the density of the fluid ρ_f. Weighing a solid sample with a balance we measure a force w^* which is the difference of the gravitational force on the solid of mass m_s and of the opposite directed buoyancy:

$$w^* = m_s g - V_s \rho_f g \tag{5.38}$$

Measurement of w^* at different density of the fluid allows us to determine both, the volume V_s and the mass m_s of the sample. Because the balance indicates the mass rather than the force that equation may be written as:

$$m^* = m_s - V_s \rho_f \tag{5.39}$$

This means that when determining the mass of a sample the weighing result m^* should be corrected by buoyancy of the sample in air and—in case of a non-symmetric balance—by the buoyancy of the balance.

Instruments for the measurement of the density by buoyancy consist of a balloon connected to a balance. Instruments for the measurement of the density by buoyancy consist of a hollow body in form of a spindle, the immersion of which gives the

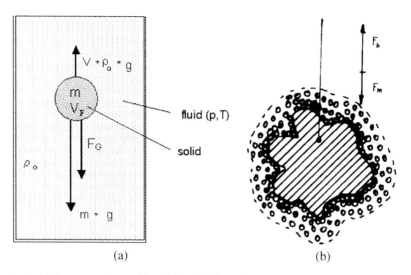

(a)   (b)

Fig. 5.94 (a) Buoyancy of a solid in a fluid. (b) Effect of an adsorbed layer on buoyancy in gas

value. Density of solids is performed using a balance which enables weighing the sample twice: in a liquid and in air.

5.9.3.1 Hydrometer

A hydrometer or areometer (Senkspindel, Senkwaage) is an instrument used to measure the density of liquids. Operation of the hydrometer is based on Archimedes' principle that a solid suspended in a fluid will be buoyed up by a force equal to the weight of the fluid displaced. Thus, the lower the density of the substance, the further the hydrometer will sink. An early description of a hydrometer appears in a letter from Synesius of Cyrene ($\sim$373–414) to his Greek scholar Hypatia of Alexandria (350/370–415). In Synesius' fifteenth letter, he requests Hypatia to make a hydrometer for him. Hypatia is given credit for inventing the hydrometer sometime in the late 4th century or early 5th century. Such a hydrometer has the shape and the size of a flute and has notches in its perpendicular line, where the density of the liquid can be read. A cone forms a lid at one of the extremities, closely fitted to the tube. The cone and the tube have one base only. This is called the baryllium. Hydrometers had been used by Abū Rayhān al-Birūnī in the 11th century and described by Al-Chazini in the 12th century. It later appeared again in the work of Daniel Gabriel Fahrenheit (1686–1736) and Jacques Alexandre César Charles (1746–1823).

A modern hydrometer is usually made of glass and consists of a cylindrical stem and a bulb weighted with mercury or lead shot to make it float upright in a graduated cylinder containing the liquid to be tested. The point at which the surface of the liquid touches the stem of the hydrometer is noted. A variety of scales exists, the one used depending on the type of fluid. In low-density liquids such as kerosene, gasoline, and alcohol, the hydrometer will sink deeper, and in high-density liquids such as brine, milk, and acids it will not sink so far. Therefore in industry and laboratories usually a set of hydrometers is available. Besides a variety of instruments exist designed for the special liquid to be controlled. Modern hydrometers usually are graduated in density units but different scales were (and sometimes still are) used in certain industries. Since density is temperature dependent some types of hydrometer enclose a thermometer in the float section (thermohydrometer). Examples of scales include:

- API gravity, universally used worldwide by the petroleum industry.
- Baumé scale, formerly used in industrial chemistry and pharmacology
- Brix scale, primarily used in fruit juice, wine making and the sugar industry
- Oechsle scale, used for measuring the density of grape must
- Plato scale, primarily used in brewing
- Twaddell scale, formerly used in the bleaching and dyeing industries

Specialised hydrometers are frequently named for their use; following are some examples:

A lactometer (or galactometer) is used to test milk. Since milk contains a variety of substances its density does not give a conclusive indication of its composition and additional tests for fat content are necessary. The instrument is graduated into a hundred parts. Milk is poured in and allowed to stand until the cream has formed, then the depth of the cream deposit in degrees determines the quality of the milk.

An alcoholometer is used for determining the alcoholic strength of liquids. It is also designed as a Proof and Traille or Proof and Tralle hydrometer named after the German physicist, Johann Tralles (1763–1822). It was graduated so that it read the percentage of alcohol by volume at 60 °F = 15.5 °C. This Proof and Tralle Hydrometer is a professional grade hydrometer that has a scale going to 200 proof. It is used to calculate proof in higher alcoholic beverages as it is not accurate for lower alcohol products like beer or wine. Alcoholmeters have scales marked with volume percent of 'potential alcohol', based on a pre-calculated specific gravity. A higher 'potential alcohol' reading on this scale is caused by a greater specific gravity, assumed to be caused by the introduction of dissolved sugars. A reading is taken before and after fermentation and approximate alcohol content is determined by subtracting the post fermentation reading from the pre-fermentation reading.

In the 1740's, the Customs and Excise and the London brewers/distillers began to use Clarke's hydrometer. In 1802 the Board of Excise held a competition to find a better instrument than Clarke's hydrometer. The winning design was that of Bartholomew Sikes. Its hydrometer was enshrined in legislation in 1816 with the *Sikes Hydrometer Act* and remained the legal standard until 1907 and was in common use until 1980 [193, 194]. The Sikes hydrometer was used to determine the strength of spirits for the distiller, victualler and tax collector alike, and works both before and after fermentation for the brewer, providing an accurate method of determining Alcohol proof, strength, percentages and duties thereof.

A saccharometer is a hydrometer used for determining the amount of sugar in a solution. It is used primarily by winemakers and brewers, and it can also be used in making sorbets and ice-creams. The first brewers' saccharometer was constructed by John Richardson in 1784. It consists of a large weighted glass bulb with a thin stem rising from the top with calibrated markings. The sugar level can be determined by reading the value where the surface of the liquid crosses the scale.

The Oechsle Scale is a hydrometer measuring the density of grape must, which is an indication of grape ripeness and sugar content used in wine-making [195]. It is named after the pharmacist and goldsmith Christian Ferdinand Oechsle (1774–1852). It is widely used in the German, Swiss and Luxembourg wine-making industries. The Oechsle scale indicates how much one litre of must is heavier than one litre water. For example, must with a density of 1088 gL^{-1} has 80° Oe. In Austria instead °KMW or Babo is used. International, but mostly in English speaking regions Brix or Balling and in France and Spain Baumé are used.

A barkometer is calibrated to test the strength of tanning liquors used in tanning leather.

The state of charge of a lead-acid battery can be estimated from the density of the sulphuric acid solution used as electrolyte. A hydrometer calibrated to read density relative to water at 60 °F = 15.5 °C is a standard tool for servicing automobile batteries. Tables are used to correct the reading to the standard temperature. Such a battery tester is equipped with bulb by which the electrolyte can be sucked.

Antifreeze tester is specialised for testing the quality of the antifreeze solution used for engine cooling. The degree of freeze protection can be related to the concentration of the antifreeze and the freezing point.

Acidometer or acidimeter, is a hydrometer used to measure the density of an acid.

The urinometer or urometer is a specialised hydrometer designed for urine analysis. As urine's density depends on its ratio of solutes (wastes) to water, it allows for quick assessment of a patient's overall level of hydration.

5.9.3.2 The Buoyancy of Solid Particles in a Liquid

The density of small particles or powders can be determined by suspension in a liquid of similar density. The sample floats at equal densities. Liquids of finely graduated density may be obtained by mixing of two liquids, one of which is of higher and the other of lower density. Suitable mixtures are made of chloroform (1.46 $g\,cm^{-3}$), bromoform (2.90 $g\,cm^{-3}$), or methylene iodide (3.30 $g\,cm^{-3}$) with benzene (0.88 $g\,cm^{-3}$), toluene (0.87 $g\,cm^{-3}$), or xylene (0.87 $g\,cm^{-3}$). Using aqueous solutions of potassium mercury iodide (Thoulet's solution) densities up to 3.2 $g\,cm^{-3}$ can be adjusted.

Also temperature variations may be applied to adjust a certain density because liquids expand more than solids with temperature increase. On the other hand that effect is applied for sensitive temperature measurement in the so-called Galileo thermometer (Fig. 5.95) described below.

The density of a liquid column in a cylinder increases continuously from top to bottom with regard to the gravitational force exerted of the higher layers to deeper ones. This allows for highly sensitive density determinations. For that purpose a liquid which has the expected density should be used and the powder sample dispersed

Fig. 5.95 Galileo
thermometer

in that liquid. The particles will be accumulated at the area in which the liquid has
the same density.

The density distribution of the liquid as a function of the height of the column
is determined by means of buoyancy bodies. Alternatively the pressure gradient as
a result of the density gradient may be measured by sensitive pressure transducers,
preferably in a differential arrangement.

In hydrometer analysis fine-grained soils or powders are graded. Hydrometer
analysis is performed if the grain sizes are too small for sieve analysis. The basis
for this test is Stoke's Law for falling spheres in a viscous fluid in which the terminal velocity of fall depends on the grain diameter and the densities of the grain in
suspension and of the fluid. The grain diameter thus can be calculated from knowledge of the distance and time of fall. The hydrometer also determines the density of
the suspension, and this enables the percentage of particles of a certain equivalent
particle diameter to be calculated.

5.9.3.3 Buoyancy in a Gas

The volume of a solid sample can be determined by weighings at different pressures using any sensitive vacuum balance. With regard to clean measuring conditions a suspension balance with separated sample vessel is favourable (Fig. 4.2b) [196]. Using a symmetrical beam balance buoyancy of the instrument can be cancelled out. For that purpose the balance is loaded with pieces of same mass but different density (e.g. quartz and gold) at opposite sides of the beam. Because difference measurements are made an error of the indication by buoyancy of the balance can be neglected, if the compensating weight is not changed. The following equation indicates a linear relation between a series of gas pressures p_i and the mass indication $m(p_i)$ influenced by buoyancy:

$$m_0 - m(p_i) = m_0 - \frac{V_s M}{RT} p_i \tag{5.40}$$

After measurement of several values a linear regression is made, the slope a_{lr} gives the volume V^* as the sum of the sample volume V_s and the volume of the parts of the balance

$$V^* = \frac{RT}{M} a_{lr} \tag{5.41}$$

The sample volume results from

$$V_s = \frac{RT}{M} a_{lr} - V_b \tag{5.42}$$

To determine V_b a calibration measurement with void balance pan is made and evaluated in the same way by linear regression. The sample mass m_s is the value indicated in vacuum m_0 corrected by the calibration term. The density of the sample ρ_s is calculated using Eq. (5.38).

Weighing of highly dispersed or porous samples may be disturbed in vacuum by adsorption of residual gases or in a gas atmosphere by impurities of the gas. It is recommended for regression analysis not to use the mass value in vacuum or that at the lowest pressure, respectively. If the sample has a large surface an adsorbed layer may contribute to mass and volume of the sample. In the vicinity of the surface the density of the gas may deviate from the density of the free gas phase and thus influence buoyancy (Fig. 5.94b).

The buoyancy pressure gauge described later can also be used for gas density measurements if the pressure is measured and kept constant with a gauge that is independent of the kind of gas used. Density measurements carried out with electronic balances (gas density balances) can be used for process checking. Gast [197, 198] used both a modified electronic beam balance equipped with a glass receptacle (Figs. 4.22a/5.96a) and a suspension system (Fig. 5.96b) to measure density changes in a gas flow.

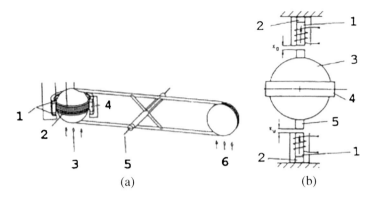

Fig. 5.96 (**a**) Microbalance system to measure density changes in a gas flow according to Gast and Talebi-Daryan. *1* field coil, *2* sensor coil, *3* glass recipient, *4* ring magnet, *5* taut band, *6* counterweight. © Theodor Gast. (**b**) Gast suspension system to measure density changes in a gas flow. *1* soft iron, *2* permanent magnet, *3* glass recipient, *4* sensor ring, *5* ferromagnetic cylinder. © Theodor Gast

5.9.3.4 Comparison of Buoyancy in Air and in a Liquid

Hydrostatic weighing with comparison of the weight of a solid sample in air and in a liquid was first described by Archimedes ($\sim$287–212 BC) [199, 200] who invented a method for determining the volume of an object with an irregular shape by immersion of the object in water (Fig. 5.97). According to Vitruvius [66], a votive crown for a temple had been made for King Hiero II, who had supplied the pure gold to be used, and Archimedes was asked to determine whether some silver had been substituted by the dishonest goldsmith. Archimedes had to solve the problem without damaging the crown, so he compared its weight in air and in water and determined the density of the material.

Abu al-Fath Abd al-Rahman Mansour al-Khāzini (Al-Chazini) (11th–12th century) [8] was a Muslim astronomer of Greek ethnicity, who taught in Merv, then in the Khorasan province of Persia (located in today's Turkmenistan). He constructed the famous 'balance of wisdom' which was an equal armed precision balance with rider weights, devoted in particular for density determinations. A model of that balance is exhibited in the Institut für Geschichte der Arabisch-Islamischen Wissenschaften at Frankfurt am Main, Germany (Fig. 3.42) [10].

The hydrostatic balance (Fig. 5.98) is equipped with two scales, one in air and the other in a liquid. By weighing the solid sample in air the mass m_1 is indicated and should be corrected by the buoyancy in air to give the sample mass m_s:

$$m_s = m_1 + V_s \rho_g \tag{5.43}$$

where V_s is the volume of the sample and ρ_g the density of the air at temperature and pressure. Then the sample is weighed in the liquid which yields the indication m_2 and the sample mass is

$$m_s = m_2 + V_s \rho_l \tag{5.44}$$

Fig. 5.97 Hydrostatic weighing according to Archimedes for determination of volume and density of an irregular shaped object by immersion of the object in water

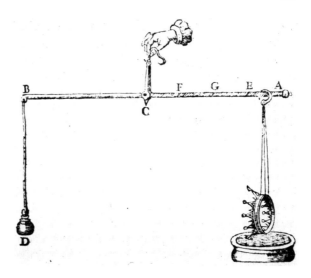

where ρ_l is the density of the liquid at measuring temperature. By combination of both equations the solid volume (including closed pores) can be calculated:

$$V_s = \frac{m_1 - m_2}{\rho_l - \rho_g} \tag{5.45}$$

From that the density of the sample ρ_s can be obtained:

$$\rho_s = \frac{m_1 + V_s \rho_g}{V_s} \tag{5.46}$$

Distilled water is generally used for hydrostatic weighings, because for all liquids considered it has the smallest volume expansion coefficient ($\alpha_v = 2.1 \times 10^{-4}$ K^{-1}) [201]. The density ρ_w of air-free water at 101325 Pa as a function of temperature t_w can be calculated using a formula given by Kell [202]. In the temperature range $19.5\,°\mathrm{C} \le t_w \le 20.5\,°\mathrm{C}$

$$\rho_w = 998.203 - 0.206(t_w - 20) \tag{5.47}$$

gives a good approximation, where t_w must be entered in °C to obtain the water density in kg m^{-3}.

A specialised hydrostatic balance with fixed counterweight has been designed by Georg Westphal in Celle [203]. It is known as Mohr's balance (Karl Friedrich Mohr 1806–1879). Mohr designed several modifications of hydrostatic balances with equal and unequal arms. At the end of the arm the solid sample is suspended by means of thin wire into the liquid of known density. In case of the measurement of the density of a liquid, a test body of known density is suspended into that fluid. A remarkable measuring error may be caused by the meniscus of the fluid at the suspension wire. This may be overcome by using a magnetic coupling (magnetic suspension balance) of the sample and of test body.

Fig. 5.98 (a) The hydrostatic balance (Senkwaage).
1 deflection sensor and compensation system,
2 balance beam, *3* balance casing, *4* balance pan in air,
5 trough with liquid,
6 balance pan in liquid.
(**b**) Mohr/Westphal balance

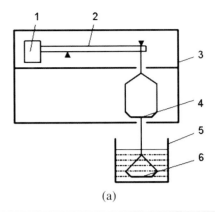

(a)

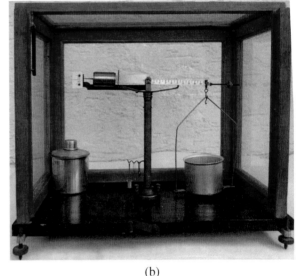

(b)

5.9.4 Radiometric Methods

A different principle is employed with the radiation method. Transmitted or scattered gamma radiation is measured; and with suitable calibration, the density of the combined gaseous-liquid-solid components of a soil mass is determined. Correction is then necessary to remove the components of density attributable to liquid and gas that are present. The radiation method is an in situ method [204]. The absorption of gamma rays in material depends on the product [205]

$$\rho l = \frac{m}{A} \tag{5.48}$$

where l is the length of the path irradiated, A the irradiated area and m the mass in the volume $A\,l$. With an initial intensity I_0 after absorption the intensity is

$$I_\gamma = I_0 e^{-\mu' \rho l} \tag{5.49}$$

where μ' is the mass adsorption coefficient. This method is used for on-line mass measurements with radiometric belt weighers (Fig. 4.133). If the coefficient, the apparatuses parameters and the density of the material are known the mass m can be calculated as follows:

$$m = \frac{A}{\mu'} \ln \frac{I_\gamma}{I_0} \tag{5.50}$$

In case of a crystalline material the true matrix density may be obtained from X-ray diffraction.

5.9.5 Acoustic Waves/Seismic Waves

The seismograph records ground movements caused by earthquakes, explosions, or other Earth-shaking phenomena. It is a pendulum fixed on the ground and the ground oscillations are observed. The propagation velocity of the waves depends on density and elasticity of the medium. Velocity tends to increase with depth, and ranges from approximately 2 to 8 km/s in the Earth's crust up to 13 km/s in the deep mantle. At so-called seismic discontinuities also the velocity of propagation of seismic waves changes. This phenomenon can be related to the change.

5.9.6 The Density of the Universe

In order to measure the density of the Universe, it is necessary to sample a region that is larger than the scale on which the Universe becomes approximately homogeneous. The volume of this region must then be measured, as well as the mass of the matter it contains. The ratio of mass to volume then gives the density. The mass of the universe can be assessed from counting the light emitting objects. However the dark mass should be included which is by far larger (see Chap. 1) and that we see only through its gravitational effects.

5.10 Particle Analysis

Particle analysis covers the investigations of the mass, size and shape of individual particles and of a multitude of particles in a surrounding fluid and its spatial distribution. From about 1000 randomly selected particles the particle size distribution can be established.

Three basic methods of particle size analysis may be distinguished: ensemble methods, counting methods, and separation methods. The ensemble methods collect data from all particles of different size present at a given time, and then digest the data to extract a size distribution. Common ensemble techniques are Low Angle Scattering, Photon Correlation Spectroscopy, Back-Scattering Spectroscopy.

Counting methods start with physical separation. Either the sample is distributed in one layer on a plane slide and scanned, or the particles are diluted in air or a liquid and fed separately through a slot. The particles then are counted and simultaneously the size of each particle is determined optically or by electrical sensing [206–208]. Then the numbers of particles in different size classes are determined in order to establish the particle size distribution. Some common counting methods are: Electrozone Counter, Light Counter, Time of Flight Counter, Light and Electron Microscopy.

Common separation and classification techniques include sieving and air separation, sedimentation in a liquid or in air and sedimentation field flow fractionation, in the gravitational field or by means of a disc centrifuge, capillary hydrodynamic fractionation. By means of holography a single particle can be inspected from all sides, a three-dimensional picture of the particle cloud can be established and the spatial distribution of particles within a fluid stream determined.

In particle analysis the particle size is never measured directly. This is true even for microscopic inspection where it is the size of an enlarged image of the particle that is assessed and not the particle itself, and where smaller parts of the particle may be hidden by larger ones.

5.10.1 Dust Concentration

Dust concentration usually is expressed in number of particles or cumulative mass per volume in which the dust is suspended. For other tasks the number of particles or their mass are related to the area on which the dust is precipitated. Measurements are made at different locations, e.g. at the place of origin of airborne dust: exhaust of engines (diesel engines, saw mill), within transport pipes for dispersed material, in the near environment of factories, in the atmosphere and in space. At high concentrations as can occur in factories and in mines the danger of dust explosion is to be taken into account. Dust particles are often porous and exhibit a large surface area ready for adsorption. As dust pollutes the environment and may be hazardous (the material itself or adsorbed species) airborne pollution has become the focus of increasing public, scientific and governmental concern in recent years [209].

At the different control locations dust concentrations are very different and therefore a variety of measuring methods are applied. Conventional air quality monitoring requires mostly sampling in advance. Dust may be collected by means of active filtration samplers or by precipitating devices. Often used are open collector samplers in connection with gravimetric analysis. Typically a small part of the fluid in which the particles are dispersed is sucked through a filter and the filter is weighed

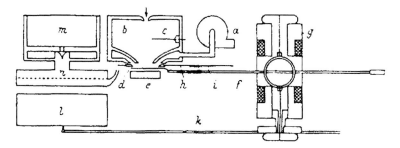

Fig. 5.99 Schematic drawing of a recording dust balance according to Gast. *a* blower, *b* inlet of airborne dust, *c* loading electrode, *d* metal plate, *e* stopper, *f* balance beam, *g* microbalance, *h* regulator, *i* velvet strip, *k* needle, pen, *l* reel, *m* cleaning blower, *n* filter

after a certain time and/or inspected by optical analysis. Favourably, the precipitating device is combined with the sensor so that the dust deposit can be monitored continuously as a function of time. Loaded particles can be collected by means of an electric field. The thermal precipitator is a cold plate at which dust particles are settled which are moved by convection of a warmed gas. Adhesives or magnetic fields assist the settling of the particles.

A recording dust balance based on an electromagnetic microbalance system of Gast is depicted in Fig. 5.99. Quartz resonators are used for monitoring the concentration of aerosols and contaminating particles. Figure 4.125 show an assembly used in space [210]. In Fig. 4.128 a diagram is depicted of a tapered element oscillation microbalance (TEOM) according to French [211]. It consists of a thin-walled tapered element, its broad end fixed to a solid base. The free end executes autonomous oscillations, the frequency of which depends on the mass and the stiffness of the element and on the additional mass connected to the free end. The oscillations are excited electrostatically, and detection is optoelectronic. Such an instrument with a measuring range from 30 mg to 330 mg is described in [212]. The specimen was taken at a constant temperature of 50 °C. The method is also suitable for air monitoring in closed and open spaces. TEOM instruments contain a filter cartridge, located on the top of the oscillating element, and an air pump regulated by a mass-flow controller unit. Inlet configuration and flow-rate restrict the intake of particulate matter with a certain average aerodynamic diameter in the millimetre range. This inducted particulate fraction is drawn through the system by the air pump, and accumulates upon the filter. The frequency of vibration of the element changes as the mass of the filter unit increases and the mass of particulate collected can be measured directly on the basis of the change in vibration of the element. One area of application is the continuous measurement of particle emissions of diesel engines. The system can also be used to provide near-real-time continuous measurement of airborne particles. Today, TEOM is the recommended and preferred method for the dust measurement in the atmosphere.

In comparison to quartz oscillators the transversally [213, 214] or longitudinally [215] oscillating belt has a considerably lower working frequency, making it possible to equip the measuring instrument with a continuously moving belt for receiving and accumulating the substance to be measured. The change in frequency of the belt is proportional to the additional mass loaded. A device for registering dust concentrations is schematically depicted in Fig. 4.127. The belt is conducted over a knife-edge air bearing, and oscillations are excited acoustically and measured capacitively. Before it is loaded in the measurement section, the belt transported forward at a constant rate runs through a comparison section.

Radiation mass sensors are applied for routine control of dust concentrations. The Beta Attention Mass Monitor (BAM) is an airborne particulate mass measurement method. BAM systems are generally cheaper than TEOM to purchase. However, BAM systems need external calibration using reference samples as they provide an indirect measure of particulate mass (beta absorption), and therefore that accuracy is a function of how representative calibration standards are of particulates in the area of monitoring. BAM monitors typically consist of a beta radiation source, a glass fibre band filter, a detector and an air pump. The beta radiation source and detector are typically situated opposite each other with one part of the glass fibre band filter situated directly between the two. Air samples are introduced into the monitor via an inlet configuration that restricts the intake of particulate matter with an average aerodynamic diameter of greater than 10 mm, and drawn across the surface of the tape. Particulate accumulating upon the surface of the tape absorbs beta radiation thereby reducing the amount measured by the detector. The difference in response is then used as a measure of the amount of particulate present. At the end of a sample period (typically between 1 and 24 hours) the tape is automatically wound on and a new sampling period begins.

Dust concentration can be measured by various optical methods. Light scatter mass estimation monitors measure the degree of light scattering occurring within a sampled volume of air. Particulate with an average aerodynamic diameter up to 10 mm can scatter light in the visible and near-infrared regions. Consequently, the degree of scatter of light within this range can provide proximate estimation of airborne particulate levels. Mass monitoring using light scatter mass estimation monitors requires external calibration.

5.10.2 Gravitational Sedimentation

A sample of particles is uniformly suspended in a fluid of known density and viscosity, and allowed to settle due to gravity. The settling of the particles is impeded by friction and this depends on size of the individual particle: smaller particles fall slower, approximately according to Stokes' law. In commercial instruments the concentration of particles remaining in the fluid is continuously monitored via extinction of X-rays or a light beam (photo-sedimentometer) at a known distance from the top of the container.

Usually the particles sediment to the bottom of the container, but if they are lower in density than the fluid, they float towards the top. Thus, the liquid should be of a lower density than that of the particles. The viscosity of the liquid is chosen so that, on the one hand, coarse particles do not drop too fast, and on the other hand, the time for the complete analysis is not too long. The formation of aggregates is impeded by means of a dispersion agent. Either in a specific layer near the bottom of the sedimentation chamber the particle concentration is measured incrementally or the cumulative particle mass at the bottom is registered as a function of time. In both methods the powder may be introduced either as a thin layer of dispersion on top of a column of a pure liquid: (the line-start technique), or uniformly dispersed: (the homogeneous technique).

Particle size determination by the investigation of the sedimentation of a powder in a liquid by gravitation or acceleration in a centrifuge may be regarded as the conversion of the observation of flow of a liquid through a porous system. The settling rate of the particles depends on size, density and shape of the particles and on density and viscosity of the liquid. At low Reynolds numbers (<0.25) the settling rate is measured and related to particle size as an equivalent settling rate diameter. For an irregular particle, the equivalent settling rate diameter is usually slightly smaller than the volume diameter. As a consequence of an adhering, immobile liquid layer, surface roughness and pores do not contribute. Highly porous particles, however, exhibit a differing apparent density, and this should be taken into account. Brownian motion of very small particles combined with long sedimentation times can lead to substantial errors in the size distribution. Large particles >25 µm do not follow Stokes law if their rate of sedimentation produces turbulent flow.

The concept of gravitational sedimentation is simple and the apparatus can be very cheap. The method can be treated well theoretically. Accuracy is excellent but measuring time is long, up to 12 hours and even more. Accuracy can be easily controlled with reference material. The range of particle diameter is quite broad: ~0.2 to >500 µm.

By gravity or in a centrifugal field the particles achieve a stationary velocity in the stagnant liquid dependent on diameter, shape, and density of the individual particles. The equivalent Stokesian diameters d_s are calculated by means of a resistance law, usually Stokes' law:

$$d_s^2 = \frac{18\eta}{\rho_s - \rho_l} \frac{v}{g} \tag{5.51}$$

where η is viscosity, ρ_s density of the particles, ρ_l density of the liquid, v velocity of sinking, and g gravitational acceleration.

Instead of monitoring with a light beam or X-rays the settling particles in gravitational sedimentation can be weighed [216]. Special sedimentation balances are offered for such measurements (Fig. 5.100).

5.11 Magnetic Susceptibility

The magnetic susceptibility is the degree of magnetisation of a material in response to an applied external magnetic field [217]. If the ratio between the induced volume

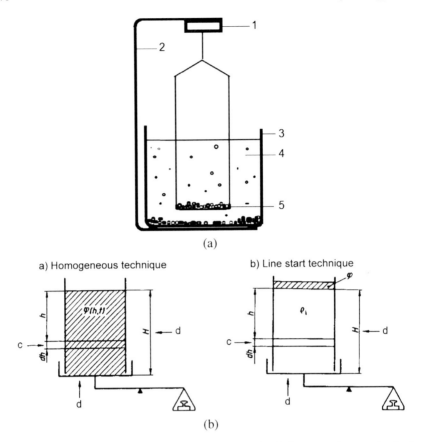

Fig. 5.100 (**a**) Diagram of a sedimentation balance. *1* load cell, *2* support, *3* measuring beaker, *4* suspension, *5* sample pan with sediment. (**b**) Principles of particle size analysis by sedimentation: (*a*) Incremental measurement with optical means. (*b*) Cumulative measurement by a sedimentation balance. (*c*) Uniformly dispersed powder sample in a liquid of lower density—the homogeneous technique, (*d*) Sample introduced as a thin layer of dispersion on top of a column of a pure liquid—line-start technique

magnetisation M and the inducing field H is expressed per unit volume, volume susceptibility χ_V is defined as

$$\chi_V = \frac{M}{H} \tag{5.52}$$

Volume magnetic susceptibility of solid samples is measured by the force change felt upon the application of a magnetic field gradient. Mass, or specific, susceptibility is defined as

$$\chi_m = \frac{\chi_V}{\rho} \tag{5.53}$$

where ρ is the density of the material. The dimension of mass susceptibility χ_m is therefore cubic meters per kilogram whereas the volume susceptibility χ_v is dimensionless.

The earliest measurement methods are from Gouy and M. Faraday. In both methods the force of the magnetic field is measured by observing the mass difference without and with the magnetic field, indicated by a balance. Early measurements were made using the Gouy balance where a sample is hung between the poles of an electromagnet. The change in mass when the electromagnet is turned on is proportional to the susceptibility. In the Gouy method one part of the sample is partially in a homogeneous magnetic field and the other part far outside. Therefore a large sample is necessary. Today, a superconductive magnet is used. An alternative is to measure the force change on a strong compact movable magnet upon insertion of the sample (Evans balance).

The Faraday method is most frequently used to measure susceptibility [218, 219]. The sample is placed on a flat quartz pan hanging on the balance (Figs. 4.58/5.47). Magnets of different strengths are arranged at the sides making it possible to interchange them. The balance is evacuated to prevent the occurrence of sorption and buoyancy effects. Starting from a position in which the sample is not influenced by the magnetic field, the magnet is lifted or lowered continuously or in discrete steps. If the sample is paramagnetic, first it is attracted to a neutral position. It is then drawn along a little until gravitational force prevails. The susceptibility of the quartz pan should be determined in a blank test:

$$F = m \chi_m H \frac{\mathrm{d}H}{\mathrm{d}x} \tag{5.54}$$

where m is the sample mass, χ_m the specific magnetic susceptibility, H the field strength and $\mathrm{d}H/\mathrm{d}x$ the field gradient. If the magnet approaches the sample from above, the total force then acting on the coil is $mg - F$, and after the magnet has passed through the neutral position it is $mg + F$. The influence of the gravitational field can thus be eliminated. The registered movement of the coil is taken at half the distance between the maximum of the deflection and compared with a standard sample to obtain the magnetic susceptibility.

Pascal used the balance according to the substitution principle, the total load always remaining constant (Fig. 5.101a). In Fig. 4.58 a susceptibility apparatus with spring balance is depicted. Today electromagnetic vacuum microbalances are used.

5.12 Gravimetric Measurement of Temperature

The so-called Galileo thermometer was invented probably by Galileo Galilei. It is based on measurement of the difference of gravitational force and buoyancy. It consists of a sealed glass cylinder containing a clear liquid and a series of test objects, usually sealed glass balls filled with a coloured liquid and with a suspended metal weight, whose densities are such that they rise or fall as the temperature changes (Fig. 5.95).

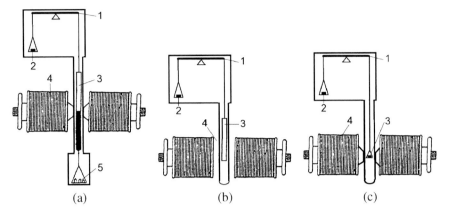

Fig. 5.101 (**a**) Balance of P. Pascal for measuring the magnetic susceptibility 1910: *1* balance in substitution mode, *2* counterweight, *3* sample, *4* electromagnet with pole shoes, *5* tare weights. (**b**) Gouy balance: *1* balance, *2* counterweight, *3* sample, *4* homogeneous magnetic field. (**c**) Faraday balance: *1* balance, *2* counterweight, *3* sample, *4* electromagnet with pole shoes

Fig. 5.102 Measurement of gas temperature with a vacuum balance. *1* gas inflow, *2* needle valve, *3* oven, *4* deflector plate, *5* ionisation gauge, *6* balance, *7* vacuum pump

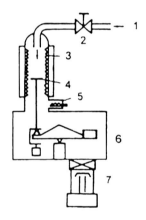

According to Wiedemann and Bayer [220] it is possible to measure the temperature of gas in high vacuum with the aid of a balance. A molecular beam is directed at a deflector plate which is attached to a balance (Fig. 5.102). If, along a sufficiently long path, the beam travels through an isothermal area, then according to Maxwell's equation the following is valid for the mean velocity at a constant pressure:

$$\bar{u} = \sqrt{\frac{3RT}{M}} \tag{5.55}$$

where R is the gas constant, T the temperature and M the molecular weight. Thus the impact force registered by the balance is proportional to the square root of the absolute temperature. The authors suggest using the method to check the temperature deviation of the sample in vacuum.

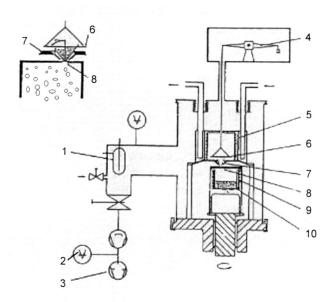

Fig. 5.103 Vapour pressure determinations by means of a Knudsen cell from which gas is evolved and blown towards a deflector plate on a balance. *1* condensation cold trap, *2* vacuum gauge, *3* vacuum pumps, *4* microbalance, *5* thermostat, *6* reflector plate, *7* terminating shutter, *8* outlet shutter, *9* sample holder with heater, *10* sample. © Netzsch Gerätebau, Selb, Germany

5.13 Gravimetric Measurement and Control of Pressure

5.13.1 Pressure Measurement

If the kind of gas used is not changed during a measurement, the buoyancy can be used to measure the pressure. The buoyancy gauge consists of a vacuum microbalance carrying a glass recipient instead of the specimen. With an instrument with a Gast balance covering the pressure range 0.1 and 10^6 Pa at a resolution of 0.1 Pa, a pressure control at 0.1 Pa was possible [44]. Exploiting the Knudsen forces by producing thermal gas flows along the buoyancy artefact, the measurement range was extended up to 10^{-3} Pa [221] (Fig. 5.45).

Vapour pressure can be controlled by thermoregulation of the liquid. To avoid the vapour unintentionally condensing in parts of the apparatus, these should be held at a higher temperature. Figure 3.9 shows an apparatus for measuring water vapour isotherms where this had been made [51].

With the aid of a Knudsen cell either standing on the balance pan [222] or blowing against a deflector on a balance, temperature-dependent vapour pressure determinations can be carried out (Fig. 5.103).

Deadweight tester (pressure balances) are the most accurate instruments for pressure measurement and the calibration of manometers for liquids and gases

Fig. 5.104 Deadweight
Tester, pressure gauge
calibrator (piston pressure
gauge, pressure balance).
1 calibrating weights,
2 rotating piston, *3* est
instrument, *4* hydraulic fluid

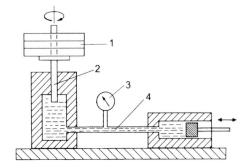

(Fig. 5.104). By means of weights a well defined force F is exerted on a vertically arranged piston of known diameter A. Usually the weights are at ambient pressure and thus overpressure (pressure—atmospheric pressure) is measured. If the absolute pressure should be determined directly the whole apparatus must be placed in a vacuum (absolute pressure piston gauge). In the fluid a pressure is generated according to the definition of pressure p.

$$p = \frac{F}{A} \tag{5.56}$$

To reduce friction the piston is rotating. By choice of piston diameter and weights a large measuring range can be covered. For precision measurements results should be corrected by the local value of gravity, atmospheric pressure and buoyancy of the weights in air. Commercial instruments may reach a discrimination of 0.005 % of the measuring range; for standardisation special instruments have a relative sensitivity of 10^{-6}.

5.13.2 Safety Valves

Pressure relief valves (PRV) are important parts in power generation plants, steam engines, chemical manufacturing, natural gas processing etc. A pressure relief valve is a safety device that relieves overpressure of a fluid in a vessel or piping exceeding their design limits. A pressure safety valve (PSV) in addition has a manual lever to open the valve in case of emergency. A vacuum pressure safety valve (VPSV) is an automatic system that relieves static pressure on a gas. It is used when the pressure difference between the vessel pressure and the ambient pressure is small, negative and near the atmospheric pressure.

Such a safety valve closes an opening in the apparatus covered with a lid and tightened by means of a gasket (O-ring). The force opposite to the working pressure can be applied either by means of a weight or by a spring. Because the spring is independent of its situation and smaller today it superseded the weight design completely. The model of a PSV with weight is depicted in Fig. 2.19. In low-pressure systems which include cryostats operated with a liquefied gas (He, N_2, O_2, Ar, air)

Fig. 5.105 Vacuum pressure safety valve of Robens and Sieglen. *1* connection to apparatus, *2* glass fibre reinforced PTFE disc, 60 mm ∅, *3* O-rings (neoprene), *4* three sickle-shaped springs, *5* springs, *6* permanent magnet, *7* reed relays, *8* PTFE piston, *9* adjusting screws, *10* exit to atmosphere

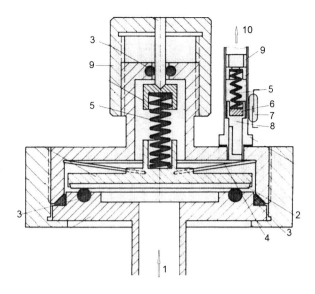

condensation of any gas can take place. Sudden expansion of the condensate can generate high overpressure. For such occurrences a special vacuum safety valve was constructed schematically depicted in Fig. 5.105 [223].

5.14 Multicomponent Systems

A balance measures besides the force of an article towards the gravity centre of Earth and simultaneously the opposite directed force of buoyancy and other disturbing forces. Usually the movement of a gravitational balance is restricted to one degree of freedom. Only forces or components of forces in vertical direction can be observed. If the design of a balance allows the free movement in more directions also lateral forces can be observed and evaluated by the same electronic sensor system.

Repeatedly the possibility has been discussed of measuring—beside mass change—other parameters, in particular temperature. Some realisations of such multicomponent systems are described in the following sections.

5.14.1 Double Platform Balance for Investigations of Single Crystals

For simultaneous determinations of condensation pressure and condensed mass or evaporation pressure and evaporated mass, the balance beam is rotatable about a point and is able to oscillate in two planes (Fig. 5.106) [224]. For mass determinations it moves in a vertical plane, and for pressure measurements in a horizontal plane. In both cases the measured quantity is electromagnetically compensated.

Fig. 5.106 Diagram of a double platform balance for simultaneous measurement of condensation pressure and condensed mass. *A* frame made of bronze wire, *B* span wire, *C* balance holder, *F* balance beam, *G* photoelements, *H* specimen, *I* horizontal Helmholtz coils, *J* vertical Helmholtz coils.© Theodor Gast

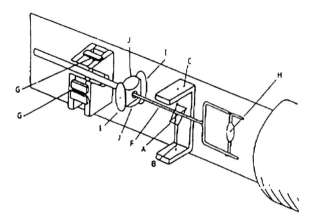

The beam consists of a glass capillary tube with a U-shaped piece for attaching the specimen. It contains a slim bar magnet to the left, beyond which it widens and is blackened. The left pole of the magnet protrudes into the magnetic field of two pairs of Helmholtz coils oriented perpendicularly to each other. These are controlled via amplifiers by two photoelectric sensors whereby the end of the beam acting as a screen influences the luminous flux of a filament to each pair of photoelements. A resolution of 10^{-8} is given for both force directions.

In another model of the balance the beam is on bearings with diamond toes under load in a concave stone. It is stabilised in two planes by the servo-loop described. In the third, it is held by the magnetic forces exerted by a small bar magnet connected to it by two parallel steel wires oriented in the direction of the beam and fixed to it.

5.14.2 Balance for Simultaneous Determination of the Mass Flux and Reaction Force of the Steam Jet of a Heated Sample

Simultaneous measurements of mass flux and the recoil of a steam jet from an aperture in the wall of a heated vessel containing the specimen enables the mean velocity of the specimen to be calculated, and from this its molecular weight, when the temperature is known [224]. For this purpose the sample can be placed in a cylindrical vessel which is centrically symmetrical with two holes drilled on opposite sides, and hung indirectly over a shank, a cardan joint, a thin-walled tube and a torsion band on the load hanger of a self-compensating balance [225]. This balance has a quartz glass framework beam with plate-shaped coil carriers joined to it by casting and two etched coils which also serve as position detectors. With a carrying force of 15 g, it achieves a resolution up to 1 μg and maintains zero-point constancy over several hours. To measure the moment of reaction by means of electrostatic compensation, the axis of the cardan joint has a wing of aluminium foil which plays between two pairs of plates. These make it possible to measure the capacitive angle and to generate the moment of compensation. The useful resolution recorded is 10^{-12} in a measurement range of 10^{-8} Nm.

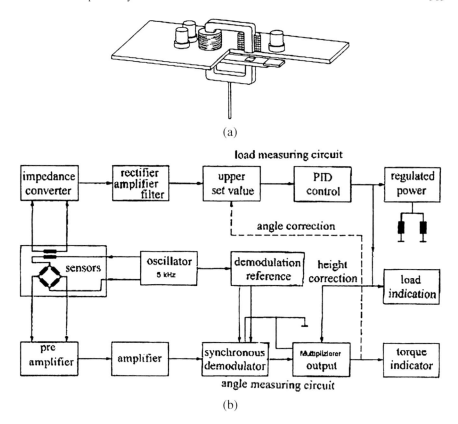

(a)

(b)

Fig. 5.107 (a) Measurement of force and torque using free suspension: Magnets and sensors. © Theodor Gast. (b) Block diagram of the double system. © Theodor Gast

5.14.3 Simultaneous Measurement of Weight and Torque Using Magnetic Suspension

On the cover plate of the receptacle, which consists of a non-magnetic poorly conducting alloy, there is a permanent horseshoe magnet with two drive windings and four inductive distance sensors mounted according to the flux displacement principle [226]. As shown in Fig. 5.107a, under the plate there is a second horseshoe magnet in a state of suspension which carries an insulating cross arm with end terminal copper lamina. This has a counter-rotational influence on two of the distance sensors if the magnet is turned about its vertical axis. Thus the diagonal voltage of the bridge contained in one of the sensors is a measure of the torsional angle. At the same time the height of the lower magnet acts uniformly on all four distance sensors. The virtual resistance of the whole measured bridge is a measure of the vertical position. For this purpose, a servo-loop can be built up (Fig. 5.107b). The bridge's virtual resistance is compared with a reference resistance. Via an amplifier, phase-dependent rectification and smoothing, the difference signal controls the cur-

Fig. 5.108 German post
office letter balance 1937:
rocker balance of A. Bizer,
Balingen, (Bizerba) Germany

Fig. 5.108 German post
office letter balance 1937:
rocker balance of A. Bizer,
Balingen, (Bizerba) Germany

rent through the drive windings of the upper magnet. In equilibrium, this current is a measure of the load. If a torque is executed on the lower magnet, it is twisted in relation to the upper one against the righting moment carried out by the tension of the lines of force. The detuning of the bridge circuit, and consequently the resulting diagonal voltage, are therefore a measure of the torsion and the torque. The dependence of the torsional sensitivity of distance requires a distance correction to be made. Correspondingly, the load signal is dependent upon the torsion of the magnets against each other. Both sources of error can be eliminated by analogue offset switching. A possible application of the system described is the simultaneous reading of density and viscosity. The sensitivity of the balance and that of the torque indication is $2.00(1)$ V g^{-1}.

5.15 Post Office Scales

An important application of weighing scales is in haulage companies. Equipment and costs depend on volume and weight of the objects to be transported. In the reform of the postal system in 1839 postal charges were changed to be based on weight rather than on distance delivered. So it became necessary to weigh letters, but appropriate balances did not exist. The first balance for post offices was developed and patented in the same year by Edwin Hill; others for commercial use by Henry Hooper and in 1840 by Robert Willis. There are special balances for nearly each type of use. Such scales have to be easily loaded, weighing should be performed fast, and reading should be easy. In many cases with letter balances instead of the mass a value is indicated e.g. of the postage. Often check weighing is applied; rocking is used as the indicator of excess weight (Fig. 5.108).

Fig. 5.109 Letter balance 'Epistola' produced by Simplicitas AB, Grävlingsbacken 2, 131 50 Saltsjö-Duvnäs, Sweden. A pencil is used as an axis stuck through one of the holes, corresponding to a mass of 20, 50 or 100 g. The letter is clamped through the slit on the tip of the device. The device can also be used as a paper knife. © Simplicitas

Fig. 5.110   Ecosol letter balance, © Ecosol, Weilheim, Germany

A great variety of letter balances have been designed. There are also very artificial designs which may be complicated to use but make decorative items used as gifts (Fig. 5.109/5.110). Decorative, but never used today, are equal-arm scales of Roberval or Béranger type (Fig. 5.111) usually with a set of dish weights. Unequal-arm steelyard-type balances as desk-balances have one or two slidable counterweights on a graduated beam (Fig. 5.112). These counterweights are moved along the beam to the equilibrium position. A smaller second weight may serve for more precise weighing. The beam has a chart of weight graduations. Hand held models of unequal type are specialised steelyards.

Simple inclination top pan balances, also referred to as pendulum scales are still widely used as letter scales. These scales are self-indicating: the mass or postage is directly read off the chart. There are pendulum scales in which the chart is fixed to the frame, and the indicator or pointer moves with the mechanism when loaded. Also, the reverse exists: the chart moves with the mechanism along a frame-fixed pointer. Some pendulum scale designs have a full circle as chart. Most of the pendulum scales have a chart with the form of a part of a circle; the size is almost a quarter. Therefore these pendulums are also called quadrant scales. There are table-

Fig. 5.111 British post office
scales of modified Roberval
principle, 19th century

Fig. 5.112 Steelyard-type
postal scale. © Wikipedia

top or desktop models and hand held models. Most of the pocket scales belong to
the latter group of models. Pure letter scales were fitted with a scale on which the
postal rates were inscribed. Sometimes the balances were equipped with change-
able charts which allow adjustments to rate changes. Many pendulum scales have a
three legged base frame (tripod). Placed on an inclined surface, the scale can be adjusted to zero with a levelling screw. The leg with this screw is situated in the plane
of movement of the scale mechanism. The counterweight of some pendulum scale
models is rotatable to a stable second position. Here the scales have two separate
weighing ranges (Figs. 5.113a/b).

Fig. 5.113 (a) Pendulum inclination letter scale. The counterweight of some pendulum scale models is rotatable to a stable second position. (b) Pocket letter scales of pendulum inclination type. (c) Columbus Bilateral letter scales, Ph.J. Maul. (d) Bilateral letter scales. Ph.J. Maul

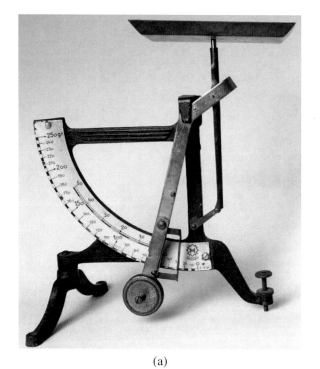

(a)

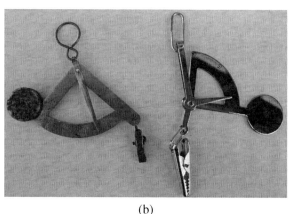

(b)

Model 'Fertig' of the German scales manufacturer Ph.J. Maul has the indicator-needle and the chart with divisions rotatable and mounted on the same axis. In this way this balance requires no separate levelling adjustment, it adjusts itself automatically to zero. The more complex design with the double pendulum, like the 'Columbus Bilateral' also made by Maul, does not require a levelling screw (Figs. 5.113c/d). This model with two contra-rotating pendulum parts as levers always indicates zero when unloaded, even in high sloping positions.

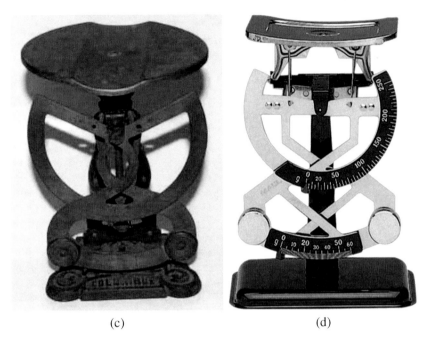

(c)   (d)

Fig. 5.113 (Continued)

Spring type balances also were used as letter scales. With a restricted load range the required sensitivity can be obtained. The mass or postage is directly read off the chart. There are table or desk models and hand held models. The simple spring balances make use of an extension or a compression coiled spring which is placed in line with the load to be determined. In England spring scales came into common usage in 1840 when Winfield developed the candlestick scale for postal use (Figs. 4.60/5.114). The vast majority of candlestick postal scales were manufactured by Robert Walter Winfield and the brothers Joseph and Edmund Ratcliff. Winfield worked in the Birmingham area from 1829 to 1860 and had a factory on the Baskerville Estate [227].

The depth of a floating body can be used as a measure of its load. This method is applied to determine the loading of ships by observing the immersion depth. A hydraulic letter balance was registered by A.F. Osler of Birmingham in 1839. Another model, the 'Eldon', came on the market in USA in the middle of the 20th century. Such a reliable letter scale can be easily made as shown in Fig. 5.115. It consists of a wide-neck bottle filled with water into which the weighing device is inserted. This latter part is made of a light piece of wood weighted at the lower end, to keep it in a stable, upright position. The wood is placed in the water and calibrated using a set of weights.

Fig. 5.114 Letter spring balance known as the 'candlestick' type, by R.W. Winfield, Birmingham, England. © Matthias Hass Germany

Fig. 5.115 Hydraulic letter balance. *1* pan with sample, *2* light wood, *3* graduation, *4* flask, *5* water, *6* metal weight

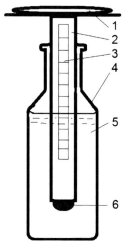

More and more, mechanical letter scales are superseded by electronic scales. Maul developed a solar powered electronic letter balance (Fig. 5.116). In letter sorting offices mass sensors are incorporated in the sorting machine. These are fast indicating electronic sensors which control the mass flow of the letters. The data are compared with data of optical sensors reading the stamps and too low franked objects are sorted out.

Fig. 5.116 Electronic letter
balance with solar loading of
the storage battery. Maul
Solar

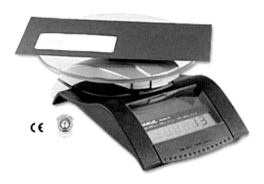

5.16 Coin Scales

Coins are derived from weights and even today—more than 2000 years after their
invention—some types of coins and likewise currency notes have similar status.
As a counter value for trading object coins are abstract items. However, initially
the value of a metal coin corresponded to its material value. The trading object was
exchanged to the more compatible coin; the mass of the coin multiplied by the value
of the metal corresponds to the value of the weighing object. So it is clear that the
mass of coins must be controlled. There are applications in weighing money:

- Control of the coin mass during production
- Control of wear and detection of counterfeit coins
- Comparison and identification of different currencies
- Sorting and counting
- Release of goods from vending machines or information on body measurements.

For the above applications, various types of coin scales have been developed.
Initially in production equal armed hand scales had been used whereby individual coins or a number of them, and likewise precious items like pearls, had been
weighed (Fig. 5.117).

In the *New Testament* use of money in trade is reported and in the Middle Ages
with its fragmented currencies this business was of high importance. For control of
wear and detection of falsifications small equal armed hand scales or small checkweighers were used (Fig. 5.118a), often stored in artificially designed cases. During
the industrial age with coin plate sorting machines underweight and overweight
coins were sorted out. Underweight coins were melted again whereas overweight
ones were reduced by mechanical treatment. Today specialised coin counting scales
are used which sort out foreign or faked coins. Banknotes are usually counted and
controlled optically, with weighing also applied. Formerly bank counters operated
on rolling of coins and use of special mechanical scales, and today electronic scales
are available (Fig. 5.118b/c). Businesses that rely heavily on cash transactions use
counting scales to speed up their cash counting processes.

(a)

(b)

Fig. 5.117 (**a**) Weighing of coins. Quentin Massys: Der Bankier (The banker) 1514. (**b**) Weighing of coins. (**d**) Weighing of coins. Several scales hang on the board. (**c**) Inspection of a coin. Weighing of coins. (**e**) Woman weighing coins. (**f**) Johannes Vermeer (1632–1675): Woman weighing pearls, 1664. National Gallery of Art, Wiener Collection, Washington, USA. (**g**) Skinflint

(c)   (d)

(e)   (f)

Fig. 5.117 (Continued)

(g)

Fig. 5.117 (Continued)

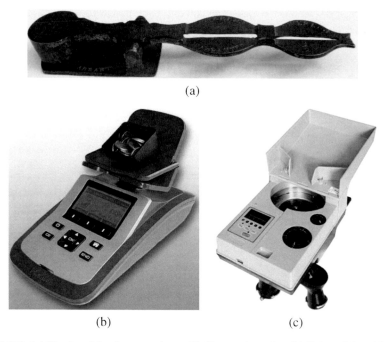

(a)

(b)   (c)

Fig. 5.118 (**a**) Check weigher for sovereign and half sovereign coins. (**b**) Coin weigher. (**c**) Coin sorting and weighing machine

Fig. 5.119 Principle of the coin operated machine for spending holy water of Heron of Alexandria (~20–70 AC)

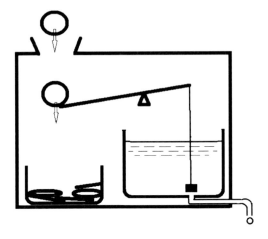

The expression 'Cotton balance' has several meanings.

- In economics it is the balance of production and export of cotton worldwide.
- In finance 'Cotton balance' designates a balance measuring the magnetic field of moved coins. In a magnetic field the mass of the coin seemingly increases and the real mass can be calculated precisely and fast. Inventor of this balance was the banker William Cotton (1786–1866).
- In physics the inventor of the 'balance de Cotton' is the French physicist Aimé Auguste Cotton (1869–1951). His balance is equipped with two beams, one of which measures the gravitational mass and the other the Laplace electromagnetic force.

A coin operated machine for spending holy water was invented already by Heron of Alexandria (~20–70 AC) [228] (Fig. 5.119). Coin operated vending scales like lollipop scales, flipper apparatus in pubs or amusement arcades, public body scales dispose on mass sensors which count different coins and reject fakes.

5.17 Body Scales

Our life starts with weighing. Often in the birth announcement the weight of the baby is indicated. Today, usually infant weight is recorded daily to monitor their health and growth.

5.17.1 Health Care

Mechanical baby scales usually are desk models of sliding weight balances: steelyards combined with the Roberval principle holding the sample pan in an upward direction. The sample pan is shaped half cylindrical so that the baby can be laid in it (Fig. 5.120). Increasingly mechanical baby scales are being replaced by those

Fig. 5.120 Baby scale: sliding weight balance, steelyard combined with the Roberval principle

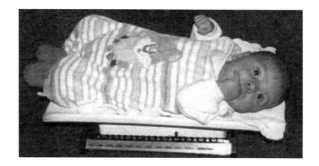

Table 5.2   Weight-estimation methods in the paediatric population

Method	Applicability	Method of calculation	
Broselow tape	46–143 cm	Read weight from corresponding colour zone	
Leffler formula	<1 year	m_{body} [kg] $= \frac{age\ [\text{months}]}{2} + 4$	(6.56a)
	1–10 years	m_{body} [kg] $= 2 \times age$ [years] $+ 10$	(6.56b)
Theron formula	1–10 years	m_{body} [kg] $= e^{0.175571 \times age\ [\text{years}]} + 2.187099^{a}$	(6.57)
So et al. Model A	<1 year	m_{body} [kg] $= (6.93 \times age$ [years]$) + 3.15 + 0.24$ if boy	(6.58a)
	<1 year, black	m_{body} [kg] $= (6.93 \times age$ [years]$) + 2.93 + 0.24$ if boy	(6.58b)
	≥1 year	m_{body} [kg] $= (3.1 \times age$ [years]$) + 5.61 + 0.65$ if boy	(6.58c)
	≥1 year, black	m_{body} [kg] $= (3.48 \times age$ [years]$) + 5.61 + 0.65$ if boy	(6.58d)
So et al. Model B	1–10 years	m_{body} [kg] $= e^{0.244 \times age\ [\text{years}]} + 1.73 + 0.05$ if boy	(6.59a)
	1–10 years, black	m_{body} [kg] $= e^{0.228 \times age\ [\text{years}]} + 1.73 + 0.05$ if boy	(6.59b)

[a] 6 digit numbers in an empiric statistical formula are senseless!

equipped with load cells based on strain gauges. The mean body weight of children may be assessed by the following methods [229] which are derived from statistics in the United States (Table 5.2). Because development and size depend on race and environmental influences those methods must be adapted when applied for other regions. They result in a mean value and deviations of the weight of an individual baby within a restricted range should not be overestimated.

For adults specialised bathroom scales are used to measure the body weight. The simplest designs are spring type platform scales (Fig. 5.121), replaced today by electronic types, each equipped with usually four load cells with strain gauges (Fig. 5.122). The problem of such bathroom scales of standing area for feet only is that for a person with a pot belly reading of the weight indicated is difficult. Therefore eye-level medical scales have been designed which are sliding weight top pan balances. The parallelogram guiding of the pan on the floor is connected by a linkage through a tube to the steelyard at chest height. Today also those accurate and expensive medical scales are replaced by electric types. With the beginning of the 20th century such mechanical scales had been put up at railway stations and other

Fig. 5.121 Spring type
platform bathroom scale.
© ADE, Hamburg, Germany

Fig. 5.122 Bathroom scale
with 4 load cells. © Efbe
Schott, Bad Blankenburg,
Germany

public places [230] (Fig. 5.123a–c). The scales were unblocked by coins during
operation. Later on the machines were equipped with ticket printer.

A very early balance to control food intake is depicted in Fig. 5.124 [231]. Indeed
bathroom scales may be used mostly to control the uptake of calories. Table 5.3
shows mean values of people's weight in some countries. It should be emphasised
that deviations from these mean values are irrelevant for the health of individual
persons. If a really ideal weight for an individual can be defined, then this is an
individual value which may deviate widely from the mean value of a population.

An obsolete recommendation for the ideal body weight [kg] is height [cm] minus
100. Because the weight of a human being depends on size the body size mass index
has been defined:

$$BMI = \frac{m_{body} \; [\text{kg}]}{(height \; [\text{m}])^2} \tag{5.57}$$

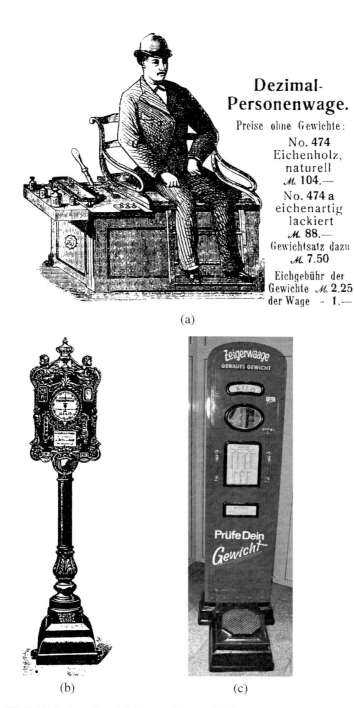

(a)

(b)   (c)

Fig. 5.123 Public body scales of G. Hartner, Ebingen 1903

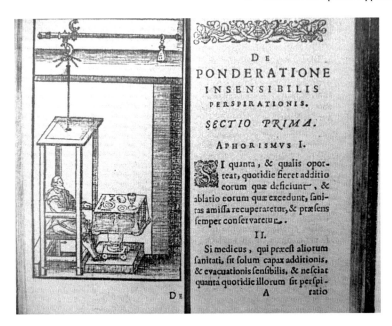

Fig. 5.124 Food control

Table 5.3 Statistical measurements of average weight of populations [232]

Country	Average male weight kg	Average female weight kg	Sample population age range	Year
Chile	77.3	67.5	>15	2009–2010
Germany	82.4	67.5	>18	2005
USA	86.6	74.4	20–74	1999–2002

Ancel Keys et al., found the *BMI* to be the best proxy for body fat percentage [233, 234]; the interest in measuring body fat being due to obesity becoming a discernible issue in prosperous Western societies. *BMI* was explicitly cited by Keys as being appropriate for population studies, and inappropriate for individual diagnosis. Nevertheless, due to its simplicity, it came to be widely used for individual diagnosis, despite its inappropriateness. *BMI* lower than 20 suggest the person is underweight while a number above 25 may indicate the person as overweight, and over 30 as obese. However, *BMI* depends on race and, as a mean value is different for different countries.

A *BMI* ≤ 25 is regarded as a not to fat body. The ratio *BMI*/25 is used as BMI Prime which is a dimensionless number. Individuals with BMI Prime <0.74 are underweight; those between 0.74 and 0.99 have optimal weight; and those at 1.00 or greater are overweight.

5.17.2 Medical Care

In the medical area gravimetry is widely applied. The requirements of accuracy in the preparation of drugs were one impetus in the development of precision balances, and until now the balance is the symbol of correctness of the pharmacist. Mass determination by means of balances or by centrifugation is practised in medical analytical laboratories. Thermal analysis may be help in clearing a wide range of problems; e.g.: diagnostics, testing of new surgical techniques, orthopaedic and traumatic phenomena [235].

In order to control the weight of patients several types of body scales are approved with regard to hygiene. Medical balances with a seat are used for patients who are unable to stand (Fig. 5.125). Also scales to weigh the patient in bed exist. Usually the bed is placed on four load cells and the total value is averaged.

In medicine mass changes are used as symptoms for diseases, e.g. diarrhoea and cancer. In hospitals weight is recorded daily as a parameter of the treatment, e.g. for the assessment of the mass of drugs to be fed. An important application is in nephrology. By means of a hosepipe blood is taken from the body, cleaned by dialysis, and recycled to the body. In this process the content of free water stored in the body is controlled and excess water is removed depending on body mass and regarding visible oedemas. The water to be removed at each treatment is assessed from weight increase in the meantime. For that purpose the patient is weighed before and after dialysis whereas the taking of water by the apparatus is controlled by the flow rate. The balance is equipped with a computer exit via data card to the dialysis apparatus (Fig. 5.125c).

A tonometry test measures the intraocular pressure (IOP) inside of the eye. This test is used to check for glaucoma, that can cause blindness by damaging the optic nerve in the back of the eye. Damage to the optic nerve may be caused by a build-up of fluid that does not drain properly out of the eye. Tonometry measures the resistance of the cornea to indentation. That has been made in the past by means of a balance. Today different methods are used, partially not involving body contact.

Weighing of the penis is depicted in a Roman wall painting (Fig. 5.126a) and on a gravestone where a woman compares the penises of two men by means of a 'trutina' (Fig. 5.126b). Results of such measurements may fail expectations.

5.17.3 Animal Scales

Also for animals special scales are on the market, equipped with a respective cage. These are sliding weight scales; today they are equipped with strain gauge sensors. Design and weighing range depend on the animals to be weighed (Fig. 5.127).

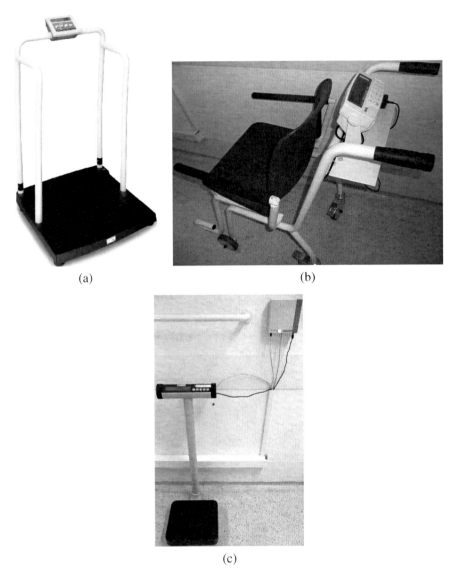

(a)   (b)

(c)

Fig. 5.125 (**a**)–(**c**) Medical body scales. (**c**) Medical body scale with data card printer. © Ebinger Waagenbau

5.18 Gravimeter

With gravimeters the local gravimetric field of the Earth is investigated. Gravimeters are used in metrology, geodesy and seismology, for prospecting minerals and oil. Absolute gravimeters measure free fall, or rise and free fall, of test bodies or atoms in vacuum. The movement usually is observed by means of a Michelson interferometer.

Fig. 5.126 (**a**) Weighing of a penis, wall painting, Pompeii. (**b**) Monument (gravestone?) Trier, Germany. A woman compares two penis' by weighing with a 'trutina'

(a)

(b)

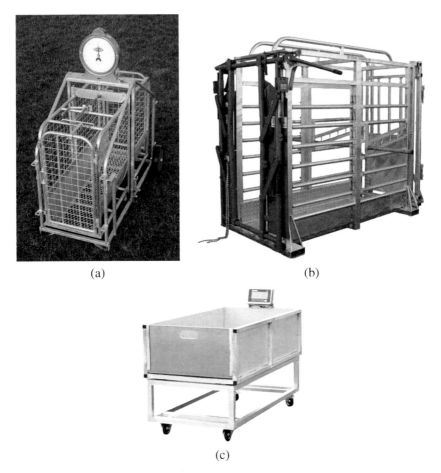

(a)   (b)

(c)

Fig. 5.127 (a) Mobile spring type animal scale. (b) Digital bull balance. © Bosche Wägetechnik, Damme, Germany. (c) Mobile piglet digital scale. © Bosche Wägetechnik, Damme, Germany

Relative gravimeters compare gravity at different places. Usually a weight on a spring is used and stretching of the spring is calibrated in a location with known gravitational force. Most accurate are superconducting gravimeters in which the position of a diamagnetic, superconducting niobium sphere in a magnetic field is observed.

5.19 Mass Determination in Astronomy

5.19.1 The Mass of the Earth

The gravitational force on a body with mass m_b on Earth's surface is given by Newton's second law (Eq. (1.1)) [236]. The distance between the centres of gravity is

Earth's radius; the mean Earth radius $r_E = 6367$ km $= 6.367 \times 10^6$ m. The mass of the Earth may be determined by combining Newton's Second law with the universal law of gravitation (Eq. (1.4)).

$$F_g = G \frac{m_b m_E}{r_E^2} = m_b g \tag{5.58}$$

Then Earth's mass m_E results in:

$$m_E = \frac{g r_E^2}{G} \tag{5.59}$$

The gravitational constant was determined first by Henry Cavendish (1731–1810) by measurement of the attractive force between two lead spheres (see Chap. 4). The actual value is $G = 6.67428(67) \times 10^{-11}$ m^3 kg^{-1} s^{-2}. The standard acceleration of gravity g_n is defined to be 9.806 65 m s^{-2}. So we obtain for the mass of the Earth $m_E = 5.9737 \times 10^{24}$ kg. The Earth mass is used as unit to describe masses of other rocky planets of our solar system and of exoplanets. The mass of the Earth increases slightly because of the cosmic debris (dust, meteorites) 10^6–10^8 kg/day.

The volume of the Earth is [237]:

$$V_E = \frac{4\pi r_E^3}{3} = \frac{4\pi \times (6.367 \times 10^6)^3}{3} = 1.1 \times 10^{21} \text{ m}^3 \tag{5.60}$$

The mean density is

$$\rho_E = \frac{3 r_E^3 m_E}{4\pi} = 5540 \text{ kg m}^3 \tag{5.61}$$

5.19.2 The Mass of the Sun

The mass of the Sun m_{Sun} may be determined by comparing the attractive gravitational force between Sun and Earth and the central force (centrifugal force) of the orbiting Earth [238]:

$$G \frac{m_E m_{Sun}}{r^2} = m_E \omega^2 r \tag{5.62}$$

with G the gravitational constant $= 6.672 59(85) \times 10^{-11}$ m^3 kg^{-1} s^{-2}, m_E mass of Earth, m_{Sun} mass of the Sun, r distance between Sun and Earth $=$ astronomical unit AU $= 1.49597870691 \times 10^{11}$ m, and ω the angular frequency. Here we assume a circular orbit and corrections should be made due to the ellipticity of the orbit. Furthermore the diurnal parallax of the Sun should be considered.

It follows

$$m_{Sun} = \frac{\omega^2 r^3}{G} = \frac{4\pi^2 r^3}{t_u G} \tag{5.63}$$

with t_u the time taken by the Earth to orbit the Sun $= 1$ a $= 365.256363$ d $= 3.1536 \times 10^7$ s.

The mass of the Sun was determined to be $1.98892(25) \times 10^{30}$ kg $= 332946 \times$ Earth mass, its volume to 1.412×10^{27} m^3, and its mean density to 1408 kg m^{-3}. In the general theory of relativity the mass units are sometimes expressed in units of length or time,

$$m_{Sun} \frac{G}{c^2} \approx 1.48 \text{ km} \qquad (5.64)$$

$$m_{Sun} \frac{G}{c^3} \approx 4.93 \text{ } \mu s \qquad (5.65)$$

5.19.3 The mass of Celestial Bodies

As discussed in Chap. 1 (1.5)–(1.8) the motion of planets and other celestial bodies in our planetary system is described by Kepler's laws. The mass of these objects can be determined with high accuracy applying Kepler's 3rd Law (Eqs. 1.10a, 1.10b). In general the motions of a planet around the sun or of a moon around a planet can be regarded as a two-body system and irregularities of the movement on account of various influences can be neglected. Likewise Kepler's 3rd Law is applied for the determination of the mass of stars. Therefore always the motion of stars in a binary system is looked at for this calculation. Even star clusters and galaxies at sufficiently large distance may be regarded as punctiform and then Kepler's laws can be applied approximately.

Measuring the orbital motion of the two objects in the binary system allows one to derive information about the mass of the objects themselves in a model-independent manner which is based solely on the application of Kepler's laws, and which does not rely on any assumptions about the state, dimensions, or evolution of the system. Often though, only information about one of the objects in the system (usually the brighter one) is known. This can be used to derive the mass function of the system, a combination of the two masses and the inclination of the orbital plane with respect to the line of sight. When similar information about the other object can be obtained, either spectroscopically or by measuring the time delays in the pulsations induced by the orbital motion, then the two masses can be calculated—if a good estimation of the inclination can be obtained. The latter can be calculated quite accurately if one of the objects is periodically eclipsed by the other. Several properties of celestial bodies (type of star, size, and irradiation) are connected with the mass of the body and these factors allow for assessment of mass.

The X-ray pulsator RX J0648.0-4418 and the subdwarf star HD49798 form a binary system with unique properties [239]. The subdwarf star is a bright object in the visible and ultraviolet wavelength range of the electromagnetic spectrum, and is well characterised. The orbital period of the system is accurately known, and the discovery in 1996 of a 13.2 s periodicity in X-rays made it clear that the companion must be either a neutron star or a white dwarf. By making use of recent XMM-Newton observations timed to coincide with the expected eclipse of the X-ray pulsator RX J0648.0-4418, the mass function of that X-ray source was accurately determined, and a well constrained estimate of the inclination of the orbital plane was obtained. Armed with these data, and applying them to the equation of orbital motion, it could be concluded that the X-ray source is a rare, ultra-massive (at least 1.2 solar masses) white dwarf. This makes RX J0648.0-4418 one of the most massive white dwarfs known to date.

Special methods can be applied for the assessment of mass of so-called 'Black Holes' which are spaces at which an extreme amount of matter is accumulated and compressed. The high gravitational field impedes the emission of electromagnetic radiation almost completely so that such objects are invisible. The mass of black holes in general is determined by observing the influence of gravity on neighbouring stars and accumulation of gas. Black holes absorb any matter in their surrounding and when this occurs jets of matter are ejected and X-rays are emitted, the intensity of which oscillates quasi-periodically. According to Nikolai Schaposchnikow and Lew Titartschuk (NASA) the frequency of those oscillations is related to the mass of the black hole and allows for its calculation [240]. The masses of the supermassive black holes found in galaxy bulges are correlated with a multitude of galaxy properties, leading to suggestions that galaxies and black holes may evolve together. The number of reliably measured black-hole masses is small. Directly measuring black-hole masses is currently possible with stellar kinematics (in early-type galaxies), ionised-gas kinematics (in some spiral and early-type galaxies) and in rare objects that have central maser emission. Davis et al. report that by modelling the effect of a black hole on the kinematics of molecular gas it is possible to fit interferometric observations of CO emission and thereby accurately estimate black-hole masses [241].

Another method is based on measuring the temperature of gas travelling in a direction towards the black hole. By the movement the gas is compressed and the peak temperature is related to the mass of the black hole. Using the X-ray telescope 'Chandra' of the University of California at Irvine the galaxy NGC 4649 was inspected and the mass of the black hole determined to be 3.4×10^9 times the mass sun in accordance with results from gravitational observations [242, 243].

According to the theory of relativity a light beam is influenced by gravity. Such a beam passing a massive object will be bent or even split simulating two sparkling sources. The effect allows for assessment of the influencing mass and also detection of dark matter.

5.19.4   The Mass of the Universe

Only about 4 % of the total energy density in the universe is connected with matter which can be seen directly. About 23 % is considered to be composed of dark matter [244] and the remaining 73 % is thought of as dark energy, distributed diffusely in space and 0.3 % of matter is neutrinos. Dark matter is a hypothetical form of matter of unknown composition that does not emit or reflect enough electromagnetic radiation to be observed directly, but whose presence can be inferred from gravitational effects on visible structures. It has vastly more mass than the 'visible' component of the universe. About 10 % of the ordinary matter is in self-lighting stars whereas 90 % is in planets, asteroids and dust. The existence of dark matter is necessary in theory because otherwise the rotating galaxies would be scattered. This concerns also our own Milky Way. The dark energy acts contrary to gravity in a similar way as Einstein's cosmological constant. It results in an increasing expansion of the volume of the universe. The density of ordinary particles and radiation in the universe is estimated to be equivalent to about one hydrogen atom per cubic metre of space.

The visible universe with a radius of 10^{28} m contains at least 10^{11} galaxies. Our galaxy contains at least 10^{11} stars and 5×10^{10} planets. The mass of the sun is about a 1.99×10^{30} kg. The mass of our galaxy amounts to about 10^{42} kg. That means the visible mass of the universe is at least 10^{53} kg. This very rough estimate is based on an approximate model, because

- We cannot see the boundaries of our Universe.
- On account of the finite value of light velocity, we look into the past.
- Dark matter is not included.

5.20 Curiosities

5.20.1 Vae Victis

The Roman historian Livius (*Livius, 5*, 47) [245, 246] wrote the following tale from 387 BC: As the Gauls tried to climb the walls of the Capitoline in Rome during the night, the tired army didn't notice and even the dogs made no sound. But the watchful, sacred geese of the Goddess Juno started with such a chatter and flapping of their wings and woke up the Roman Marcus Manilus and his troops. Therefore the citadel was saved from the invading Gaul's. The Romans honoured Juno with the name Moneta, out of gratitude for the deed that her sacred geese had performed. The dogs were crucified because they failed to bark when the Gaul's made their sneak attack.

Nevertheless, in the following negotiations the Gallien chieftain Brennus promised retreat of the soldiers if the Romans would pay 1000 pounds of gold. The Romans accepted but when weighing that tribute the Roman tribune Quintus Sulpicius complained that the weights used to calibrate that tribute were faked. In response Brennus threw his sword on the balance in addition, crying the famous "Vae Victis" (woe to the vanquished).

5.20.2 The Mass of the Soul

The Egyptians believed in a complementary spiritual world and on a life after death. Human beings and also animals had invisible attributes: Ka and Ba, which correspond approximately to soul and spirit and which survived the deceased. When a

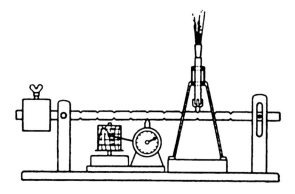

Fig. 5.128 Max Valier's rocket test bench with decimal scale to measure the thrust (1928)

human being died he was embalmed and the dead started a voyage along the underground Nile. At the end he was examined, not Ka or Ba, but his heart or the dead itself or his embryo was weighed against an ostrich feather in order to ascertain whether he or she was worthy to enter eternity. Of course this was a symbolic act and therefore discussed in a following chapter. However, the scales depicted are exactly of man's height real scales. Such scales could resolve the mass of an ostrich feather of about 2 g [247].

Bernhard Wälti tried recently to measure the weight of the astral body when leaving a sleeping person [248]. The results are inconclusive. As reported by Muldoon and Carrington [249], Dr. Malta und Dr. Zaalberg von Zelst, Den Haag calculated a mass of about 69.5 g of the astral body.

In 1906 William McDougall, Head of Massachusetts General Hospital and Joseph Banks Rhine [250] registered the weight of six dying patients. They used a balance with 2 g resolution on which the bed with the dying person was placed. Immediately following death they observed a sudden mass loss between 8 and 35 g. Similar experiments with dying dogs had no significant results.

Similar experiments with several dying patients by the Swedish physician and psychiatrist Nils-Olof Jacobson [251] resulted in a mass loss of around 21 g. This gave the suggestion for an American drama film in 2003 with title "21 grams".

5.20.3 Dangerous Weighing

In general, the balance is a safe and harmless instrument. Indeed the authors are aware of only one weighing procedure with deadly end. The rocket developer Max Valier (1895–1930), born in Bozen, Southern Tyrol, was killed when an alcohol-fuelled rocket exploded on May 17, 1930 on his test bench in Berlin. The rocket combustion chamber was placed on a decimal balance used to measure its thrust (Fig. 5.128). It exploded and a small metal fragment hit his pulmonary artery [252, 253].

Fig. 5.129 Weighing the bride against coins

5.20.4 Gravitational Waves

Gravitational waves, predicted by Einstein, could not be observed until now (see Chap. 1).

In physics, a gravitational wave is a fluctuation in the curvature of spacetime which propagates as a wave, travelling outward from the source. Sources of gravitational waves include binary star systems composed of white dwarfs, neutron stars, or black holes. Gravitational waves transport energy as gravitational radiation. Although gravitational radiation has not yet been directly detected, there is indirect evidence for its existence. Various gravitational wave detectors of impressive size exist. However, to date they have been unsuccessful in detecting such phenomena.

Although not yet observed with certainty, applications of gravitational waves have been sold successfully. Hartmut Müller claimed to transport data by a new technique named 'global scaling' [254]. Here gravitational waves should be modulated by the information signal. For this project a fraudulent company collected about €9 million and deposited the money at a Cypriot bank, but recently the inventor and the directors were accused of fraud.

5.20.5 Counterbalancing of Persons

In the Middle Ages sometimes the bride was weighed against coins in order to determine the dowry (Fig. 5.129).

In India it is still a practice at certain festivities to weigh the ruling Marahaja against gold or silver. The Maharaja then sits under a canopy and is weighed on a huge scale against pieces of treasure—gifts from his people. Minted coins are given away to Brahmins (a priestly caste) or to the poor.

References

1. E. Robens, H.R. Jenemann, Remarks on the notion "Microbalance", in *Proceedings of the XXVIth International Conference on Vacuum Microbalance Techniques*, ed. by M.b. Ben-Chanaa (Faculté des Sciences Semlalia, Université Cadi Ayyad, Marrakech, 1995), pp. 7–12
2. T. Gast, E. Robens, Review on modern vacuum microbalances, in *Proceedings of the XXVIth International Conference on Vacuum Microbalance Techniques*, ed. by M.b. BenChanaa (Faculté des Sciences Semlalia, Université Cadi Ayyad, Marrakech, 1995), pp. 151–158
3. C.A. Russell (ed.), *Recent Developments in the History of Chemistry* (The Royal Society of Chemistry, London, 1985)
4. J.A. Chaptal, *Elements de Chymie. Trois volumes* (Deterville, Paris, 1796)
5. L. Žagar, Ermittlung der Größenverteilung von Poren in feuerfesten Baustoffen. Arch. Eisen-hüttenwes. **26**(9), 561 (1956)
6. A.M. Basedow, H.R. Jenemann, Waage und Wägung, in *Quantitative organische Elemen-taranalyse*, ed. by F. Ehrenberger (VCH, Weinheim, 1991), pp. 79–108
7. H.R. Jenemann, Zur Geschichte der Mikrochemischen Waage, in *Acta Metrologiae Histori-cae II*, ed. by H. Witthöft (Linz, Trauner, 1989)
8. Al-Chazini, A. a.-R, *Kitab mizan al-hikma* (Buch der Waage der Weisheit) (Merw 1122) (In Arabian)
9. N. Khanikoff, Analysis and extract of the book of the balance of Wisdom—an Arabic work of the water balance, written by Al-Chazini in the 12th century. J. Am. Orient. Soc. **6**, 1–128 (1860)
10. F. Sezgin, Al-Chazini's Balance of Wisdom. Institut für Geschichte der Arabisch-Islamischen Wissenschaften, Johann Wolfgang Goethe-Universität, Frankfurt am Main (2000)
11. J. Reidy (ed.), *Thomas Norton's Ordinal of Alchemy* (Oxford University Press, London, 1975)
12. G. Agricola, *De Re Metallici Libri XII. Zwölf Bücher vom Berg- und Hüttenwesen*, 4th edn. (VDI-Verlag, Düsseldorf, 1556)
13. Wikipedia, Joseph Black. http://en.wikipedia.org/wiki/Joseph_Black (2013)
14. H.R. Jenemann, Die frühe Geschichte der Mettler-Waage, in *Siegener Abhandlungen zur Entwicklung der materiellen Kultur* (St. Katharinen, 1992)
15. E. Salvioni, *Misura di Masse Compresa fra g* 10^{-1} *e g* 10^{-6} (University of Messina, Messina, 1901)
16. B.B. Cunningham, J. Am. Chem. Soc. **71**, 1521 (1949)
17. J.S. Stas, Nouvelles recherches sur les lois des proportions chimiques et sur les poids atom-iques. Mém. Acad. R. Sci. Lett. B.-Arts Belg. **35**(3), 1–311 (1865)
18. A.A. Fomkin, T.S. Jakubov, V.V. Serpinskii, Adsorption Equilibrium at High Pressure, Ex-perimental Technique and Discussion. 29.11., in *Statusseminar 1991: Morphologie und Aus-tauschmechanismen poröser Stoffe (POST)* (Universität Siegen, 1991)
19. H.R. Jenemann, A.M. Basedow, E. Robens, Die Entwicklung der Makro-Vakuumwaage. PTB-Berichte. Vol. TWD-38. Braunschweig, Bremerhaven, Wirschaftsverlag NW. 70 (1992)
20. H.R. Jenemann, Das Kilogramm der Archive vom 4. Messidor des Jahres 7: Konform mit dem Gesetz vom 18. Germinal des Jahres 3? in *Genauigkeit und Präzision*, ed. by D. Hoff-mann, H. Witthöft (Physikalisch-Technische Bundesanstalt, Braunschweig, 1996), pp. 183–213
21. L.-J. Deleuil, J.-A. Deleuil, *Balance de Deleuil* (Bibiothèque Polytechnique, Paris, 1874)
22. A. de Lapparent, Biographie de Victor Regnault, in *(Ecole Polytechnique) Livre du Cente-naire* (Gauthier-Villars et fils, Paris, 1897), p. 326. et suiv
23. L. Loewenherz, Apparate für Messen und Wägen, in *Bericht über die wissenschaftlichen Ap-parate auf der Londoner Internationalen Ausstellung im Jahr 1876* (Vieweg, Braunschweig, 1878), pp. 207–278
24. K. Scheel, *Grundlagen der praktischen Metronomie*. Die Wissenschaft, vol. 36 (Vieweg, Braunschweig, 1911), pp. 119–123, exp. 122

25. T.J. Quinn, New techniques in manufacture of platinum-iridium mass standards. Platin. Met. Rev. **30**, 74–79 (1986)
26. T.R. Quinn, The kilogram: the present state of our knowledge. IEEE Trans. Instrum. Meas. **40**(4), 81–85 (1991)
27. S. Ikeda et al., Surface analytical studyof cleaning effects and the progress on contamination on prototypes of the kilogram. Metrologia **30**, 133–144 (1993)
28. L.R. Pendrilll, Microscopic and macroscopic properties of gas and buoyancy in precision weighing, in *La Massa e la sua Misura*, ed. by L. Grossi (Bologna, CLUEB, 1995), pp. 153–160
29. Y. Kobayashi et al., Prototype kilogram balance II of NRLM. Bull. NRLM **33**(2), 7–18 (1984)
30. Y. Kobayashi et al., Prototype kilogram balance II of NRLM. Bull. NRLM **35**(2), 143–158 (1986)
31. M. Gläser, R. Schwartz, M. Mecke, Experimental determination of air density using a 1 kg mass comparator in vacuum. Metrologia **28**, 45–50 (1991)
32. W. Nax, Meilenstein in der Wägetechnik: TU Ilmenau und Sartorius AG entwickeln genaueste Waage der Welt. idw/Technische Universität Ilmenau. Informationsdienst Wissenschaft (2009)
33. M. Gläser, Mass comparators, in *Comprehensive Mass Metrology*, ed. by M. Kochsiek, M. Gläser (Wiley/VCH, Berlin, 2000), pp. 441–478
34. H.R. Jenemann, E. Robens, History of vacuum macrobalances, in *Microbalance Techniques*, ed. by J.U. Keller, E. Robens (Multi-Science Publishing, Brentwood, 1994), pp. 13–23
35. F. Emich, Ein Beitrag zur quantitativen Mikroanalyse. Monatsh. Chem. **36**(6), 407–440 (1915)
36. J.W. McBain, A.M. Bakr, J. Am. Chem. Soc. **48**, 690–695 (1926)
37. J.W. McBain, H.G. Tanner, A robust microbalance of high sensitivity, suitable for weighing sorbed films. Proc. R. Soc. Lond., A **125**, 579–586 (1929)
38. E. Robens et al., Considerations on the planned use of a scientific balance on mars. Part II: Choice of the balance. J. Therm. Anal. Calorim. **86**(1), 27–29 (2006)
39. A.W. Czanderna, S.P. Wolsky, *Microweighing in Vacuum and Controlled Environments* (Elsevier, Amsterdam, 1980)
40. C. Eyraud, E. Robens, P. Rochas, Some comments on the history of thermogravimetry. Thermochim. Acta **160**, 25–28 (1990)
41. E. Robens, C. Eyraud, P. Rochas, Some comments on the history of vacuum microbalance techniques. Thermochim. Acta **235**, 135–144 (1994)
42. E. Robens, K.K. Unger, Vacuum for balances. J. Therm. Anal. **55**, 431–439 (1999)
43. S. Weber, Bemerkungen über die Gleichgewichtsbedingungen der Gase. Commun. Kamerlingh Onnes Lab. Univ. Leyden **22**(246 d) (1936/38)
44. E. Robens, G. Sandstede, Anordnungen zur präzisen Druckmessung und -regelung im Bereich von 0,1 bis 760 Torr. Vak.-Tech. **16**, 125–130 (1967)
45. C.H. Massen et al., Application of micro balances to the measurement of gas pressure over eight decades, in *Thermal Analysis*, ed. by H.G. Wiedemann (Birkhäuser, Basel, 1962), pp. 225–233
46. W. Kollen, A.W. Czanderna, Dynamic vacuum in microbalance chambers, in *Vacuum Microbalance Techniques*, ed. by C.H. Massen, H.J. van Beckum (Plenum, New York, 1970), pp. 145–159
47. G. Hakvoort, TG measurement of solid-gas reactions in aggresive gas atmospheres, in *Microbalance Techniques*, ed. by J.U. Keller, E. Robens (Multi-Science Publishing, Brentwood, 1994), p. 89
48. A.W. Czanderna, H. Wieder, An apparatus for the simultaneous measurement of the optical transmission and mass changes of thin films, in *Vacuum Microbalance Techniques*, ed. by R.F. Walker (Plenum, New York, 1962), pp. 147–164
49. M. Wutz et al., in *Theory und Practis of Vacuum Technology* (Vieweg, Braunschweig, 1989)
50. E. Robens, Vacuum systems for vacuum microbalances. Vacuum **35**(1), 1–4 (1985)

51. H.H. Willems, *Creep Behaviour and Microstructure of Hardened Cement Pastes* (Technische Universiteit, Eindhoven, 1985)
52. J.W. Clark, An electronic analytical balance. Rev. Sci. Instrum. **18**, 915–918 (1947)
53. J.W. Beams, Magnetic suspension for small rotors. Rev. Sci. Instrum. **21**, 182–184 (1950)
54. T. Gast, Registrierendes Wägen im Milligrammbereich und seine Anwendung auf die Staubmessung, in *DECHEMA-Monographien* K. Bretschneider, K. Fischbeck (DECHEMA, Frankfurt am Main, 1960), pp. 1–19
55. T. Gast, Microweighing in vacuo with a magnetic suspension balance, in *Vacuum Microbalance Techniques*, ed. by K.H. Behrndt (Plenum, New York, 1963), pp. 45–54
56. W. Baran, Overview of applications of permanent magnets, in *Proceedings of the 8th International Workshop on Rare-Earth Magnets*, ed. by K.J. Strnat (Dayton, Ohio, 1985), pp. 1–14
57. T. Gast, G. Luce, A directly weighing suspension balance with frequency variant output, in *Mechanical Problems in Measuring Force and Mass*, ed. by H. Wierings (Nijhoff, Dordrecht, 1986)
58. T. Gast, H. Jakobs, G. Luce, Transmission of DTA-values from a magnetically suspended sample. Thermochim. Acta **82**, 1–14 (1984)
59. W. Pahlke, Berlin. Technische Universität (1991)
60. T. Gast, W. Pahlke, Magentic coupling for a microbalance. J. Therm. Anal. **37**, 1933–1941 (1991)
61. T. Gast, A new magnetic coupling for the separation of microbalance and reaction vessel in experiments with controlled atmospheres. Thermochim. Acta **24**, 247–250 (1978)
62. F. Dreisbach, H.W. Lösch, Adsorption equilibria of the pure gases CH4 and H2S and of the mixture CH4/H2S on the zeolite DAY at t = 25°C. J. Therm. Anal. Calorim. **62**(2), 515–521 (2000)
63. F. Dreisbach, H.W. Lösch, Magnetic suspension balance for simultaneous measurement of a sample and the density of the measuring fluid. J. Therm. Anal. Calorim. **62**(2), 515–521 (2000)
64. H.W. Lösch, R. Kleinrahm, W. Wagner, Neue Magnetschwebewaagen für gravimetrische Messungen in der Verfahrenstechnik, in *Verfahrenstechnik und Chemieingenieurwesen, Jahrbuch 1994* (VDI-Verlag, Düsseldorf, 1994), pp. 117–137
65. W.F. Hemminger, H.K. Cammenga, *Methoden der thermischen Analyse* (Springer, Heidelberg, 1989)
66. M. Vitruvius Pollio, *De Architectura*, vol. II/III (Rome, 33–14 BC)
67. R.C. Mackenzie, Thermochim. Acta **75**, 251–306 (1984)
68. C. Duval, *Inorganic Thermogravimetric Analysis* (Elsevier, Amsterdam, 1953)
69. S. Iwata, *Über die Entwicklung der Thermowaage, Besonders in Japan* (Chemischen Institut der Universität Bonn, Bonn, 1961)
70. S. Iwata, *Soil-Water Interaction*, 2nd edn. (Dekker, New York, 1995)
71. C. Keattch, *An Introduction to Thermogravimetry* (Heyden/Sadtler, London, 1969)
72. C.J. Keattch, *The History and Development of Thermogravimetry* (University of Salford, Salford, 1977)
73. C.J. Keattch, Studies in the history and development of thermogravimetry. J. Therm. Anal. Calorim. **44**(5) (1995)
74. J. Šesták, P. Hubík, J.J. Mareš, Historical roots and development of thermal analysis and calorimetry, in *Glassy, Amorphous and Nano-Crystalline Materials: Thermal Physics, Analysis, Structure and Properties*, ed. by J. Šesták, P. Hubík, J.J. Mareš (Springer, Dordrecht, 2011)
75. C. Eyraud, P. Rochas, Thermogravimetry and silk conditioning in Lyons. A little known story. Thermochim. Acta **152**, 1–7 (1989)
76. W.F. Hemminger, K.-H. Schönborn, A nineteenth century thermobalance. Thermochim. Acta **39**, 321–323 (1980)
77. K. Honda, On a thermobalance. Sci. Rep. Tôhoku Univ., Sendai Serie **1**(4), 97–103 (1915)
78. K. Honda, Kinzoku No Kenkyū **1**, 543 (1924)

79. M. Guichard, Bull. Soc. Chim. Fr. **33**, 258 (1923)
80. P. Dubois, Bull. Soc. Chim. Fr. **3**, 1178 (1935)
81. P. Chevenard, X. Waché, R. de la Tullaye, Bull. Soc. Chim. Fr. **10**, 41 (1944)
82. C. Eyraud, I. Eyraud, Catalogue, in *50e Expos. Soc. Fr. Physique* (1953)
83. C. Eyraud, I. Eyraud, Laboratoires **12**, 13 (1955)
84. W.A. de Keyser, Nature **172**, 364 (1953)
85. J. Paulik, F. Paulik, Thermal analysis, part A: Simultaneous thermoanalytical examinations by means of the derivatograph, in *Thermal Analysis*, ed. by W.W. Wendlandt (Elsevier, Amsterdam, 1981)
86. J. Paulik, F. Paulik, Simultaneous thermoanalytical examination by means of the derivatograph, in *Comprehensive Analytical Chemistry*, ed. by W.W. Wendlandt (Elsevier, Amsterdam, 1981)
87. F. Paulik, *Special Trends in Thermal Analysis* (Wiley, Chichester, 1995)
88. G. Liptay, *Atlas of Thermoanalytic Curves* (Akadémiai Kiadó/Heyden, Budapest/London, 1971–1977)
89. E. Robens, G. Walter, Thermogravimetrische Arbeitsmethoden. Sprechsaal **104**(10), 426–428 (1971)
90. E. Robens, G. Walter, Thermogravimetrische Arbeitsmethoden. Sprechsaal **104**(11), 489–492 (1971)
91. DIN, *DIN 51006: Thermische Analyse (TA) – Thermogravimetrie (TG) – Grundlagen* (Beuth, Berlin, 2005)
92. A.W. Coats, J.P. Redfern, Nature **201**, 68–69 (1964)
93. H. Barkia, H.L. Belkbir, S.A.A. Jayaweera, Thermal analysis studies of oil shale residual carbon. J. Therm. Anal. Calorim. **76**, 615–622 (2004)
94. T. Berrajaa et al., Thermal analysis studies of the pyrolysis of Tarfaya oil shale, in *Carbon '88*, ed. by B. McEnaney, T.J. James (1988), pp. 567–569
95. P. Staszczuk, D. Glazewski, Study of heterogeneous properties of solids by means of a special thermal analysis techniques. J. Therm. Anal. Calorim. **55**, 467–481 (1999)
96. P. Staszczuk et al., Total heterogeneity of Al2O3 surface. J. Therm. Anal. Calorim. **71**(2), 445–458 (2003)
97. N. Gérard, Thermogravimetric investigation of the decomposition kinetics of a gaseous aluminum hydride compound, in *Microbalance Techniques*, ed. by J.U. Keller, E. Robens (Multi-Science Publishing, Brentwood, 1994)
98. R. Berger, J. Gutmann, R. Schäfer, Scanning probe methods: from microscopy to sensing. Bunsenmagazin **2**, 42–53 (2011)
99. S.J. Gregg, K.S.W. Sing, *Adsorption, Surface Area and Porosity*, 2nd edn. (Academic Press/Mir, London/Moscow, 1982/1984)
100. R.S. Mikhail, E. Robens, *Microstructure and Thermal Analysis of Solid Surfaces* (Wiley, Chichester, 1983)
101. F. Rouquerol, J. Rouquerol, K. Sing, *Adsorption by Powders & Porous Solids* (Academic Press, San Diego, 1999)
102. S. Lowell et al., *Characterization of Porous Solids and Powders: Surface Area, Pore Size and Density* (Springer, Dordrecht, 2004)
103. DIN, *Partikelmeßtechnik*, 4th edn. DIN Taschenbuch/CD, vol. 133 (Beuth, Berlin, 2004)
104. E.L. Fuller Jr., Volumetric and gravimetric methods of determining monolayer capacities. Thermochim. Acta **29**, 315–318 (1979)
105. E. Robens et al., Sources of error in sorption and density measurements. J. Therm. Anal. Calorim. **55**(2), 383–387 (1999)
106. E. Robens, A.W. Czanderna, J.A. Poulis, Surface area and pore size determination on finely divided or porous substances by adsorption measurements: comparison of volumetric and gravimetric methods. PMI, Powder Metall. Int. **12**(4), 201–203 (1980)
107. G. Sandstede, E. Robens, Automatisierte Apparatur zur gravimetrischen Bestimmung der spezifischen Oberfläche und der Porengröße. Chem. Ing. Tech. **34**(10), 708–713 (1962)

108. P. Klobes, E. Robens, *Standardization of the Pore Size Distribution*. Particle & Particle Systems Characterization (2011)
109. S. Brunauer, P.H. Emmett, E. Teller, Adsorption of gases in multimolecular layers. J. Am. Chem. Soc. **60**(2), 309–319 (1938)
110. ISO, *ISO 9277: Determination of the Specific Surface Area of Solids by Gas Adsorption Using the BET Method* (Beuth, Berlin, 2007)
111. V.V. Kutarov, E. Robens, *The Pickett equation analytical continuation*. Adsorption (2010)
112. V.V. Kutarov, E. Robens, B. Kats, Universal function for the description of multilayer adsorption isotherms. J. Therm. Anal. Calorim. (2006)
113. J. Adolphs, Excess surface work—a modelless way of getting surface energies and specific surface areas directly from sorption isotherms. Appl. Surf. Sci. **253**, 5645–5649 (2007)
114. J. Adolphs, M.J. Setzer, A model to describe adsorption isotherms. J. Colloid Interface Sci. **180**, 70–76 (1996)
115. J. Adolphs, M.J. Setzer, Energetic classification of adsorption isotherms. J. Colloid Interface Sci. **184**, 443–448 (1996)
116. ISO, ISO 15901-3 Evaluation of pore size distribution and porosimetry of solid materials by mercury porosimetry and gas adsorption—Part 3: Analysis of micropores by gas adsorption. ISO, Genève
117. E.P. Barrett, L.G. Joyner, P.H. Halenda, The determination of pore volume and area distribution in porous substances. I. Computation from nitrogen isotherms. J. Am. Chem. Soc. **73**, 373–380 (1951)
118. ISO, ISO 15901-2 Evaluation of pore size distribution and porosimetry of solid materials by mercury porosimetry and gas adsorption—Part 2: Analysis of mesopores and macropores by gas adsorption. ISO, Genève
119. W. Thomson, Lord Kelvin of Largs, On the equilibrium of vapour at a curved surface of liquid. Philos. Mag. **42**(282), 448–452 (1871)
120. P. Pfeifer, D. Avnir, Chemistry in noninteger dimensions between two and three: I. Fractal theory of heterogeneous surfaces. Chem. Phys. **79**(7), 3558–3565 (1983)
121. P. Pfeifer, D.J. Avnir, Chemistry in noninteger dimensions between two and three. Fractal surfaces of adsorbents. Chem. Phys. **79**(7), 3566–3571 (1983)
122. D.D. Farin, D. Avnir, The fractal nature of molecule-surface interactions and reactions, in *The Fractal Approach to Heterogeneous Chemistry*, ed. by D. Avnir (Wiley, Chichester, 1989), pp. 271–293
123. A.V. Neimark, Thermodynamic method of calculating surface fractal dimension. JETP Lett. **51**(10), 607–610 (1990)
124. A.V. Neimark, A new approach to the determination of the surface fractal dimension of porous solids. Physics A **191**(1–4), 258–262 (1992)
125. A.V. Neimark, Determination of surface fractal dimension from adsorption experimental data. Russ. J. Phys. Chem. **64**(10), 2593–2605 (1990)
126. A.V. Neimark, Calculating surface fractal dimension of adsorbents. Adsorp. Sci. Technol. **7**(4), 210–219 (1990)
127. A.V. Neimark, E. Robens, K.K. Unger, Berechnung der Fraktaldimension einiger poröser Feststoffe aus der Stickstoff-Adsorptionsisotherme. Z. Phys. Chem. **187**, 265–280 (1994)
128. E. Robens et al., Water vapour sorption and humidtity—a survey on measuring methods and standards, in *Humidity Sensors: Types, Nanomaterials and Environmental Monitoring*, ed. by C.T. Okada (Nova Science, Hauppauge, 2011), pp. 1–87
129. E. Gerland, F. Traumüller, Geschichte der physikalischen Experimentierkunst, Leipzig (1899)
130. N. Cusanus, Idiota de Staticis Experimentis, Dialogus. Codex Cusanus 1456/64 1450, Straßburg Folio 135r
131. N. Cusanus, Nicolai de Cusa opera omnia. Gesamtausgabe der Heidelberger Akademie. Felix Meiner, Hamburg (2008)
132. L.B. Alberti, L'architettura. Padua, Firenze (1483/1485)
133. L.B. Alberti, M. Theurer, Zehn Bücher über die Baukunst (Heller, Wien, 1912)

134. L. da Vinci, Codex atlanticus—Saggio del Codice atlantico, ed. by Aretin. Vol. fol. 249 verso-a + fol. 8 verso-b Milano (1872)
135. L. da Vinci, in Catalogue "Les Mots dans le Dessin" of the Cabinet des Dessin. Paris, Louvre (1986)
136. M.C. Cantu', T. Settle, The Antique Instruments at the Museum of History of Science in Florence. Arnaud, Firenze (1973)
137. C.W. Meyer et al., Automated continuous-flow gravimetric hygrometer as a primary humidity standard, in *Proceedings for the International Symposium on Humidity and Moisture* (National Institute of Metrology, Standardization and Industrial Quality, Rio de Janeiro, 2006), pp. 1–6
138. V.M. Mecea, J.O. Carlsson, R.V. Bucur, Extension of the quartz-crystal-microbalance technique. Sens. Actuators **53**, 371–378 (1996)
139. T. Brokate et al., Survey on mass determination with oscillating systems. Part II: Instruments and weighing of matter from gaseous environment. J. Therm. Anal. Calorim. **71**(1), 25–29 (2003)
140. S. Mintova, S. Mo, T. Bein, Chem. Mater. **13**, 901 (2001)
141. H. Römpp, Römpp Chemie Lexikon. CD 1.0 ed. Stuttgart, Thieme (1995)
142. H.-D. Isengard, Bestimmung von Wasser in Lebensmitteln nach Karl Fischer. ZFL, Int. Z. Lebensm.-Technol. -Verfahr. Tech. **42**, 1–6 (1991)
143. H.-D. Isengard, How to determine water in foodstuffs? Analytix **3**, 11–15 (2003)
144. J.B. Hannay, J. Chem. Soc. **32**, 381 (1877)
145. E. Robens, K. Rübner, Gravimetrische Wasserdampfsorptions- und Feuchtemessung an Feststoffen. GIT Z. Labortechnik **47**, 1046–1050 (2003),
146. E. Robens et al., Measurement of water vapour sorption and humidity. A survey on measuring methods and standards. ICTAC News **38**(1), 39–46 (2005)
147. K. Rübner, E. Robens, D. Balköse, Methods of humidity determination. Part I: Hygrometry. J. Therm. Anal. Calorim. **94**(3), 669–673 (2008)
148. Y.M. Volfkovich et al., Techniques of standard porosimetry and possible areas of their use in electrochemistry (Review). Sov. Electrochem. **16**(11), 1325–1353 (1981)
149. A. Brenner et al., Application of a gravimetric sorption system for the investigation of sorption kinetics exemplified at the uptake of hexyne-3 on silikalit-1, in *Microbalance Techniques*, ed. by J.U. Keller, E. Robens (Multi-Science Publishing, Brentwood, 1994), pp. 73–78
150. W.H. Kuhn, G. Walter (eds.), *Microgravimetric Investigation into the Mechanisme of Corrosion of Reactor Materials in the Presence of Nuclear Radiation*. Euratom Report, vol. 1474e (Presses Académiques Européennes, Brussels, 1964)
151. B. Franklin, Of the stilling of waves by means of oil. Philos. Trans. **64**, 445–460 (1774)
152. A. Pockels, Surface tension. Nature **43**, 437–439 (1891)
153. Wikipedia, Langmuir-Blodgett trough. http://en.wikipedia.org/wiki/Langmuir-Blodgett_trough (2013)
154. P. Lecomte du Noüy, A new apparatus for measuring surface tension. J. Gen. Physiol., 522–524 (1919)
155. P. Lecomte du Noüy, An interfacial tensiometer for general use. J. Gen. Physiol. **7**, 625–633 (1925)
156. G. Binning, H. Rohrer, Scanning tunnelling microscopy. IBM J. Res. Dev. **38**, 4 (1986)
157. Wikipedia, Scanning tunneling microscope. http://en.wikipedia.org/wiki/Scanning_tunneling_microscope (2013)
158. P. Poncheral et al., Electrostatic deflections and electromechanical resonances of carbon nanotubes. Science **283**, 1513–1516 (1999)
159. J. Wood, Mass detection finds new resonance. Mater. Today **4**, 20 (2004)
160. B. Ilic et al., Attogram detection using nanoelectromechanical oscillators. J. Appl. Phys. **95**(7), 3694–3703 (2004)
161. S. Gupta, G. Morell, B.R. Weiner, Electron field-emission mechanism in nanostructured carbon films: a quest. J. Appl. Phys. **95**(12), 8314–8320 (2004)

162. LaHaye et al., Science **304**, 74 (2004)
163. C. Sealy, Probing the quantum world with uncertainty. Mater. Today **6**, 9 (2004)
164. R. Berger, J. Gutmann, Polymer brushes on micromechanical cantilevers. Nanopticum **1**, 4–5 (2005)
165. H.J. Jodl, F. Glas, Millikan's experiment. http://millikan.edu.hel.fi/eng/index.htm. RCL (2008)
166. H. Hörstermann, I. Jandt, Der Millikan'sche Öltröpfchenversuch. http://home.wtal.de/i-jandt/Physik/Millikan/Millikan.html. Ross Moore, Mathematics Department, Macquarie University, Sydney (2002)
167. U. Kilian, C. Weber, *Lexikon der Physik*, vol. 4 (Spektrum Akademischer Verlag, Heidelberg, 2000)
168. R.A. Millikan, On the elementary electrical charge and the Avogadro constant. Phys. Rev. **2**, 109–143 (1913)
169. H. Straubel, Naturwissenschaften **42**, 506 (1955)
170. H. Straubel, Z. Elektrochem. **60**, 1033 (1956)
171. H. Straubel, Acta Phys. Austriaca **13**, 265 (1960)
172. H. Straubel, Phys. Bull. **28**, 56 (1972)
173. G. Böhme et al., Determination of relative weight changes of electrostatically suspended particles in the sub-microgram range, in *Progress in Vacuum Microbalance Techniques*, ed. by S.C. Bevan, S.J. Gregg, N.D. Parkyns (Heyden, London, 1973), pp. 169–174
174. G. Böhme et al., Messungen von Gewichtsänderungen an im elektrischen Feld frei schwebenden Teilchen. Sprechsaal **106**, 184–188 (1973)
175. H. Straubel, Chem. Ing. Tech. **43**, 853 (1971)
176. H. Straubel, Aerosol-Rep. **6**, 77 (1967)
177. W. Heide, Berührungslose Messung von Dichte, Viskosität und Oberflächenspannung kleiner Probenvolumina mit Hilfe akustischer Probenpositionierung und -anregung. Frankfurt am Main Battelle-Institut e.V (1990)
178. C.E. Wieman, D.J. Wineland, D.E. Pritchard, Atom Cooling, Trapping, and Quantum Manipulation. Rev. Mod. Phys., S253–S352 (2000) (Centennial Issue)
179. D. Rodríguez et al., MATS and LaSpec: high-precision experiments using ion traps and lasers at FAIR. Eur. Phys. J. Spec. Top. **183**, 1–123 (2010)
180. F. DiFilippo et al., Accurate atomic mass measurements from penning trap mass comparisons of individual ions, in *Atomic Physics 14*, ed. by D.J. Wineland, C.E. Wieman, S.J. Smith (AIP, Boulder/New York, 1995), pp. 149–175
181. M. Block et al., Direct mass measurements above uranium bridge the gap to the island of stability. Nature **463**, 785–788 (2010)
182. Wikipedia, Mass spectrometry. http://en.wikipedia.org/wiki/Mass_spectrometry (2008)
183. O.D. Sparkman, *Mass Spectrometry Desk Reference* (Global View Pub., Pittsburgh, 2000)
184. J.J. Thomson, *Rays of Positive Electricity and Their Application to Chemical Analysis* (Longman's Green, London, 1913)
185. W. Paul, H. Steinwedel, Ein neues Massenspektrometer ohne Magnetfeld. Z. Naturforsch. A **8**(7), 448–450 (1953)
186. Frühgeschichte der Gastechnik, http://www.dvgw.de/dvgw/geschichte/geschichte-des-dvgwF/fruehgeschichte-der-gastechnik/. DVGW, Deutscher Verein des Gas- und Wasserfaches e.V., Bonn (2011)
187. H. Recknagel, B. Sprenger, E.-R. Schramek, *Taschenbuch für Heizung + Klimatechnik 11/12*, 75th edn. (Oldenbourg Industrieverlag/Vulcan, München, 2010)
188. W. Fritsche, Elster – Stationen der Geschichte 1848–1998. http://www.elster-instromet.com/de/downloads/EI_Stationen_Geschichte_1998.pdf. Elster (1998)
189. J. van Keulen, Density of porous solids. Matér. Constr. **6**(33), 181–183 (1973)
190. J.U. Keller, R. Staudt, *Gas Adsorption Equilibria. Experimental Methods and Adsorptive Isotherms* (Springer, Heidelberg, 2004)
191. S. Bohn, Untersuchung der Adsorption von Helium an Aktivkohle Norit R1 Extra und dem Molekularsieb 5A., in *Inst. f. Fluid- & Thermodynamik* (Siegen, 1996)

192. Archimedes, *§7: About Swimming Bodies*, vol. 1 (Wissenschaftliche Verlagsbuchhandlung, Frankfurt am Main, 1987)
193. D. Fage, *Tablles of Temperature for Sikes's Hydrometer* (Robins, London, 1855)
194. J.S. Donovan, History of Sikes Hydrometer. http://www.promash.com/sikes/history.html (2002)
195. H.R. Jenemann (ed.), *Zur Geschichte der Dichtebestimmung von Flüssigkeiten insbesondere des Traubenmostes in Oechsle-Graden*. Schriften zur Weingeschichte, vol. 98 (Gesellschaft für Geschichte des Weins e.V, Wiesbaden, 1990)
196. K. Brachthäuser et al., *Entwicklung eines neuen Dichtemeßverfahrens und Aufbau einer Hochtemperatur-Hochdruck-Dichtemeßanlage*. Fortschrittsberichte VDI-Z, Reihe 8, vol. 371 (VDI Verlag, Düsseldorf, 1993)
197. T. Gast, Elektrische Mikrowägung mit Hilfe trägerfrequenter Regelkreise. ETZ, Elektrotech. Z., Ausg. A **87**, 9–13 (1966) (Sonderheft zum 70. Geburtstag von R. Vieweg)
198. T. Gast, R. Talebi-Daryani, *Selbstkompensierend Waage zur kontinuierlichen Messung der Gasdichte*. Meß- u. Regelungstechnik, vol. 7010-11-TUB (Technische Universität, Berlin, 1985)
199. Archimedes, *The Works of Archimedes, §7. About Swimming Bodies*, vol. 1 (Wissenschaftliche Verlagsbuchhandlung, Frankfurt am Main, 1987)
200. Archimedes, *Abhandlungen*. Ostwalds Klassiker der exakten Wissenschaften, vol. 201 (Harri Deutsch, Frankfurt am Main, 2003)
201. F. Kohlrausch, *Praktische Physik*, 23rd edn., vol. 3 (Teubner, Stuttgart, 1986)
202. G.S. Kell, Density, thermal expansivity and compressibility of liquid water from $0°C$ to $150°C$. J. Chem. Eng. Data **20**, 97–105 (1975)
203. H.R. Jenemann, Die Waagenkonstruktionen von Georg Westphal, in 125 Jahre 1860–1985 Westphal-Mechanik/Westphal-Augenoptik, Westphal Mechanik-Augenoptik Celle (1985), pp. 35–45
204. G.R. Blake, K.H. Hartge, Bulk density, in *Methods of Soil Analysis. Part i. Physical and Mineralogical Methods: Agronomy Monograph*, ed. by A. Klute (1986)
205. M. Gläser, Methods of mass determination, in *Comprehensive Mass Metrology*, ed. by M. Kochsiek, M. Gläser (Wiley/VCH, Berlin, 2000), pp. 184–231
206. T. Allen, *Particle Size Measurement*, 5th edn. (Chapman & Hall, London, 1997)
207. M. Deleuil, Powder technology and pharmaceutical processes, in *Handbook of Powder Technology*, ed. by D. Chulia, M. Deleuil, Y. Pourcelot (Elsevier, Amsterdam, 1993)
208. G. Parfitt, Powder Technol. **17**(2), 157–162 (1977)
209. K. Ropkins, R.N. Colville, Airborne pollutants: current practices in air quality monitoring programs. LabPlus Int. **14**(5), 22–23 (2000)
210. A.P.M. Glassford, Application of the quartz crystal microbalance to space system contamination studies, in *Applications of Piezoelectric Quartz Crystal Microbalances*, ed. by C. Lu, A.W. Czanderna (Amsterdam, Elsevier, 1984), pp. 281–305
211. C.C.J. French, Advanced techniques for engine research and design. J. Automot. Eng. **203**(D3), 169–183 (1989)
212. H. Patashnick, E.G. Rupprecht, Continuous measurement using the tapered element oscillating microbalance. J. Air Waste Manage. Assoc. **41**(8), 1079–1083 (1991)
213. D. Büker, T. Gast, Kontinuierliche gravimetrische Staubmessung durch mechanische Resonanz. Chem. Ing. Tech. **39**(16), 963–966 (1967)
214. H. Bahner, Th. Gast, Massebestimmung disperser Stoffe mit Hilfe transversal schwingender Filterbänder. Tech. Mess. **50**(1), 3–13 (1983)
215. K.U. Kramm, *Bestimmung von Massen, Viskositäten und E-Moduln mit dem longitudinal schwingenden Band* (Technische Universität, Berlin, 1985)
216. K. Leschonski, W. Alex, B. Koglin, Teilchengrößenanalyse. Chem. Ing. Tech. **43**, 23–26 (1974)
217. Wikipedia, Electromagnetism. http://en.wikipedia.org/wiki/Electromagnetism (2013)
218. F.E. Senftke et al., Quartz helix susceptibility balance using the Curie-Cheneveau principle. Rev. Sci. Instrum. **29**(5), 429–432 (1958)

219. A. van den Bosch, Static magnetic susceptibility measurements on NBS-SRM aluminium, in *Progress in Vacuum Microbalnce Techniques*, ed. by C. Eyraud, M. Escoubes (Heyden, London, 1975), pp. 398–408

220. H.G. Wiedemann, G. Bayer, Comparison of temperature measurements in the range of 400–2500 k by use of a thermobalance, in *Progress in Vacuum Microbalance Techniques*, ed. by C. Eyraud, M. Escoubes (London, Heyden, 1975), pp. 103–107

221. C.H. Massen et al., Application of micro balances to the measurement of gas pressure over eight decades, in *Thermal Analysis*, ed. by H.G. Wiedemann (Birkhäuser, Basel, 1972), pp. 225–233

222. H.G. Wiedemann, Application of thermogravimetry for vapor pressure determination. Thermochim. Acta **1**(3), 355–366 (1972)

223. E. Robens, R. Sieglen, Overpressure protection valve for high-vacuum apparatus. Vacuum **21**(10), 484 (1971)

224. T. Gast, T. Brokate, E. Robens, Vacuum weighing, in *Comprehensive Mass Metrology*, ed. by M. Kochsiek, M. Gläser (Wiley/VCH, Weinheim 2000), pp. 296–399

225. T. Gast, A device for simultaneous determination of mass and reaction force of a gas stream from a heated sample, in *Thermal Analysis*, ed. by H.G. Wiedemann (Birkhäuser, Basel, 1972), pp. 235–241

226. T. Gast, Simultaneous measurement of weight and torque and torque by magnetic suspension, in *Microweighing in Vacuum and Controlled Environments*, ed. by A.W. Czanderna, S.P. Wolsky (Elsevier, Amsterdeam, 1980), pp. 393–395

227. M. Hass, Scales and Weights. A collection of historical scles and weights from different periods of the past 3000 years. http://www.s-a-w.net/ (2010)

228. B. Woodcroft (ed.), *The Pneumatics of Hero of Alexandria* (Taylor/Maberly, London, 1851)

229. T.-Y. So, E. Farrington, R.K. Absher, Evaluation of the accuracy of different methods used to estimate weights in the pediatric population. Pediatrics **123**(6), e1054–e1051 (2009)

230. H. Homann, Die Personenwaage der Wiesbadener Curverwaltung. Mass & Gewicht **97**, 2390–2391 (2011)

231. S. Santorio, M. Lister, G. Baglivi, De statica medicina aphorismorum. Sectionis septem. De ponderatione insensibilis perspirationis (Venezia, 1759)

232. Wikipedia, Body weight. Human_weight. http://en.wikipedia.org/wiki/ (2011)

233. A. Keys et al., Indices of relative weight and obesity. J. Chronic. Dis. 1 **25**(6), 329–343 (1972)

234. Body mass index. http://en.wikipedia.org/wiki/Body_mass_index (2011)

235. D. Lörinczy (ed.), Thermal Analysis in Medical Application Wide Diversity in Thermal Analysis and Calorimetry, ed. by J. Simon (Akadémiai Kiadó, Budapest, 2011)

236. S. Dong, Mass of Earth, in *The Physics Factbook*. http://hypertextbook.com/facts/2002/SamanthaDong2.shtml, ed. by G. Elert (2002)

237. Wikipedia, *Earth*. http://en.wikipedia.org/wiki/Earth (2013)

238. Wikipedia, *Solar mass*. http://en.wikipedia.org/wiki/Solar_mass (2011)

239. S. Mereghetti et al., An ultra-massive fast-spinning white dwarf in a peculiar binary system. Science **4.9** (2009)

240. NASA, *Apollo*. http://www.nasa.gov/mission_pages/apollo/index.html. National Aeronautics and Space Administation (2007)

241. T.A. Davis et al., A black-hole mass measurement from molecular gas kinematics in NGC4526. Nature, 11819 (2013)

242. F. Gastaldello et al., Probing the dark matter and gas fraction in relaxed galaxy groups with x-ray observations from chandra and XMM-Newton. Astrophys. J. **669**(1), 158–183 (2007)

243. D.A. Buote et al., The x-ray concentration-virial mass relation. Astrophys. J. **664**(1), 123–134 (2007)

244. D.B. Cline, The Search for Dark Matter. Sci. Am. **3** (2003)

245. T. Livius, Ab urbe condita. Lib. 1-5. 10, Rom

246. T. Livius (ed.), The history of Rome (Ab urbe condita. Lib. 1-5). Electronic Text Center, University of Virginia Library, ed. by E.t.b.R.C. Roberts. E.P. Dutton and Co., New York (1912)

247. C.H. Massen et al., Investigation on a model for a large balance of the XVIII Egyptian dynasty, in *Microbalance Techniques*, ed. by J.U. Keller, E. Robens (Multi-Science Publishing, Brentwood, 1994), pp. 5–12
248. B. Wälti, in Ein metaphysikalisches Experiment zur Frage ob der Astralkörper ein Gewicht hat und mit einem Hinweis auf die dunkle Materie. http://www.beniwaelti.ch/inline_libra. htm (1997/2005)
249. S.J. Muldoon, H. Carrington, *Die Aussendung des Astralkörpers* (Herm. Bauer, Freiburg, 1996)
250. J.B. Rhine, *Extra-Sensory Perception After Sixty Years* (1940)
251. N.-O. Jacobson, Leben nach dem Tod? (Liv efter döden?) Bastei-Lübbe-Taschenbuch, vol. 10086. Bergisch Gladbach Bastei-Lübbe (1937/1979)
252. W.-H. Hucho, Vorstoß in den Weltraum – Max Valier – Pionier der Raumfahrt, wurde vor 100 Jahren in Bozen geboren. VDI-Nachr., Mag. **5**, 34 (1995)
253. M. Wade, Valier, http://www.astronautix.com/astros/valier.htm, in Encyclopedia Astronautica (2011)
254. EsoWatch, Global scaling. http://www.esowatch.com/ge/index.php?title=Global_Scaling EsoWatch (2011)

Chapter 6
Balance as Symbol and Object of Art

After its invention, more than 5 millenniums ago, for a long period the balance was a mysterious instrument and was difficult to handle. Its working mode could hardly be understood without knowledge of the natural forces in its operation. Until recently the entity, mass, was a mysterious property. It is not surprising that soon after its invention the instrument was mystified and thought of as a divine gift [1].

The balance is an instrument very often depicted as a symbol and operated in symbolic procedures initially by gods and saints and later by a monkey [2, 3]. For that symbolic weighing, always a real instrument was depicted. This allows us to study the constructional features of the ancient instruments. The idea of a symbolic significance of the balance in religion first appeared in Egypt and then spread through the whole of Europe [4]. The impartial judgement of the weighing operation was well known ca. 2000 B.C., as evidenced by the adoption of the balance as a symbol of social justice. However, a Budhistic Tibetan fate balance is also known (Fig. 6.1) [5]. Indeed there are basic differences in the different cultures and the respective religions.

Obviously 'weighing' is a fundamental method of human thinking, as demonstrated from very different branches of the arts [6]. We *weigh* our financial state as well as our ideal enterprises in order to calculate prospective gains and losses. Georg Groddeck (1866–1934), the founder of psychosomatic medicine, says that "the existence is filled from beginning to end with the *weighing* of guilt and the *balancing* of effort and worry against joy as a fruit of our actions" [7]. It was a theologian who mentioned that an individual's freedom is given only by the fact that he can *weigh* his actions in advance and that this freedom is limited by the actions being *weighed* by others against the existing moral code. And the philosopher Baruch Spinoza (1632–1677) assumed that we can live only because of the fear of death and the desire to live would *balance* it.

In modern times, the instrument is depicted as an object of art and as an objective technical symbol. It has become a popular token in comics. However, also antique examples have been found of balances as souvenirs or toys.

E. Robens et al., *Balances*, DOI 10.1007/978-3-642-36447-1_6,
© Springer-Verlag Berlin Heidelberg 2014

6.1 Mythology and Religion

In the mythology of the civilisations and in the holy scriptures of religions we often find symbolic notations in connection with a balance or its use in weighing. There are also historical reports included on weighing procedures of daily life. This concerns in particular the Egyptian, Hellenistic and Roman world of deities, the Jewish part of the Bible, the Old Testament, and the Koran. The original scripts of the New Testament contain only one remark on the balance and the Book Mormon only the notation 'weigh down' which has hardly to do with weighing. Christian civilisation, however, started very early with legends, in which the balance plays an important role.

6.1.1 Egypt

In the Old Kingdom of Egypt, divine qualities were attributed to the Pharaoh and it was believed that he enjoyed eternal life after his death. During the Middle Kingdom the idea of a future life was transferred to all members of the society, resulting in lavish funeral rites [8]. Wealthy citizens established voluminous vaults fitted out with numerous objects they presumed to need for a pleasant life in the next world. The dead had been provisioned with some real commodities, whereas servants, domestic animals, tools and food were replaced by figurative representations. In vaults the walls often are covered with scenes from the daily life and occasionally the use

Fig. 6.1 Tibetan Budhistic painting with fate balance of steelyard type, 18th century. © Kretzenbacher

(a)

(b)

Fig. 6.2 (a) Weighing of the heart. 9 Steps, Champdor, A.: Das Ägyptische Totenbuch. S. 114 [12]. (b) Book of the death of Neferini, papyrus ptolmeic 4.–1. century BC. (c) Weighing the heart, Book of the death of Neferini. (d) Weighing the heart, papyrus Cairo. (e) Weighing the heart, Papyrus Anhai. Thebes XXIst or XXIInd dynsty, ~1110 BC. (f) Weighing the heart, Papyrus Anhai. XXIInd dynasty. (g) Weighing the heart, papyrus similar to TT69, 18. Dynasty. (h) Weighing of the heart. Papyrus Hunefer 9901. (i) Weighing of the heart. Papyrus Ani, Chap. 300, British Museum London. (j) Weighing of the heart. Papyrus Berlin. (k) Weighing the heart, papyrus Ryerson, Oriental Institute Chicago. (l) Weighing the heart. Papyrus Milbank Chicago (Oriental. Inst.). (m) Weighing the heart, papyrus Paris. (n) Weighing the heart, papyrus. (o) Weighing the heart, pBM 10470, nach Le Page Renouf, in PSBA 17 (1985) Tf. 36 G.C. Seeber: Untersuchungen zur Darstellung des Totengerichts im Alten Ägypten. 55/269. (p) Weighing the heart, papyrus. (q) Weighing the heart, papyrus. (r) Weighing the heart, p München BSB ptol. C. Seeber: Untersuchungen zur Darstellung des Totengerichts im Alten Ägypten. /268. (s) Weighing of the heart. Papyrus Djed-Hor. Eggebrecht, Arno: Suche nach Unsterblichkeit. S. 101. (t) Weighing of the heart. Papyrus Nodjmet. James, T.G.H.: Egyptian Painting, S. 62. (u) Weighing of the heart. Papyrus Hor. Faulkner, R.O.: Book of the Dead. S. 31

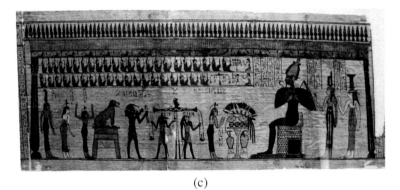

(c)

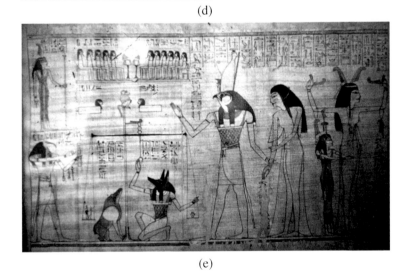

(d)

(e)

Fig. 6.2 (Continued)

(f)

(g)

Fig. 6.2   (Continued)

(h)

(i)

(j)

Fig. 6.2 (Continued)

(k)

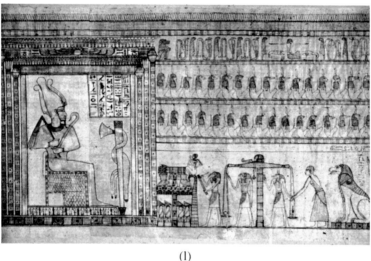

(l)

Fig. 6.2 (Continued)

of a balance is depicted. The corpse was embalmed and prepared for a voyage on the underground Nile. The dead person got a papyrus in his hand, the so-called *Book of Death*, a papyrus scroll containing instructions on how to succeed the dangerous voyage on the underground Nile to pass the examination by Osiris [9]. During this travel, the dead person had to pass dangerous adventures and finally he was examined in a tribunal. In front of 42 judge gods he presented his declaration of innocence (negative confession), documented in Spell 125 of the *Book of the Dead* [10]. After-

(m)

(n)

Fig. 6.2 (Continued)

wards his ka (~soul), represented by his heart, by his embryo or by his corpse, was weighed in order to decide whether the dead is worthy to achieve eternity or should be annihilated by the devourer. The heart weighing ceremony may be explained by a short form representation in Figs. 6.2–6.9: The deceased stands before the balance. His/her heart is placed on the balance pan and its mass is compared with an ostrich feather representing Maat, the goddess of truth, justice and harmony. The jackal-headed Anubis, tutelary god of the embalmers, calms the balance by grasping the suspension cords and touching the plummet. On top of the balance stand a baboon (identical with Thoth) is sitting as a symbol of measure. He controls the operation. (Sometimes his penis seems o be part of the indicator system.) The bal-

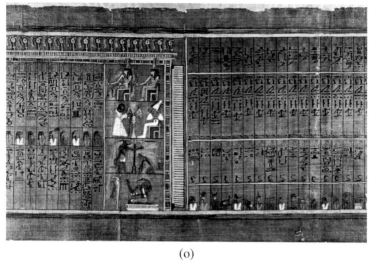

(o)

(p)

(q)

Fig. 6.2   (Continued)

(r)

(s)

Fig. 6.2 (Continued)

ance beam must come to equilibrium in horizontal position, thus showing that the
life of the dead person had been in harmony with the world. Each sin detected in the
questioning by the judges would result in a deviation of the balance. In this case,
the Devourer, a combination of hippopotamus and crocodile pounces upon the deceased. (Indeed, only one example of a painting is known, in which that happened.)
The ibis-headed Thoth, Lord of the Moon, god of learning and inventor of writing,
reports the result. A favourable Osiris, the emperor of the West, decides to enter

(t)

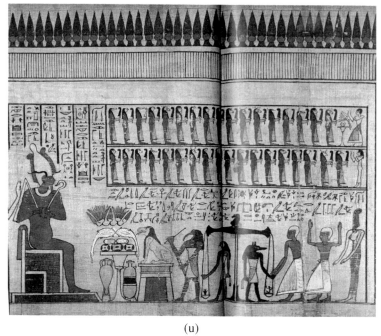

(u)

Fig. 6.2 (Continued)

Fig. 6.3 Weighing the heart.
Temple of Deir el-Medina
20th dynasty

eternity. Behind Osiris stand his sister and wife Isis. This ceremony is accompanied by Spell 30B of the Book of the Dead [10]:

"O my heart which I had from my mother! O my heart which I had from my mother!
O my heart of my different ages! Do not stand up as a witness against me, do not be opposed to me in the tribunal, do not be hostile to me in the presence of the keeper of the balance, for you are my ka which was in my body, the protector who made my members hale. Go forth to the happy place whereto we speed; do not make my name stink to the entourage who make man. Do not tell lies about me in presence of the god; it is indeed well that you should hear"!

That says Thoth, judge of truth, in the Great Ennead, which is in the presence of Osiris: "Hear the word of very truth. I have judged the heart of the deceased, and his soul stands as a witness for him. His deeds are righteous in the great balance, and no sin has been found in him. He did not diminish the offerings in the temples, he did not destroy what had been made, and he did not go about with deceitful speech while he was on earth".

Thus says the Great Ennead to Thoth who is in Hermopolis: "This utterance of yours is true. The vindicated Osiris *N* is straightforward, he has no sin, there is no accusation against him before us, Ammit shall not be permitted to have power over him. Let there be given to him the offerings which are issued in the presence of Osiris, and may a grant of land be established in the Field of Offerings as for the Followers of Horus".

Thus says Horus son of Isis: "I have come to you, O Wennefer, and I bring *N* to you. His heart is true, having gone forth from the balance, and he has not sinned against any god or any goddess. Thoth has judged him in writing which has been told to Ennead, and Maat the great has witnessed. Let there be given to him bread and beer which have been issued in the presence of Osiris, and he will be for ever like the Followers of Horus".

Thus says *N*: "Here I am in your presence, O Lord of the West. There is no wrong-doing in my body, I have not wittingly told lies, and there has been no second fault. Grant that I may be like the favoured ones who are in your suite, O Osiris, one greatly favoured by the good god, one loved of the Lord of the Two Lands, *N*, and vindicated before Osiris".

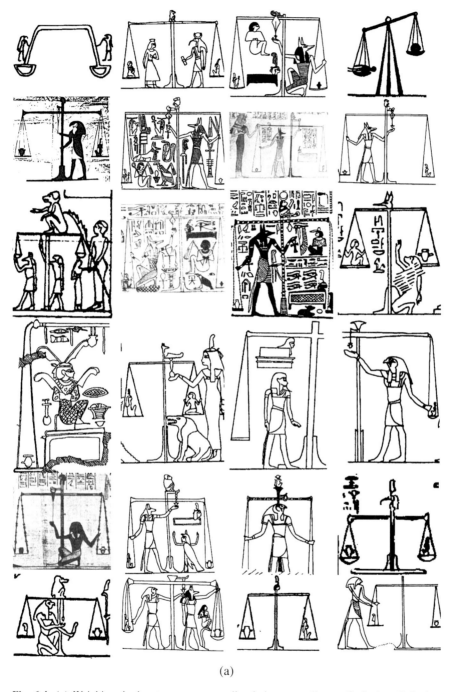

(a)

Fig. 6.4 (**a**) Weighing the heart; papyrus or wall painting, according to C. Seeber. © Seeber. (**b**) Weighing the heart; papyrus or wall painting, according to C. Seeber. © Seeber

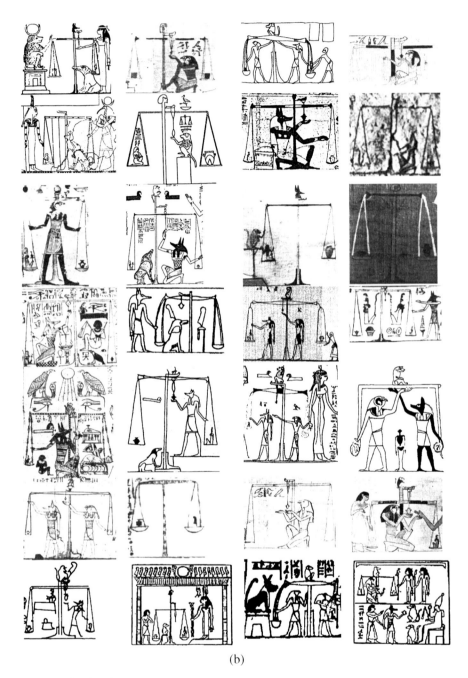

(b)

Fig. 6.4 (Continued)

Fig. 6.5 Weighing the heart; papyrus, according to C. Seeber. © Seeber

Fig. 6.6 (**a/b**) Weighing the heart. Coffin Seshepenmehyt XXVI–XXX dynasty, ~450 BC Brit. Museum London 22814. (**c**) Coffin of a woman, Teuris, 2nd century AD, Brit. Museum London. (**d**) Coffin Seshenmehyt 22814 XXVI–XXX 450 BC. (**e/f**) Weighing the heart. Coffin 24906??, 24986 XXII dyn. 900 BC Brit. Museum London Coffin Pasenhor. James, T.G.H.: Egyptian Painting. (**g**) Coffin 6667, XXI Dyn. 700 BC, XXI Dyn. 700 BC. (**h**) Coffin, © Seeber. (**i**) Coffin C. Seeber: Untersuchungen zur Darstellung des Totengerichts im Alten Ägypten. /268. (**j**) Coffin Berlin. (**k/l**) Weighing of the heart. Coffin Penju, 22/23 Dynasty ~800 BC. Roemer- & Pelizaeusmuseum Hildesheim, Germany. (**m**) Death tribunal on coffin, Muzeum Książąt Czartoryskich, Kraków, Poland

(e)

(f)   (g)

Fig. 6.6 (Continued)

(h)   (i)

(j)

Fig. 6.6 (Continued)

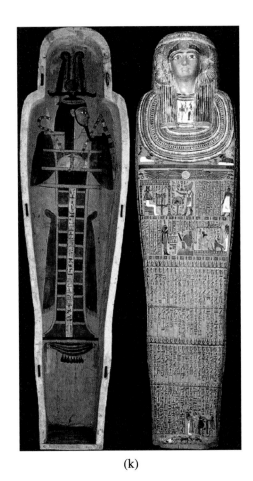

(k)

(l)   (m)

Fig. 6.6 (Continued)

Fig. 6.7   Table with pharaoh.
On the left side hieroglyph of
balance

Fig. 6.8   Weighing the heart. Painted wooden case, Louvre, Paris

(a)

(b)

Fig. 6.9 (**a/b**) Shroud, Roman period 200 BC. British Museum, London, UK (**c**) Shroud Hoffmann 325 nach Legrain: Collection Hoffmann III. Catalogue des Antiquités Egyptiennes Paris 1984, 100

(c)

Fig. 6.9 (Continued)

This scenery is depicted very often during the New Empire (1550–1070 BC), in particular from the middle of the 18th Egyptian dynasty (1550–1320 BC) onwards till the end of the Empire in several modifications in tombs, on coffins or shrouds. In 30 BC Octavian conquered Egypt and overturned the Ptolemaic dynasty. The country became a Roman province between 30 BC and 641 AD. The Egyptian religion decayed, however its basic ideas survived and the burial rites had been maintained for a long period of time [11], of course influenced and changed by the Greek and Roman cultures and to a less extent by Christian cultures.

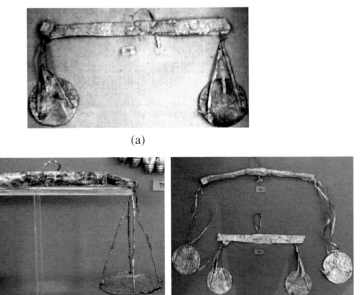

(a)

(b)   (c)

Fig. 6.10 (**a**) Old Chinese jewellery: small balance made of gold. (**b/c**) Symbolic balances made of gold foils for weighting of the soul (= butterfly) in the Underworld. Arch. Museum Athen, 16. Cent. BC Mycenae, Greece

6.1.2 China

The overestimation of the importance of the balance as a consequence of its obscure history and the lack of knowledge of natural laws is reflected in the rather lyrical Chinese descriptions in the Huai-Nan-Tsu book from the second century BC [9]. Six standard measuring methods are described which include the equal-beam and the unequal-beam balances. The equal-beam balance is associated with winter, the unequal-beam steelyard with summer, and the latter is described as follows:

> The steelyard, as a measuring instrument, moves with deliberation, but not too slowly. It equalises without inducing resentments, extends benefits for right-doing without ostentatious virtue, and expresses sorrow for wrong-doing without ostentatious reproof. It takes care to equalise the people's means, prolonging thereby the lives of those who would otherwise suffer want. Glorious and majestic it is, in its operations never unvirtuous. It nourishes, gives growth, transforms and develops, so that the thousand things flourish exceedingly, the five grains come to fruition, and the fields and fiefs bring forth their produce. Its administration is without error, so that sky and earth are brightened thereby.

A balance as a gold item of jewellery (Fig. 6.10a) may have had a symbolic character like the Greek soul balance (Figs. 6.10b/c).

6.1.3 Israel

In the *Tanach* (the *Old Testament*) we have several descriptions of events with weighing procedures and warnings for betraying (Table 6.1). Only a few illustrations were allowed in the *Tanach*, therefore we have no original pictures of scenes in which scales are included.

Table 6.1 Balance and weighing in the Bible

Genesis	Several trading negotiations are reported including weighing of money.
23:16	Abraham agreed to Ephron's terms and weighed out for him the price he had named in the hearing of the Hittites: four hundred shekels of silver, according to the weight currently in use among merchants.
24:21–23	Without saying a word, the man watched her closely to learn whether or not the Lord had made his journey successful. When the camels had finished drinking, the man took out a gold nose ring weighing a beka and two gold bracelets weighing ten shekels. Then he asked, "Whose daughter are you? Please tell me, is there room in your father's house for us to spend the night?"
43:20–22	"Please, sir," they said, "we came down here the first time to buy food. But at the place where we stopped for the night we opened our sacks and each of us found his silver—the exact weight—in the mouth of his sack. So we have brought it back with us. We have also brought additional silver with us to buy food. We don't know who put our silver in our sacks."
Exodus	
30:13	Each one who crosses over to those already counted is to give a half shekel, according to the sanctuary shekel, which weighs twenty gerahs. This half shekel is an offering to the Lord.
Leviticus	The book of Leviticus, among others, contains laws and priestly rituals [13]. In Hebrew the book is called *Vayikra*, literally "and He called"; it was written most probably by priests. This suggests that priests kept watch also on use of weights and measures and performed standardisation and calibration.
19:35–36	Do not use dishonest standards when measuring length, weight or quantity. Use honest scales and honest weights, an honest ephah and an honest hin. I am the Lord your God, who brought you out of Egypt.
26:26	When I cut off your supply of bread, ten women will be able to bake your bread in one oven, and they will dole out the bread by weight. You will eat, but you will not be satisfied.
Numbers	The Book of Numbers was written between 1440 and 1400 BC probably by Moses. It includes measures to organise the Israelitic tribes during the wandering in the wilderness. There are also included lists of offerings at the dedication of the tabernacle installed by Moses. Many gifts are registered with indication of their mass and conversion of weights or money is presented.
3:46–48	To redeem the 273 firstborn Israelites who exceed the number of the Levites, collect five shekels for each one, according to the sanctuary shekel, which weighs twenty gerahs. Give the money for the redemption of the additional Israelites to Aaron and his sons.
3:50	From the firstborn of the Israelites he collected silver weighing 1.365 shekels, according to the sanctuary shekel.

Table 6.1 (Continued)

7:13	His offering was one silver plate weighing a hundred and thirty shekels, and one silver sprinkling bowl weighing seventy shekels, both according to the sanctuary shekel, each filled with fine flour mixed with oil as a grain offering; one gold dish weighing ten shekels, filled with incense;
7:19–20	The offering he brought was one silver plate weighing a hundred and thirty shekels, and one silver sprinkling bowl weighing seventy shekels, both according to the sanctuary shekel, each filled with fine flour mixed with oil as a grain offering; one gold dish weighing ten shekels, filled with incense;
7:2–26	His offering was one silver plate weighing a hundred and thirty shekels, and one silver sprinkling bowl weighing seventy shekels, both according to the sanctuary shekel, each filled with fine flour mixed with oil as a grain offering; one gold dish weighing ten shekels, filled with incense;
7:31–32	His offering was one silver plate weighing a hundred and thirty shekels, and one silver sprinkling bowl weighing seventy shekels, both according to the sanctuary shekel, each filled with fine flour mixed with oil as a grain offering; one gold dish weighing ten shekels, filled with incense;
7:37–38	His offering was one silver plate weighing a hundred and thirty shekels, and one silver sprinkling bowl weighing seventy shekels, both according to the sanctuary shekel, each filled with fine flour mixed with oil as a grain offering; one gold dish weighing ten shekels, filled with incense;
7:43–44	His offering was one silver plate weighing a hundred and thirty shekels, and one silver sprinkling bowl weighing seventy shekels, both according to the sanctuary shekel, each filled with fine flour mixed with oil as a grain offering; one gold dish weighing ten shekels, filled with incense;
7:49–50	His offering was one silver plate weighing a hundred and thirty shekels, and one silver sprinkling bowl weighing seventy shekels, both according to the sanctuary shekel, each filled with fine flour mixed with oil as a grain offering; one gold dish weighing ten shekels, filled with incense;
7:55–56	His offering was one silver plate weighing a hundred and thirty shekels, and one silver sprinkling bowl weighing seventy shekels, both according to the sanctuary shekel, each filled with fine flour mixed with oil as a grain offering; one gold dish weighing ten shekels, filled with incense
7:85–86	Each charger of silver weighing an hundred and thirty shekels, each bowl seventy: all the silver vessels weighed two thousand and four hundred shekels, after the shekel of the sanctuary: The golden spoons were twelve, full of incense, weighing ten shekels apiece, after the shekel of the sanctuary: all the gold of the spoons *was* an hundred and twenty shekels.
1. Samuel	
2:3	Do not keep talking so proudly or let your mouth speak such arrogance, for the Lord is a God who knows, and by him deeds are weighed.
17:4–7	A champion named Goliath, who was from Gath, came out of the Philistine camp. He was over nine feet tall. He had a bronze helmet on his head and wore a coat of scale armour of bronze weighing five thousand shekels; on his legs he wore bronze greaves, and a bronze javelin was slung on his back. And the staff of his spear was like a weaver's beam; and his spear's head weighed six hundred shekels of iron: and one bearing a shield went before him.
2. Samuel	
14:26	And when he polled his head, (for it was at every year's end that he polled it: because the hair was heavy on him, therefore he polled it: he weighed the hair of his head at two hundred shekels after the king's weight.

Table 6.1 (Continued)

21:16	And Ishbi-benob, which *was* of the sons of the giant, the weight of whose spear weighed three hundred shekels of brass in weight, he being girded with a new sword, thought to have slain David.
1. Chronicles	
20:2	And David took the crown of their king from off his head, and found it to weigh a talent of gold, and there were precious stones in it; and it was set upon David's head: and he brought also exceeding much spoil out of the city.
Ezra	
8:25–33	... and I weighed out to them the offering of silver and gold and the articles that the king, his advisers, his officials and all Israel present there had donated for the house of our God. I weighed out to them 650 talents of silver, silver articles weighing 100 talents, 100 talents of gold, 20 bowls of gold valued at 1,000 darics, and two fine articles of polished bronze, as precious as gold. I said to them, "You as well as these articles are consecrated to the Lord. The silver and gold are a freewill offering to the Lord, the God of your fathers. Guard them carefully until you weigh them out in the chambers of the house of the Lord in Jerusalem before the leading priests and the Levites and the family heads of Israel. Then the priests and Levites received the silver and gold and sacred articles that had been weighed out to be taken to the house of our God in Jerusalem. On the twelfth day of the first month we set out from the Ahava Canal to go to Jerusalem. The hand of our God was on us, and he protected us from enemies and bandits along the way. So we arrived in Jerusalem, where we rested three days. On the fourth day, in the house of our God, we weighed out the silver and gold and the sacred articles into the hands of Meremoth son of Uriah, the priest. Eleazar son of Phinehas was with him, and so were the Levites Jozabad son of Jeshua and Noadiah son of Binnui. Everything was accounted for by number and weight, and the entire weight was recorded at that time.
Job	
6:1–4	Then Job replied: "If only my anguish could be weighed and all my misery be placed on the scales! It would surely outweigh the sand of the seas—no wonder my words have been impetuous. The arrows of the Almighty are in me, my spirit drinks in their poison; God's terrors are marshalled against me."
28:15	It cannot be bought with the finest gold, nor can its price be weighed in silver.
31:5–6	If I have walked in falsehood or my foot has hurried after deceit—let God weigh me in honest scales and he will know that I am blameless.
37:16	Do you know how the clouds hang poised, those wonders of him who is perfect in knowledge?
Psalms	
62:8–10	Lowborn men are but a breath, the highborn are but a lie; if weighed on a balance, they are nothing; together they are only a breath.
Proverbs	Several times merchants are called to weigh honestly.
11:1	The Lord abhors dishonest scales, but accurate weights are his delight.
16:11	Honest scales and balances are from the Lord; all the weights in the bag are of his making.
20:23	The Lord detests differing weights, and dishonest scales do not please him.
Isaiah	describes how Godfather used the balance during the creation of the world [14]. Later he reports on idolatry [15].

Table 6.1 (Continued)

40:12–15	Who has measured the waters in the hollow of his hand, or with the breadth of his hand marked off the heavens? Who has held the dust of the earth in a basket, or weighed the mountains on the scales and the hills in a balance? Who has understood the mind of the Lord, or instructed him as his counsellor? Whom did the Lord consult to enlighten him, and who taught him the right way? Who was it that taught him knowledge or showed him the path of understanding? Surely the nations are like a drop in a bucket; they are regarded as dust on the scales; he weighs the islands as though they were fine dust.
46:6	Some pour out gold from their bags and weigh out silver on the scales; they hire a goldsmith to make it into a God, and they bow down and worship it.
Jeremiah	
32:9–10	"I knew that this was the word of the Lord; so I bought the field at Anathoth from my cousin Hanamel and weighed out for him seventeen shekels of silver. I signed and sealed the deed, had it witnessed, and weighed out the silver on the scales."
Ezekiel	
5:1–3	"Now, son of man, take a sharp sword and use it as a barber's razor to shave your head and your beard. Then take a set of scales and divide up the hair. When the days of your siege come to an end, burn a third of the hair with fire inside the city. Take a third and strike it with the sword all around the city. And scatter a third to the wind. For I will pursue them with drawn sword. But take a few strands of hair and tuck them away in the folds of your garment."
45:9–11	'This is what the Sovereign Lord says: You have gone far enough, O princes of Israel! Give up your violence and oppression and do what is just and right. Stop dispossessing my people, declares the Sovereign Lord. You are to use accurate scales, an accurate ephah and an accurate bath. The ephah and the bath are to be the same size, the bath containing a tenth of a homer and the ephah a tenth of a homer; the homer is to be the standard measure for both.
Daniel	The metaphor of weighing up of bad and good deeds by god [16] probably is taken over from Egypt. Proverbial has become the "Mene tekel" in Daniel's prophecy to Belshazzar.
5:25–28	This is the inscription that was written: Mene, Mene, Tekel, Parsin. This is what these words mean:
	Mene: God has numbered the days of your reign and brought it to an end.
	Tekel: You have been weighed on the scales and found wanting.
	Peres: Your kingdom is divided and given to the Medes and Persians.
Hosea	
12:7	The merchant uses dishonest scales; he loves to defraud.
Amos	
8:4–6	Hear this, you who trample the needy and do away with the poor of the land, saying, "When will the New Moon be over that we may sell grain, and the Sabbath be ended that we may market wheat?"—skimping the measure, boosting the price and cheating with dishonest scales, buying the poor with silver and the needy for a pair of sandals, selling even the sweepings with the wheat."
Micah	
6:11	Shall I acquit a man with dishonest scales, with a bag of false weights?
Zechariah	
11:12	I told them, "If you think it best, give me my pay; but if not, keep it." So they weighed for me thirty pieces of silver.

6.1.4 Greece

The balance lost some of its bad publicity when Aristotle (384–322) defined the action of forces and Archimedes (285–212 BC) explained the mode of operation of levers [17, 18]. On the other hand, in Greece the balance symbol achieved a poetic dimension. The idea of weighing the soul of a deceased (Figs. 6.10b/c), the balance of fate and of fairness and justice, already based in pre-antique oriental thinking, was developed. In ceramic ware, gods are frequently represented operating the balance.

In Hellenic mythology, Goddess Moira, the Goddess of fate, or Godfather Zeus, himself used the fate balance in order to decide about victory or death of Achilles and Hector by weighing the two heroes (Fig. 6.11a). He whose balance pan sinks is doomed to die. During the Trojan War, Zeus weighed the fates of the two armies in the balance, and that of the Greeks sinks down (Figs. 6.11b/c). This idea has been repeatedly taken up poetically by Homer and others. Also Kairos, the opportune moment, sometimes is personalised as a young man, with wings and equipped with a balance (Fig. 6.12). Always symmetric beam balances are shown in the pictures.

Aischylos wrote [19, 20]:

> "The awe of majesty (of kings) once unconquered, unvanquished, irresistible in war, that penetrated the ears and heart of the people, is now cast off (with death). But there is still fear. And Eutykhia (prosperity)–this, among mortals, is a God and more than a God. But the balance of Dike (Justice) keeps watch: swiftly it descends on those in the light; sometimes pain waits for those who linger on the frontier of twilight; and others are claimed by strengthless night."

From an Anonymous source, we have the sentence [21]:

> "Tykhe (Fortune), beginning and end for mankind, you sit in Sophia's (Wisdom's) seat and give honour to mortal deeds; from you comes more good than evil, grace shines about your gold wing, and what the scale of your balance gives is the happiest; you see a way out of the impasse in troubles, and you bring bright light in darkness, you most excellent of Gods."

Themis was the embodiment of divine order, law, and custom, in her aspect as the personification of the divine rightness of law. Together with Zeus she had a daughter Dike, the Goddess of justice. In a fragment of a book of the poet Bacchylides, (5th century BC) we find [22, 23]:

> "If some God had been holding level the balance of Dike ..."

(a)

(b)   (c)

Fig. 6.11 (**a**) Encaustic painting on a Greek vase: Gods weighing the heroes in order to decide which one will win and which will be killed. (**b/c**) Kerostasy on a Greek vase. Zeus weighs the fates of the two armies in the balance, and that of the Greeks sinks down

Fig. 6.12 The Greek god Kairos, the opportune moment, with balance. Reconstruction of A. Krstulovic, Trogir. (Kretzenbacher)

6.1.5 Rome

The Romans modified the balance symbolism to a technical and identification icon, often with legislative significance.

In Roman mythology, Aequitas, also known as Aecetia, was the minor Goddess of fair trade and honest merchants. She is depicted with a cornucopia, representing wealth from commerce and holding an equilibrated balance, representing equity and fairness. During the Roman Empire, Aequitas was sometimes worshipped as a quality or aspect of the emperor, under the name Aequitas Augustus. Aequitas is often depicted on coin's reverse [24] (Fig. 6.13).

Juno Moneta (Fig. 6.14) is the Roman Goddess of Good Counsel, whose name means 'Advisor' or 'Warner', said to give good advice to the people in general and those about to be married in particular. She had a large and famous temple on the Arx, a height on the Capitoline Hill in Rome. Some believe that her epithet 'Moneta' is derived from "mons", but is usually assumed to derive from the Latin word for 'warn', monere. In 273 BC the Romans established a mint at Juno Moneta's temple, and for the next four hundred years or so all the silver coins of Rome were made there, up to the time of the Emperor Augustus in the 1st century AD. Juno Moneta then accordingly became associated with money and minting coins, and both the English words 'money' and 'mint' trace back to her epithet Moneta. Oddly enough

Fig. 6.13 (**a**) Aequitas on the reverse of this antonianus = 2 denarii struck under Roman Emperor Claudius II. The goddess is holding her symbols, the balance in equilibrium and the cornucopia. (**b**) Roman coins with Goddess Aequitas with balance and sceptre or cornucopia on the reverse. (**c**) Coin Rom, Titus Flavius Vespasianus; Caesar Vespasianus Augustus, 9–79 As 72, Æ10.49 g. Laureate head r. Rev. Aequitas standing l., holding scales and sceptre. (**d**) Coin Rom Hadrian, 117–138 Sestertius 134–138, Æ 35.59 g. Laureate head r. Rev. Aequitas standing l., holding scales and sceptre. (**e**) Coin Rom Lucius Verus, 161–169 Aureus 168, AV 6.94 g. Laureate, draped and cuirassed bust r. Rev. Aequitas seated l., holding scales and cornucopiae. (**f**) Coin Rom Titus Fulvius Iunius Macrianus (†261), Aequitas holding scales and cornucopiae. (**g**) Coin Rom Macrinus, 217–218 Denarius 217–218, AR 3.04 g. Laureate and draped bust r. Rev. Aequitas standing l., holding scales and cornucopiae. (**h**) Coin Rom Philip II Augustus, 246–249 Antoninianus, Antiochia 246–249, AR 4.50 g. Rev. Aequitas standing l., holding scales and cornucopiae

(a)

(b)

(c)

(d)

Fig. 6.13 (Continued)

(e)

(f)

(g)

(h)

Fig. 6.14 Jupiter monuments (copies), with balance in the hand of goddess Moneta, Saalburg, Bad Homburg vdH/Mainz

for a temple of such importance and fame, no trace of it survived and no one is even sure now where it was located, though it is suspected to lie beneath the Church of Aracoeli. Juno Moneta is depicted on coins, holding a pair of scales with which she controlled the correctness of the money lying in a pile at her feet. She also frequently holds a cornucopia as an emblem of abundance (Fig. 6.15).

Iustitia's attributes parallel those of the Hellenic deities Themis and Dike. Themis was the embodiment of divine order, law, and custom, but her daughter Dike was imaged carrying scales. Of all deities and personifications appearing on Roman imperial coins, Iustitia is among the rarest only of the regency of Hadrian. Iustitia with diadem, sword, balance and patera has become symbol of justice throughout millenniums until now and therefore is discussed in a separate chapter. A modern coin and a medal of the Vatican with Iustitia are shown in Fig. 6.16.

Fig. 6.15 (a) Roman coin. (b) As of Domitianus. (c) On the reverse Moneta holding cornucopia and balance. (d) Coin Rom Pescenius Niger, 193–194 Denarius, Antiochia 193–194, AR 2.81 g. Laureate head r. Rev. Moneta standing l., holding scales and cornucopiae

(a)   (b)

(c)   (d)

(e)

Fig. 6.16 (**a**) Medal of pope Innocence XII with Justitia holding balance and sword. (**b**) Coin with balance of Italy and coin with Justitia holding balance and sword of Vatican 1952. (**c**) 1818 Hanumanji Balance East India Company. (**d**) Half Anna coin, reverse. (**e**) Coin with balance, France 1793

6.1.6 The Christian Civilisation

In the gospels several sceneries could include a balance though not expressly mentioned. A painting of Jan van Hemessen 'Evocation of Matthew by Jesus' presents the customs officer when exchanging currency (Fig. 6.17). On the desk there is a balance, the scales filled with coins.

In the New Testament we have only one remark on a balance in the Book of Revelation [25]. Apparently here the balance is a symbol for misery, in connection with the announcement of imminent inflation:

> When the Lamb opened the third seal, I heard the third living creature say, "Come!" I looked, and there before me was a black horse! Its rider was holding a pair of scales in his hand. Then I heard what sounded like a voice among the four living creatures, saying, "A quart of wheat for a day's wages, and three quarts of barley for a day's wages, and do not damage the oil and the wine!

The Apocalypse had been illustrated very often, e.g. within the famous Apocalypse Tapestry of Angers from 14th century. Some examples of the Four Horseman of are shown in Fig. 6.18 [26–28].

In the brotherly community of the first Christians with their hopeful belief, there was no room for a symbolic balance. That changed when at the beginning of the fourth century Christianity became a state-supported religion. Now we find it as the attribute of the archangel Michael and of St. Anthony of Florence (1489–1459). Sometimes saints are depicted with a balance with reference to their profession. So we have pictures of the physician Pantaleimon (San Pantaleone) with drug balance (Fig. 6.19). Only two cardinals have a balance in their coat of arms, probably from their descent.

Both, the Egyptian significance as measuring instrument of good deeds and the Roman as a judicial symbol was adopted by the Christian culture. Occasionally we see Jesus or Mary operating a balance (Fig. 6.20a) or Mary assisting in weighing at the Last Judgement, but usually Michael Archangel is the operator. While the devils put millstones and themselves on the scale as a counterweight, Mary or the child sometimes falsify the result in favour of the poor wretch in a manner setting an unsavoury example to laboratory assistants (Figs. 6.20b–d). With Jesus as a pharmacist (Fig. 6.21) the symbolism of the balance degenerated into an advertising icon. Most of those balances depicted are symmetric hand balances made of metal.

At the time of selling indulgences, the balance as the weighing instrument for vows and indulgences was used in a rather perverted form: at places of pilgrimage in the Alpine countries, Belgium, and the Netherlands, the weight of the sinner was counterbalanced by coins (Fig. 6.22), precious metals, wax or natural products which had to be consecrated to obtain absolution or health [29]. Usually this was a symbolic invitation to spent money for savinner the deceased sinner from purgatory.

(a)

(b)

Fig. 6.17 (**a**) Vocation of Matthäus, Egbert-Codex 980 AC, Stadtbibliothek Trier, Germany. (**b**) Jan van Hemessen 1536: Vocation of customs officer Matthew as apostle by Jesus. On the desk a balance, the scales filled with coins. Alte Pinakothek, München

(a)

Fig. 6.18 (**a**) Four Horsemen of Apocalypse, Beatus von Osman-Codex 11th century. (**b**) Albrecht Dürer: Four Horsemen of Apocalypse, woodcut (1498). (**c**) Wiktor Wasnezow: Four Horsemen of Apocalypse (1887). (**d**) An Orthodox Christian: Die apokalyptischen Reiter. Four Horsemen of Apocalypse (2009). (**e**) Carl Burger: Four Horsemen of Apocalypse relief. (1931) Memorial, Rheinbrohl, Germany

(b)

Fig. 6.18 (Continued)

(c)

(d)

Fig. 6.18 (Continued)

(e)

Fig. 6.18 (Continued)

6.1.6.1 Michael the Archangel

The central personality operating a balance is Archangel Michael. In Jewish, Christian and Islamic tradition he is viewed as the field commander of the Army of God. In the last struggle he will defeat the devil, represented by a dragon. Michael's name means "Who is like El" (Who is like God?) and he is regarded as the patron of Israel, France, Germany and of the City of Brussels. St Michael is also considered in many Christian circles as the patron saint of the warrior, policeman and soldiers, paratroopers and fighter pilots. Whereas in the Jewish and Islamic societies, imaging is tabooed, in Christianity we have many drawings and statues representing St. Michael almost in arms. Because he is regarded as a powerful protector of castles many chapels in medieval castles are devoted to St. Michael and we find there representations of him.

Fig. 6.19 The Greek physician Pantaleimon (San Pantaleone) died as martyr during the Diocletian persecution of 303 AD. The icon shows him with drug balance. © Ikonen-Museum, Frankfurt am Main, Germany

Fig. 6.20 (**a**) Mary as Justitia in the last justice. Flemish devotion tablet 12th century. (**b**) Mary and Jesus falsify the weighing result in favour of the poor wretch. (**c, d**) Michael weighing the deceased, Mary with child influencing the balance, (**e**) Weighing of good deads in presenc of a moibund, (**f**) Mary weighing and pouring holy water on the scale. Codex 14th century, München

(a)

(b)

(c)

Fig. 6.20 (Continued)

(d)

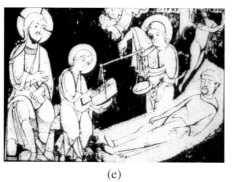

(e)

(f)

Fig. 6.20   (Continued)

Fig. 6.21 (**a**) Jesus as a
pharmacist. (**b**) Jesus as a
pharmacist. (**c**) Jesus as a
pharmacist. (**d**) Jesus as a
pharmacist. (**e**) Jesus as
Pharmacist, PZF 'CEFARM'
Apteka Muzeum, Lublin,
Poland

(a)

(b)

The idea that Michael is the Charon of individual souls, which is common among
Christians, is not found in Jewish sources, but that he is in charge of the souls of the
just appears in many Jewish writings. Michael accompanies the souls of the pious
and helps them to enter the gates of the heavenly Jerusalem. Michael's function is
to open the gates also of justice to the just and, therefore, also today his assistance is
evoked in the catholic funeral mass. In Christianity St. Peter is watching Heaven's

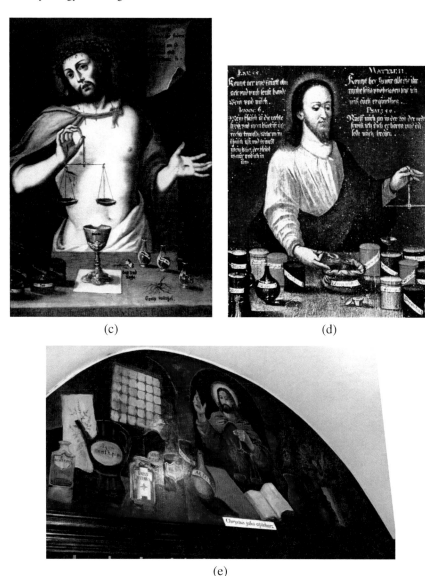

(c)

(d)

(e)

Fig. 6.21   (Continued)

door. However, Michael archangel escorts the deceased into the Heaven's light. Before the throne of God he weighs the deceased or rather his good deeds against the bad ones (Figs. 6.23–6.31) [17, 18]. He weighs in order to find out whether he is worthy of entering eternity near God. Simultaneously by means of a sword or a lance he defends the deceased against the devil who tries to influence the balance to the detriment of the poor sinner.

Fig. 6.22 Colin de Coster
1455, Michael archangel. The
money in the balance pan (as
an invitation for the
collection) should rescue the
sinner from purgatory,
Brussels, Belgium

There had been several appearances of the archangel inspiring the erection of a shrine, the foundation of a pilgrimage church and/or of a monastery. The earliest shrine from 4th century at Chonae near Colossae, east of Laodicea (modern Khonas, east of Denizli) on the Lycus in Phrygia, was dedicated to Michael as healer of a mute girl; it is still the site of a miraculous church of St. Michael. In Byzantine and Russian art (Fig. 6.31), the theme of the *Miracle of the Archangel Michael at Chonae* is intimately linked with the site. Eastern Orthodox tradition tells that the pagans directed the stream of a river against the sanctuary of St. Michael there to destroy it, but Michael the Archangel appeared and split the rock by lightning to give a new bed to the stream, diverting the flow away from the church and sacrificing forever the waters which came from the gorge. The Orthodox Church celebrates a feast in commemoration of this event on 6 September. Likewise in the Coptic (Fig. 6.31) and Ethiopian Church (Fig. 6.31e) Archangel Michael is worshiped.

(a)

(b)

Fig. 6.23 (a) During the Last Judgement Michael archangel weighs before the throne of God the deceased or rather his good deeds against the bad ones. Hospices de Beaune, France. (**j**) Meditation picture of the Last Judgement with archangel Michael weighing the soul. (**k**) Michael Prepositus Paradisi. (**m**) Archangel Michael, statue painting Kiedrich, Germany. (**n**) Michael archangel, Liebighaus Frankfurt am Main. Painting of a statue by Mechthild Kreuter. The balance is added

The first apparitions of the archangel Michael in Western Europe were granted to the Bishop of Sipontum in Apulia in 490. After Chonae become insignificant, as a substitute, at Monte Sant'Angelo sul Gargano in Apulia, Italy, Michael's veneration "his original glory as patron in war was restored to him." The legend of the Archangel's apparition at Gagano is related in the Roman Breviary for May 8. At

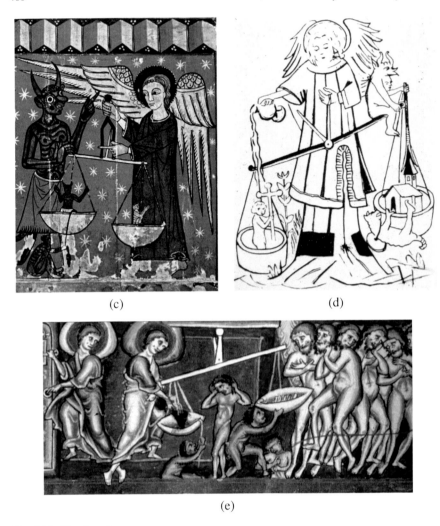

(c)   (d)

(e)

Fig. 6.23   (Continued)

this oldest shrine in Western Europe dedicated to the archangel Michael also the oldest sculpture dated to the 5th century representing Michael archangel with sword but without balance is exhibited in the grotto of the chapel [30, 31].

Mont-Saint-Michel is a rocky tidal island in Normandy, France, at the border with Bretagne (Fig. 6.32). According to legend, St. Michael the Archangel appeared to St. Aubert, bishop of Avranches, in 708 and instructed him to build a church on the rocky islet. Aubert repeatedly ignored the angel's instruction, until St. Michael burned a hole in the bishop's skull with his finger. In the cathedral a statue of Michael archangel is seen with sword and balance (Fig. 6.33).

The wealth and influence of the abbey extended for many daughter foundations, including St Michael's Mount in Cornwall, UK.

(f)   (g)

(h)

Fig. 6.23 (Continued)

(i)

(j)

(k)

Fig. 6.23   (Continued)

(l)

(m)

(n)

Fig. 6.23 (Continued)

Fig. 6.24 (**a**) Last Judgement
with St.Michael, wall
painting. (**b**) Archangel
Michael, Bayeux cathedral,
wall painting. (**c**) Painting in
Archangle Michael's Church,
Debno Podhalanskie.
(**d**) Hans Memling:
Archangel Michael with
balance of the Last
Judgement, Tryptichon,
Muzeum Narodowe Gdansk
1467–71. (**e, f**) Michael
weighing at the last
judgement

(a)

(b)

6.1.6.2 The Witch Balance

During the witch prosecutions in Europe between the fifteenth and eighteenth centuries (last dating in Europe 1782, in Tennessee 1833) the weighing of people came into fashion. As witches were supposed to fly on their brooms, they should be light. A person lighter than an average of around 50 kg in weight was assumed to be able fly. In 1545 there was such a witch trial in the village of Polsbroek (near Oudewater, The Netherlands). Emperor Charles V, who happened to pass by, refused to believe that the prosecuted woman really was too light as the corrupt weigh master was trying to make everybody believe (Fig. 6.34). He gave Oudewater the privilege to a fair weighing trial (Figs. 6.35, 6.36 and 6.37). It is and was the only place in Europe where you can get yourself weighed and get your 'Cerjpgicaet van Weghinge'. Nobody was ever convicted as a witch here.

Fig. 6.24 (Continued)

Fig. 6.25 St.Michael
painting on a small wooden
box. Kiev, Ukraine, 2000

Fig. 6.26 St.Michael with
balance, statue, ivory,
Elfenbeinmuseum Erbach

(a)   (b)

(c)   (d)

Fig. 6.27   (**a**) St.Michael with balance, statue. (**b**) St.Michael with balance, statue. (**i**) Erhart Kün: Archangel Michael (1485). Main entrance of the minster of Bern, Switzerland. (**m**) Hans Holbein d.J.g: Der Seelenwäger (Weigher of the soul). Early 16th century. (**o**) St.Michael statue, Neunkirchen, Germany. (**p**) St.Michael pedestal of wood, Allgäu. (**q**) Archangel Michael with balance statue on the Miltenberg market place, with C.H. Massen

(e)

(f)

(g)

(h)

Fig. 6.27 (Continued)

(i)

(j)

(k)

(l)

Fig. 6.27 (Continued)

Fig. 6.27 (Continued)

(m)

(n)

(o)

(p)

(q)

(a)

(b)

(c)   (d)

Fig. 6.28 (**a–d**) Last Judgement with Archangel Michael, frieze of statues. (**d**) Archangel Michael with sword and scales

(a)   (b)

Fig. 6.29 (**a**) Cathedral of Otranto, Spain. Mosaic in the central nave of Grazio Gianfreda: St.Michael with scales. (**b**) St.Michael, modern mosaic, Wiesbaden

Fig. 6.30 Archangel
Michael, cathedral Bayeux,
France. Painted windows

(a)   (b)

(c)   (d)

Fig. 6.31 (**a**) Archangel Michael with balance of the Last Judgement, Russian. (**b**) Michael archangel. Coptic, contemporary. (**c**) Michael archangel. Coptic, contemporary. (**d**) Michael archangel. Coptic, contemporary. (**e**) Michael archangel. Ethiopia, modern

Fig. 6.31 (Continued)

(e)

Fig. 6.32 Mont-Saint-Michel is a rocky tidal island in the Normandy, France

6.1.7 The Islamic Civilisation

In answering questions of society at the time of proclamation of the Koran (610–632) the balance is mentioned frequently (Table 6.2) [32]. The tradition of the symbolic character of justice and the idea of the use of the balance to compare good

Fig. 6.33 Sculpture of St. Michael with balance in the monastery of Mont-Saint-Michel, Normandy, France

Fig. 6.34 Weighing of a witch

and bad deeds, in particular in the last judgement, was adopted. Obviously the balance at that time was an important measuring instrument of trade and commerce, and mass or weight, respectively, had been important quantities. So frequently it is admonished at Allah's behest to weigh correctly.

Fig. 6.35 Gezicht op
Oudewater by Willem
Koekkoek (1839–1893).
© National Gallery London

Fig. 6.36 Museum de
Heksenwaag, Oudewater, The
Netherlands

Fig. 6.37 Scales dated 1842
as used for weighing a witch,
Oudewater, The Netherlands

Fig. 6.38a Balance
monument made of stone,
Sanaa, 70 Refer-Road,
Jemen, © Dr. Theuerkauf

Some Koran interpreters assume that the balance was Allah's gift, brought to Noah by the archangel Gabriel (*Sure* 42:17, 55:7 and 57:25). Because in Islam illustrations are disapproved we have no pictures.

In modern times Islamic countries adopted some western symbols, in particular the balance of justice. Furthermore, fair weighing as demanded in the Koran is represented. As an example a monument at Sanaa, Yemen is depicted in Fig. 6.38a.

6.1.8 India

In Hindu (Indian) philosophy and as notified in holy books weighing up of good deeds with evil will decide on a place in heaven or hell after our death.

In an old Indian parable the balance is operated by an ape (Fig. 6.38b): Two cats got one piece of bread and started quarrelling with each other. A monkey saw them and very cleverly offered to divide it equally. He divided the bread in two pieces and compared it on a balance. The cats patiently waited for the justice by an impartial balance while the monkey misused it. He bit off a small piece from the heavier sample and he repeated weighing. Now the balance inclined to the other side because the ape had removed too large a piece. Again he bit off a small piece

Table 6.2 Balance and weighing in the Koran

Al-Nisa	*(The Women)*
4:40	Surely Allah does not do injustice to the weight of an atom, and if it is a good deed He multiplies it and gives from Himself a great reward.
Al-Anam	*(The Cattle)*
6:153	And do not approach the property of the orphan except in the best manner until he attains his maturity, and give full measure and weight with justice.
Al-Araf	*(The Elevated Places)*
7:8–10	The scales will be set on that day, equitably. Those whose weights are heavy will be the winners. As for those whose weights are light, they will be the ones who lost their souls as a consequence of disregarding our revelations, unjustly. We have established you on earth, and we have provided for you the means of support therein. Rarely are you appreciative.
7:85	And to Madyan (We sent) their brother Shu'aib. He said: O my people! serve Allah, you have no God other than Him; clear proof indeed has come to you from your Lord, therefore give full measure and weight and do not diminish to men their things, and do not make mischief in the land after its reform; this is better for you if you are believers.
Jonah	*(Jonah)*
10.81	And you are not (engaged) in any affair, nor do you recite concerning it any portion of the Quran, nor do you do any work but We are witnesses over you when you enter into it, and there does not lie concealed from your Lord the weight of an atom in the earth or in the heaven, nor any thing less than that nor greater, but it is in a clear book.
Hud	*(The Holy Prophet)*
11:84–85	And to Madyan (We sent) their brother Shu'aib. He said: O my people! Serve Allah, you have no god other than He, and do not give short measure and weight: surely I see you in prosperity and surely I fear for you the punishment of an all-encompassing day. And, O my people! Give full measure and weight fairly, and defraud not men their things, and do not act corruptly in the land, making mischief:
Bani-Israil	*(The Children of Israel)*
17:35	And give full measure when you measure out, and weigh with a true balance; this is fair and better in the end.
Al-Kahf	*(The Cave)*
18:105	These are they who disbelieve in the communications of their Lord and His meeting, so their deeds become null, and therefore We will not set up a balance for them on the day of resurrection.
Al-Anbiya	*(The Prophets)*
21:47	And We will set up a just balance on the day of resurrection, so no soul shall be dealt with unjustly in the least; and though there be the weight of a grain of mustard seed, (yet) will We bring it, and sufficient are We to take account.

Table 6.2 (Continued)

Al-Mu'minun	*(The believers)*
23:103–104	As for those whose weights are heavy, they will be the winners. Those whose weights are light are the ones who lost their souls; they abide in Hell forever.
Al-Schuara	*(The Poets)*
26:182	And weigh (things) with a right balance,
Al-Casas	*(The Narratives)*
28:76	Surely Qaroun was of the people of Musa, but he rebelled against them, and We had given him of the treasures, so much so that his hoards of wealth would certainly weigh down a company of men possessed of great strength. When his people said to him: Do not exult, surely Allah does not love the exultant;
Luqmaan	*(Lugman)*
31:16	O my son! surely if it is the very weight of the grain of a mustard-seed, even though it is in (the heart of) rock, or (high above) in the heaven or (deep down) in the earth, Allah will bring it (to light); surely Allah is Knower of subtleties, Aware;
Sheba	*(The Saba)*
34:3	And those who disbelieve say: The hour shall not come upon us. Say: Yea! by my Lord, the Knower of the unseen, it shall certainly come upon you; not the weight of an ant becomes absent from Him, in the heavens or in the earth, and neither less than that nor greater, but (all) is in a clear book.
34:22	Say: Call upon those whom you assert besides Allah; they do not control the weight of an ant in the heavens or in the earth nor have they any partnership in either, nor has He among them any one to back (Him) up.
Al-Fatir	*(The Originator)*
35:18	And a burdened soul cannot bear the burden of another and if one weighed down by burden should cry for (another to carry) its burden, not aught of it shall be carried, even though he be near of kin. You warn only those who fear their Lord in secret and keep up prayer; and whoever purifies himself, he purifies himself only for (the good of) his own soul; and to Allah is the eventual coming.
Al-Schura	*(The Counsel)*
42:17/18	Allah it is Who revealed the Book with truth, and the balance, and what shall make you know that haply the hour be nigh?
Al-Rahman	*(The Beneficent)*
55:7	And the heaven, He raised it high, and He made the balance.
55:9	He constructed the sky and established the law. (scales). You shall not transgress the law. And keep up the balance with equity and do not make the measure deficient.

Table 6.2 (Continued)

Al-Hadid	*(The Iron)*
57:25	Certainly We sent Our apostles with clear arguments, and sent down with them the Book and the balance that men may conduct themselves with equity; and We have made the iron, wherein is great violence and advantages to men, and that Allah may know who helps Him and His apostles in the secret; surely Allah is Strong, Mighty.
Al-Muzzammil	*(The Wrapped Up)*
73:5	Surely We will make to light upon you a weighty Word.
Al-Tatfif	*(The Deceivers in Measuring)*
83	In the name of Allah, the Beneficent, the Merciful.
	Woe to the defrauders, Who, when they take the measure (of their dues) from men take it fully. But when they measure out to others or weigh out for them, they are deficient. Do not these think that they shall be raised again for a mighty day, the day on which men shall stand before the Lord of the worlds? Nay! most surely the record of the wicked is in the Sijjin. And what will make you know what the Sijjin is? It is a written book. Woe on that day to the rejecters, Who give the lie to the day of judgement. And none gives the lie to it but every exceeder of limits, sinful one When Our communications are recited to him, he says: Stories of those of yore. Nay! Rather, what they used to do has become like rust upon their hearts. Nay! most surely they shall on that day be debarred from their Lord. Then most surely they shall enter the burning fire. Then shall it be said: This is what you gave the lie to. Nay! Most surely the record of the righteous shall be in the Iliyin. And what will make you know what the highest Iliyin is? It is a written book, Those who are drawn near (to Allah) shall witness it. Most surely the righteous shall be in bliss, On thrones, they shall gaze; You will recognise in their faces the brightness of bliss. They are made to quaff of a pure drink that is sealed (to others). The sealing of it is (with) musk; and for that let the aspirers aspire. And the admixture of it is a water of Tasnim, A fountain from which drink they who are drawn near (to Allah). Surely they who are guilty used to laugh at those who believe. And when they passed by them, they winked at one another. And when they returned to their own followers they returned exulting. And when they saw them, they said: Most surely these are in error; And they were not sent to be keepers over them. So today those who believe shall laugh at the unbelievers; On thrones, they will look. Surely the disbelievers are rewarded as they did.
Al-Zilzal	*(The Quaking)*
99:7–8	So, He who has done an dust particle's weight of good shall see it. And he who has done an ant's weight of evil shall see it.
Al-Qaare`ah	*(The Shocker)*
101:6–10	Then as for him whose scale of good deeds is heavy, he shall live a pleasant life. And as for him whose scale of good deeds is light, his abode shall be the abyss. And what will make you know what it is? A burning fire.

Fig. 6.38b The parable of the monkey with the balance. © Aggarwal Book Depot, Delhi

and repeated the procedure of weighing and eating until the bread was removed ultimately and the cats did not get any share [33].

6.2 Icon and Arts

Looking at the oldest written or pictorial reports of mankind, the Egyptian papyri, it appears that balances were used for weighing anything rather than physical masses and forces. To balance and to weigh are ambiguous words. Almost daily, we find the words balance and weighing used in symbolic sense in the literature.

6.2.1 Zodiac Libra

6.2.1.1 Astronomy

The zodiac is the ring of constellations that lines the ecliptic, which is the apparent path of the Sun across the sky over the course of the year, seen from Earth [34]. The zodiac encompasses also the paths of the Moon and the planets corresponding to the band of about eight arc degrees above and below the ecliptic. The zodiac is divided into 12 sectors corresponding to the 12 months and these sectors are labelled by star constellations, symbolised mostly by animals from which the notation zodiac is derived, meaning circle of animals (Figs. 6.39/6.40). Indeed a lot of imagination is necessary to see any resemblance between the star constellations and the respective notation. Constellations are rather arbitrary because the respective stars are very far from one another and there are no interactions between them. Furthermore, because of distances of light years the star lights observed do not present simultaneous events.

For medical purposes in the middle-ages the zodiac was also applied to characterise regions of the human body (Fig. 6.41).

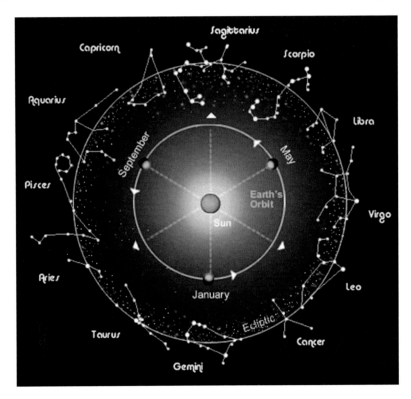

Fig. 6.39 Illustration of the 'zodiac band' with a few of the constellations depicting the objects they represent. © Lunar and Planetary Institute, Houston, TX 77058

The ecliptic zodiac coordinates are still used today besides equatorial coordinates and the twelve signs of star constellations are still a help for a vivid description of the sky. This ring-shaped coordinate system was developed by Babylonian astronomers in the middle of the 1st millennium BC most probably from earlier systems. The construction of the zodiac is described in Ptolemy's *Almagest* [35, 36]. It was originally described in *Rigveda* [37], an Indian collection of Vedic Sanskrit hymns dated between 1700 and 1100 BC.

Libra, also known as 'The Scales' or 'Balance', is the only symbol of the zodiac that is represented by an inanimate object, rather than an animal or human (Fig. 6.42). The sign of libra (Fig. 3.3) is an Egyptian hieroglyph representing 'horizon'. The sign shows schematically the sun leaving the horizon or disappearing below it. Egyptian hieroglyphs for a balance are quite different.

6.2.1.2 Astrology

As a sign of the zodiac, the balance has acquired a superstitious meaning. Libra is the seventh astrological sign in the Zodiac, corresponding to its astronomical sig-

Fig. 6.40 (a) Ethiopian scheme of zodiac. (b) Lund calendar clock with signs of the zodiac. (c) Lund calendar clock, sign of the zodiac Libra. (d) Gate tower with clock with signs of the zodiac. München (Munic), Germany. (e) Cathedral of Otranto, Spain. Central nave with mosaic of Grazio Gianfreda: tree of life with zodiac signs. (f) Libra. Aratos manuscript, Codex 902, 9th century. Libra depicted as Justitia with balance and with palm frond of Pax. Library of the monastery St. Gallen, Switzerland

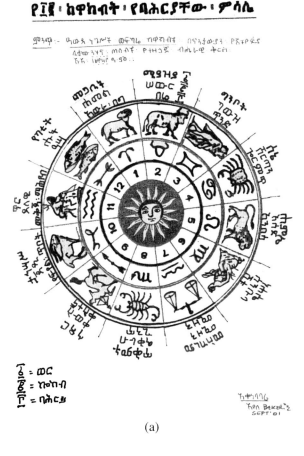

(a)

nificance. In western astrology, this sign is no longer aligned with the constellation as a result of the precession of the equinoxes. In astrology, Libra is considered a "masculine', positive (extrovert) sign. It is also considered an air sign and is one of four cardinal signs (Figs. 6.43/6.44). Libra is ruled by the planet Venus. Being the seventh sign of the zodiac, Libra has been associated with the astrological seventh house.

Individuals born when the Sun was in this sign are considered Libra individuals. Under the tropical zodiac, the Sun enters Libra at the moment of autumnal equinox by definition on September 23, leaving it on October 22. Under the sidereal zodiac, it is currently there roughly between October 18 and November 16. The Sanskrit name of Libra in Hindu astrology is Tula.

The Trutina Hermetis (Hermes balance) is based on the exchange of the horoscopes of conception and birthday.

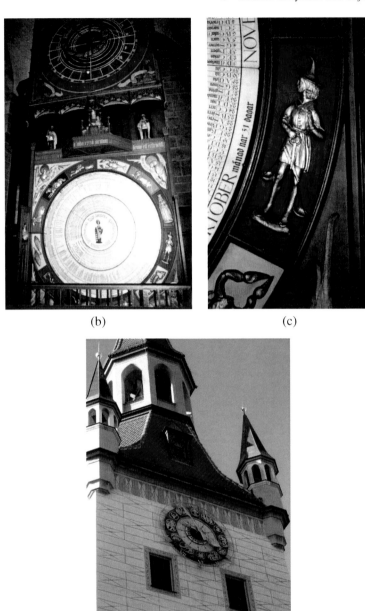

(b)   (c)

(d)

Fig. 6.40 (Continued)

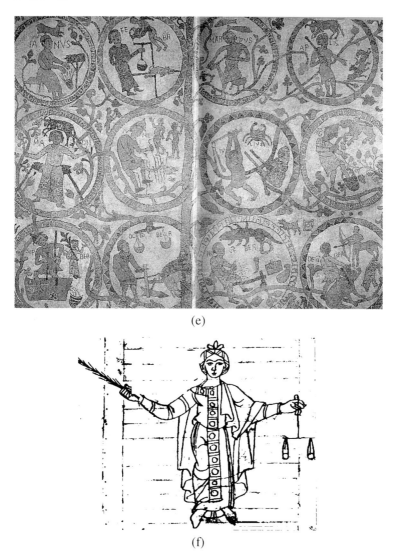

(e)

(f)

Fig. 6.40   (Continued)

6.2.2 Lady Justice

Weighing of a (criminal) action in comparison to law is a very old symbolic as we
know from representation of Egyptian death tribunals. Thus the balance became a
token for justice (Fig. 6.45a), usually in the hands of Lady Justice, but sometimes
also in the hand of the judge (Fig. 6.45b).

Lady Justice which frequently adorns courthouses and courtrooms is an allegorical personification of the moral force in judicial systems. That modern iconography

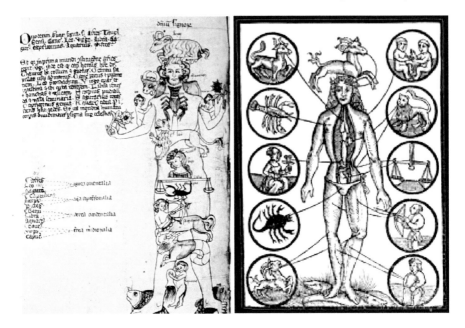

Fig. 6.41 The zodiac-man

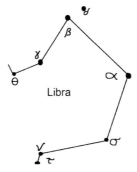

Fig. 6.42 Libra (the balance) is a constellation of the zodiac. Libra is seen along the ecliptic between Virgo and Scorpius. The brightest star α_{Lib} in Libra is Zuben Elgenubi (Southern scissors), distance 78 ly (light years), the second-brightest star β_{Lib} is Zuben Elschemali (Northern scissors) 160 ly, the third-brightest star γ_{Lib} is Zuben Elakrab (Scorpion scissors) 150 ly, and the fourth-brightest star δ_{Lib} is Zuben Elakribi

represents the Roman Goddess Iustitia the goddess of Justice who is equivalent to the Greek Goddess Dike, both with diadem, sword and balance. Iustitia conflates attributes of several Goddesses who embodied right rules of Greeks and Romans: Greek Themis (divine rightness of law), Tyche (fate) and sword carrying Nemesis (vengeance).

Fig. 6.43 Tarod card Libra

Fig. 6.44 Luis Ricardo
Falero (1851–96), The
Balance of the Zodiac

Justitia is most often depicted with a set of scales typically suspended from her
left hand, upon which she measures the strengths of a support and opposition in a
case (Fig. 6.46) Whereas initially the balance was in equilibrium today the beam
occasionally is tilted symbolising 'In dubio pro reo' (Giving the defendant the ben-
efit of the doubt). Justitia is also often seen carrying a double-edged sword in her
right hand, symbolising the power of Reason and Justice, which may be wielded
either for or against any party. Indeed it would have been the sword of execution.
Occasionally we see a combination of Justitia with the zodiacal sign Libra and with
different attributes (Fig. 6.47).

(a)

(b)

Fig. 6.45 (**a**) Relief of a balance at the tomb of duke François II and his wife Marguerite de Foix in the cathedral of Nantes, France. (**b**) Court hearing and judgement

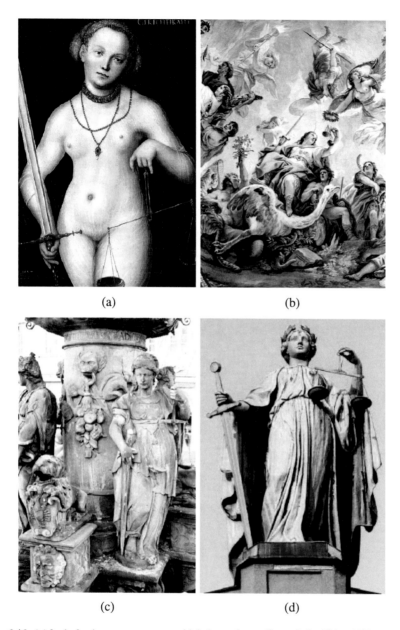

(a)   (b)

(c)   (d)

Fig. 6.46 (a) Lady Justice, young woman with balance, Lucas Cranach the Elder, 1537. (b) Justitia. Luca Giordano, Palazzo Medici Riccardi in Florence, 1684–1686. (c) Lady Justice without blindfold, statue. (d) Statue of Justitia, residence of the government of Upper Bavaria at Munich, Germany, Lady Justice without blindfold. (e) Justitia without blindfold, Römer, Frankfurt am Main, Germany. (f) Justitia without blindfold on the roof of a courthouse. (h) Justitia, stadhuis (town hall) Bolsward, The Netherlands. (g) Justitia Simeonsstift Trier, Germany (Hans Ruprecht Hofmann, 1545). (i) Justitia without blindfold, statue on the spire of Old Bailey courthouse in London

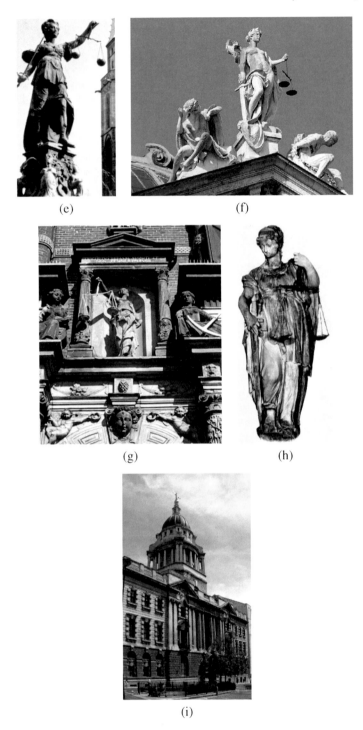

(e)   (f)

(g)   (h)

(i)

Fig. 6.46   (Continued)

Fig. 6.47 Justitia with and
without blindfold. © Kissel
p. 53

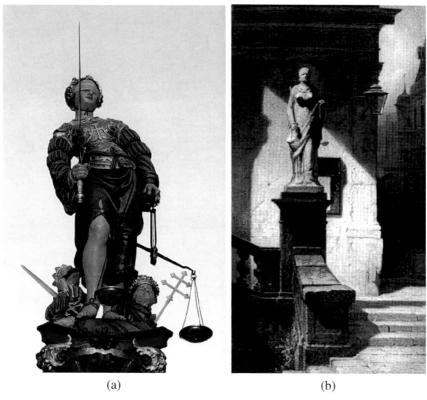

(a)   (b)

Fig. 6.48 (**a**) Hans Gieng's 1543 statue on the Gerechtigkeitsbrunnen (Fountain of Justice) in Bern (Switzerland). First Justitia with blindfold. (**b**) Carl Spitzweg: 'Fiat Justitia' 1857. The painting was stolen by the Nazis and then exhibited in the German Bundespräsidialamt. Recently it was given back to the heir of the Jewish owner. (**c**) Ludwig Richter 1864: Market scene with commercial balance and Lady Justice with balance. (**d**) Justitia with blindfold. (**e**) Justitia with blindfold on a Netherlands town hall, (**f**) Justitia with blindfold, stadhuis Hindelopen, The Netherlands. (**g**) Inciralti Sea Museum, Izmir, Turkey: Justice girl. (**h**) Lady Justice's attribute: scales, cornucopia and blindfold. Exhibition: 'Justitia is a woman' 2009 Erbach, Germany. (**i**) Justitia, Bavaria Studios at München (Munich, Germany). (**j**) Justitia Shelby County Courthouse in Memphis, Tennessee. Blindfolded lady weighing with his hands. (**k**) Justitia Stemer. (**l**) St.Michael with sword and balance, Relief

(c)   (d)

(e)   (f)

Fig. 6.48 (Continued)

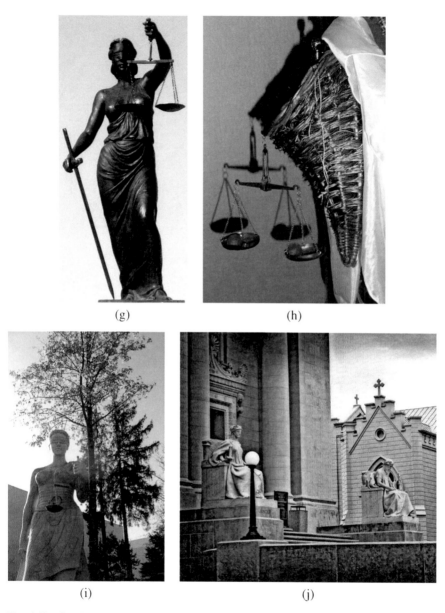

(g)   (h)

(i)   (j)

Fig. 6.48   (Continued)

(k)   (l)

Fig. 6.48 (Continued)

The blindfold is taken from the Roman Goddess Fortuna. However, it appeared
first in the early 16th century as a mocking symbol for blindness of justice. Today the blindfold indicates that justice should be meted out objectively and im-
partially, without fear or favour, regardless of identity, money, power, or weakness (Fig. 6.47). The first known representation of blind Justice is Hans Gieng's
1543 statue on the Gerechtigkeitsbrunnen (Fountain of Justice) in Bern (Switzerland) (Fig. 6.48a). The blindfold is still disputed, because in a legal procedure all
things should be observed carefully. Furthermore, a blind personality with sword
seems to be dangerous. Therefore, also today occasionally, Lady Justice is sometimes unblindfolded (Fig. 6.46). Atop the Old Bailey courthouse in London, a statue
of Lady Justice stands without a blindfold; the courthouse brochures explain that
this is because Lady Justice was originally not blindfolded, and because her 'maid-
enly form' is supposed to guarantee her impartiality which renders the blindfold
redundant (Fig. 6.46h). Another solution to the conundrum is to depict a blindfolded Lady Justice as a human scale, weighing competing claims in each hand;
this is done, for example, at the Shelby County Courthouse in Memphis, Tennessee (Fig. 6.48k). With the ship 'Tribuna' lawyers come to remote places in
the Amazonas delta (Fig. 6.49a). Also in theatre Lady Justice appears occasionally
(Fig. 6.49b).

Fig. 6.49 (**a**) With the ship
'Tribuna' lawyers visit
remote places in the
Amazonas delta. At the front
of that Ship of Justice an
abstract Balance of Justice.
(**b**) Scene of 'Meine
Schwester und ich' (My sister
and I) of Ralph Benatzky and
Robert Blum: Justitia with
sunglasses, balance and
sword. Volkstheater Frankfurt
am Main (Germany) 2008

(a)

(b)

6.2.3 Icons

It is obvious for an instrument manufacturer to include in the company logo the
balance, typical parts or weights, however mostly in an abstract form (Fig. 6.50).

The balance is the emblem of the merchant as a symbol of fair measure
(Fig. 6.51). Drug stores and laboratories use the balance as a token because it is
its most used measuring instrument and it demonstrates accuracy and trustworthi-
ness in their counsel. The town of Onstmettingem (Germany) as a centre of balance
makers has the instrument in his arms (Fig. 6.52). Organisations concerned with
balance work have a respective logo (Fig. 6.53). Besides Lady Justice with balance
is also the icon of lawyers, judges and judicial offices. Fair advice and trustworthi-
ness may also be the background for the use of Justitia with balance by insurance
companies, whereas publishers of scientific books demonstrate the seriousness. For
stimulating selling some products are decorated with zodiacal signs including Libra
(Fig. 6.52).

Fig. 6.50 (**a**) Logo Waagen Kissling GmbH, Bahnhofstraße 17A, D-64668 Rimbach, Germany. (**b**) Banner, Precisa Gravimetrics AG, Moosmattstrasse 32. (**c**) Logo Ados Pesatura Electronica Via Bersaglio, I-20–22015 Gravedona (CO)—Lombardy, Italy. (**d**) Banner Itin Scale Company, Inc. 4802 Glenwood Rd., Brooklyn, NY 11234, USA. (**e**) Logo Dibal S.A. Astintze Kalea, 24, I-48160 Derio, Spain. (**f**) Logo Ebinger Waagenbau GmbH, Obere Riedwiesen 15, D-74427 Fichtenberg, Germany. (**g**) Logo Häfner Gewichte GmbH, Hohenhardtsweiller Str. 4, D-74420 Oberrot, Germany. (**h**) Logo Radwag Wagi Elektroniczne Bracka 28, PL-26–600 Radom, Poland. (**i**) Logo Chronos Richardson, Hennef, Germany/Chronos BTH, Rivière-du-Loup (Québec) Canada. (**j**) Logo Waagen Steitz, Kaiserslautern, Germany

(a)   (b)   (c)

(d)

Fig. 6.51 (**a**) Sign-board of a drugstore. (**b**) Logo Dallmayr on sugar bag. (**c**) Sign-board of the Kur- & Therapiehaus Nanzer (natural phsician), Lenzerheide, Schweiz. (**d**) Corner house 'Goldene Waage' (Golden Scales), destroyed in war but in planning for reconstruction, Frankfurt am Main, Germany

Fig. 6.52 Flag with arms of
Onstmettingen, Germany

Fig. 6.53 (**a**) Logo UK
Weighing Federation.
(**b**) Logo International
Conferences on Vacuum
Microbalance Techniques.
(**c**) Logo Unie výrobcùvah
Ćeské republiky (UVV).
(**d**) Logo IUPAC

(a)   (b)

(c)   (d)

Fig. 6.54 Cube sugar and bag decorated with Libra

6.2.4 Painting and Sculpture

Paintings depicting a balance by itself or symbolising weighing are rare. An exceptional example is Paul Klee's "Gewagt wägend" (risky weighing) (Fig. 3.1). Others show applications (Figs. 6.57a–e) or its symbolic use (Figs. 6.57f–h). Often depictions of balances are committed in logos, icons, emblems, etc. (Figs. 6.50–6.53). Above we mentioned representations of saints and gods with balances, both as statue and paintings. The esoteric Libra as depiction of the zodiacal sign, as an icon or as an allegorical image, is often used (Fig. 6.54). In religions but likewise in political (Fig. 6.55a) and scientific reports (Fig. 6.55b) the balance is used as an allegoric instrument. Very popular is the usage of the symbolic balance or of the Lady Justice in cartoons, and Figs. 6.56–6.57 give some examples. Series of pictures with or concerning balances are printed as postal stamps (Fig. 6.58).

(a)

(b)

Fig. 6.55 (**a**) Joseph Motier Marquis de Lafayette (2. from right) 1789. Allegory of human rights. (**b**) Comparison of the geocentric (Ptolemaic) and the heliocentric (Copernicanic) model by weighing of the solar system

Fig. 6.56 (**a**) Chiquita Superfruit Smoothies. Fruits on seesaw © Spiegel 2009. (**b**) Prüf den Prof. (Test the professor) with Prof. Jürgen U. Keller, Siegen, after a cartoon in the 'Demokratische Blätter'. (**c**) An example of barter: A sample labour note of Carpenter's workshop for Joseph Peters, with Justitia (1846). (**d**) Justitia mit Sprenggürtel. (**e**) Justitia from back. Old sin often produces new shame. (**f**) Cartoon: 'Der Handel', 'Dealing', Pfundsmuseum, Kleinsassen, Germany

(a)

(b)

Fig. 6.57 (**a/b**) Paintings: pharmaceutical tools. (**c**) Relief: cheese balance. Alkmaar, The Netherlands. (**d/e**) Rosa Bonleitner, Hundham-Fischbachau, Germany. Staines with beer balance. (**f**) Horseman with two persons in a scale. (**g**) Gabriel von Max: Vivisektor 1883, Lenbachhaus München, Galerie Konrad Bayer. Justitia compares on a balance the heart of a dog with its brain whereas the vivisector is waiting with the scalpel. (**h**) Sybille Smolak, Kärnten 2010: Dream of Nicolaus Cusanus: The hearts = love weigh more than any other thing other in the world. The symbolic blue Jacob's ladder leads over the balance of justice directly to the Saviour. The balance represents simultaneously science. The nuns pray for the salvation of the world. (**i**) Russian grocer with balance, Pfundsmuseum, Kleinsassen, Germany. (**j**) James Sidney Ensor (1860–1949), Dance of Death. (**k**) Albrecht Dürer: Melancholia I, copperplate engraving

Fig. 6.57 (Continued)

(c)

(d)   (e)

Fig. 6.57 (Continued)

(f)

(g)

(h)

Fig. 6.57 (Continued)

(i)

(j)

Fig. 6.57 (Continued)

(k)

Fig. 6.58 Postal stamps

Fig. 6.59 (a) Karl May: Durch das Land der Skipetaren. (b) Heinrich Böll (1917–1985): Die
Waage der Baleks (The Scales of the Baleks)./Joseph Roth (1894–1939): Das falsche Gewicht
(The faked weight). © Pfundsmuseum, Kleinsassen, Germany

6.2.5 Literature and Music

Besides technical books there exists a vast astrological literature on scales. In fiction
balance and imbalance play an important role. In the following some examples are
provided from serious literature and music.

6.2.5.1 Karl May (1842–1912): Durch das Land der Skipetaren (1862)

In this novel describing a tour through Albany Kara ben Nemsi and his companions
survive many dangerous events. A female judge plays a role and is depicted on the
cover of the first edition [38] (Fig. 6.59a).

6.2.5.2 Franz Kafka (1883–1924): Der Prozess (The Trial)

Der Prozess is a novel by Franz Kafka, first published in 1925. It tells the story of
Joseph K. who was arrested and prosecuted by a remote, inaccessible authority, with
the nature of his crime revealed neither to him nor to the reader [39].

When K. visited the studio of a painter, he asked: 'Are you working on a picture
just now?' 'Yes' said the painter, and threw the shirt hanging on the easel onto
the bed with the letter. 'It is a portrait. A good piece of work, but it's not quite
finished.'

This was quite fortunate for K.; it gave him a clear opportunity to mention the court, for it was obviously the portrait of a judge. ... There was a large figure hovering over the back of the chair that he could not make out, and he asked the painter what it was. He replied that it still needed to be worked on; he fetched a pastel crayon from a small table and drew a few strokes at the edges of the figure, but this made it not clearer to K. 'It is Justice,' the painter said finally. 'Ah, now I recognise it,' said K., 'this is the blindfold, and these are the scales. But aren't those wings on her ankles, and is she not flying?' 'Yes,' said the painter. 'I was commissioned to paint it like that; actually it's Justice and the goddess Victory all in one.' 'That is not a very happy combination,' said K., smiling. 'Justice must be still, otherwise the scales will waver and no just verdict is possible.'

6.2.5.3 Joseph Roth (1894–1939): Das falsche Gewicht (The Faked Weight) (Fig. 6.59)

This is one of the most impressive novels [40] in which balances are at the centre: Anselm Eibenschütz was a weights and measures inspector in a little Galician town near the eastern borderline of the ancient Austrian kingdom. Because he was incorruptible and very hard in his decisions he had many enemies among the merchants. Personally he was discontented because he had no friends and he was deceived by his wife, who got a child with a subordinated colleague. When he fell in love with a young gipsy girl he got more and more problems to decide between right and wrong. At the end he was killed by a co-partner.

6.2.5.4 Heinrich Böll (1917–1985): Die Waage der Baleks (The Scales of the Baleks) (Fig. 6.59)

In this narration [41], the fraudulent use of the fruit scales of a feudal family is revealed by a little boy. The scales, or more precisely, the gauging-mark of the magnificent instrument simultaneously stood for fairness and fair measure, and the revelation of the deception produced a local revolution with a bloody end resulting in the expulsion of the small boy's family from their home.

6.2.5.5 Siegfried Obermeier (*1936): Die Hexenwaage (The Witch Balance) [42]

A crime fiction from the 17th century. A respectable virgin should marry her old, rich uncle. As she refused she was suspected to be a witch. Dressed as a horseman she escaped to the Netherlands in order to be weighed in a witch balance, which demonstrates her innocence. However her vengeful uncle sent persecutors.

6.2.5.6 Frederic Lenoir (*1962), Violette Cabesos: La promesse de l'ange (The Angel's Promise)

In this thriller an important role play representations of Michael Archangel equipped with sword and balance [31, 43]. Rising out of the sea on a rocky isle off the coast of France, the church and abbey of Mont-Saint-Michel have withstood the harsh northern weather, and withheld its darkest medieval mysteries, for one thousand years. A young, brilliant archaeologist, Joanna, began to uncover and explore secrets that have long lain buried at the sacred site. In ancient ritual, in foreboding visions, in an expedition, simultaneously metaphysical and archaeological, Johanna ardently researches the past only to discover that its bizarre, murderous history has begun to be repeated in the present. Reality becomes an elusive, dangerous place, and it leaves Joanna with no one she can fully trust.

6.2.5.7 Sabine Marienberg: Still and Again

Weighing still against silence	Weighing still against silence
Crystallised time in a haze	Blank eyed, in ceaseless visions
Motionless, sensing my senses	Lastingly held in unrest
In vigil, unguarded	My days are my nights now
My insight made of echoes	Though could be turned by one touch
Ruled by an uncertain pulse	And sounds, and absent voices
Delusion or distant response	Are seeping into my skin

That canto of Penelope was set to music 2011 by Hanspeter Kyburz for his opera ΟΥΤΙΣ [44, 45].

6.2.5.8 Carl Orff (1895–1982): Carmina Burana (1936) cantoribus et choris cantandae comitantibus instrumentis atque imaginibus magicis

The scenic cantata composed by Carl Orff in 1935/36 [46–48] is based on poems found in the medieval collection Carmina Burana [49]. Michel Hofmann, assisted Orff in the selection and organisation into a libretto in Latin, Middle High German and Old Provencal verses covering fickleness of fortune and wealth. The poem 'In trutina' describes a young girl's feeling to be trapped between love and chastity. She decides to fall in love rather than to become a nun.

In trutina mentis dubia	In the wavering balance of my mind
Fluctuant contraria	hesitating in reverse
Lascivus amor et pudicitia	Lascivious love and chastity
Sed eligo quod video	But I choose what I see
Collum iugo prebeo	And offer my neck to the yoke
Ad iugum tamen suave transeo	I submit to the yoke with a kiss.

Fig. 6.60 Three-armed
balance of sustainability:
Economy, ecology, and social
engagements should be in
equilibrium. BASF promises
to work successfully in
economics, agreeable in
ecology, and well balanced in
social responsibility. © BASF,
Ludwigshafen, Germany

6.2.5.9 Patti Smith (*1946): Kodak

The book includes a poem 'Balance' [50].

6.2.5.10 Popular Music

'Balance' was an early 1980s pop rock group based out of New York City and
fronted by Peppy Castro, formerly of Blues Magoos [51]. Several album's include
pieces of music connected to the balance: Van Halen album, Kim-Lian album, Akro-
batik album, Leo Kottke album, Swollen Members album, and a song by Axium
from 'The Story Thus Far'.

6.3 Economy and Ecology

In Chap. 2 the close connection between money and weights is mentioned. Coins
and weights had (and have in some examples also today) identical designations, and
coins had been used as weights. Consequently, many expressions in economics had
been adopted from weighing techniques. Values and risks are weighed up carefully.
Since double-entry book-keeping is usual practice, the result is drawn up in a balance. In cartoons occasionally economic and financial events are visualised using a

balance. An example of assessing economic, ecologic and social influences using a three-armed balance is depicted in Fig. 6.60.

References

1. F. Szabadváry, History of analytical chemistry, in *Classics in the History and Philosophy of Scienve Series*, ed. by R. Hahn (Pergamon Press, Oxford, 1993)
2. E. Robens, Michael archangel and balance. J. Therm. Anal. Calorim. **62**, 579–580 (2000)
3. E. Robens, S.P. Garg, Strange weighers. J. Therm. Anal. Calorim. **71**, 3–18 (2003)
4. J.H. Breasted, *The Dawn of Conscience* (Charles Scribners's Sons, New York, 1933)
5. L. Kretzenbacher, *Die Seelenwaage* (Landesmuseum für Kärnten, Klagenfurt, 1965)
6. E. Robens, Some remarks on the balance as a symbol. Thermochim. Acta **24**, 205–217 (1978)
7. G. Groddeck, *Das Buch vom Es. Geist und Psyche*, vol. 2040 (Kindler, München, 1923)
8. C. Seeber, Untersuchungen zur Darstellung des Totengerichts im Alten Ägypten, in *Münchner Ägyptologische Studien*, vol. 35, ed. by H.W. Müller (Deutscher Kunstverlag, München, 1976)
9. S. Needham, W. Ling, K.G. Robinson (eds.). *Science and Civilisation in China*, vol. 4/1 (University Press, Cambridge, 1962)
10. R.O. Faulkner, C. Andrews, *The Ancient Egyptian Book of the Dead* (British Museum Publications, London, 1985)
11. K. Lembke (ed.), *Ägyptens späte Blüte. Die Römer am Nil* (Philipp von Zabern, Mainz, 2004). Zaberns Bildbände zur Archäologie
12. A. Champdor, *Das Ägyptische Totenbuch*, 3rd edn. (Droemer Knaur, München/Zürich, 1977)
13. Leviticus, Vayikra, in *Bible* (550–400 BC), pp. 19:34–36
14. Isaiah, Scriptures of Isaiah, in *Bible* (740–701 BC), pp. 40:11–13
15. Isaiah, Scriptures of Isaiah, in *Bible* (740–701 BC), p. 46:6
16. Daniel, Book of Daniel, in *Bible*, pp. 5:25–28
17. Archimedes, About swimming bodies, in *The Works of Archimedes, §7*, vol. 1 (Wissenschaftliche Verlagsbuchhandlung, Frankfurt am Main, 1987)
18. Archimedes, *Abhandlungen*. Ostwalds Klassiker der exakten Wissenschaften, vol. 201 (Harri Deutsch, Frankfurt am Main, 2003)
19. Aischylos, *The Libation Bearers (The Choephoroi)* (458 AC)
20. Aischylos (ed.) *Aischylos. Tragödien und Fragmente*. Reihe Tusculum ed. Werner and Oskar. München (2005)
21. A.J. Atsma, Theoi Greek Mythology. http://www.theoi.com/, The Theoi Project
22. Wikipedia, Bacchylides. http://en.wikipedia.org/wiki/Bacchylides (2013)
23. A.J. Atsma, Dike. http://www.theoi.com/Ouranios/HoraDike.html. Theoi Greek Mythology (2013)
24. Reppa, Lexikon numismatischer Fachbegriffe. https://www.reppa.de/. Pirmasens (2012)
25. John, Book of Revelation, 6, in *Bible, New Testament*, pp. 69–96
26. Wikipedia, English Apocalypse manuscripts. http://en.wikipedia.org/wiki/English_Apocalypse_manuscripts (2012)
27. J.B. Lohmann, Das Leben unsers Herrn und Heilandes Jesus Christus nach den vier Evangelien (1911)
28. A.O. Christian, Das Leben Jesu. http://vitajesu.wordpress.com/2009/12/31/116/ (2009)
29. I. Höfer, K. Rahner, *Lexikon für Theologie und Kirche* (Herder, Freiburg, 1965)
30. A. Kayserling, *Graf von, Monte Gargano. Europas ältestes Michaelsheiligtum*, 3rd edn. (Urachhaus-Verlag, Stuttgart, 1987)
31. F. Lenoir, V. Cabesos, *La promesse de l'ange* (Editions Albin Michel, Paris, 2004)
32. M.H. Shakir (ed.), The Holy Qur'an. http://quod.lib.umich.edu/k/koran/. Tahrike Tarsile Qur'an, Inc. (1983)
33. Anonymus, *Hindi Sachitra Priemer* (Aggarwal Book Depot, Delhi, 1957)

34. Anonymus, Zodiac. Education (2012)
35. C. Ptolemaeus, Almagest (Alexandria, 150)
36. G.J. Toomer (ed.), *Ptolemy's Almagest* (Princeton University Press, Princeton, 1998)
37. R.P. Arya, K.L. Joshi (eds.), *Ṛgveda Samhitā: Sanskrit Text, English Translation, Notes & Index of Verses*. Parimal Sanskrit Series, vol. 45 (Parimal Publications, Delhi, 2001/2003)
38. K. May, Durch das Land der Skipetaren, in *Gesammelte Werke*, vol. 5, ed. by E.A. Schmid (Karl May Verlag, Bamberg, 1862/2002)
39. F. Kafka, J.T. Williams, in *The Trial*, ed. by T. Griffith. Wordsworth Classic of World Literature (Wordsworth, London, 2008)
40. J. Roth, *Das falsche Gewicht. Die Geschichte eines Eichmeisters* (Querido, Amsterdam, 1937)
41. H. Böll, Die Waage der Baleks, in *Leserunde*, vol. 8 (Matthiesen, Lübeck, 1970)
42. S. Obermeier, *Die Hexenwaage* (Econ, Düsseldorf, 1997)
43. F. Lenoir, V. Cabesos, The Angel's Promise (Pegasus Books, 2006)
44. H. Kyburz, S. Marienberg, Still and again. Partitur (Breitkopf & Härtel, Wiesbaden, 2011)
45. K. Allihn, Ensemble Modern, Programm (Alte Oper, Frankfurt, 2011)
46. C. Orff, Carmina Burana (B. Schott's Söhne, Mainz, 1936)
47. B. Bischoff (ed.), *Carmina Burana. Faksimile-Ausgabe der Handschrift Clm 4660 und Clm 4660a* (Prestel, München, 1967)
48. F. Willnauer, Carmina Burana von Carl Orff. Entstehung, Wirkung, Text (Schott, Mainz, 2007)
49. Anonymus, Carmina Burana. Kloster Benediktbeuren, Benediktbeuren, 11th century
50. P. Smith, Kodak (Middle Earth, Philadelphia, 1972)
51. Wikipedia, Balance. http://en.wikipedia.org/wiki/Balance (2011)

Chapter 7
Documentation and Archiving

This chapter is concerned with the history of the conventional balance its documentation in museums, in the literature, and by scientific societies. "Like religions, the natural sciences have relics, and among the relics pertaining to chemistry there are balances made famous either because of the men who used them or because these instruments played an important part in outstanding researches" [1]. For the history of weights see Chapter 'Weights'. The development of balances comprises many more or less important technical and theoretical steps. The following may be regarded as the most important steps:

- Invention of the symmetrical beam or counterweight balance (prehistoric).
- Invention of steelyard and besemer (~500 BC).
- The theory of levers providing the basis for the design of unsymmetrical balances (Archimedes of Syracuse (~287 BC–212 BC) (Fig. 7.1a)).
- The use of a rod parallelogram in which two parallel balance beams are connected, enabling top loaders (Gilles Personne de Roberval (1602–1675) (Fig. 7.1b)).
- Invention of the spring balance (Robert Hooke (1635–1703) (Fig. 7.1c)).
- Introduction of electricity for force compensation, measurement and control of balances (Anders Jonas Ångström (1814–1874) (Fig. 7.1d)).
- Invention of the electric resistance strain gauge and application on flexural bodies (load cell) (Edward E. Simmons (1911–2004) (Fig. 7.1e) and Arthur Claude Ruge (1905–2000) (Fig. 7.1f)).
- Theory and development of all types of modern balances (Theodor Gast (1916–2010) (Fig. 7.1g)).
- Invention of the oscillating balance (Günter Sauerbrey (1933–2003) (Fig. 7.1h)).

Roger Bacon (1220–1292) was the first medieval scholar expressing the view that experiments are the most vital source of knowledge: "Oported ergo omnia certificari per viam experientiae" [2]. For that statement, he was imprisoned from 1277 for the rest of his life by his Franciscan fellows. Nicolaus Cusanus (Nicholas of Cusa, Germany) (1401–1464) regarded the use of scales in medicine, physics and alchemy as the most accurate and reliable method of measurement [3, 4]. With Antoine

E. Robens et al., *Balances*, DOI 10.1007/978-3-642-36447-1_7,
© Springer-Verlag Berlin Heidelberg 2014

(a)   (b)

(c)   (d)

Fig. 7.1 (a) Archimedes of Syracuse (~287 BC–212 BC). (b) Gilles Personne de Roberval (1602 –1675). (c) Robert Hooke (1635–1703). (d) Anders Jonas Ångstrøm (1814–1874). (e) Edward E. Simmons (1911–2004). (f) Arthur Claude Ruge (1905–2000). (g) Theodor Gast (1916–2010). (h) Günter Sauerbrey (1933–2003)

Lavoisier (1743–1794) and other scientists in his environment, systematic weighing in chemistry started [5, 6].

7.1 History of the Balance

Neolithic society's people settled in the eastern Mediterranean regions during about 9000 BC and townships were founded at about 5000 BC. Commerce and trade over long distances started in those societies. Trade was first purely based on barter and initially the exchange of single objects like domestic animals, tools and weapons or small batches of materials like victuals may have been predominant (Fig. 7.2). Besides length and volume, the mass is an important parameter characterising objects

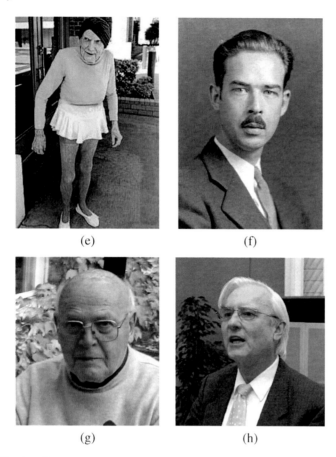

(e)   (f)

(g)   (h)

Fig. 7.1 (Continued)

to be traded or treated. Rulers and calibrated or sometimes graduated vessels are the oldest measuring instruments. In barter weighing was hardly required and we have drawings of Old Egyptian market scenes without any balance (Fig. 7.2). Weighing was invented when and where a need existed and this had been towards the end of the above mentioned settling period [7–9]. "The existence of man-made alloys of copper with lead for small ornaments and alloys of copper with varying amounts of tin for a wide variety of bronzes implies an ability to make accurate measurements with a weighing device during ca. 3000 BC and perhaps earlier" [10]. We can localise the invention of the balance in the eastern Mediterranean area up to the Indus valley region (today Pakistan), between the Sahara and the Gobi desert, where the modern civilisation of mankind developed first. Very early, in Mesopotamia a highly developed standardised measuring system exited which covered most probably weighing, but we have no findings. Similar developments in China in the Yangtse-kiang and Huangho valleys and in Middle and South America are clearly dated later [7].

Fig. 7.2 Grave of Ipui at Thebes 13th century BC. Sailors carrying sacks containing grain. A woman is selling bread and possibly beer (*top left*), beside her a sailor is exchanging grain for fish. Another is acquiring some vegetables. All by barter; no balance is used

7.1.1 Egypt

After the ice-age the climate changed dramatically. In Egypt the Nile decreased, turned to its present size and became enclosed by almost impassable deserts on both sides, concentrating the human population, forcing them to settle in the few kilometres wide valley. The favourable geographic situation protected the country against enemies and allowed for a period of more than 4000 years of relatively stable development of its culture. Besides agriculture in the extremely fertile area, specialisation of the production of goods, trade and traffic in the modern sense began. Such business as well as new techniques like metallurgy required measuring instruments. The dry climate preserved artefacts and thus we have most findings from Egypt [11–13].

7.1.1.1 The Pre-dynastic Era

Already before the dynastic era (∼3000 BC) systems of measures and weights had been established. Balances were used in the exchange of commodities; its type and mass gave the value since money did not yet exist. Besides trade important tasks had been weighing of tributes, controlling inventory and handling out of precious metals to be treated [14]. Proportioning of the constituents of mixtures in general was made volumetrically. It is not known that balances were used for that purpose.

Probably the oldest weighing artefacts, in particular weights, have been found in Upper Egypt first by Sir Flinders Petrie at the turn of the 20th century (Fig. 7.3) [15–17]. Besides weights he described a balance beam:

> "A small balance beam is made of hard pink-brown limestone, a material often used in prehistoric work, but seldom later. The beam is 3.35 inches long, 0.16 to 0.20 inches wide, 0.17 to 0.20 inches deep. The middle hole for suspension is 0.08 inches wide; the end holes for the pans are 0.06 inches wide. The arms between the holes are 1.595 and 1.600 inches long, a difference of 1 to 320; but on actual trial a difference of 1 to 120 was found; a change of 1 to 500 was visible in the level of the beam. The strings shown in the photograph are modern" [16].

Many of those artefacts including the balance beam were bought from various persons during excavations [15, 18, 19]. Therefore, the site of discovery is unclear and

Table 7.1 Measures of the oldest Egyptian balance beam of reddish limestone (Figs. 3.39/5.1/7.2)

	inch	mm
Length	3.35	85
Width	0.16–0.20	4–5
Depth	0.17–0.20	4–5
Arms between the holes	1.595/1.600	40.5/40.6
Middle hole ∅	0.08	2
End holes ∅	0.06	1.5

a dating could be made only by appearance of the object. Regarding the type of the object and its elaboration Petrie assigned the balance beam conclusively to the Amratic period of Naqada (Negade) I, or to Naqada II [17, 18, 20]. Petrie's absolute dating of the Egyptian history, however, was incorrect as proved using the radiocarbon dating method [21]. The beginning of the Egyptian dynasties is about 3000 BC and not 5000, as he assumed [8, 9] and thus his dating of the balance beam was also incorrect. Today the Naqada I period is dated to 4000–3500 BC and Naqada II to 3500–3200 BC. Skinner dated that beam to 3300 BC [14]. Unfortunately, many others accepted Petrie's dating, and we still find erroneous dating in some museums.

On account of the collar around the central hole the balance was a rather insensitive and inaccurate instrument, but of an already advanced type (Figs. 3.39/5.1/7.2, Table 7.1). The relative sensitivity of a model was 10^{-2}. Doubts whether this object is really a balance beam were overcome, when two beams of similar type were depicted in the tomb of Hesire 1000 years later.

7.1.1.2 The Old Kingdom

A drawing in the tomb of Hesire at Saqqara, built about 2750 BC (Figs. 3.37/4.8), encloses two balance beams of the same type, and sets of weights and cylinders [14]. This wall painting represents an early standard of instruments and measures. During the time of pharaoh Djoser (enthronement between 2691 and 2625 BC) [22, 23] of the 3rd dynasty Hesire was the most prominent scientist in Egypt and most celebrated dentist [24]. There are no hints that any other type than equal armed balances existed in Egypt during the Old Kingdom era.

7.1.1.3 The New Kingdom

The equal armed balance was used exclusively until the middle of the New Kingdom (1567–1085 BC). Possibly it was derived from the supporting beam (Fig. 7.4a). All early depictions show the equal armed balance operated in equilibrium, though with draw wells and in construction engineering unequal armed lever had been applied (Fig. 7.4b). A few Egyptian drawings of scales with only one pan are regarded as incomplete [25]. Balances with only one pan and with variable lever arm, like scales

Fig. 7.3 Table XLVI of Flinder Petrie's book: Prehistoric Egypt [16]. 36 balance beam; strings are added

with sliding weight (steelyard) or with sliding suspension of the fulcrum (besemer or bismar), should be dated later, some hundred years BC [17]. Most likely the thousand year's earlier attribution of such instruments to the Egyptians [26] resulted from the wrong interpretation of the plummet of the indicator system as a sliding weight. Variable lever arm balances are described first in 350 BC [27, 28]. To derive satisfactory sensitivity and reproducibility using such scales, knife edges made of hard metal are necessary. Thus the steelyard was invented most probably by people familiar with iron work, as was the case in the Northern and Eastern Mediterranean

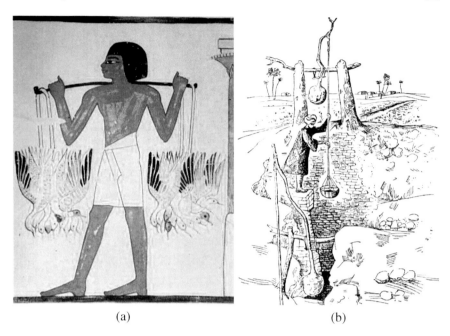

(a)   (b)

Fig. 7.4 (**a**) Duck shooting. Transport of ducks with supporting beam. Egypt, New Empire. (**b**) Egyptian draw well, 19th century

era, rather than in Egypt. Steelyards found in the Nile delta are dated in the Roman period and may have been imported [29].

In the Egyptian Museum of Cairo several findings from the New Kingdom era of small hand- and stand-balances as well as weights of different mass and shape are exhibited (Figs. 7.5). Besides scales of simple design balances had been improved to a better relative resolution [30] (the quotient of resolution threshold to maximum load) up to 10^{-4}. The load capacity ranges from a few grams to a kilogram. The purpose of gram balances [31] is unclear since the ingredients of pharmaceuticals and cosmetics were proportioned by volumetric techniques as we may conclude from the findings of measuring vessels and spoons. In medical papyri we find amounts in volume rather than in mass [32].

Two small balances in the Egyptian Museum at Cairo should be discussed in detail. We learned that these had been reconstructed by Ducros [33, 34] using fragments which he found besides other things in a case stored in that Museum. From the 'hand balance' (Fig. 7.5c/d) only a round wooden beam and metal pans existed. He assumed that it was not equipped with pointer and plummet and served as a rough instrument only [34].

Only the beam, two pans and the upright form of an arm with fist of the second instrument was found and he assumed that the arm was the shaft of a small stand balance (Fig. 7.5e). The beam was made of rolled copper or bronze sheet. Ducros supplemented the arm with a right-angled hook in form of an ostrich feather and added to the beam a wire pointer passed through a central bore of the beam. At the

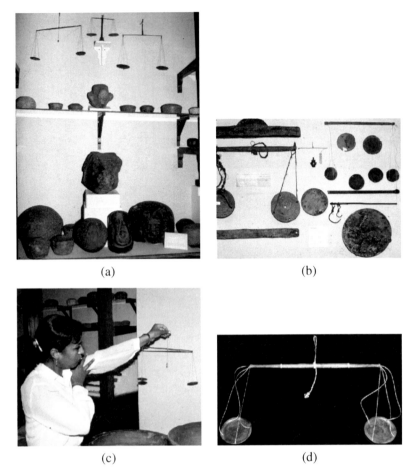

(a)   (b)

(c)   (d)

Fig. 7.5 (**a/b**) Scales and weights. Egyptian Museum, Cairo, Egypt. (**c/d**) Hand balance, Egyptian Museum, Cairo, Egypt. (**e**) Stand balance, Egyptian Museum, Cairo, Egypt

upper end he formed that wire to a ring which he used as fulcrum suspension. As a result of that construction the axis of rotation of balance is far above the connecting line of the bearings of the pan suspension making the balance largely insensitive. We assume indeed that this balance had no pointer and the bore-hole was used for a textile thread as a fulcrum suspension. Both ends of the beam have a biased boring for the threads of the pan suspensions. In Table 7.2 the dimensions of these parts according to Ducros' measurements are summarised.

Whereas findings of small balances exist the construction of large balances has to be derived from the paintings in tombs. Some of the mural paintings or drawings in the 'Book of Deaths' are accurate like constructional drawings giving a good impression of the instruments and of the weighing techniques of this time (Figs. 6.8/7.6).

Fig. 7.5 (Continued)

(e)

Fig. 7.6 Egyptian wooden
case decorated with painting
of the death tribunal dated to
the New Empire. Musée du
Louvre, Paris

However, on account of the development of more precise rules for the way to
eternity for everyone a wholesale manufacturing of those papyri came into existence. They were prefabricated leaving place for the name of the dead. Some of
the copying second-class artists had obviously never seen a balance. Thus, we have
more inaccurate drawings. But even the accurate representations are difficult to un-

Table 7.2 Dimensions of scales exhibited in the Cairo Egyptian Museum, reconstructed by Ducros. Values with * are obtained by measurements of Robens and Jenemann made with models, both with thread suspension of the fulcrum. The sensitivity was defined as the mass producing a beam deflection of $1°$

	Hand balance	Stand balance	Unit
Catalogue number	31489		
Material of shaft	–	Copper	
Height of standard shaft		158	mm
Material of beam	Mahagony	Copper or bronze	
Length of beam	277	138	mm
Diameter of beam	6–8	2–4	mm
Mass of beam	13.50	4.85	g
Distance of lateral borings	137	78	mm
Material of pans	Metal	Copper	
Diameter of pans	61	58	mm
Height of pans	3	3	mm
Mass of pan A	7.05	7.76	g
Mass of pan B	7.03	8.00	g
Maximum load*	200	60	g
Sensitivity at 0.5 g load*	0.05	0.10	g
Sensitivity at 30 g load*	0.15	0.12	g
Sensitivity at 60 g load*	0.30	0.15	g

derstand on account of the Egyptian method of drawing [12]. The Old Egyptians never used any type of perspective. Each object is shown as a silhouette, either in front view or in side view. Furthermore each part of the object is shown from the most impressive side. This results in a combination of front and side view: if necessary some parts are twisted by $90°$ within the figure whereby some details must be distorted. The face is shown from the side, the eye, however, from front. Arms and legs are represented from the side, the breast from the front. Hidden details were shifted to become visible. With regard to the operation of scales we made some schematic sketches showing Anubis weighing the heart at the left hand side against Maat's ostrich feather on the right hand side. Anubis is kneeling in front of the balance showing his back and calming down the oscillating balance and the pans by grasping a suspension cord (Fig. 7.7a). The side view (Fig. 7.7b) shows Anubis operating the balance but indeed that view is not very informative. In the Egyptian mode of drawing (Fig. 7.7c) we see clearly that he calms with the left hand the plummet and we make out a pointer fastened at the beam.

From the middle of the 18th Egyptian dynasty (1567–1320 BC) onwards, scenes of a death tribunal are depicted as mural paintings in tombs and in the 'Books of Death'. The 'Book of Death' was given in the hand of the deceased as a script helping him to pass the examination by Osiris. The death tribunal includes the weighing of the heart by means of a man's height symmetric beam scales of special advanced

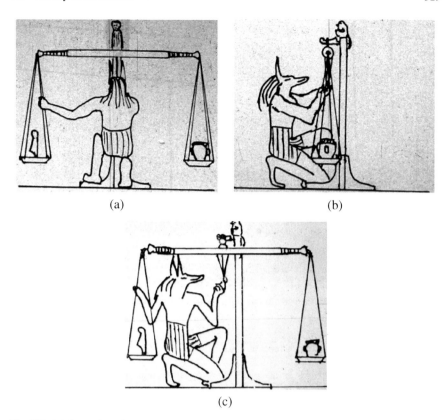

Fig. 7.7 (**a**) Operation of a large Egyptian balance from the New Kingdom. The master of weighing (here Anubis) is kneeling in front of the balance showing his back and calming down the oscillating balance and the pans by grasping a suspension cord, front view. (**b**) The side view is hardly informative. (**c**) Egyptian method of representation

type (Figs. 6.2–6.9). The beam is round and tapered to the ends, and there widened in the form of a lotus or papyrus flower. The beam of large balances was made of wood. Because native wood was hardly available, this was imported from southern (e.g. ebony) or Mediterranean countries (conifers). The beam is suspended centrally by a string or chain. Sometimes the central part and the ends seem to be reinforced by metallic furniture. The ends of the beam were axially drilled to give slightly ascending tubes penetrating the beam at the top. Four strings carrying the balance pans come out laterally of the bore-holes at the beam ends and are dragged down by the balance pans. Thus, the ends of the beam serve as bearings. Most balances are equipped with an indicator system always including a plummet and often a pointer [9]. Paintings representing scenes of daily life show similar balances of different size (Figs. 7.8/7.9). This type of balance (without or with modified indicator system) was used until modern age in Mediterranean countries.

In order to find out the features of this type of balance and to test some speculations concerning the indicator system we built a model with a beam length of

Fig. 7.8 Kom-Obmo temple, table with surgical instruments including balance

1.8 m [30, 35]. The balance beam (Figs. 3.40/7.10a) was turned from oak. Because turning of wood of such a length is difficult, the beam was made of three parts, mortised and glued. The maximum diameter of the beam for the model balance is 73 mm ~ 1 palm. The beam length l_b equals 4 short cubits ~ 1.8 m. The beam is suspended by a rope fed through a vertical hole, thus, the fulcrum is on the top of the beam. The data of the components and of the balance model itself are summarised in Table 7.3.

Starting from the expression for the momentum

$$M = \frac{P l_b}{4} \tag{7.1}$$

using

$$\sigma = \frac{M}{W}, \quad \sigma = 1.1 \times 10^7 \, \text{N} \, \text{m}^{-2} \text{ for oak} \tag{7.2}$$

and inserting $W = 3.8 \times 10^{-5} \, \text{m}^3$ as the section modulus for bending load for a diameter of 73 mm, the breaking load at each side of the beam results to

$$P = \frac{4 W \sigma}{l_b} = 930 \, \text{N} \tag{7.3}$$

The carrying load of the sisal rope used for the beam suspension amounts to 10^3 N and corresponds with the carrying load of the beam.

In order to obtain an impression of the accuracy of such a balance we measured the angular deflection of the beam as a response to little mass changes m_α for different loads m_l. The observation of the deflection of the beam was made by means of a little pointer at one end of the beam opposite to a fixed ruler. Combined with the beam length l_b this yields the angle of deflection α. The results are depicted in Fig. 7.10b. As can be seen in the graph, proportionality between and m_α is satisfactory.

The handling of the model balance turned out to be easy on account of the slow natural frequency of the unloaded beam of 0.2 Hz. It may be assumed that in the

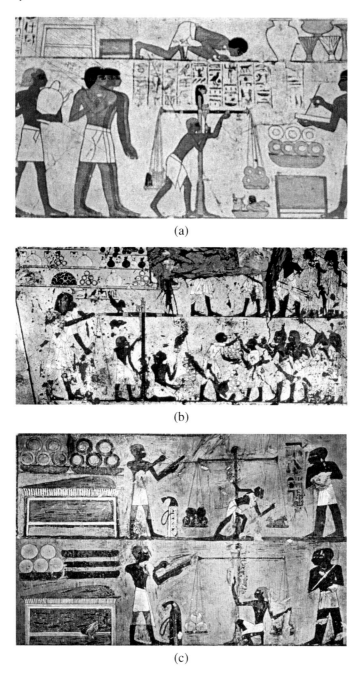

(a)

(b)

(c)

Fig. 7.9 (a) Gold balance, Workshop, Wall painting in the tomb of Benya, called Pahequamen TT343: weighing of gold rings. (b) Forge. © Wreszinski. (c) Forge with two gold balances. © Wreszinski. (d) Weighing of tributes. Tomb of Neferronpet (Ramses). (e) Forge, gold weighing, tomb of Rechmere (Rekhmara) 1500 BC. (f) Workshop with balance. © Wreszinski. (g) Forge with two gold balances. © Wreszinski. (h) Tribute weighing. © Wreszinski

(d)

(e)

(f)

Fig. 7.9 (Continued)

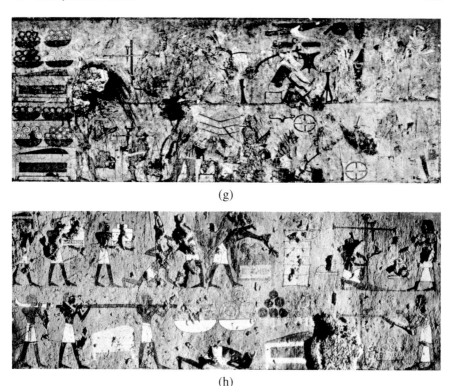

(g)

(h)

Fig. 7.9 (Continued)

drawings some time compression is made: at first the balance and the pans loaded with weighing object and weights were calmed down, then the indicator system was stabilised by manipulating the plummet.

The indicator system of such a large balance was rather unclear and several theories of its construction and operation existed. Therefore some experiments were made concerning the indicator system. Because the balance beam is shown from the front, the console of the standard, however, from the side, some Egyptologists supposed that the balance was equipped with a rider weight. This idea was realised to be wrong on account of clear evidence in many drawings. Wilkinson's speculation [29] of a ring-shaped bench-mark is rather insensitive and had already been rejected by Petrie. The abstruse idea of a three string plummet as proposed by Neuburger [36] and Spiegler [37] could be disproved by our experiments. Even for very large deflections slackening of threads could not be observed. Comparison of the situation of such a large beam with a horizontal reference line is very effective. We have, however, no hints in the drawings, that in this way equilibrium was checked. Especially, when the balance was equipped with a plummet and a pointer it is evident, that this was used to verify equilibrium, as already suggested by Petrie [15]. The triangular pointer was made probably in most cases of metal, according to the colour of this object in some paintings. Suspension of the plummet at two strings and the

Fig. 7.10 (**a**) Dimensions of
the balance model of a large
Egyptian balance of the
XVIII Egyptian dynasty. The
arrangement is shown in
Fig. 3.40. (**b**) Model of a
large scales of the XVIII
Egyptian dynasty. Reciprocal
sensitivity as a function of the
load of the balance beam

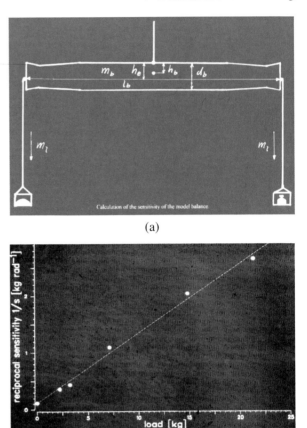

(a)

(b)

pointer between could be assumed only with regard to view drawings. Jenemann
applied consequently the Egyptian mode of drawing on the representation of the
pointer and of the fulcrum suspension. To compensate for parallax either the pointer
could be provided with a vertical line (Fig. 7.11a) or it could be arranged crosswise
to the beam (Fig. 7.11b) as proposed by Jenemann, regarding the Egyptian drawing techniques [38, 39]. In the front view it would appear as a line in the fulcrum
centre, covered by the plummet thread. Because a separate line is never depicted at
the pointer, we assume that the Egyptians turned this object in the drawing by 90°.
They did with the suspension hook similarly.

From an estimate of the bending of the beam it follows that for a beam of the
length 1.8 m and a diameter of 0.073 m with a load of 20 kg corresponding erroneous angular deflection amounts to less than 0.0004 rad which is small enough
to ignore. Careful observation allows for the reading of a deflection of 0.01 rad
($= 0.6°$), corresponding to a mass increment of about 30 g at a load of 25 kg. At
such sensitivity differences in the arm length should be regarded. Furthermore, when
weighing gold in comparison to stone-weights the influence of buoyancy might be

Table 7.3 Data of the model balance and symbols used

Parameter	Symbol		Unit
Mass of the beam (oak)	m_b	3.8	kg
Density of oak		0.6–0.9	$kg\,m^{-3}$
Beam length	l_b	1.8	m
Difference in armlength		3	mm
Maximum diameter of the beam	d_b	0.073	m
Frequency of the balance beam with or without buckets		0.2	Hz
Distance between centre of gravity of the beam and beam fulcrum	h_b		m
Distance between the beam fulcrum and the line connecting the bucket fulcrums	h_e		m
Breaking load at each side of the beam in central suspension	P	10^3	N
Diameter of the suspension rope		5	mm
Maximum carrying load of the suspension rope		10^3	N
Mass of each of the load buckets		2.1	kg
Mass of bucket + load at each side of the balance	m_l	∅ 2.1	kg
Mass of an ostrich feather		10^{-3}	kg
Additional mass at one side of the balance	m_α	0–0.3	kg
Angular deflection of the beam	α		rad
Sensitivity	s		$rad\,kg^{-1}$
Acceleration due to gravity	g	9.81	$m\,s^{-2}$
Material constant for oak	σ	1.1×10^7	$N\,m^{-2}$

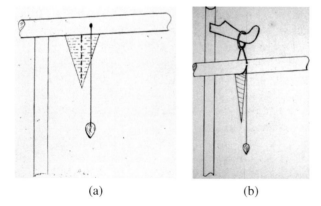

(a)   (b)

Fig. 7.11 (a) Suspension of the beam and arrangement of pointer and plumb line of a large balance from the Egyptian New Kingdom according to Hans Jenemann. Pointer with line. (b) Pointer arranged crosswise to the beam

noticeable. Most probably these errors were unknown at that time. In the case of weighing the heart in comparison to a ostrich's feather the detection limit is about

1 g and assuming the density of both objects to be comparable, weighing in the death tribunal could have been rather accurate.

7.1.2 The Near East

In 2600 BC Gudea was king of Babylon. Some weights in the form of geometric or animal figures have been discovered [40]. Such stone weights were marked with their mass and they had also sacerdotal seals. In 2300 BC Dungi was king of the town of Ur in Sumeria. He founded an institute, presided by priests, for testing of measures.

7.1.3 Greece

On a Hittite grave column (~1000 BC) at Marash (today Kahramanmaraş, Turkey) a man is depicted as a bas-relief with a small balance in his hand. This statue is now at the Louvre, Paris. Details cannot be seen. This balance was once thought to be a coin balance, but coins did not yet exist at that time.

First in Lydia, during about 650–500 BC, the use of metallic coins as an equivalent for a traded object became familiar. These coins were pieces of molten electrum, which is a naturally occurring alloy of gold and silver, with trace amounts of copper and other metals. Initially without any mark (see Chap. 2), coins were stamped later, on one side the value and on back any object, figure or the portrait of the emperor. On the back of many coins a goddess with a balance is depicted, usually with cornucopia and balance. Now, balances found a new application for the control of the weight of coins [41] (see Chap. 5, Figs. 5.117/5.118).

A Greek dish of ~500 BC from Laconia, the so-called Archesilaus dish, shows a weighing scenery with a balance similar to the model used in the Egyptian New Empire, but the suspension and indication devices are different (Fig. 7.12). A number of other pictures of that time show the original Egyptian model. Figure 7.13 shows a weighing procedure on a vase.

Greek philosophers developed science as a basis of techniques. Aristotle Αριστοτέληζ (384–322 BC) and his pupils formulated the theoretical basis of science, described in the general physics textbooks: *Mechanics*, *On the Heavens*, and *The Nature*, *Mechanics* [28] (Μηχανικά—*Mekhanika*), probably not written by him [42] but by Archytas of Tarentum (428–350) [43]. Archytas of Tarentum was a Greek mathematician, political leader and philosopher, active in the first half of the fourth century BC. Archytas was born in Tarentum, Magna Graecia (now in Italy) [44]. In Mechanics the different kinds of levers are described with technical applications, including balances [28]. He described a special balance with variable ratio of arms, whereby the central suspension was movable, later called bismar or besemer. Similarly steelyards came into use (Figs. 7.17).

Fig. 7.12 The so-called
Archesilaus (Arkesilaos) dish.
King Arkesilaos of Laconia
watches on weighing (of
wool?). ~560 BC

Fig. 7.13 Greek vase
tradesmen measuring
merchandise, the Taleides
Amphora, 6th century BC
© Metropolitan Museum of
Art

The lever principle was then deduced in mathematical form by Archimedes (285–212 BC). Archimedes of Syracuse (Greek: 'Αρχιμήδηζ; ~287 BC–~212 BC) was the leading scientist in Greece [45]. Most of his books are lost but may be reconstructed on the basis of reports of Heron [46] and Pappos of Alexandria. Besides his survived works *Deplanorum aequilibris*, he wrote most probably a book on balances which has been lost. He explained the Law of the Lever, stating, "Magnitudes are

Fig. 7.14 Steelyard weight
with a bust of a Byzantine
Princess, 400–450

Fig. 7.15 Model of the
balance of Archimedes with
sliding sample hook

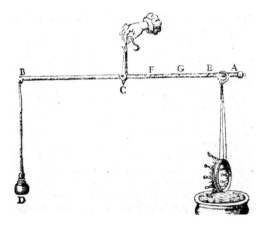

in equilibrium at distances reciprocally proportional to their weights". Archimedes
uses the principles derived to calculate the areas and centres of gravity of various geometric figures. He developed a general theoretical treatment of hydrostatics and of
buoyancy in liquids [47]. He invented the density determination of solids by comparing the weight of an object in air and in water. Here he used a symmetrical balance
with sliding weight or sliding sample hook on one arm (Figs. 5.3/7.15/7.17). This
was a forerunner of the analytical balance with rider weight.

Most probably the Greeks had been the first to use the balance not only in trade
and for weighing components of mixtures but also in science in order to determine
parameters like mass, density, and force.

Fig. 7.16 Metal beam of a
Roman trutina. The upwards
directed pointer is broken off.
The bearings are round holes

7.1.4 Rome

The Romans had been less interested in science than in techniques. They established
a unique system of weights and measures in their empire. The centre of minting was
in the Capitol, most probably in the temple of Juno Moneta. A close network of
surveillance of scales, weights and measures was incorporated in both the administration and the military. During the 1st to the 3rd century different types of scales
had been improved for their practical use. Whereas large scales were made of wood,
for smaller stand and hand balances besides wood improved metallic materials were
available. The general notion 'trutina' was used in particular for the hand held balance as depicted often in the hand of the goddesses Aequitas, Juno Moneta (without
sword) and later (seldom) of Justitia. Small equal-armed scales were fitted with an
upwards directed pointer to the equilibrium position, some controlled by a diop-
tre device. The bearings in general were round holes (Fig. 7.16). Occasionally the
beam was graduated for the use of a sliding weight. The bismar, a lever scale with
variable beam was described by Marcus Vitruvius Pollio (~70–10 BC). Hard metals like bronze or steel allowed for the construction of complicated lever balances,
in particular of the statera (steelyard) (Figs. 7.14/7.16/7.17). According to Isidor of
Sevilla (560–636) such scales had been applied first in the 2nd century in the Cam-
pania and were called therefore 'trutina campera', today 'Roman scales'. Designs
of that type with up to three ranges for sliding weights had been constructed. Either
a hook for the weighing object or a scale pan was balanced on the shorter side of
the fulcrum. The longer side was divided into lengths equal to the short arm side,
and then again subdivided. A weight was moved along the longer arm. For heavier
objects, the user could turn the bar over and hang the steelyard from the other hook
nearer the end of the bar. Such scales of modest sensitivity allowed for fast weighing
without use of a set of loose weights and were called therefore also 'trutina momen-
tana'. Statera were in common use in the Roman world, especially by shopkeepers
and traders, for weighing food, other commodities and coins and have been found
by excavations in the whole Roman Empire, e.g. in Pompeii.

Presumably the first record of the gravimetric observation of a chemical process
can be read in Vitruvius' report on lime-burning, dated to 27 BC [48, 49].

7.1.5 Northern Europe

First, Celtic tribes brought scales to northern regions [50]. They settled in about
600 BC in Middle Europe and established a network of handicraft centres, connected by trade-routes which reached from the Mediterranean countries to the

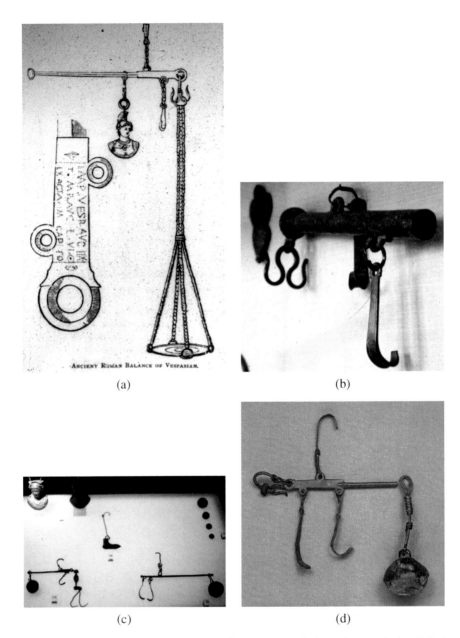

Fig. 7.17 (**a**) Roman statera of Vespasian. (**b**) Roman statera, the beam is a metal tube. © Saalburg, Bad Homburg vdH, Germany. (**c**) Roman scales, © Römisch-Germanisches Museum Köln, Germany. (**d**) Roman statera, length 186 mm, capacity > 8 pounds, © British Museum, London, UK. (**e**) Roman statera and weight from Caernarfon. Copper alloy. © National Museums & Galleries of Wales, UK. (**f**) This statera is a copy of an example excavated from the Roman town of Pompeii in Italy. © Science Museum, London, UK

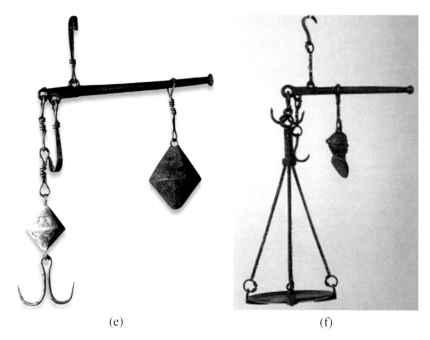

(e)   (f)

Fig. 7.17 (Continued)

British Isles. In the third century BC they extended their sphere of influence to Asia Minor where they learned commerce with money and the use of coin balances. Celts were driven away by the Teutons who had a lower state of civilisation. Thus, when the Romans conquered the countries in the north of the Alps they soon introduced the balance. They probably produced even scales in the colonies but most of them may have been imported from the mother-country. The Franconians took over the basic models of the Romans: equal-armed scales and steelyards, however less advanced types [51].

7.1.6 Islamic Countries

From Byzantine sources the Arabs learned manufacturing and handling of balances as well as its theory. Between the 8th and 15th centuries AD, in the golden age of the caliphate, the Islamic World formed centres of science [52]. During the caliphates of Harun al-Rashid (786–809) and his son al-Ma'mun (809–833) an ambitious programme of construction was carried out in Baghdad that included an observatory, a library, and an institution for research named 'Bayt al-Hikma' ('House of Wisdom'). This project brought groups of scholars together from the Mediterranian, Persia, Indus region up to today's Turkestan. But also in other important towns, science and techniques has been taught. Works were translated from Greek, Sanskrit, Pahlavi and Syriac sources into Arabic. The scientists of the Islamic world developed original theories in mathematics, astronomy, physics, medicine and engineering. Their knowledge came over Spain to the medieval Middle and Northern Europe. They were the first group of scientists who relied on experiment and observation as well as theory, and if the data they gathered did not support the theories of Aristotle, Galen or Ptolemy, they went with the empirical results. The spirit of 'Mutazilism', or critical thinking and rationalism, prevailed in this culture at a time that predated the European era of Enlightenment by about a thousand years.

They also improved weighing skills, in particular Archimedes' method of density determination [53]. Abd ar-Rahmān al-Chāzinï or Abu al-Fath Abd al-Rahman Mansour al-Khāzini was a Muslim astronomer of Greek ethnicity from Merv, then in the Khorasan province of Persia (located in today's Turkmenistan). He constructed the famous 'Balance of Wisdom' and described it in his book which appeared in 1120 together with density values of several solids and liquids (Fig. 5.3) [54]. A model of that balance exposed in the Institut für Geschichte der Arabisch-Islamischen Wissenschaften at Frankfurt am Main, Germany (Fig. 3.42) has a relative sensitivity of 10^{-5} [55]. Most probably European and in particular Italian scientists of the Renaissance period took advantage of Islamic knowledge. However, the Islamic literature of that time was not evaluated exhaustively until now.

7.1.7 The Middle Ages

In Europe in the Middle Ages (6th to 15th century) the Roman system of currency and weight were adopted. However, since the central institution for minting and control of weights and measures had collapsed each government made its own coins and weights and could not compare it with a common standard. As a result of weakening of the central power, the weights and measures system became increasingly fragmented. In the Middle Ages almost each town had its own weighing and measuring system. It was necessary to control the coins and a large variety of coin balances had been designed, usually small hand balances which had been stored together with a series of copper weights, the mass of which corresponding to that of the most common coins used in trade (Fig. 5.118). Underweight had been compensated by means of an 'ass' and its value was regarded in the transaction. Weighing coins became a popular subject of paintings of rich people (Fig. 5.117). After 1200 the Cologne Mark became a standard and likewise the Cologne pound (467.7 g). From this derivates were defined in the whole of Europe and elsewhere. Special coin balances are still used until today (see Chap. 6).

High sensitivity was required for druggist scales [56]. In middle age pharmacies and until today balances have become symbols of correctness. Many instruments had been designed very artificially (Figs. 5.5/7.18). The pharmacists had their own weight system and we still have relics today.

The so-called Renaissance (14th to 17th centuries) may be regarded as a transitional period from middle ages to modern ages. A new way of thinking took place.

Fig. 7.18 Rococo pharmacy of the Julius Hospital at Würzburg, Germany. It was crafted by Johann Gorg Oegg in 1762. At the wrought-iron lattice on the dispensary table hang scales, balls of string, and bags

Besides philosophy and theology, natural sciences came into the centre of interest. New techniques and their scientific basis were developed. Speculations were tested by experiments. In the centre of investigations was mechanics and the most important scientific instrument was the balance.

In the 14th century assay-balances were developed for the investigations in ore prospection and preparation of alloys (Fig. 5.6). This may be regarded as the beginning of systematic chemical analysis and of weighing in chemistry. In 1343 a decree of King Philippe of France covered a standard regulation for cupellation. This oxidizing process, known already in antiquity, is applied to separate lead from precious metals. The decree is a first example of standardisation of an analytical method including weighing [57]. In Agricola's (1494–1555) books [58] more analytical methods are described and in particular execution of weighing (Fig. 5.6) [59]. Balances got their special place in the laboratory and the most sensitive of which were protected by casings.

Nicolaus Chrypfs (Cryffiz) of Cusa (Germany), later known as Cardinal Nicolaus Cusanus (1401–1463) assumed that measurements are the basis of science and medicine. He wrote the first books in which weighing techniques are described in detail [3] and recommended weighing of blood and urine. The aim of the first gravimetric adsorption measurements was weather forecasting by measurement of atmospheric humidity. In Nicolaus Cusanus book, *Idiota de Staticis Experimentis* (Figs. 7.19) [4, 60, 61] a gravimetric hygrometer is described. Some twenty years later the Italian architect, painter and philosopher Leon Battista Alberti (1404–1472) described a similar device in the book *L'architettura* [62, 63] and from Leonardo da Vinci (1452–1519) we have three sketches of inclination balances [64, 65] loaded with a sponge or with cotton (see Chap. 5). At the end of that period optics and astronomy come into the centre of interest but all the time mechanics held its important place.

Simon Stevin (Stevinus) of Bruges (1548–1620) was a great mathematician, physicist and engineer, inventor of the decimal system. He distinguished stable from unstable equilibria. He derived the condition for the balance of forces on inclined planes using an ingenious and intuitive diagram. He demonstrated the resolution of forces before Pierre Varignon, which had not been remarked previously, even though it is a simple consequence of the law of their composition. He wrote a comprehensive work on balances and their use [66–68]. For his experiments he used equal-armed scales with open beam ends, similar to those of the Egyptians. This mathematician and physicist is less known because he published in Flemish.

In 1586, Galileo (1564–1642) wrote a short treatise entitled *La Bilancetta* (*The Little Balance*) [69–71]. He proved Archimedes' experiment and presented his own theory based on laws of lever and buoyancy. He described in detail an improved balance with sliding weight and its operation (Fig. 7.20) designed for density determinations according to Archimedes' Principle. Most important in this field, however, had been his investigation of gravity (see Chap. 1).

7.1.8 Modern Age

Many improvements can be recorded with the beginning of the Age of Enlightenment (17th to 18th century). With *Philosopha naturalis principia mathematica* (Isaac Newton, 1643–1727) classical mechanics was founded [72, 73]. He introduced a new terminology in order to develop the general theory of attracting forces and defined the quantities, mass and force. The mass of a body is determined by

(a)

Fig. 7.19 (**a**) Nicolaus Cusanus: De docta ignorata. (**b**) Nicolaus Cusanus: *Idiota de Staticis Experimentis*

comparison of forces observed with a balance. The acceleration of a body is proportional to the force exerted on it and has the same direction as this force; the mass being a measure of the body's inertia. Thus, Newton made a distinction between two properties: the inertial and the gravitational mass, which, however, are equal. Also today, the gravitational mass in a state of rest is regarded as the real property

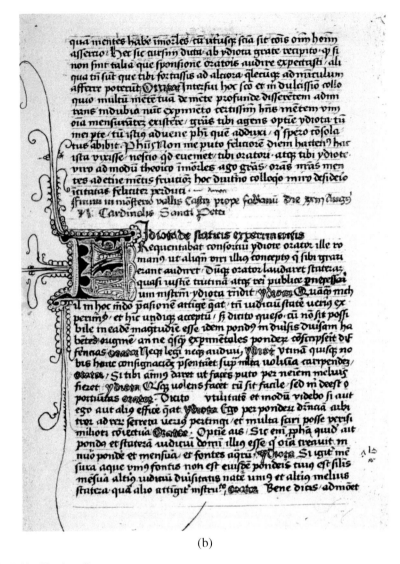

(b)

Fig. 7.19 (Continued)

of a body. Furthermore, he claimed that gravitational laws are valid in the whole universe, as assumed already by Nicolaus Copernikus (1473–1534).

In 1726 Jacob Leupold (1674–1727) [74] in his *Theatrum Staticum* summarised the technical improvements on balances and their use. High performance and assay balances as well as commercial scales are described in detail (Fig. 7.21). He demonstrated the conditions that must be fulfilled to obtain high sensitivity. The beam and likewise the pan's suspension should be pivoted by wedge-shaped bearings. Hereby friction was reduced allowing the balance to oscillate freely and not be 'lazy'. The

Fig. 7.20 Improved
Archimedes' density balance
of Galileo Galilei, in *La
Bilancetta*

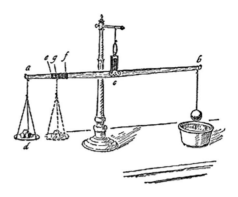

three knife-edges should be arranged at the same level. By that means he obtained
highest sensitivity and the sensitivity remains constant over the whole weighing
range. If the pan suspensions are at a lower level than the pivot a 'lazy' balance
with reduced sensitivity results, whereas, if they are at a higher level the balance
become unstable. With such knowledge the relative sensitivity of balances could be
improved to 10^{-5}, much better than the Roman scales and the Middle Ages' assay
balances.

An almost complete theory of the balance was developed by Leonhard Euler
(1707–1783) in 1737 [75]. He demonstrated that sensitivity is directly proportional
to the length of the beam, but inversely proportional to its mass and the distance
between beam's axis of rotation and its centre of gravity. This centre of gravity must
be slightly lower than the axis of rotation. Otherwise the balance would be instable.
The beam of an equal-armed balance should be fairly long, but as light as possible.
These conflicting conditions required a compromise in the design of the instrument.
Euler's theory may be summarised in the following equation:

$$\tan \alpha = k \frac{l}{dm_b} \Delta m \qquad (7.4)$$

where k is the proportionality factor (dependent also on the system's friction),
l length of the balance beam, measured between the knife-edges, d distance between axis of rotation and centre of gravity, m_b mass of the oscillating system, in
particular of the beam, Δm additional small mass placed on the pan, resulting in the
angle of deflection α.

In 1669 the French mathematician Gilles Personne de Roberval (1602–1675)
presented a mechanical parallelogram device which could be applied in a top pan
balance. Though this allowed for the construction of special kind of top pan scales,
which are very favourable in trade, such scales were made first in the early 19th
century [76].

In 1678 Robert Hooke (1635–1705) published his investigations on elastic bodies
[77–79]. Several basic types of spring balance had been invented by Robert Hooke:

- Vertical elastic string
- Horizontal metal rod firmly clamped on one side

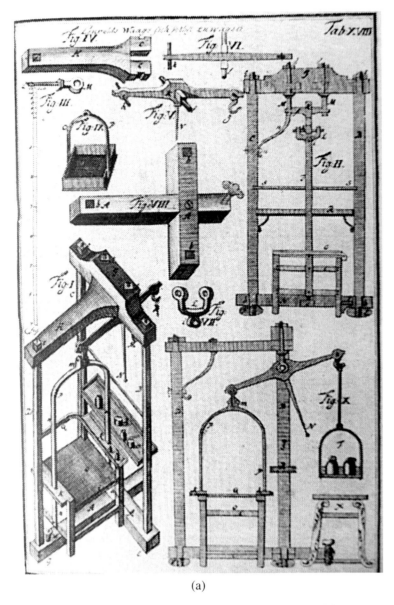

(a)

Fig. 7.21 (**a/b**) Design of sensitive balances according to Jacob Leupold, 1736

- Spiral-shaped spring
- Helical coiled spring.

Many variants of such elastic elements have been designed afterwards. On account of such very cheap and simple elements, springs are widely used for force measurements and weighing and even today many manufacturers produce spring and torsion

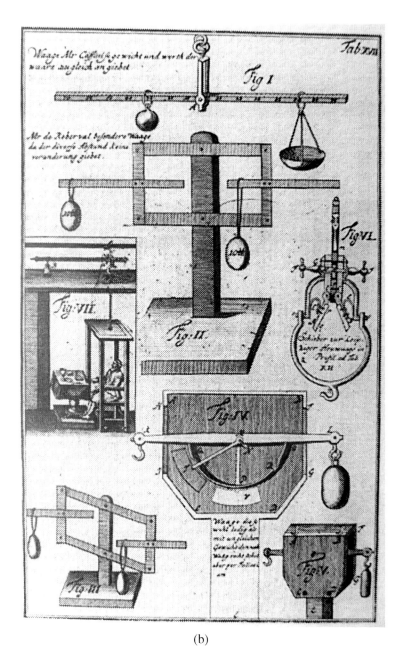

(b)

Fig. 7.21 (Continued)

type balances. Load cells equipped with electric strain gauges are applied in many modern scales.

Some historians assume that the inclination or deflection balance was known for a long time, because the inclined lever for lifting of heavy loads and in the draw well was known already in antiquity. Indeed we have not any reports and so we may conclude that initially only the equal armed balance in equilibrium was used. Earliest sketches of inclination scales are from Leonardo da Vinci (1452–1519). Designs of deflection scales for practical use appeared in the first decades of the 18th century; e.g. a Nuremberg coin balance and a hydrostatic balance of Johann Adam Cass [74]. Johann Heinrich Lambert (1728–1777) developed a deflection balance with quadrant reading on the basis of the pendulum theory and discussed several possibilities of the construction of such devices (Fig. 7.22).

Disregarding few exceptions the relative resolution was 10^{-4} until the end of the 18th century. Such an exception was the Netherlands physics professor Willem Jacob's Gravesande (1688–1742) [80]. He constructed a hydrostatic balance with a load capacity of about 100 g and improved its sensitivity by observing the deflection to a relative resolution of 10^{-5}. At this time the Netherlands had been leading in instrument manufacturing. Later on, in England industry became predominant, besides sea faring. For marine navigation efficient observation instruments, e.g. chronometers were necessary. Some London manufacturers produced also high sensitive balances.

Important impetus for further development were given by chemistry (detection of the law of conservation of mass) and by the work for the metric convention [81]. As the most important personality Lavoisier (1743–1794) should be mentioned. In Paris instrument factories were founded, which were able to construct balances meeting the demands of the scientists (Fig. 7.23). The appearance of assay balances as used in ore prospection changed. The new analytical balances had been mostly protected by casings with glass windows. The balance beam was not suspended in a shear but supported with a knife-edge made of steel, agate or cornelian mounted on a metal block at the upper end of a brass column. Special devices enabled the knife-edges to be separated from that mounting by leverage in order to arrest the scales when loaded with the object and weights.

To improve the sensitivity, as a consequence of Euler's recommendations fairly long-armed balances were constructed [82]. With a long arm simultaneously the error of unequal arm length could be reduced (Fig. 7.24). Such balances, however, oscillated at low frequency due to the large mass of the beam, and thus caused a long measuring time [53, 83]. To reduce beam mass it was made of hollow cones (Fig. 7.25) or metal framework (Figs. 7.22/7.24). By trial and error it was found that indeed short beams had a better accuracy and allowed fast weighing (Fig. 7.26). The error in arm length could be cancelled out by transposition or substitution weighing. Instead of placing rider weights manually, balances were fitted out with devices for external operation. In order to reduce the measuring time damping methods were introduced. At the high-end of that development metrological balances had been constructed with relative sensitivity of 2.5×10^{-9} with mechanical remote control and the possibility of operation in a vacuum.

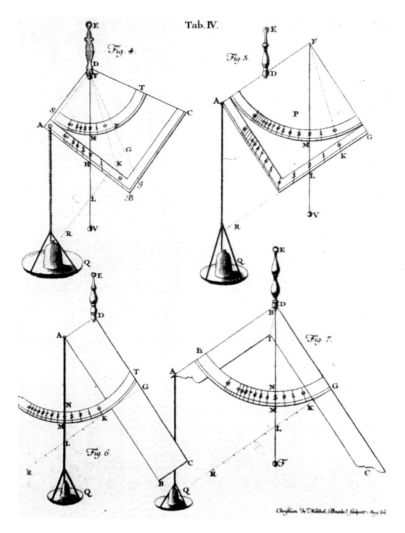

Fig. 7.22 Johann Heinrich Lambert's derivation of the theory of the deflection or inclination balance, 1758

In the 20th century many attempts had been made for easier operation of analytical weighing e.g. by mechanisms for weight changing and indication of results. The solution was a single-armed balance operated in the substitution mode and realised first by Erhard Mettler (*1917) and Hans Meier (*1914) (Figs. 5.9/7.27). The balance was equipped with air damping. Mechanical weight changing was indicated digitally; the deflection range was projected optically.

Simultaneously in the 19th and 20th centuries sensitivity and stability of scales for household trade and industry had been improved and its operation simplified. This concerns the various types of inclination balances, top pan balances according

Fig. 7.23 Balance of Jean-Adrien Deleuil, Paris (1874). © Musée des Arts et Métiers, Paris

Fig. 7.24 Laboratory balance with long symmetric framework beam

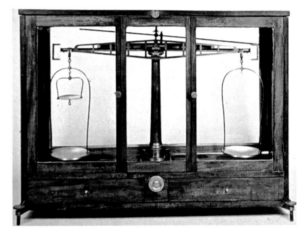

to the Roberval and similar principles, platform scales, and—for modest demands of sensitivity—spring type balances. Their history is mentioned in the Chaps. 4 and 5.

Although no other instrument had such a high sensitivity and such an extended measuring range, at the end of the 19th century scientists were not satisfied. Ångstrøm (1895) constructed a microbalance with electromagnetic compensation [84, 85]. Warburg and Ihmori constructed a microbalance with a quartz tube beam. As the central knife-edge they used a razor blade. Salvioni used a quartz fibre as a spring. Such designs, indeed, could not be invented by mechanics. Most important, however, was the application of electrical methods for indication and control of balances [84, 85]. Ångstrøm, Emich and others compensated the beam deflection electromagnetically and this was automated by Eyraud. The most important developer in this field, Theodor Gast, designed a variety of types of electrodynamically compensating balances as well as magnetic suspension balances in which the

Fig. 7.25 Laboratory balance, beam of long hollow cones

Fig. 7.26 Laboratory balance with short beam

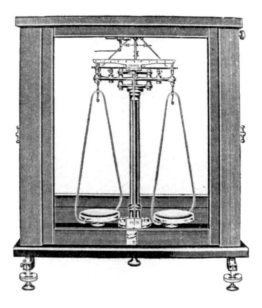

sample is separated from the balance [86, 87]. Self-compensating control, first developed for laboratory balances, today is applied for the majority of scales of any size and application.

Concerning microbalances we depend on the detailed reviews of Emich [88], Gorbach [89], Cunningham [90] and Behrndt [91]. The history of thermogravimetry and thermobalances we find in Duval's book [92] and by Iwata [93, 94] and Keattch [95–97]. The invention of thermogravimetry, however, must be antedated, as discovered by Eyraud [98–100]. In 1833 Talabot equipped a laboratory at Lyon with thermobalances for quality control of Chinese silk (Fig. 2.18). Such 'conditioning apparatus' were widely used for moisture determination of any materials. The first thermogravimetric assembly seems to have been described by Nernst and Riesenfeld in

Fig. 7.27 Mettler
single-armed top pan balance
operated in the substitution
mode

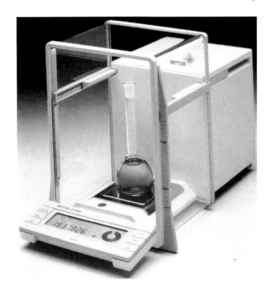

1903: a quartz beam balance with electric oven. In 1915 Honda seems to be the very first using the expression 'thermobalance' for his instrument [101, 102] (Fig. 2.19). Soon afterwards such instruments were used to investigate the metabolism of plants (Fig. 2.20). Further developments on thermogravimetry were made in particular in France by Guichard, Dubois, Chevenard and Duval and also in Japan. In 1953 de Kaiser developed the differential thermogravimetric method.

Another line of development was started as a consequence of the application of gravimetry to sorption measurements. One of the most essential steps forward in this development was the introduction of vacuum, which affected the construction of the balances. A prerequisite for sorption measurements was the development of sensitive microbalances, the first one attributed to Warburg and Ihmori in 1886. The spring balance was invented in 1665 by Robert Hooke. The first spring microbalance is probably of Salvioni [103]. The first vacuum microbalance with helical spring was designed by Emich [104]. In 1912 he made also an electromagnetic compensating beam microbalance which may be regarded as prototype for modern instruments (Fig. 2.21). Petterson published very sensitive observations of adsorption in 1914. McBain and Bakr [105–107] reported on a series of sorption measurements by means of helical spring balances. Further important developments were made by Rhodin, Gulbransen, McBain and Bakr, Gregg and Wintle, Cahn and Gast. In 1953 the company A.R.A.M. brought C. and I. Eyrauds electrodynamic recording balance on the market (Fig. 2.22). In 1949 van Nostrand claimed a patent application of a gravimetric apparatus for surface area and pore determination. An automatic gravimetric sorption measuring instrument was realised by Sandstede and Robens (Fig. 5.72). The most progressive concept for mass determination in vacuo was realised by Gast using electromagnetic suspension of the sample, the balance being outside the sample vessel (Fig. 5.58).

Mass and force determination based on mechanical oscillators is a fairly new measuring technique though the influence of tension on a string was known since lutes, harps and fiddles existed. Already 200 years ago elastic waves were studied experimentally and theoretically [108]. However, in the 19th century mechanical beam balances achieved the extraordinary relative sensitivity (sensitivity/maximum load) of 10^{-9}. So there was no motive to concern with other principles for mass determination apart from the very simple spring balance.

The effect of added mass on crystal frequency has been known since the early days of radio when frequency adjustment was accomplished by a pencil mark on the controlling quartz crystal. Sauerbrey was the first to investigate theoretically and experimentally the quartz crystal's suitability for determining mass [109, 110]. Probably independently, Warner and Stockbridge developed a quartz crystal balance based on long years of research on application of quartz crystals [111]. In the same year Wade and Slutsky reported on adsorption measurements of water vapour and hexane at a vibrating quartz crystal [112]. A period of lively activity and enthusiastic research and development was initiated, in the course of which the quartz crystal microbalance (QCM) underwent considerable improvements and became very widely used. Already in 1971 King Jr. had been asked to present a survey [113] and in 1984 the applications were reviewed in a book [114].

Besides, cheaper materials were applied in the form of strings, ribbons and thin-walled hollow bodies. The suitability of a horizontally arranged rotary pendulum for adsorption measurements was investigated. Using micromechanical cantilevers as developed for scanning atomic force microscopes it is possible to investigate samples with nanogram mass and mass changes in the picogram range [115]. Such sensors are widely used in air e.g. for dust concentration measurements. In a molecular beam apparatus it was used to measure the beam's profile and properties of clusters with sensitivity of 6×10^{-12} g. Liquids that surround the cantilever lead to significant damping. Nevertheless a mass resolution of femtograms of biologically relevant species could be obtained [116].

In the appended table (Table 7.4) the early history of balances, important development steps and their first applications are listed.

7.2 The Balance in the Museum

A museum is a "permanent institution in the service of society and of its development, open to the public, which acquires, conserves, researches, communicates and exhibits the *tangible and intangible* heritage of humanity and its environment, for the purposes of education, study, and enjoyment", as defined by the International Council of Museums [252]. The UK Museums Association definition (adopted 1998) is: "Museums enable people to explore collections for inspiration, learning and enjoyment. They are institutions that collect, safeguard and make accessible artefacts and specimens, which they hold in trust for society". For technical museums that 'inspiration and learning' aspect is especially important in modern times

Table 7.4 Chronological table concerning the development of balances

Year	Invention/investigation	Inventor, researcher	Reference
<−3000	Oldest balance beam, 85 mm lime stone, Egypt, Negade period, relative sensitivity 10^{-2}	Petrie	[17, 38]
<−3000	Finding of weights, Eastern Mediterranean		[117]
−2600	Weights in the shape of geometric or animal figures, Babylon	King Gudea	[40]
−2650	Standardisation table of weights and balances, Egypt	Hesire	[118]
−2300	Institute for the testing of measures, Ur in Sumeria	King Dungi	[119]
−1800	Relative sensitivity 10^{-4}		[95]
−500	Steelyard, Roman balance	Greece	[120]
−400	Bismar, Besemer	Greece	[121]
−350	Theory of lever	Aristotle	[28]
−250	Theory of lever and of the balance. Gravimetric density determination of solids	Archimedes	[47, 122]
−200	Ganchang (steelyard)	China	[123]
−27	Mass loss measurement when burning limestone	Vitruvius Pollio	[48, 49]
850	Definition of specific mass	Al Pindi	[124]
1120	Gravimetric density determination of solids and liquids	Al Chazini	[125, 126]
1343	Standardisation of cupellation	King Philipp VI	[57]
1450	Gravimetric hygrometer with wool	Nicolaus Cusanus	[4, 60]
1470	Gravimetric hygrometer with a sponge	Alberti	[62, 63]
1490	Gravimetric hygrometer with cotton. Buoyancy and air density measurements	Leonardo da Vinci	[65]
1550	De mensuris et ponderibus	Agricola	[59]
1650	Measurement of the mass of air	Guericke	[64]
1669	Toploader balance	Roberval	[76]
1678	Spring balance	Hooke	[77–79]
1685	Gravimetric hygrometer with vitriolic acid	Gould	[127, 128]
1798	Torsion apparatus, weighing of the Earth, gravitational constant	Michell, Cavendish	[129, 130]
1822	Decimal scale	Schwilgué, Quintez	[131]
1833	Dessicateurs Talabot-Persoz-Rogeat, thermogravimetry of silk	Talabot	[99, 100]
1837	Torsion balance	Weber	[132]
1849	Top loaders	Béranger, Pfanzeder	
1854	Susceptibility measuring method	Plücker	[133]
1854	Susceptibility measuring method	Gouy	[134]

Table 7.4 (Continued)

Year	Invention/investigation	Inventor, researcher	Reference
1860	Metrological vacuum balance	Deleuil, Regnault	[135, 136]
1875	Vacuum balance	Bunge	[137]
1881	First recording and automatic weight switching	Stückrath	[138]
1886	First microbalance, glass beam	Warburg, Ihmori	[139]
1887	Torsion balance	Kent	[140]
1895	Electro mechanical balance	Ångstrøm	[141, 142]
1901	Glass fiber balance	Salvioni	[103]
1903	Inclination torsion microbalance	Nernst	[143, 144]
1903	Thermogravimetric apparatus	Nernst, Riesenfeld	[145]
1903	Adsorption measurement	Giesen	[146]
1905	Dissociation of carbonates	Brill	[147]
1907	Weighing of pyrite and pyrrhotite decomposition	Truchot	[148]
1909	Triangular quartz rod buoyancy balance	Steele, Grant	[149, 150]
1909	Electrostatic suspension of particles, 10^{-15} g	Ehrenhaft, Millikan	[151]
1910	Suszetibility measurement	Pascal, Curie, Chéneveau	[152, 153]
1910	First photographic recording of weight changes	Kuhlmann, Bunge	[154, 155]
1912	Electromagnetic compensation vacuum microbalance	Urbain, Boulanger	[156]
1912	Triangular quartz rod beam zero balance	Gray, Ramsay	[157]
1912	Quartz fiber beam	Steele, Grant	[150]
1912	Electromagnetic compensating vacuum microbalance, quartz spring vacuum balance	Emich	[104]
1914	Quartz fiber suspension, sorption measurement 10^{-10} g	Petterson	[158, 159]
1914	Quartz rod beam	Aston	[160]
1915	Thermobalance, force compensation by a spring	Honda	[101, 102]
1920	Coefficient of expansion	Honda	[161]
1923	Recording thermobalance	Guichard	[162]
1924	Electromagnetic force compensation, electro-mechanical weight switching	Odén, Keen	[163]
1925	Thermobalance with scale on top	Saito	[164, 165]
1926	Quartz spring thermobalance	McBain, Bakr	[107]
1927	Thermobalance with electromagnetic force compensation	Shibata, Fukushima	[166]
1927	Etching	Inamura	[167]

Table 7.4   (Continued)

Year	Invention/investigation	Inventor, researcher	Reference
1929	Suscebtibilty measurement in vacuum	Aharoni, Simon	[168]
1929	Quartz fiber suspension balance	Strømberg	[169]
1930	Quartz spring thermobalance with electromagnetic force compensation	Oshima, Fukuda	[170]
1931	Quartz balance with pointer	Donau	[171, 172]
1933	First photographic recording on a thermobalance	Skramovsky	[173]
1934	Thermobalance with oven on top	Rigollet	[174]
1935	Thermoelement in the sample	Dubois	[175]
1936	Vapour pressure	Mochida	[176]
1936	Vapour pressure, Knudsen method	Shibata, Niwa	[177]
1936	Photographic registration	Chevenard	[178]
1937	Magnetic suspension	Holmes	[179]
1940	Electromagnetic suspension balance	Beams, Clark	[180]
1940	Compensating electrodynamic vacuum microbalance with Helmholtz coils, permeation measurements	Gast, Vieweg	[181]
1940	Quartz balance with rider	Barret, Birnie, Cohen	[182]
1942	High temperature NiCr-spring balance	Oyama, Nakaji	[183]
1943	Permeation	Gast	[86]
1944	Automatic quartz rod balance	Gulbransen	[184, 185]
1945	ADAMEL photographic recording thermobalance	Chevenard	[178]
1946	Glass beam balance with electromagnetic suspension	Gregg, Wintle	[186]
1948	Automatic electromagnetic compensating vacuum microbalance, measurement of dielectric properties	Gast, Alpers	[187]
1948	Electromagnetic suspension	Clark	[188]
1949	Particle size distribution	Hara	[189]
1949	Vacuum torsion balance	Rhodin	[190, 191]
1950	Thermoelement in the sample	Kinjo, Iwata	[192, 193]
1950	Suspension in alternative electrical field	Straubel	[194]
1950	Gravimetric surface area and pore size determination	Glasson	[195]
1950	Electromagnetic suspension balance	Beams	[180, 196]
1953	Electromagnetic registration	Kinjo, Iwata	[197]
1953	Sartorius differential thermogravimetry	de Keyser	[198]
1953	Inorganic thermogravimetric analysis	Duval	[92]
1953	ARAM/DAM/SETARAM vacuum thermobalance	Eyraud	[199, 200]
1954	Magnetic susceptibility	Hirone, Maeda, Tsuya	[201]
1955	Low/high temperature thermobalance	Gregg	[202]
1956	Fusion point determination	Shikazone, Kanayama	[203]

Table 7.4   (Continued)

Year	Invention/investigation	Inventor, researcher	Reference
1956	MOM Derivatograph	Paulik, Erdey	[204]
1957	DTA in the balance pan	Powell, Blazek	[205, 206]
1957	AMINCO high pressure/vacuum gravimetric sorption measuring instrument with spring balance	Campbell, Gordon, Smith	[207, 208]
1957	Quartz crystal balance	Sauerbrey	[109, 110]
1958	Derivatograph	Paulik, Erdey	[204]
1958	Combination of DTA/TGA	Splitek	[209]
1958	Automatic vacuum thermobalance	Tomonari, Takahashi, Arakawa, Togawa	[210]
1958	Order of reactions	Freeman, Carroll	[211]
1959	Sartorius vacuum microbalance	Gast	[212, 213]
1959	Electromagnetic suspension	Gast	[214]
1959	Netzsch thermogravimetric apparatus	Netzsch GmbH	[215]
1959	Viscosity	Kunugi, Yamate	[216]
1959	High pressure thermobalance	Rabatin, Card	[217]
1960	Fractionating thermogravimetric analysis	Waters	[218]
1960	Liquid density measurements	Schöneck, Wanninger	[219]
1960	Cahn vacuum microbalance	Cahn, Schultz	[220, 221]
1962	Sorption measuring apparatus Sartorius Gravimat	Sandstede, Robens	[222]
1963	Fast balance reading	Poulis	[223]
1963	Electromagnetic suspension	Gast	[214, 224]
1967	Gravimetric pressure control	Robens, Sandstede	[225]
1967	Vibrational belt dust balance	Büker, Gast	[226]
1970	Vibrational belt vacuum balance	Gast	[227]
1970	Compensation by light pressure	Zinnow, Dybwad	[228]
1971	Automatic quartz fiber ultramicrobalance	Rodder, Czanderna	[229, 230]
1971	Simultaneous measurement of mass and torque	Gast	[231]
1972	Weighing of a particle suspended in an alternative electrical field 10^{-13} g	Böhme, Robens, Straubel, Walter	[232, 233]
1972	Gravimetric pressure measurement	Massen, Schubart, Knothe, Poulis	[234]
1972	Gravimetric vapour pressure measurement	Wiedemann	[235]

Table 7.4   (Continued)

Year	Invention/investigation	Inventor, researcher	Reference
1975	Gravimetric temperature measurement	Wiedemann, Bayer	[236]
1975	Wireless transmission of values from sample	Gast	[237]
1978	Glass ceramic beam	Lotmar, Ulrich	
1982	Simulation of high heating rates	Dunn, Jayaweera, Stevens, Davies	[238]
1985	Gravimetric density measurement	Gast, Talebi-Daryani	[239]
1989	Tapered element oscillating balance	French	[240]
1997	Fast reading of weighing cells	Horn	[241–243]
1999	Fast balance reading	Poulis, Massen	[244, 245]
1999	Oscillating carbon tubes	Poncheral	[246]
2006	Quartz crystal microbalance is really a mass sensor	Mecea	[247, 248]
2010	Penning trap mass measurement	Block	[249, 250]
2011	Micromechanical device for thermogravimetry	Berger, Gutmann, Schäfer	[115, 251]

with its rapid change of techniques. Otherwise, for example, the knowledge of construction of mechanical balances gained over thousands of years may be forgotten soon, because such instruments today are replaced by electronic balances [253].

7.2.1 Science Museums and Technical Libraries

The Museion of Alexandria, Egypt and some similar institutions of Greece may be regarded as the first museums. Most probably the Museion was founded at the beginning of the third century BC and it was conceived and opened during the reign of Ptolemy I Soter (367–283 BC) (Fig. 7.28) or his son Ptolemy II (309–246 BC). Museion means sanctuary of the muses. Carved into the wall above the shelves, a famous inscription read: "The place of the cure of the soul". Such institutions had more the character of an academy or university and they had been connected with large libraries. The included Royal Library of Alexandria [254], Egypt, was also home to a host of international scholars, well-patronised by the Ptolemaic dynasty with travel, lodging and stipends for their whole families. Most probably there were also scientific instruments, though standardisation work for scales and weights was carried out in the temples by priests. Heron Alexandrinus (the mechanicus) was a mathematician, physicist and engineer who lived in 10–70 AD. From Heron's

Fig. 7.28 Tetra drachma with
head of Ptolmaios I, Soter,
Pharao of Egypt 367–283

writings it is reasonable to deduce that he taught at the Museion in Alexandria and
most probably he presented there his inventions.

During the caliphates of Harun al-Rashid (786–809) and his son al-Ma'mun
(809–833) in Baghdad the 'Bayt al-Hikma' ('House of Wisdom') was founded with
observatory and library. This institution brought groups of scholars together from
the Mediterranian, Persia, Indus region up to today's Turkestan.

In the Roman Republic stolen Hellenistic statues provided the basis for art collections of rich people. Later, beginning in the 14th century, European dynasties made
similar collections. In the second half of the 16th century interest in unusual and cu-
rious things caused the initiation of 'Cabinets of arts and curiosities' in which also
old and newly invented instruments found their place. The earliest prototype was
the famous 'Kunst- und Wunderkammer' in the Castle Ambras, Tirol (Fig. 7.29)
of the archduke Ferdinand II of Austria and Tirol (1564–1595) [255]. The collection is scattered and scientific objects are exhibited today in the Kunsthistorisches
Museum, Wien, Austria and in the Amabras Castle. In the 17th and 18th centuries
at almost every residence such collections were established. However, these collections had been reserved for aristocrats and privileged people and were not open for
the public.

Since the 18th century the museum developed to be an institution. The British
Museum (Fig. 7.30) was established in 1753, as the first national, public and secular
museum in the world [256]. Also such 'public' museums were often accessible only
by the middle and upper classes. It could be difficult to gain entrance. In London
for example, prospective visitors to the British Museum had to apply in writing
for admission and to wait for a reply for some weeks. Some Egyptian scales and
weights are exhibited there and in particular many papyri containing drawings of
scales. As a result of the French Revolution in France the museums were opened for
the public, e.g. the Museum Français in the Louvre (Fig. 7.31). This was the first
truly public museum, opened in 1793, which enabled for the first time in history free
access to the former French royal collections for people of all stations and status.
In the Louvre are exhibited probably the oldest bas-relief of a balance in the hand
of a Hittite weigher (Fig. 5.2) and a small case with the most accurate painting
of a large Egyptian death balance. Furthermore Leonardo's sketches of gravimetric
hygrometers are stored there.

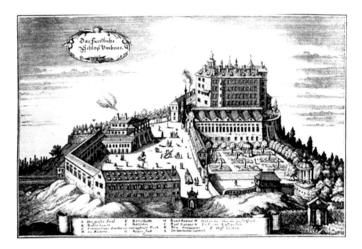

Fig. 7.29 Castle Ambras, Tirol, near Innsbruck, copper engraving of Matthäus Merian

Fig. 7.30 British Museum, London, UK, opened 1753, first national, public and secular museum in the world

In the 19th century many specialised museums were founded. The Smithsonian Institution founded 1846 at Washington, USA, began to establish several museums, e.g. the National Museum of American History which includes collections of scientific instruments. In 1857 the Science Museum, London, started with technical objects shown at the Great Exhibition held in the Crystal Palace. There are exhibited several Old-Egyptian balances and a copy of the oldest balance. In the Musée des Arts et Métiers, Paris, the balances of Lavoisier are shown and also the first vacuum balance of Deleuil and Regnault.

7.2.2 The Present Situation

The Deutsches Museum at Munich was opened in 1906 (Fig. 7.32) [257]. It houses a large collection of scales including Gast's electromagnetic and suspension balances.

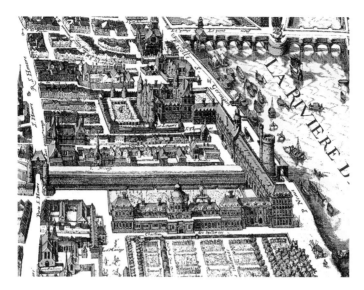

Fig. 7.31 Louvre, Paris, 1815. First truly public museum, opened in 1793

Fig. 7.32 Expositon of metrological balances taken out of service in the Federal Office of Metrology (METAS), Bern, Switzerland

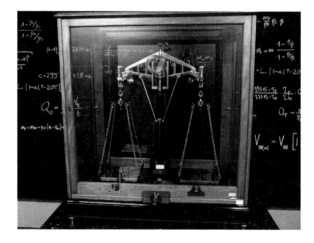

In the 20th century many specialised collections were installed. All standardisation authorities possess collections of old instruments and they present their standard equipment e.g. the exposition of the Bureau International des Poids et Mesures at Sèvres and of METAS at Bern, Switzerland (Fig. 7.33). Several scale & weight museums [258] and numerous pharmaceutical museums were installed [259]. Today nearly every village has a local museum or a collection in the town hall and in each of those small and lovely museums you will find some household and commercial scales. Many drug stores exhibit small collections of balances stored after their use sometimes for more than a hundred years. A recent development, with the expansion of the web, is the establishment of virtual museums [260].

Fig. 7.33 Deutsches Museum, München (Munich), Germany, opened 1906

The progress in mechatronics caused a rapid change from pure mechanical scales to electronically controlled designs which conquered the market of balances for any purpose. The sensitivity of force cells with strain gauge was improved so that they satisfied the requirements of industrial and household scales. With oscillating sensors the range was extended below the microgram. The main advantage of these new types of balances was easy and fast operation and recording of the result which allowed for online evaluation of results and its use for process control. Therefore suddenly a large number of obsolete laboratory and other scales submerged into the antique market. A lively scene of collectors of historical instruments and respective societies arose. In particular weights and scales were regarded by people interested in history of techniques. Some of these collections are of scientific importance besides the historical aspect. The balance collection of the Deutsches Museums, Munich, Germany (Fig. 7.33) documents the progress in science in instrumental techniques.

The Avery Historical Museum is a private museum opened in 1927. It houses probably the finest collection of weighing artefacts in the world. The museum is located at the Avery Weigh-Tronix headquarters, close to Birmingham city centre, England. The site Smethwick, West Midlands UK is an historic one. Known for more than 200 years as Soho Foundry, it was originally established by Birmingham's three most famous industrial pioneers—Matthew Boulton, James Watt and William Murdoch—to build commercial steam engines at the start of the Industrial Revolution.

The Museum für Waage und Gewicht at Balingen, Germany (Fig. 7.34) owes its existence to Prof. Wilhelm Kraut's (1875–1957) passion to collect. As a start in 1943, he gave his private collection on loan to the Museum für Waage und Gewicht. The display of about 450 items shows almost the complete development steps of scales as a manifold measuring element, from the simple beam scale up to the intelligent price computing scale and highly sensitive electronic scales of the present. Another important weight and scales museum covering the whole field of mass determination is the 'Museo della Bilancia' at Campogalliano (Modena), Italy. The balance collection of the Egyptian Museum, Cairo, gives a survey of the early devel-

Fig. 7.34 Museum für
Waage und Gewicht,
Balingen, Germany

Fig. 7.35 Philipp-
Mathäus-Hahn-Museum,
Im Kasten,
Albstadt-Onstmettingen,
Germany

opment of such instruments. Because much Egyptian artwork is displaced and scattered throughout the world we find in particular depictions of the Egyptian 'Death
balance' on papyri and on sarcophagi in many museums in every country.

Other collections are centred on inventors, e.g. the Philipp-Mathäus-Hahn-
Museum at Onstmettingen, Germany (Fig. 7.35). Hahn (1739–1790) was for several years pastor of Onstmettingen. As a pious priest he wrote theological papers
and translated the Bible and Books of Sermons, and he saw himself as a "Prophet
of the upcoming kingdom of Jesus Christ." Besides he was an ingenious inven-

Fig. 7.36 Pfundsmuseum, Kleinsassen (Rhön), Germany. The private museum houses several thousand measuring instruments and accessories, in particular balances and weights. On the left the founder Reinhard Kremer

tor of astrolabes, watches, balances, calculators and he is regarded as the initiator of precision mechanical industries in the Swabian Alb. The large scale collection of the Förderverein Philipp-Matthäus-Hahn-Museum, a group of people promoting the museum for the pastor Hahn, is housed in a former school, the Riedschule. There are over 1000 scales on display from about 50 manufacturers. It comprises various collections and foremost those from Jenemann and Danell. Max Danell of Bensheim, Germany, collected scales and documents of scales manufacturers. He restored old books and documents in this field. The Jenemann collection [261] includes laboratory scales and provide a survey on the development of mechanical balances. A part of the Jenemann collection is exhibited at the DECHEMA, Frankfurt, Germany. The Pfundsmuseum (pound museum) at Kleinsassen (Röhn), Germany is situated in a 16th century farmhouse (Fig. 7.36). It includes the collection of Reinhard Kremer of several thousand instruments, in particular of scales and weights.

Furthermore many laboratory scales are included in pharmaceutical museums and in historical drugstores which are found everywhere in the world. The 'Museo de la Farmacia Hispana' Facultad de Farmacia University Complutense of Madrid exhibits several splendid baroque drugstores equipped with scales, splendid as well. In contrast the small Apteka Muzeum, Lublin, Poland presents in a medieval town house a nice collection of instruments (Fig. 7.37).

The development of scales is well documented by artefacts of instruments and weights everywhere in museums. Extensive, but nevertheless incomplete lists of museums which exhibit large collections of balances are given in refs [258, 259]. They include museums in the following countries: Argentina (1), Australia (5), Austria (6), Belgium (3), Brazil (2), Canada (4), Chile (1), China (2), Costa Rica (1), Croatia (2), Czech (4), Denmark (13), Egypt (1), Finland (25), France (35), Ger-

Fig. 7.37 PZF 'CEFARM'
Apteka Muzeum, Lublin,
Poland

many (61), Greece (1), Hungary (8), Iceland (2), Italy (14), Japan (1), Lithuania
(2), Luxemburg (1), Netherlands (6), Norway (7), Poland (5), Portugal (1), Puerto
Rico (1), Romania (2), South Africa (3), Spain (13), Switzerland (6), Slovakia (1),
Sweden (26), Ukraine (2), United Kingdom (8), USA (29), Virgin Islands (1).

7.2.3 Museum Crime Fiction

Hans R. Jenemann (1920–1996) worked at Schott Glassworks in Mainz as head of
the analytical laboratory [262]. He established education and training of the laboratory staff and was chief instructor until his retirement in 1982. In 1975 together
with his wife Irene (Inis) [263] he started to collect several hundred discarded analytical balances with respect to technical peculiarity and repaired them carefully.
A part of the balance collection is exhibited at the DECHEMA in Frankfurt am Main
(Germany). A large part they assigned to the Mettler-Toledo company at Giessen
(Germany). From the profit and from his own funds he established the Jenemann
foundation which grants every two years the Paul Bunge award of about 10 000
€ for completed research on the history of scientific instruments. In spring 1996

Fig. 7.38 Move 2008 of the Jenemann collection from Hochheim (Taunus) to Albstadt-Onstmettingen. The 3 tons truck is overloaded. From left to right: Fritz Brenner, Erwin Kiefer. Behind Inis Jenemann: Rudi Keinath, Ludwig Bosch, Susanne Kiefer, Erich Robens. The photo was made by Fabian Jenemann

Hans Jenemann was appointed as an honorary member by the Scientific Instrument Society.

After his death his wife decided to transfer the remaining 150 balances (which covered the walls of the small townhouse) to the Philipp-Mathäus-Hahn-Museum at Onstmettingen, Germany. She trusted the curator, Susanne Kiefer, who organised a truck and some men which cleaned and packed all the balances carefully; it was somewhat more than a load of 3 tons of the truck (Fig. 7.38). The truck started to Onstmettingen, but instead of going to the Philipp-Mathäus-Hahn-Museum the men (all former workers or heads of ancient scale workshops) unloaded the balances in their own museum in the Riedschule. That museum now houses more than 1000 scales, carefully repaired and excellently exhibited. The curator of the Philipp-Mathäus-Hahn-Museum handed in her notice. We never told this history to Inis Jenemann who died in the same year [263].

7.3 Literature on Balance Techniques

The Royal Library of Alexandria [254], Egypt, was once the largest library in the ancient world containing about 500 000 papyrus scrolls. Mark Antony (83–30 BC) supposedly gave Cleopatra (69–30 BC) over 200 000 scrolls taken from the great Library of Pergamum for the library as a wedding gift. As a research institution, the

Table 7.5 Scales exhibitions. The list includes museum in which scales are exhibited: Specialised balance collections [258], museums of weights and measures, science museums, medical and pharmacy museums [259], antique pharmacies, etc. The list is incomplete!

Argentina	
Buenos Aires	Museo de Farmacia 'Dra. Rosa D'Alessio de Carnevale Bonino'
Australia	
Boulder	Boulder Pharmacy Museum
Chiltern, Victoria	Dows Pharmacy Museum
Coolgardie	Coolgardie Pharmacy Museum (formerly: Dwellingup)
Sydney	Sydney Powerhouse Museum
Melbourne	Medical History Museum
Austria	
Eisenstadt	Eichkundliche Sammlung im Eisenstädter Eich- und Vermessungsamts
Graz	Stadtmuseum
Köflach	Waagen und Gewichte Museum Pichling
Leoben	Josefee Apotheke
Trofaiach	Heimatmuseum Schloß Stibichhofen
Vorau	Freilichtmuseum
Belgium	
Brügge	Oude Apotheek—Memling Museum
Brussels	Le Musée de Pharmacognosie
Ghent	Museum of Folklore—Alijns House
Brazil	
Ouro Preto	Centro de Memória da Escola de Pharmacia de Ouro Preto
Rio de Janeiro	Museo Histórico Nacional
Canada	
London, Ontario	Banting House National Historic Site
Niagara, Ontaria	Niagara-on-the-Lake Pharmacy Museum
Québec	Musée des Augustines de l'Hôtel-de-Dieu de Québec
St. John's, NF	James J. O'Mara Pharmacy Museum
Chile	
Santiago de Chile	Museo de Historia de la Farmacia
China	
Beijing	Museum of Chinese History
Beijing	Palace Museum
Costa Rica	
San José	Botica Solera

Table 7.5 (Continued)

Croatia	
Dubrovnik	Museum Collection and Old Pharmacy of the Friars Minor
Klostar Ivanic	The Old Franciscan Pharmacy
Czech	
Cesky Krumlov	County museum
Kuks	Hradec Kralove: Czech Pharmacy Museum
Praha	The Museum of Czechoslovakian Pharmaceutical Industry
Praha	Exhibition of Historical Pharmacies, Lékarna Dittrrich U zlatého lva
Denmark	
Aalborg	Jens Bang's Stenhus' Apotekerloft
Århus	Apoteket i 'Den Gamle By', Danmarks Købstadsmuseum
Århus	Steno Museet—Denmark's Museum for the Hystory of Science
Ærøskøbing	ÆrøMuseum
Hillerød	Study Collection for the History of Pharmacy
Højerup	Stevns Museum
København	Københavns Universitets Medicinsk-Historiske Museum
København	Den Farmakognostiske Samling ved Danmarks Farmaceutiske Højskole
København	Fonden til Bevarelse af Gammelt Dansk Apoteksinventar
Farum	Apotekslaboratoriet
Rudkøbing	Langelands Museum
Store-Heddinge	Det Gamle Apothek (Østsjællands Museum)
Tønder	Apotekermuseet 'Det Gamle Apotek'
Egypt	
Cairo	Egyptian Museum
Estonia	
Tallinn	Raeapteek (The Town Hall Pharmacy)
Tartu	The University Museum
Finland	
Alajärvi	Alajärven apteekki/The Alajärvi Pharmacy
Åland	Ahvenmaan apteekkimuseo/The Åland Pharmacy Museum
Carlsrontie/Carlsrovägen	Apteekkarinnurkkaus Carlsron kartanossa
Esbo	Lääketehdas Orinin museo/The collections of the Orion pharmaceutical company
Hauho	Hovinkartanon apteekkimuseo/The Hovinkartano Pharmacy Museum
Helsinki/Helsingfors	Arppeanum—Helsingin yliopistomuseo/University Museum
Helsinki/Helsingfors	Apteekkariliiton esinekokoelma/The Historic Collections of the Apothecaries' Association
Helsinki/Helsingfors	Farmasian tiedekunnan vitriinit/The Faculty of Pharmacy Showcases

Table 7.5   (Continued)

Helsinki/Helsingfors	Helsingin XII Joutsen apteekki ja sen apteekkimuseo/The Helsinki Swan Pharmacy and its Museum
Helsinki/Helsingfors	Tekniikan museon
Ilmajoki	Ilmajoen apteekkimuseo/The Ilmajoki Pharmacy Museum
Korpo	Korppoon apteekkimuseo/Korpo Pharmacy Museum
Kouvola	Kouvolan apteekkimuseo
Koupio	Kuopion korttelimuseo
Liperi	Liperin apteekkimuseo/The Pharmacy Museum in Libelits
Muonio	Mounion kotiseutumuseo/Muonio Local Museum
Närpiö/Närpes	Öjskogparkenin museoalue/Öjskogsparkens museiområde
Oulu	Pohjois-Pohjanmaan museo/North-East Bothnia Museum
Raahe	Raahen museo/The Raahe Museum
Simola	Simolan Apteekkimuseo
Seinäjoki	Etelä-Pohjanmaan maakuntamuseo/South-East Bothnia County Museum
Tohmajärvi	Tohmajärven pitäjämuseo/The Parish Museum of Tohmajärvi
Turku	Turku County Museum, Apteekimuseio Turussa ja Queselin talo
Ylivieska	Ylivieskan kotiseutumuseo/Ylivieska Local Museum.
Tampere	Tamperen museoapteekin kokoelmat
France	
Alise-Ste-Reine	Hôpital Sainte-Reyne & apothicairie
Baugé	L'Apothicairerie de l'Hôtel-Dieu
Belleville-sur-Saône	L'Apothicairerie de l'Hôtel-Dieu
Besançon	La Pharmacie 'de Gascon', l'apothicairerie de l'Hôpital Saint-Jacques
Bourg-en-Bresse	L'Apothicairerie de Bourg-en-Bresse
Chagny	Apothicairerie
Chalon-sur-Saône	L'Apothicairerie de l'Hôtel-Dieu
Charlieu	L'Apothicairerie du Musée Hospitalier
Chatillon-sur-Chalaronne	L'Apothicairerie
Cluny	L'Apothicairerie de l'Hôtel-Dieu, Hôpital de Cluny
Crémieu	Apothicairerie de l'ancien couvent des Visitandines
Lons-le-Saunier	L'Apothicairerie de l'Hôtel-Dieu
Louhans	L'Apothicairerie de l'Hôtel-Dieu
Lyon	Musée des Hospices Civils de Lyon
Lyon	Musée d'Histoire de la Médecine et de la Pharmacie Université Claude Bernard
Macon	L'Apothicairerie de l'Hôtel-Dieu
Montluel	L'Apothicairerie de l'ancien hospice de Montluel
Montpellier	Musée Albert Ciurana
Montpellier	Oeuvres de la Misèricorde
Nice	Palais Lascaris

Table 7.5 (Continued)

Paris	Pharmacie Ordener
Paris	Musée des arts et métiers—CNAM
Paris	Musée de l'Assistance Publique de Paris, Hôtel de Madame de Miramion
Paris	Musée de Boulaq
Pont-de-Veyle	L'Apothicairerie de l'Hôpital de Pont-de-Veyle
Rouen	Le Musée Flaubert et d'Histoire de la Médecine
Saint-Amour	Les Apothicaireries Hospitalières de St.-Amour
Saint-Germain-en-Laye	Apothicairerie royale—Bibliothèque municipale
Salins-les-Bains	Apothicairerie de l'Hôpital
Seurre	Hôtel-Dieu
Sèvres	Bureau International des Poids et Mesures
Thoissey	L'Apothicairerie de l'Hôpital
Tournus	Hôtel-Dieu, Musée Greuze
Trevoux	L'Apothicairerie de Trevoux
Troyes	Apothicairerie de l'Hôtel-Dieu-le Comte
Yssingeaux	Musée de la Pharmacie à Tence
Germany	
Albstadt-Onstmettingen	Philipp-Mathäus-Hahn-Museum
Albstadt-Onstmettingen	Riedschule: Sammlung Jenemann, Sammlung Danell
Amberg	Museum für Metrologie
Annaberg-Buchholz	Adam-Ries-Museum
Bad Münstereifel	Apotheken-Museum
Bad Sulza	Salinen- und Heimatmuseum
Bad Winsheim	Fränkisches Freilandmuseum
Balingen	Museum für Waage und Gewicht
Bechhofen	Museum für Wiegen und Messen
Bickendorf	Museum von freher
Bönnigheim	Die Arzney-Küche in Bönnigheim
Brekendorf	HAGET collection
Burg	Burger Museumsapotheke
Clausthal-Zellerfeld	Die Zellerfelder Bergapotheke
Cottbus	Brandenburgisches Apothekenmuseum
Dresden	Dresdner Pharmaziegeschichte im GEHE-Dauerausstellungsbereich
Dresden	Mathematisch- Physikalischer Salon im Zwinger
Dortmund	Apothekenmuseum in Ausbüttel's Adler-Apotheke
Emmerich	Museum für Prüf- und Dampfmaschinen
Frankfurt am Main	Geldmuseum der Deutschen Bundesbank
Frankfurt am Main	Historische Museum
Frankfurt am Main	Museum der Hoechst AG

Table 7.5 (Continued)

Frankfurt am Main	Sammlung Jenemann in der DECHEMA
Gardelegen	Stadtmuseum
Geithain	Heimatmuseum
Giessen	Liebig-Museum
Göttingen	Städtisches Museum
Göttingen	Waagensammlung des Physikalischen Instituts der Universität
Halberstadt	Städtisches Museum im Spiegelschen Palais
Hamburg	Deutsches Zollmuseum
Heidelberg	Deutsches Apothekenmuseum
Heidelberg	Kurpfälzisches Museum
Hennef	Hennefer Waagenwanderweg [264, 265]
Hofbiber-Kleinsassen	Pfundsmuseum
Hofgeismar	Apothekenmuseum
Ingolstadt	Deutsches Medizinhistorische Museum
Kaiserslautern	Waagen-Steitz
Kiel	Medizin- und Pharmaziehistorische Sammlung der Christian-Albrechts-Universität
Köln	Kölnisches Stadtmuseum
Konz	Freilichtmuseum Roscheider Hof
Leipzig	Sächsisches Phamaziemuseum
Leverkusen	Bayer Pharma-Museum
Molfsee	Schleswig-Holsteinisches Freilichtmuseum
München	Deutsches Museum
München	Pharmazie im Klinikum rechts der Isar
München	Bayrisches Landesamt für Maß und Gewicht
Nastätten	Heimatmuseum 'Blaues Ländchen'
Neuötting	Stadtmuseum
Nürnberg	Germanisches Museum
Obernkirchen	Berg- und Stadtmuseum
Osterode	Museum im Ritterhaus
Oschatz	Stadt- & Waagenmuseum
Plön	Museum des Kreises Plön
Rhede	Rhede Museum für Geschichte der Medizin und Pharmazie
Schiltach	Apotheken-Museum
Seligenstadt	Ehemalige Benediktinerabtei
Solingen	Bergisches Museum Schloss Burg an der Wupper
Untermerzbach	Collection Peter Ulrich
Wachenheim	Waagenmuseum
Wismar	Das Schabbelhaus
Würzburg	Apotheke der Stiftung Juliusspital

Table 7.5 (Continued)

Greece	
Chorio/Symi	The old pharmacy
Hungary	
Budapest	Pharmacy museum
Budapest	Golden Eagle Pharmacy Museum
Budapest	Semmelweis medical historical Museum
Eger	Telekessy Pharmacy Museum
Gyôr	Museum Pharmacy
Köszeg	Köszegi Patikamúseum
Sopron	Pharmacy museum
Székesfehérvár	The Fekete Sas Pharmacy Museum
Iceland	
Seltjarnarnes	The Pharmacy Museum—Lyfjafraedisafnid
Seltjarnarnes	The History of Medicine Museum—Nesstofusafn
Italy	
Brixen/Bressanone	Pharmaziemuseum Brixen
Campogalliano MO	Museo della Bilancia
Faenza	Museo Internationale della Ceramiche
Figline Valdarno	'La Spezeria' Serristori Hospital
Firenze	The Museum of History of Science—The Evolution of Pharmacy
Loreto	Museo Pinacoteca della Santa Casa
Parma	Storica Spezeria di San Giovanni Evangelista
Pavia	Farmacie storiche di Pavia
Roccavaldina	16th century Pharmacy
Roma	Historical Museum of Medical Art
Sansepolcro (AR)	Museo delle Erbe
Torino	La farmacia chimica Algostino e Demichelis
Trastevere	Farmacia di S. Maria della Scala
Villacidro (Cagliari)	Il Farmamuseo di Villacidro
Japan	
Gifu	The Naito Museum of Pharmaceutical Science and Industry
Lithuania	
Kaunas	The Museum of the History of Lithuanian Medicine and Pharmacy
Vieksniai	Museum of the Pharmacy
Luxemburg	
Luxembourg	Musée National d'Histoire et d'Art

Table 7.5 (Continued)

The Netherlands	
Amsterdam	Rijksmuseum
Haarlem	Frans Halsmuseum
Leeuwarden	Fries Museum
Leeuwarden	Museum het Princessehof
Leiden	Museum Boerhaave
Leiden	De Lakenhal
Naarden-Vesting	Weegschaal Museum
Norway	
Elverum	Glomdal Museum
Halden	Halden Historiske Samlinger
Hamar	The Hedmark Museum and Domkirkeodden
Lillehammer	The Sandvig Collections
Oslo	Norsk Folkemuseum, Norsk Farmasihistorisk Museum
Svelvik	Svelvik Museum
Trondheim	Tröndelag Folkemuseum
Poland	
Kraków	Jagellonian University—Muzeum Farmacji
Kraków	Muzeum Książąt Czartoryskich
Lublin	PZF 'CEFARM' Apteka Muzeum, Lublin, Poland
Poznan	Pharmacy Museum
Warszawa	Glówny Urzad Miar
Portugal	
Lisboa	Museo da Farmácia da ANF
Puerto Rico	
San Juan	Pharmacy Museum
Romania	
Cluj	The Pharmacy History Collection
Sibiu	The Museum of History of Pharmacy
South Africa	
Durban	Old Court House Museum
Johannesburg	The Adler Museum of the History of Medicine
Ladybrand	The Catharina Brand Museum
Spain	
Barcelona	Museu farmàcia catalana
Burgos	Botica de la abadía cisterciense de Santo Domingo de Silos
Castellón	Museo Provincial de Farmacia—Colegio Oficial de Farmacéuticos

Table 7.5   (Continued)

Girona	Museo-farmacia del Hospital de Santa Caterina
Leoia	Museo Vasco de Historia de la medicina 'José Luis Goti'
Llivia	Museu de Farmàcia de Llivia
Madrid	Museo de la Farmacia Hispana, Universitad Complutense
Madrid	Museo de Farmacia Militar
Madrid	Palacio Real
Valldemosa/Mallorca	Museum pharmacy in the Carthusian monastery
Vitoria-Gasteiz	Museo de Farmacia
Virtual museum	Las colecciones del Museo de la Farmacia Hispanica
Virtual museum	Museo Virtual de la Sanidad
Switzerland	
Basel	Historisches Museum, Pharmaziemuseum of the University
Engelberg	Collection Thomas Meierhofer
Luzern	Alte Suidtersche Apotheke
Zürich	Schweizerisches Landes-Museum
Zürich	Medizinhistorische Sammlung
Slovakia	
Bratislava	Pharmaceutic exhibition of the City Museum
Sweden	
Borås	History of Medicine museum
Gothenburg	The History of mdicine Museum in Oterdal House
Helsingborg	The Pharmacy Museum, Fredriksdal Friluftsmuseum
Helsingborg	Leo Nostalgicus
Köping	The Köpings Pharmacy Museum—Scheele's Memory
Linköping	The Pharmacy Museum in Old Linköping
Lund	History of medicine Museum at Lund
Lund	The Swan Pharmacy
Malmö	The Lion Pharmacy
Mariestad	Vadsbo Museum
Mönsterås	The Pharmacy Museum
Skara	Museum of the History of Veterinary Medicine
Skara	The Skara Pharmacy
Stockholm	The History of Pharmacy Museum
Stockholm	Friluftsmuseet Skansen, The Crown Pharmacy
Stockholm	Eugenia History of Medicine Museum
Stockholm	The Stork pharmacy
Stockholm	The Raven pharmacy
Stockholm	The Owl pharmacy
Stockholm	Scales collection of the Academy of Sciences

Table 7.5   (Continued)

Uppsala	History of Medicine Museum
Uppsala	Linné's Hammarby
Vadstena	The Vadstena pharmacy
Vindeln	The Vindeln pharmacy
Västerås	County History of Medicine Museum
Västerås	Vallby Friluftsmuseum

Ukraine

Kyiv	Pharmacy Museum
Lvov	Antique Pharmacy Museum

United Kingdom

Birmingham	Avery Berkel Museum
Bradford on Avon	Bradford on Avon Museum
Edinburgh	Victorian Pharmacy Museum
London	Science Museum
London	Museum of the Royal Pharmaceutical Society of Great Britain
London	Chelsea Physics Garden
Lydney Gloucester	The Scales Museum
Norwich	The Bridewell Museum

USA

Ada, OH	Pharmacy Museum, Ohio Northern University
Albany, NY	Throop Pharmacy Museum
Alexandria, VA	The Stabler-Leadbetter Apothecary Museum
Amarillo, TX	Texas Pharmacy Museum, Texas Tech School of Pharmacy
Ashland, NE	Ashland Pharmacy Museum
Bellingham, WA	Fairhaven Pharmacy Museum
Buffalo	Pharmacy museum, Cooke Hall, Univ. of Buffalo North Campus
Canterbury, NH	Shaker-by med 'Physician's Botanical Garden'
Chokoloskee, FL	The Store and Museum (incl. the old pharmacy)
Dover, DE	Delaware Museum of Small Town Life
Guthrie, OK	Oklahoma Frontier Drug Store Museum
Indianapolis, IN	Hook's Historic Drug Store and Pharmacy Museum
Manchester, CT	Pharmacy Museum
Minneapolis, MN	Owen H. Wangenstein Historical Library of Biology and Medicine
Morgantown, WV	Cook-Hayman Pharmacy Museum
Monroe, NY	Museum Village
New Orleans, LA	New Oleans Pharmacy Museum—La Pharmacie Française
Oklahoma City, OK	Lemuel Dorance Museum, University of Oklahoma
Owing Mills, MD	The B. Olive Cole Pharmacy Museum
Philadelphia, PA	Pennsylvania Hospital, History of Medicine Museum

Table 7.5 (Continued)

Philadelphia, PA	Kendig Memorial Museum
Pittsburgh, PA	The Elmer H. Grimm Sr. Pharmacy Museum
Quincy, IL	Pharmacy Museum of Quincy & Adams County
Sacramento, CA	The Donald Salvatori California Pharmacy Museum
St. Peter, MN	The Drugstore Museum
Tucson, Arizona	The History of Pharmacy Museum, University of Arizona
New Braunfels, TX	Sophienburg Museum & Archives
Washington D.C.	Armed Forces Medical Museum
Weimar, TX	Heritage Society Pharmacy Museum of Weimar
Virgin Islands	
Christiansted	Christiansted's Apothecary Hall

library filled its stacks with new works in mathematics, astronomy, physics, natural sciences and other subjects. Each ship attending the harbour of Alexandria was obliged to deliver all its scripts to the library which made copies which then were included in the inventory. Plutarch (AD 46–120) wrote that Caesar accidentally burned the library down during his visit to Alexandria in 48 BC. However, this version is not confirmed in contemporary accounts. It has been reasonably established that the library or parts of the collection were destroyed on several occasions. Finally the whole building was destroyed, there exist no image and it is still unknown where it was situated.

Figure 7.39 shows the new Bibliotheca Alexandrina opened in 2002 near the site of the Old Library. In contrast to the old Royal Library the acquisition of books is censored. Publications from Israel or those which are not in agreement with Islam are not included.

Prior to the 17th century, the most significant works on mechanics were written in the 4th century BC by the Greek philosopher Aristotle (Ἀριστοτέληζ) (384–322 BC): *Mechanics*, *On the Heavens*, and *The Nature*. Indeed, Aristotle's Μηχανικά—*Mekhanika* [28] probably was not written by him [42] but by Archytas of Tarentum (428–350) [43]. Archytas of Tarentum was a Greek mathematician, political leader and philosopher, active in the first half of the 4th century BC. He was born in Tarentum, Magna Graecia (now Italy) [44].

In the section on the lever both the force and the load are seen as weights and accordingly as directly comparable variables. The relationship of moved weights to moving weights is in reverse proportion to the relationship of the lengths of the corresponding lever arms to each other. This means that the further a force is from the pivot of the lever, the easier it can move something. Similarly he had a rough and tacit notion of centre of gravity. Besides the use of levers as a tool or a primitive machinery also tasks are discussed like "Why is a balance beam harder to move when laden than when empty?"

The theory of the balance originates from Archimedes [266]. Archimedes of Syracuse (Ἀρχιμήδηζ; ∼287 BC–∼212 BC) was a Greek mathematician, physi-

Fig. 7.39 The new
Bibliotheca Alexandrina,
Alexandria, Egypt

cist, engineer, inventor, and astronomer [45]. Only few details of his life are known. He proved experimentally the statements in the *Mekhanika*. Most of his books are lost but may be reconstructed on the basis of reports of Heron [46] and Pappos of Alexandria. In the book *On the Equilibrium of Planes* (two volumes) he explains the *Law of the Lever*, stating, "Magnitudes are in equilibrium at distances reciprocally proportional to their weights." Archimedes uses the principles derived to calculate the areas and centres of gravity of various geometric figures including triangles, parallelograms and parabolas.

From Byzantine sources the Arabs learned manufacturing and handling of balances as well as its theory. There are hints that Middle Ages scientists learned mechanics from Arabic books.

In 1586 at the age of 22, Galileo (1564–1642) wrote a short treatise entitled *La Bilancetta* (*The Little Balance*) [69, 70]. He was sceptical of Vitruvius's account of how Archimedes determined the fraud in Hiero's crown and in this treatise presented his own theory based on Archimedes' Law of the Lever and Law of Buoyancy. He included a description of a hydrostatic balance that determined the precise composition of an alloy of two metals.

Today history and progress of development of scales are documented in the large popular and scientific literature. As recent standard works may be regarded Kochsiek and Gläser's *Comprehensive Mass Metrology* [267] and Jenemann's *The Chemist's Balance* [83]. The Proceedings of the International Conferences on Vacuum Microbalance Techniques (Table 7.3) [268, 269] register the progress in the development of that special type of balances. In a variety of books the application of such balances is described [86, 270–273] Oscillating crystal balances require special experimental techniques [114]. Several detailed reviews on microbalances are published by Emich [88, 274], Gorbach [89], Cunningham [90], Behrndt [91], and Benedetti-Pichler [275]. Today many books are already digitalized by Google and several new ones are published directly in the Internet. In Wikipedia much information on balances is available, though, of course, some are also erroneous.

In many countries national libraries had been founded in which the complete literature published in books is gathered of the respective country or language. Besides libraries exist which are specialised on techniques worldwide, e.g. the Technische Informationsbibliothek—Universtiätsbibliothek Hannover (Germany) [276].

H.R. Jenemann gathered catalogues and other documents in particular of German precision scales manufacturers [267–269]. That substantial documentation together with his specialised library on the history of balances is now in the Riedschule, Albstadt-Onstmettingen (Germany). Similarly Max Danell collected documents in particular of the precision scales companies in the Swabian Alp (Germany) (Hartner, Kern, Sauter, Kistler, Ernst & Schaal, Boss, Haigis, Bosch). Copies are now in the Wirtschaftsarchiv Baden-Württemberg at the University of Hohenheim and in the Riedschule, Albstadt-Onstmettingen. Richard Vieweg (1896–1972), formerly president of the Physikalisch-Technische Bundesanstalt, Braunschweig, Germany, photographed balances elsewhere. His collection of slides of scientific instruments is now in the archive of the PTB library.

Literature lists are depicted in the Appendix.

7.4 Scientific Societies

Societies concerned with mass determination are listed in Table 7.4. ISWM, the International Society of Weighing and Measurement are the trade association for weighing and measurement industry professionals. It has nearly 800 members in the United States and 27 countries around the world. ISWM members include manufacturers and dealers/distributors of weighing equipment, state weights and measures officials and industry end users.

The International Confederation for Thermal Analysis and Calorimetry (IC-TAC) has the aim of promoting international understanding and cooperation in these branches of science through activities within its Scientific Committees, congresses to exchange information, and coordinating efforts towards standardisation and nomenclature.

It was not only necessary to adapt the balance to the vacuum technique, but the vacuum apparatus had also to be modified for the requirements of the incorporated highly sensitive force transducer. Many disturbances together with the sophisticated microbalance techniques gave rise to the desire for a forum for discussions. Thus, the International Conferences on Vacuum Microbalance Techniques (ICVMT) (Table A.6) were started in 1960 by K.H. Behrndt, W.E. Boggs, C.N. Cochrane, A.W. Czanderna, J. Efimenko, E.A. Gulbransen, R.D. Hampson, J.M. Honig, BC Johnson, O.M. Katz, S. Kosiba, A.D. Magnuson, D.B. Medved, T.N. Rhodin, P.M. Rodriguez, R. Schwoebel, R.F. Walker, S.P. Wolsky, E. Zdanuk and A.C. Zettelmoyer. Since that time, ad-hoc conferences were organised by interested participants, without sponsorship by a scientific society or an interested commercial enterprise. During the 1972 Conference, a Steering Committee was formed to provide continuity, with A.W. Czanderna and E. Robens as co-chairman. Since 1970, the conferences were organised alternatively in Europe and the USA. At the 20th Conference at Plymouth (which was planned to be definitively the last one [277]) Al Czanderna retired; up to that time the conferences were organised alternatively in Europe and Northern Africa. Following the 29th Conference at Middlesbrough in 2001, S.A.A. Jayaweera was elected as co-chairman and the Steering Committee re-organised. In 2011 E. Robens resigned.

The UK Weighing Federation act on regulatory matters emerging at UK, EC and International level for scales manufacturers. For that purpose it holds relations and continuous dialogue with all levels of Government agencies and professional bodies. The society promotes and improves the understanding of modern weighing technology e.g. by regional meetings.

The principal objective of CECIP is to represent at European level the weighing industry and its member organisations. Founded in 1958, CECIP has today 15 member associations: Austria, Czech Republic, France, Germany, Hungary, Ireland, Italy, Netherlands, Poland, Romania, Russia, Slovak Republic, Spain, Switzerland and United Kingdom. Together these associations represent around 400 companies. The secretariat of CECIP is located in Brussels Several societies unify scale collectors.

Table 7.6 Societies concerned with balance work

AECIP	C/Viladomat, 174, E-08015 Barcelona, Spain	www.aecip.es/
AFCAT	Association Francaise de Calorimétrie et d'Analyse Thermique	http://www.afcat.org/
AICAT	Associazione Italiana di Calorimetria e Analisi Termica	http://www.aicat.net/IT/
ASIP	Association Suisse pour les instruments de pesage, c/o Mettler Toledo AG, Im Langacher, CH-8606 Greifensee, Switzerland	
ASTM	Committee E37 on Thermal Measurements	http://www.astm.org/commit/committee/E37.htm
ATAS	Australian Thermal Analysis Society	http://www.ictac.org/ stephen.clarke@flinders.edu.au
ABRATEC	Brazilian Association for Thermal Analysis and Calorimetry	valterjr@gasol.com.br, vjfj@superig.com.br
	Arbeitskreis Waagenbau Nord	http://www.waagenbau-nord.de/ waagenbau-nord@metallhandwerk.net
	Association Interregional Manufacturers of Weighing Equipments. 15A, Pirogovskay nab., Saint Petersburg, 194044, Russia	www.zerkalospb.ru/
AZET	Unia Vaharov SR, Sladkovicova 532/55955 01 Topolcany, Slovak Republic	www.azet.sk/
BIPM	Bureau International des Poids et Mesures, France	http://www.bipm.org/
BSTAC	Bulgarian Society of Thermal Analysis and Calorimetry, Bulgary	pelovsky@uctm.edu
CALCON	American Calorimetry Conference, USA	www.calcon.org
CECIP	CECIP, Diamant Building, Boulevard Auguste Reyers 80, BE-1030 Brussels, Belgium	http://www.cecip.eu/
CERN	European Organization for Nuclear Research, Meyrin, Switzerland	www.cern.ch
China	Chinese Group for Thermal Analysis, China	cqy@csu.edu.cn xhli@mail.csu.edu.cn
CTAS	Canadian Thermal Analysis Society, Canada	http://www.ctas.org/
Czech	Czech Working Group for Thermal Analysis	kovarova@imc.cas.cz
DAkkS	Deutsche Akkreditierungsstelle	http://www.dakks.de/
EA	The European co-operation for Accreditation	http://www.european-accreditation.org/
Egypt	Egyptian ICTAC group	a_rehim@hotmail.com

Table 7.6 (Continued)

EURAMET	European Association of National Metrology Institutes	http://www.euramet.org/
EVITERM	European Virtual Institute for Thermal Metrology	www.evitherm.org
GECAT	Grupo Español de Calorimetría y Análisis Térmico	falisar1@usc.es
CECIP	European Federation of Weighing Machine Manufacturers	http://www.cecip.eu/
GEFTA	Gesellschaft für Thermische Analyse e.V.	http://www.gefta.org/
GMVV	Gewichten en Maten Verzamelaars Vereniging, Netherlands	www.gmvv.org
HSTA	Hellenic Society for Thermal Analysis	lalia@chem.auth.gr
HUNTAC	Hungarian Thermoanalytical Group	cs-novak@mail.bme.hu
ICVMT	International Conferences on Vacuum Microbalance Techniques	erich.robens@t-online.de saajayaweera@yahoo.co.uk
IFAC	International Federation of Accountants	http://ifac.org/; http://www.vdi.de/
ISASC	International Society of Antique Scale Collectors	http://www.isasc.org/index.html
ISO	International Organisation for Standardization	www.iso.ch/
IGTA	Israel Group for Thermal Analysis	yarivs@vms.huji.ac.il
IWA	Irish Weighing Association Limited c/o ISME 17, Kildare Street, Dublin 2	http://www.isme.ie/
ITAS	Indian Thermal Analysis Society	http://www.itasindia.org/
ICTAC	International Confederation for Thermal Analysis and Calorimetry	http://www.ictac.org/
ISASC	International Society of Antique Scale Collectors	http://www.isasc.org/index.html
ISWM	International Society of Weighing and Measurement	http://www.iswm.org/
JSCTA	Japan Society of Calorimetry and Thermal Analysis	http://wwwsoc.nii.ac.jp/jscta/e/index.html
M&G	Maß & Gewicht, Gemany	http://www.mass-und-gewicht.de
NATAS	North American Thermal Analysis Society	http://natasinfo.org/
NOSTAC	The Nordic Society for Thermal Analysis & Calorimetry	P.H.Larsen@risoe.dk
OIML BIML	International Organization of Legal Metrology, Paris, France	http://www.oiml.org/
PTB	Physikalisch Technische Bundesanstalt	www.ptb.de/
Poland	Section on Thermal Analysis of the Polish Mineralogical Society	alan@pgi.waw.pl
PTKAT	Polish Society of Calorimetry and Thermal Analysis	malecki@agh.edu.pl

Table 7.6 (Continued)

Romania	Romanian Thermal Analysis Group	popescu@dwi.rwth-aachen.de
Russia	National Committee of Russia on Thermal Analysis and Calorimetry	
SATAS	Southern African Thermal Analysis Society	cstrydom@postino.up.ac.za
Slovakia	Slovak Group For Thermal Analysis & Calorimetry	simon@cvt.stuba.sk
SFT	Société Française de Thermique	www.sft.asso.fr
Singapore		ChinHow@sp.edu.sg
STK SGTK	Schweizerische Gesellschaft für Thermische Analyse und Kalorimetrie	http://www.stk-online.ch/
	Swiss Society for Thermal Analysis & Calorimetry	danielle.giron@pharma.novartis.com
SPWAG	Stowarzyszenie Producentów Wag w Polsce	http://spwag.org.pl/
TAWN	Dutch Society for Thermal Analysis, The Netherlands	http://home.wanadoo.nl/tawn/home-en.htm http://www.chem.uu.nl/ctg
TMG	Royal Society of Chemistry, Thermal Methods Group, UK	http://www.thermalmethodsgroup.org.uk/
UCISP/ Anima	Italian Association of Scale and Weighing Equipment Manufacturers Via A. Scarsellini 13, I-20161 Milano, Italy	www.ssuu.com/federazioneanima/
UFID— AMC	Mures Street n° 63, Bucharest, Romania	
UKWF	UK Weighing Federation, Federation House, 10 Vyse Street, B18 6LT, UK	http://www.ukwf.org.uk/
UVV	Unie výrobcùvah Ćeské republiky	www.uvvcr.cz/
VDMA	VDMA/Fachverband Waagen, Lyoner Strasse 18, D-60498 Frankfurt am Main, Germany	http://machines-for-testing.com/
WKO	Mechatroniker, Schaumburgergasse 20/4, A-1040 Wien, Austria	http://portal.wko.at/
VLW	Vereniging Nederlandse Leveranciers van Weeginstrumenten, Dodeweg 6B, NL-3830 AK Leusden	www.weeginstrumenten.nl

References

1. R.E. Oesper, Some famous balances. J. Chem. Educ. **17**, 312–323 (1940)
2. *Roger Bacon*, in *Stanford Encyclopedia of Philosophy*. http://plato.stanford.edu/entries/roger-bacon/. Stanford University (2007)
3. N. Cusanus, Nicolai de Cusa opera omnia. Gesamtausgabe der Heidelberger Akademie (Felix Meiner, Hamburg, 2008)

4. N. Cusanus, Idiota de Staticis Experimentis, Dialogus. Codex Cusanus 1456/64 1450, Straßburg Folio 135r
5. A.L. de Lavoisier, Traité élémentaire de Chimie. Paris (1789)
6. J.A. Chaptal, *Elements de Chymie. Trois volumes* (Deterville, Paris, 1796)
7. K.E. Haeberle, *10 000 Jahre Waage* (Bizerba, Balingen, 1966)
8. H.R. Jenemann, Zehntausend Jahre Waage? Teil 1. Maß und Gewicht. Zeits. Metrologie **21**, 470–487 (1992)
9. H.R. Jenemann, Zehntausend Jahre Waage? Teil 2. Maß und Gewicht. Zeits. Metrologie **22**, 509 (1992)
10. G.V. Childe, The prehistory of science: archaeological documents, Part 1, in *The Evolution of Science, Readings from the History of Mankind*, ed. by G.S. Metraux, F. Crouzet (New American Library/Mentor Books, New York, 1963), pp. 66–67
11. E. Robens, R.S. Mikhail, The ancient Egyptian balance. Thermochim. Acta **82**(1), 63–80 (1984)
12. E. Robens, The ancient Egyptian balance—Part II. Thermochim. Acta **152**, 243–248 (1989)
13. S.R.K. Glenville, Weights and balances in ancient Egypt. Nature **137**, 890–892 (1936)
14. F.G. Skinner, *Weights and Measures—Their Ancient Origins and Their Development in Great Britain up to AD 1855* (Science Museum, London, 1967)
15. W.M.F. Petrie, A Season in Egypt. London (1888)
16. W.M.F. Petrie, *Prehistoric Egypt* (British School of Archaeollogy in Egypt and Bernhard Quaritch, London, 1920)
17. W.M.F. Petrie, *Ancient Weights and Measures*. London (1926)
18. R. Hall, Assistant curator of the Petrie Collection. Personal communication. London, 1980
19. W.M.F. Petrie, *History of Egypt—from the earliest Kings to the XVIth Dynasty*, 11th edn. London (1924)
20. T.G.H. James, *An Introduction to Ancient Egypt* (British Museum, London, 1979)
21. F.A. Hassan, J. Near East. Stud. **39**, 203–207 (1980)
22. T. Schneider, *Lexikon der Pharaonen* (Albatros, Düsseldorf, 2002)
23. C.B. Ramsey et al., Radiocarbon-based chronology for dynastic Egypt. Science **328**, 1554 (2010)
24. T. Willwerding, *History of Dentistry 2001* (Creighton University, Creighton, 2001)
25. *Handbuch der Ägyptologie*, vol. 6 (Harrasowitz, Wiesbaden, 1966)
26. F.M. Feldhaus, Die Technik der Vorzeit, der geschichtlichen Zeit und der Naturvölker. Leipzig (1914)
27. W.S. Hett (ed.), *Aristoteles, Minor Works* (Heinemann, London, 1836)
28. Aristoteles, a.t. (ed.), Questiones mechanicae. *Kleine Schriften zur Physik und Metaphysik*, ed. by P. Gohlke, Paderborn (1957)
29. J.G. Wilkinson, *The Manner and Customs of the Ancient Egyptians*, vol. 2. London (1879)
30. C.H. Massen et al., Investigation on a model for a large balance of the XVIII Egyptian dynasty, in *Microbalance Techniques*, ed. by J.U. Keller, E. Robens (Multi-Science Publishing, Brentwood, 1994), pp. 5–12
31. H.R. Jenemann, Über altägyptische Kleinwaagen und artverwandte Wägeinstrumente. Technikgeschichte **62**, 1–28 (1995)
32. W. Westendorf, *Erwachen der Heilkunst* (Artemis & Winkler, Zürich, 1992)
33. H. Ducros, Étude sur les balances égyptinnes. Ann. Service des antiquités de l'égypte **9**, 32–53 (1908)
34. H. Ducros, Deuxième étude sur les balances égyptiennes. Ann. Service des antiquités de l'égypte **10**, 240–253 (1910)
35. E. Robens, C.H. Massen, J.A. Poulis, Untersuchungen an einem Modell für eine große Waage der XVIII. Ägyptischen Dynastie, in *Ordo et Mensura*, ed. by D. Ahrens, R.C.A. Rottländer (Scripta Mercaturae, St. Katharinen, 1995), pp. 130–143
36. A. Neuburger, Die Technik des Altertums. Leipzig (1919)
37. O. Spiegler, Die Bestimmung der Gleichgewichtslage an ägyptischen Balkenwaagen. Bull. Soc. Hist. Metrol. Jpn. **1**(2), 59–63 (1979)

38. H.R. Jenemann, Über die Aufhänge- und Arretierungsvorrichtung der ägyptischen Waage der Pharaonenzeit. Berichte zur Wissenschaftsgeschichte **11**, 67–82 (1988)
39. H.R. Jenemann, E. Robens, Indicator system and suspension of the old Egyptian scales. Thermochim. Acta **152**, 249–258 (1989)
40. F. Szabadváry, in *History of analytical chemistry*, ed. by R. Hahn. Classics in the History and Philosophy of Scienve Series (Pergamon Press, Oxford, 1993)
41. H.R. Jenemann, Über Ausführung und Genauigkeit von Münzwägungen in spätrömischer und neuerer Zeit. Trier. Z. **48**, 163–194 (1985)
42. G. Elert, The Physics Hypertextbook—A Work in Progress. http://hypertextbook.com/physics/ (1998–2008)
43. T.N. Winter, The Mechanical Problems in the Corpus of Aristotle. Classics and Religious Studies, Faculty Publications, Classics and Religious Studies Department. University of Nebraska, Lincoln (2007)
44. C.A. Huffman, *Archytas of Tarentum: Pythagorean, Philosopher, and Mathematician King* (Cambridge University Press, Cambridge, 2005)
45. T.L. Heath, L.G. Robinson, The Works of Archimedes. http://www.archive.org/details/worksofarchimede029517mbp (2001)
46. A.G. Drachmann, Fragments from Archimedes in Heron's mechanics. Centaurus **8**, 91–146 (1963)
47. Archimedes, *The Works of Archimedes, §7. About Swimming Bodies*, vol. 1 (Wissenschaftliche Verlagsbuchhandlung, Frankfurt am Main, 1987)
48. M. Vitruvius Pollio, De Architectura, vol. II/III. Rome (33–14 BC)
49. R.C. Mackenzie, Thermochim. Acta **75**, 251–306 (1984)
50. H. Birkhan, *Kelten – Versuch einer Gesamtdarstellung Ihrer Kultur* (Verlag der Österreichischen Akademie der Wissenschaften, Wien, 1997)
51. H.R. Jenemann, Die Geschichte der Waage im Mittelalter. NTM Internationale Zeitschrift für Geschichte und Ethik der Naturwissenschaften, Technik und Medizin **3**, 145–166 (1995)
52. J. Al-Khalili, in *The Golden Age of Arabic Science*, ed. by P. Books (Allen, Harlow, 2010)
53. H.R. Jenemann, The development of the determination of mass, in *Comprehensive Mass Metrology*, ed. by M. Kochsiek, M. Gläser (Wiley/VCH, Berlin, 2000), pp. 119–163
54. H. Bauerreiß, *Zur Geschichte des spezifischen Gewichts im Altertum und Mittelalter* (Universität Erlangen, Erlangen, 1913/1914)
55. F. Sezgin, Al-Chazini's Balance of Wisdom. Institut für Geschichte der Arabisch-Islamischen Wissenschaften, Johann Wolfgang Goethe-Universität, Frankfurt am Main (2000)
56. L. Phryesen, O. Brunfels, *Spiegel der artzney* (Balhasar Beck, Strassburg, 1532)
57. F. Hoefer, Histoire de la Chimie. Paris (1866)
58. G. Agricola, *De Re Metallici Libri XII. Zwölf Bücher vom Berg- und Hüttenwesen*, 4th edn. (VDI-Verlag, Düsseldorf, 1556)
59. G. Agricola, De mensuris et ponderibus. Basel (1550)
60. H. Menzel-Rogner, *Nikolaus von Cues: der Laie über Versuche mit der Waage*, 2nd edn. Philosophische Bibliothek, vol. 220 (Meiner, Leipzig, 1944)
61. E. Robens, C.H. Massen, J.J. Hardon, Studies on historical gravimetric hygrometers. Thermochim. Acta **235**, 125–133 (1994)
62. L.B. Alberti, L'architettura. Padua, Firenze (1483/1485)
63. L.B. Alberti, M. Theurer, *Zehn Bücher über die Baukunst*. Wien, Heller (1912)
64. E. Gerland, F. Traumüller, Geschichte der physikalischen Experimentierkunst. Leipzig (1899)
65. L. da Vinci, Codex atlanticus—Saggio del Codice atlantico, ed. by Aretin. Vol. fol. 249 verso-a + fol. 8 verso-b. Milano (1872)
66. S. Stevin, De Beghinselen der Weeghconst (Principles of the Art of Weighing). Leyden (1585)
67. S. Stevin, *De Waegdaet*. Leyden (1586)
68. S. Stevin, *De Beghinsel des Waterwichts*. Leyden (1586)
69. G. Galilei, *La Bilancetta*. Pisa (1588)

70. L. Fermi, G. Bernardini, *Galileo and the Scientific Revolution* (Basic Books, New York, 1961)
71. F. Brunetti (ed.), Opera di Galileo Galilei. Unione tipografico-editrice torinese (UTET). Torino (1980)
72. I. Newton, *Philosophiae Naturalis Principia Mathematica*. London (1687)
73. I. Newton, *The Principia: Mathematical Principles of Natural Philosophy Trans. I* (University of California Press, Berkeley, 1999)
74. J. Leupold, Theatrum staticum – das ist: Schauplatz der Gewichtskunst. Theatrum staticum universale. Leipzig (1726)
75. L. Euler, Disquisitio de Billancibus. Commentari Acadimiae Scientiarium Imperialis Petropolitanae **10** (1738/1747)
76. L. Auger, *Un Savant Méconnu, Gilles Personne de Roberval* (Blanchard, Paris, 1962)
77. R. Hooke, *De Potentia Restitutiva or of Spring, Explaining the Power of Springing Bodies.* Lectiones Cutleriana or a Collection of Lectures: Physical, Mechanical, Geographical, & Astronautical. Early Science in Oxford, vol. VIII (Gunther, London, 1678/9)
78. R. Hooke, Lectures de potentia restitutiva or of spring. Explaining the power of springing bodies, Tract VI of: R. Hooke: Lectiones Cutleriana or a Collection of Lecturs Physical, Mechanical, Geographical, & Astronomical. Early Science in Oxford: The Cutler Lectures of Robert Hooke, ed. by R.T. Gunther, vol. VIII. Oxford & London (1679)
79. H.R. Jenemann, Robert Hooke und die frühe Geschichte der Federwaage. Ber. Wiss.gesch. **8**, 121–130 (1985)
80. W.J. 's Gravesande, Physices elementa mathematica, experimentis confirmata, sive introductio ad philosophiam Newtonianam. Leiden (1720)
81. H.R. Jenemann, Das Kilogramm der Archive vom 4. Messidor des Jahres 7: Konform mit dem Gesetz vom 18. Germinal des Jahres 3? in *Genauigkeit und Präzision*, ed. by D. Hoffmann, H. Witthöfft (Physikalisch-Technische Bundesanstalt, Braunschweig 1996), pp. 183–213
82. H.R. Jenemann, Zur Geschichte des langarmigen Balkens von Präzisionswaagen. Maß und Gewicht - Zeitschrift für Metrologie **29**(3), 672–687 (1994)
83. H.R. Jenemann, *Die Waage des Chemikers—the Chemist's Balance* (DECHEMA, Frankfurt am Main, 1997)
84. H.R. Jenemann, The early history of balances based on electromagnetic or electrodynamic force compensation, in *La Massa e la Sua Misura—Mass and Its Measurement—Proceedings of International Congress 1993*, Modena, Italy, ed. by L. Grossi (CLUEB, Bologna, 1995), pp. 9–20
85. H.R. Jenemann, The early history of balances based on electromagnetic and elektrodynamic force compensation, in *Microbalance Techniques*, ed. by J.U. Keller, E. Robens (Multi-Science Publishing, Brentwood, 1994), pp. 25–53
86. T. Gast, T. Brokate, E. Robens, Vacuum weighing, in *Comprehensive Mass Metrology*, ed. by M. Kochsiek, M. Gläser (Wiley/VCH, Weinheim, 2000), pp. 296–399
87. T. Gast, The Impact of control on the development of recording balances. J. Thermal Analysis (1998)
88. F. Emich, Einrichtung und Gebrauch der zu chemischen Zwecken verwendbaren Mikrowaagen, in *Handbuch der biochemischen Arbeitsmethoden*, ed. by E. Abderhalden. Berlin/Wien (1919), pp. 55–147
89. G. Gorbach, Die Mikrowaage. Mikrochemie **20**(2/3), 236–254 (1936)
90. B.B. Cunningham, Microchemical methods used in nuclear research. Nucleonics **5**(5), 62–85 (1949)
91. K.H. Behrndt, Die Mikrowaagen in ihrer Entwicklung seit 1886. Z. Angew. Phys. **8**(9), 453–472 (1956)
92. C. Duval, *Inorganic Thermogravimetric Analysis* (Elsevier, Amsterdam, 1953)
93. S. Iwata, *Über die Entwicklung der Thermowaage, besonders in Japan* (Chemischen Institut der Universität Bonn, Bonn, 1961)
94. S. Iwata, *Soil-Water Interaction*, 2nd edn. (Dekker, New York, 1995)

95. C. Keattch, *An Introduction to Thermogravimetry* (Heyden/Sadtler, London, 1969)
96. C.J. Keattch, *The History and Development of Thermogravimetry* (University of Salford, Salford, 1977)
97. C.J. Keattch, Studies in the history and development of thermogravimetry. J. Therm, Anal. Cal. **44**(5) (1995)
98. C. Eyraud, E. Robens, P. Rochas, Some comments on the history of thermogravimetry. Thermochim. Acta **160**, 25–28 (1990)
99. C. Eyraud, P. Rochas, Thermogravimetry and silk conditioning in Lyons. A little known story. Thermochim. Acta **152**, 1–7 (1989)
100. W.F. Hemminger, K.-H. Schönborn, A nineteenth century thermobalance. Thermochim. Acta **39**, 321–323 (1980)
101. K. Honda, On a thermobalance. Sci. Rep. Tôhoku Univ., Sendai Ser. **1**(4), 97–103 (1915)
102. K. Honda, Kinzoku no Kenkyu **1**, 543 (1924)
103. E. Salvioni, *Misura di masse compresa Fra g* 10^{-1} *e g* 10^{-6} (University of Messina, Messina, 1901)
104. F. Emich, Ein Beitrag zur quantitativen Mikroanalyse. Monatsh. Chem. **36**(6), 407–440 (1915)
105. J.W. McBain, Z. Phys. Chem. **68**, 471–497 (1909)
106. J.W. McBain, *Sorption of Gases by Solids*. London (1932)
107. J.W. McBain, A.M. Bakr, J. Am. Chem. Soc. **48**, 690–695 (1926)
108. E.H. Weber, E.W. Weber, *Wellenlehre, auf Experimente gegründet*. Leipzig (1825)
109. G. Sauerbrey, Wägung dünner Schichten mit Schwingquarzen. Phys. Verhandl. **8**, 193 (1957)
110. G. Sauerbrey, Verwendung von Schwingquarzen zur Wägung dünner Schichten und zur Mikrowägung. Z. Phys. **155**, 206–222 (1959)
111. A.W. Warner, C.D. Stockbridge, Mass and thermal measurements with resonating crystalline quartz, in *Vacuum Microbalance Techniques*, ed. by R.F. Walker (Plenum, New York, 1962), pp. 71–92
112. W.H. Wade, L.J. Slutsky, Adsorption on quartz single crystals, in *Vacuum Microbalance Techniques*, ed. by R.F. Walker (Plenum, New York, 1962), pp. 115–128
113. W.H. King Jr., Applications of the quartz crystal resonator, in *Vacuum Microbalance Techniques*, ed. by A.W. Czanderna (Plenum, New York, 1971), pp. 183–200
114. C. Lu, A.W. Czanderna, *Applications of Piezoelectric Quartz Crystal Microbalances* (Elsevier, Amsterdam, 1984)
115. R. Berger, J. Gutmann, R. Schäfer, Scanning probe methods: from microscopy to sensing. Bunsenmagazin **2**, 42–53 (2011)
116. T.P. Burg et al., Weighing of biomolecules, single cells and single nanoparticles in fluid. Nature **446**(4), 1066–1069 (2007)
117. T. Ibel, *Die Wage Im Altertum und Mittelalter* (Friedrich-Alexander-Universität, Erlangen, 1908)
118. J.E. Quibell, *Excavations at Saqqara 1911–1912. The Tomb of Hesy* (Institut Français d'Archéologie Orientale, Cairo, 1913)
119. P. Walden, *Maß, Zahl und Gewicht in der Chemie der Vergangenheit*. Stuttgart (1931)
120. M.J. Schiefsky (ed.), Even without math, ancients engineered sophisticated machines. http://www.fas.harvard.edu/home/news_and_events/releases/math_10012007.html. Archimedes Project. Havard University, Cambridge, MA (2009)
121. Avery-Weigh-Tronix, The History of Weighing. http://www.wtxweb.com/. Avery Weigh-Tronix (2008)
122. Archimedes, *The Works of Archimedes*. Dover
123. China-Window, Chinese Steelyard—Gancheng. http://www.china-window.com/china_culture/china_culture_essentials/chinese-steelyard-ganchen.shtml. China Window (2008)
124. K. Simonyi, *Kulturgeschichte der Physik*, 2nd edn. (Harry Deutsch, Thun, 1995)
125. Al-Chazini, Buch der Waage der Weisheit. Merw (1120)
126. N. Khanikoff, Analysis and extract of the book of the balance of Wisdom—an Arabic work of the water balance, written by Al-Chazini in the 12th century. J. Am. Orient. Soc. **6**, 1–128

(1860)

127. W. Gould, Philos. Trans. **220**, 156, 304 (1683/4)
128. W. Gould, Acta Erud., 317 (1685)
129. H. Cavendish, Experiments to determine the density of the Earth. Philos. Trans. R. Soc. Lond. (Part II) **88**, 469–526 (1798)
130. H. Cavendish, Experiments to Determine the Density of the Earth, in *Scientific Memoirs*, vol. 9: The Laws of Gravitation, ed. by A.S. MacKenzie (American Book Co., 1900), pp. 59–105
131. Wikipedia, Dezimalwaage. http://de.wikipedia.org/wiki/Dezimalwaage (2011)
132. W. Weber, Über drei neue Methoden der Konstruktion von Waagen, in *Wilhelm Weber's Werke*. Berlin (1892), pp. 489–496
133. J. Plücker, Poggendorfs Annalen **91**, 1 (1854)
134. L.G. Gouy, C. R. Acad. Sci. Paris **109**, 935 (1889)
135. Wagen im luftabgeschlossenen Raume, Bericht über die wissenschaftlichen Apparate auf der Londoner Internationalen Ausstellung (1876), p. 223 ff.
136. H.V. Regnault, Morin, Brix, Rapport sur les comparaisons qui ont été faites à Paris en 1859 et 1860 de plusieurs kilogrammes en platine et en laiton avec le kilogramme prototype en platine des Archives Impériales. Berlin (1861)
137. P. Bunge, Centr. Ztg. Optik u. Mechanik **5**, 220–225 (1884)
138. P. Stückrath, Waage zur graphischen Aufzeichnung veränderlichen Gewichtes. Z. Instrum.kd. **3**, 95–99 (1883)
139. E. Warburg, T. Ihmori, Über das Gewicht und die Ursache der Wasserhaut bei Glas und anderen Körpern. Ann. Phys. **263**, 481–507 (1886)
140. W. Kent, Kent's torsion balance. Sci. Am. **24**(601) (1887). Supplement
141. K. Ångström, Tvåmetronomiska hjälpapparater. Oefversigt af Kongl. Vetenskap-Akademiens Foerhandlingar **52**, 643–655 (1895)
142. K. Ångström, in *Eine Wage zur Bestimmung der Stärke magnetischer Felder*, ed. by P. Carl. Repertorium der Physik, vol. 25 (1889), pp. 383–387
143. W. Nernst, Z. Elektrochem. **9**, 622 (1903)
144. N. Nernst W. Nachr. Königl. Ges. Wiss. Göttingen (1903)
145. W. Nernst, E.H. Riesenfeld, Chem. Ber. **36**, 2086 (1903)
146. J. Giesen, Ann. Phys. **10**(4), 830 (1903)
147. O. Brill, Z. Anorg. Chemie **45**, 275 (1905)
148. P. Truchot, Rev. Chim. Pure Appl. **10**(2) (1907)
149. D. Steele, K. Grant, Proc. R. Soc. Lond., A **82**, 580 (1909)
150. D. Steele, K. Grant, Proc. R. Soc. Lond., A **86**, 270 (1912)
151. D. Konstantinowsky, Naturwissenschaften **6**, 429, 448, 473, 488 (1918)
152. P. Pascal, Mesure des susceptibilités magnétiques des corps solides. C. R. Hebd. Séances Acad. Sci. **150**, 1054–1056, 1514 (1910)
153. C. Chéneveau, Philos. Mag. **20**, 357 (1910)
154. W.H.F. Kuhlmann, Neue Wage zum automatischen photographischen Registrieren des Gewichtsverlustes. Der Mechaniker **18**(13), 146–147 (1910)
155. P. Bunge, Katalog der Kollektiv Ausstellung der deutschen Präzisions-Mechnik und Optik. Brüssel (1910)
156. G. Urbain, C. Boulanger, Sur une balance-laboratoire à compensation électromagnetique à l'étude des systèmes qui dégagent des gaz avec une vitesse sensible. Compt. Rend. **154**, 347–349 (1912)
157. R.W. Gray, W. Ramsay, Proc. R. Soc. Lond., A **86**, 270 (1912)
158. H. Pettersson, A new micro-balance and its use. Diss. Stockholm, in *Göteborg's Vet. of Vitterh. Samhalle's Handlinger*. Göteborg (1914)
159. H. Pettersson, Experiments with a new micro-balance. Proc. Phys. Soc. Lond. **32**, 209–221 (1919)
160. F.W. Aston, Proc. R. Soc. Lond., A **80**, 439 (1914)
161. K. Honda, Proc. Phys. Math. Soc. Jpn. **2**, 92 (1920)
162. M. Guichard, Bull. Soc. Chim. Fr. **33**, 258 (1923)

163. J.R.N. Coutts et al., An automatic and continuous recording balance (The Odén-Keen-balance). Proc. R. Soc. Lond., A **106**, 33–51 (1924)
164. H. Saito, in *Records of Proc. Imp. Acad. of Japan on the 188. General Meeting* (1925)
165. H. Saito, Sci. Rep. Tohoku University **16**, 37 (1927)
166. Z. Shibata, M. Fukushima, Kinzoku no Kenkyu **4**, 108 (1927)
167. K. Inamura, Kinzoku no Kenkyu **4**, 25 (1927)
168. J. Aharoni, F. Simon, Z. Phys. Chem., B **4**, 175 (1929)
169. R. Strömberg, Adsorptionsmessungen mit einer verbesserten Mikrowaage. K. Sven. Vetensk.akad. Handl. **3**(6), 33–122 (1928)
170. Y. Oshima, Y. Fukuda, J. Chem. Soc. Jpn, Ind. Chem. Sect. **33**, 733 (1930)
171. J. Donau, Über eine neue Mikrowaage. Mikrochemie **9**(1), 1–14 (1931)
172. J. Donau, Zur neuen Mikrowaage. Mikrochemie **13**(1), 155–164 (1933)
173. Skramowsky, Collect. Trav Chim. Czechoslov. **5**, 6–9 (1933)
174. C. Rigollet, Diploma of Higher Studies, vol. 552. Paris (1934)
175. P. Dubois, Bull. Soc. Chim. Fr. **3**, 1178 (1935)
176. T. Mochida, Kinzoku no Kenkyu **13**, 415 (1936)
177. Z. Shibata, Z. Niwa, J. Chem. Soc. Jpn., Pure Chem. Sect. **57**, 1309 (1936)
178. P. Chevenard, X. Waché, R. de la Tullaye, Bull. Soc. Chim. Fr. **10**, 41 (1944)
179. F.T. Holmes, Rev. Sci. Instrum. **8**, 444 (1937)
180. J.W. Beams, Magnetic suspension for small rotors. Rev. Sci. Instrum. **21**, 182–184 (1950)
181. R. Vieweg, T. Gast, Registrierende Mikrowaage für Diffusionsmessungen an Kunststoff-Membranen. Kunststoffe **34**, 117–119 (1944)
182. H. Barret, W. Birnie, M. Cohen, J. Am. Chem. Soc. **62**, 2839 (1940)
183. H.M.A. Oyama, Y. Nakaji, Rep. Elechtrochemical Lab. 50 Commeration Number (1942), p. 158
184. E.A. Gulbransen, K.F. Andrew, An enclosed physical chemistry laboratory: the vacuum microbalance, in *Vacuum Microbalance Techniques*, ed. by M.J. Katz (Plenum, New York, 1961), pp. 1–21
185. E.A. Gulbransen, Rev. Sci. Instrum. **15**, 201 (1944)
186. S.J. Gregg, M.F. Wintle, J. Sci. Instrum. **23**, 259 (1946)
187. T. Gast, E. Alpers, Ponderometrische Bestimmung dielektrischer Größen. Z. Angew. Phys. **1**, 228–232 (1948)
188. J.W. Clark, An electronic analytical balance. Rev. Sci. Instrum. **18**, 915–918 (1947)
189. Z. Hara, Seisan Kenkyu **1**, 94 (1949)
190. T.N. Rhodin, Discuss. Faraday Soc. **5**, 213 (1949)
191. T.N. Rhodin, Adv. Catal. **5**, 39, 53 (1953)
192. K. Kinjo, S. Iwata, Records at the 2. Meeting. The Agency of Ind Sci. and Techn. (1950)
193. K. Kinjo, S. Iwata, J. Chem. Soc. Jpn., Pure Chem. Sect. **74**, 642 (1953)
194. H. Straubel, Naturwissenschaften **42**, 506 (1955)
195. D.R. Glasson, J. Chem. Soc., 1506–1510 (1956)
196. J.W. Beams, Phys. Rev. **78**, 471 (1950)
197. K. Kinjo, S. Iwata, J. Chem. Soc. Jpn., Pure Chem. Sect. **72**, 958 (1951)
198. W.A. de Keyser, Nature **172**, 364 (1953)
199. C. Eyraud, I. Eyraud, Laboratoires **12**, 13 (1955)
200. C. Eyraud, I. Eyraud, Catalogue, in *50e Expos. Soc. Fr. Physique* (1953)
201. T. Hironne, S. Maeda, N. Tsuya, Sci. Res. Inst. Tohoku University **6**, 67 (1954)
202. S.J. Gregg, J. Chem. Soc., 1438 (1955)
203. N. Shikazone, T. Kanayama, Rep. Gov. Chem. Ind. Res. Inst. Tokyo **51**, 275 (1956)
204. F. Paulik, J. Paulik, L. Erdey, Z. Anal. Chem. **160**, 241 (1958)
205. D.A. Powell, Rev. Sci. Instrum. **34**, 225 (1957)
206. A. Blazek, Hutnické Listy **12**, 1096 (1957)
207. C. Campbell, S. Gordon, C. Smith, Anal. Chem. **3**(7), 1188 (1959)
208. R.H. Müller, Anal. Chem. A, 77–80 (1960)
209. R. Splitek, Hutnickè Listy **8**, 697 (1958)

210. T. Tomonari et al., J. Electrochem. Soc. Jpn. **26**, 485 (1958)
211. E.S. Freeman, B. Carroll, J. Phys. Chem. **62**, 394 (1958)
212. T. Gast, Registrierendes Wägen im Milligrammbereich und seine Anwendung auf die Staubmessung, in *DECHEMA-Monographien*, ed. by K. Bretschneider, K. Fischbeck (DECHEMA, Frankfurt am Main, 1960), pp. 1–19
213. T. Gast, Bull. Schweiz. Elektrotech. Ver. **53**, 1061–1069 (1962)
214. T. Gast, Microweighing in vacuo with a magnetic suspension balance, in *Vacuum Microbalance Techniques*, ed. by K.H. Behrndt (Plenum, New York, 1963), pp. 45–54
215. W.-D. Emmerich et al., Comments on the papers: "Some comments on the history of vacuum microbalance techniques" and "Some comments on the history of thermogravimetry" by Eyraud, C., Robens, E., Rochas, P. Thermochim. Acta **254**, 391–392 (1995)
216. M. Kungi, T. Yamate, Yoyuen **2**(1), 99 (1959)
217. J.G. Rabatin, C.S. Card, Anal. Chem. **31**, 1689 (1958)
218. P.L. Waters, Anal. Chem. **32**, 852 (1960)
219. H. Schöneck, W. Wanninger, Chem. Ing. Tech. **32**, 409 (1960)
220. L. Cahn, H.R. Schultz, in *Vacuum Microbalance Techniques*, ed. by R.F. Walker (Plenum, New York, 1962), pp. 7–18
221. L. Cahn, H.R. Schultz, The Cahn recording GRAM electrobalance, in *Vacuum Microbalance Techniques*, ed. by K.H. Behrndt (Plenum, New York, 1963), pp. 29–44
222. G. Sandstede, E. Robens, Automatisierte Apparatur zur gravimetrischen Bestimmung der spezifischen Oberfläche und der Porengröße. Chem. Ing. Tech. **34**(10), 708–713 (1962)
223. J.A. Poulis, J.M. Thomas, Sensitivity of analytical balances and relevance of fluctuation theory, in *Vacuum Microbalance Techniques*, vol. 3, ed. by K.H. Behrndt (Plenum, New York, 1963), pp. 1–14
224. T. Gast, Vak.-Tech. **16**, 41 (1965)
225. E. Robens, G. Sandstede, Anordnungen zur präzisen Druckmessung und -regelung im Bereich von 0, 1 bis 760 Torr. Vak.-Tech. **16**, 125–130 (1967)
226. D. Büker, T. Gast, Kontinuierliche gravimetische Staubmessung durch mechanische Resonanz. Chem. Ing. Tech. **39**(16), 963–966 (1967)
227. T. Gast, Microweighing in vacuo with the aid of vibrations of a thin band, in *Vacuum Microbalance Techniques*, ed. by C.H. Massen, H.J. van Beckum (Plenum, New York, 1970), pp. 105–107
228. K.P. Zinnow, J.P. Dybwad, Pressure of light used as restoring force on an ultramicrobalance, in *Vacuum Microbalance Techniques*, ed. by A.W. Czanderna (Plenum, New York, 1971), pp. 147–153
229. J. Rodder, in *Vacuum Microbalance Techniques*, ed. by A.W. Czanderna (Plenum, New York, 1971), pp. 43–53
230. A.W. Czanderna et al., Photoelectrically automated, bakeable, high-load ultramicrobalance. J. Vac. Sci. Technol. **13**, 556–559 (1976)
231. T. Gast, A device for simultaneous determination of mass and reaction force of a gas stream from a heated sample, in *Thermal Analysis*, ed. by H.G. Wiedemann (Birkhäuser, Basel, 1972), pp. 235–241
232. G. Böhme et al., Determination of relative weight changes of electrostatically suspended particles in the sub-microgram range, in *Progress in Vacuum Microbalance Techniques*, ed. by S.C. Bevan, S.J. Gregg, N.D. Parkyns (Heyden, London, 1973), pp. 169–174
233. G. Böhme et al., Messungen von Gewichtsänderungen an im elektrischen Feld frei schwebenden Teilchen. Sprechsaal **106**, 184–188 (1973)
234. C.H. Massen et al., Application of micro balances to the measurement of gas pressure over eight decades, in *Thermal Analysis*, ed. by H.G. Wiedemann (Birkhäuser, Basel, 1962), pp. 225–233
235. H.G. Wiedemann, Application of thermogravimetry for vapor pressure determination. Thermochim. Acta **1**(3), 355–366 (1972)
236. H.G. Wiedemann, G. Bayer, Comparison of temperature measurements in the range of 400–2500 K by use of a thermobalance, in *Progress in Vacuum Microbalance Techniques*, ed. by

C. Eyraud, M. Escoubes (Heyden, London, 1975), pp. 103–107

237. T. Gast, in *Progress in Vacuum Microbalance Techniques*, ed. by C. Eyraud, M. Escoubes (Heyden, London, 1975), pp. 108–113

238. J.G. Dunn, S.A.A. Jayaweera, S.G. Davies, Proc. - Australas. Inst. Min. Metall. **290**, 75–82 (1985)

239. T. Gast, R. Talebi-Daryani, Selbstkompensierend Waage zur kontinuierlichen Messung der Gasdichte. Meß- u. Regelungstechnik Vol. 7010-11-TUB. Berlin, Technische Universität (1985)

240. C.C.J. French, Advanced techniques for engine research and design. J. Automob. Eng. **203**(D3), 169–183 (1989)

241. K. Horn, Verfahren und Meßeinrichtung zur Bestimmung mechanischer Meßgrößen, insbesondere eines unbekannten Gewichts, Patentschrift DE 3743897 C2. Germany (1987)

242. K. Horn, To make load-cell-scales settle faster, in *XI IMEKO Congress*, vol. 2 (1988), pp. 217–243

243. H. Wente, *Ein Beitrag zur intelligenten Dämpfung schwingungsfähiger elastischer Systeme unter besonderer Berücksichtigung elektromechanischer Waagen*. Fortschritts-Berichte VDI, vol. 11 (VDI-Verlag, Düsseldorf, 1992)

244. C.H. Massen et al., Computer simulation of balance handling. J. Therm. Anal. Calorim. **55**(2), 367–370 (1999)

245. C.H. Massen et al., Optimizing of balances of the second generation. J. Therm. Anal. Calorim. **55**(2), 449–454 (1999)

246. P. Poncheral et al., Electrostatic deflections and electromechanical resonances of carbon nanotubes. Science **283**, 1513–1516 (1999)

247. V.M. Mecea, Fundamentals of mass measurement. J. Therm. Anal. Calorim. **86**(1), 9–16 (2006)

248. V.M. Mecea, Is quartz crystal microbalance really a mass sensor? Sens. Actuators A, Phys. **128**(2), 270–277 (2006)

249. M. Block et al., Direct mass measurements above uranium bridge the gap to the island of stability. Nature **463**, 785–788 (2010)

250. D. Rodríguez et al., MATS and LaSpec: high-precision experiments using ion traps and lasers at FAIR. Eur. Phys. J. Spec. Top. **183**, 1–123 (2010)

251. G. Binning, H. Rohrer, Scanning tunnelling microscopy. IBM J. Res. Dev. **38**, 4 (1986)

252. Wikipedia, Museum. http://en.wikipedia.org/wiki/Museum (2009)

253. E. Robens, S. Kiefer, The balance in the museum. J. Therm. Anal. Calorim. **101**(2), 737–769 (2010)

254. Wikipedia, Library of Alexandria. http://en.wikipedia.org/wiki/Ancient_Library_of_ Alexandria (2009)

255. A. Primisser, *Die kaiserlich-königliche Ambraser-Sammlung* (Heubner, Wien, 1819)

256. M. Caygill, Creating a Great Museum: Early Collectors and The British Museum. http:// www.fathom.com/course/21701728/index.html. The British Museum (2009)

257. F. Klemm, Geschichte der naturwissenschaftlichen und technischen Museen. Abhandlungen und Berichte - Deutsches Museum **41**(2) (1973)

258. Maß und Gewicht, http://www.mass-und-gewicht.de/, in *Maß und Gewicht – Zeitschrift für Metrologie*

259. PerBos Farmacihistoriska Sidor, http://home.swipnet.se/PharmHist/

260. M. Hass, Scales and Weights. A collection of historical scles and weights from different periods of the past 3000 years. http://www.s-a-w.net/ (2010)

261. E. Robens, S. Kiefer, Die Sammlung Jenemann ist nun in Onstmettingen. Mass Gewicht **89**, 2195–2196 (2009)

262. E. Robens, Hans Richard Jenemann. J. Therm. Anal. Calorim. **55**(2), 707 (1999)

263. E. Robens, Inis Jenemann (20.10.1933–19.11.2008). J. Therm. Anal. Calorim. **101**(2), 807 (2010)

264. W. Euler, Hennefer Waagen-Wanderweg. chronos begleitheft-online.pdf. Hennef: METAP Metrology/Bürgermeister von Hennef (2008)

265. E. Gareis, W. Euler, The Hennef Weigher-Walking-Way. chronos begleitheft-online.pdf. Hennef: METAP Metrology/Bürgermeister von Hennef (2008)
266. A. Czwalina (ed.), *Über das Gleichgewicht ebener Flächen; Über schwimmende Körper.* Archimedes Werke, ed. by A. Czwalina. Darmstadt (1983)
267. M. Kochsiek, M. Gläser (eds.), *Comprehensive Mass Metrology* (Wiley/VCH, Berlin, 2000)
268. E. Robens, A.W. Czanderna, The conferences on vacuum microbalance techniques. Thermochim. Acta **51**(1), IX–XI (1981)
269. G.W. Chądzyński, E. Robens, S.A.A. Jayaweera, Preface. J. Therm. Anal. Calorim. **86**(1), 1–6 (2006)
270. S.P. Wolsky, E.J. Zdanuk, *Ultra Micro Weight Determination in Controlled Environments* (Wiley, New York, 1969)
271. A.W. Czanderna, S.P. Wolsky, *Microweighing in Vacuum and Controlled Environments* (Elsevier, Amsterdam, 1980)
272. R.S. Mikhail, E. Robens, *Microstructure and Thermal Analysis of Solid Surfaces* (Wiley, Chichester, 1983)
273. T. Gast, T. Brokate, E. Robens, Vakuumwägung, in *Massebestimmung*, ed. by M. Kochsiek, M. Gläser (VCH, Weinheim, 1996), pp. 294–399
274. F. Emich, Einrichtung und Gebrauch der zu chemischen Zwecken verwendbaren Mikrowaagen, in *Handbuch der biologischen Arbeitsmethoden*, ed. by E. Abderhalden. Berlin/Wien (1921), pp. 183–269
275. A.A. Benedetti-Pichler, Waagen und Wägungen, in *Handbuch der Mikrochemischen Methoden*, ed. by F. Hecht, M.K. Zacherl (Springer, Wien, 1959)
276. Technische Informationsbibliothek – Universitätsbibliothek. http://www.tib.uni-hannover. de/, Hannover
277. S.A.A. Jayaweera (ed.), *Proceedings of the International Conference on Vacuum Microbalance Techniques, Plymouth 1983.* Thermochim. Acta (Elsevier, Amsterdam, 1984)

Chapter 8
Weighing Scales Manufacturers

The balance was already a widely used instrument in antiquity and standardised types had been elaborated. Therefore specialised workshops for the fabrication of instruments and tools including balances must have existed. We have no names of those factories, but a representation was found in Pompeii (destroyed 24. August 79 AD) (Fig. 8.1). It is not known to the authors whether any Roman and other early European scales had been labelled by the manufacturer, and the same applies worldwide. With beginning of the 14th century, precision balances were used for the analysis of ores, causing increasing demand.

In the 18th century specialised workshops for balances existed (Fig. 8.2). Usually both, balances for trade and industry and precision balances, had been side products. Precision balances were made by instrument manufacturers founded often by clockmakers. Best balances were offered by famous instrument workshops located in the Netherlands, London, and Paris and later in Berlin. Important impetus for further development was given by chemistry and adoption of the metric convention. By co-operation of mechanics and scientists the relative resolution was improved to 10^{-5}.

In the 19th century in Western Europe a large number of factories were known and new ones were established which produced scales for trade and industry, as well as high performance balances for laboratories and standardisation bureaus. Centres of manufacture were established e.g. at London, Amsterdam, Vienna, Berlin, Hamburg Cologne, in the Schwäbische Alb (Suabian Jura) and in the Bergisches Land.

With the beginning of the industrial age a large number of balance types and force sensors were in operation. A considerable number of factories specialised in the production of measuring instruments and some of them only with balances and force sensors. Today, on the world market manufacturers from Japan, the USA, China and India compete with European companies. Most of them produce balances with a variety of other instruments and other technical products. Only few companies cover the whole range of demands on mass determination.

It should be emphasised that even at the turn towards the 20th century balances had been handmade articles, built in relatively small instrument workshops, which usually had other workshops (joiner, glazier) and home worker as supplier. Centrally driven transmission for lathes had not been installed. Later on electrically powered

E. Robens et al., *Balances*, DOI 10.1007/978-3-642-36447-1_8,
© Springer-Verlag Berlin Heidelberg 2014

(a)

(b)

Fig. 8.1 (**a**) Frieze with Erotes, Casa Vettii, house of Aulus Vettius Conviva and Aulus Vettius Restitus, Pompeii, Italy 79 AD. Several types of balances are depicted and putti manufacturing and applying balances. (**c**) Roman smith with balance

Fig. 8.1 (Continued)

(c)

machines came into use. Assembly lines had been installed in large, specialised factories first in the middle of the 20th century.

8.1 Centres of Balance Manufacturing

In the following section examples of some regions with concentrated precision balance production and some balance manufacturing companies there are reviewed.

8.1.1 England

For centuries, shipping made up to about 95 % of transport for the entire world trade. Since the 18th century nations of the British Commonwealth were leading in shipping. The shipping industry is still the underlying supporter of almost all other industries. In accordance with the leading power of British industry also British instrument makers had a high level of output [1, 2].

Jesse Ramsden (1735–1800) was one of the most prominent manufacturers of scientific instruments in the latter half of the eighteenth century. To own a Ramsden instrument, was to own not only a high level instrument of great practical use, but also a thing of beauty [3]. Besides instruments for surveyors, barometers etc. he

Fig. 8.2 (**a**) Johannes en
Casparus Luiken, Amsterdam
1694: 'De Balanse maker',
Middle Ages scales factory.
Pfundsmuseum, Kleinsassen,
Germany. (**b**) Arte del
'bilanciere', Diderot:
Encyclopédie ou dictionnaire
raissonné 1751

(a)

produced precision scales at his London workshop. At this time the experience was
gained that a lighter beam gives greater sensitivity. However, the beam bends when
the balance is loaded and then the sensitivity is diminished. Ramsden designed a
beam consisting of two hollow cones joined at their base (Fig. 7.22). Ramsden-
beam balances were produced by a number of London instrument makers including
Edward Troughton (1784–1826), Robert Fidler [4]. One example of that time is
Henry Wood—Scale, Weight and Weighing machines Manufacturer, London 1740
(Fig. 8.3a). Later on the firm name was changed to Herbert & Sons. The company
still produces electronic cash balances for warehouses and its history is told in more
detail in the next section.

In the second and third decade of the 19th century Thomas Charles Robinson
(1792–1841) was reputed to be the best mechanic of London [5]. He made balances
with long, punched rhombic beam and maximum load up to kilograms. The balances
of Robinson apparently all had knife edges also for their end bearings. A reticulated
balance beam was described first by J. Hyacinthe Magellan (1722–1790) (a descen-
dent of the Portuguese world sailor Magelhaes) who emigrated to England in 1746
[6]. Magellan also first described optical aids for the observation of the equilibrium.

Arte del «bilanciere»

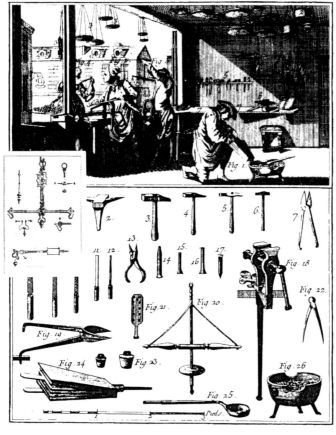

Tavole da L'Encyclopédie o Dictionnaire raisonné des sciences, des arts et des métiers, 1751.

(b)

Fig. 8.2 (Continued)

In the 19th century the firm of L. Oertling at Orpington, Kent, UK was the biggest maker of precision balances in Great Britain [7]. Ludwig Oertling (1818–1891), a younger brother from the second marriage of the Berlin August Oertling's father, set up his workshop in London around 1846, which quickly specialised in the manufacture of fine balances. At the London world exhibition of 1851 L. Oertling displayed a balance with a beam of 3 foot length for 56 lbs. (0.91 m/25.4 kg) capacity, for which he was awarded the Council Medal. A rider weight could be placed on the graduated beam. For this reason L. Oertling is named in English publications as the first designer of a rider mechanism. Indeed a similar device was exhibited already in 1844 by W. Kleiner at Berlin. The company was considered to be one of the best in providing accurate measuring apparatus in the world. This may be demonstrated by an electrodynamic Oertling balance for the measurement of Weston Standard

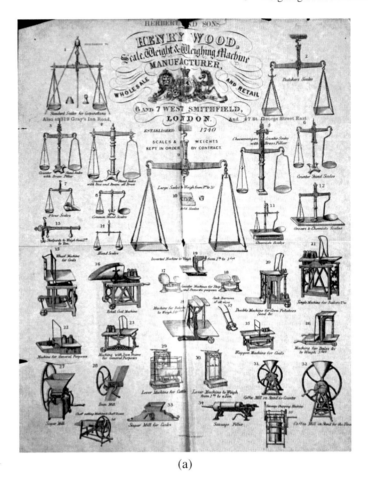

(a)

Fig. 8.3 (**a/b**) Henry Wood, leaflet—Scale, Weight and Weighing Machines Manufacturer, London, UK 1740. (**c**) Rocking horse kitchen scale with Tooth rack for the kilogram region and micrometer screw for movement of the counterweight in the gram region. © Döft, Sontheim, Germany

Cell (Fig. 4.93). (Until recently the Normal Weston Cadmium Cell was used as the standard of voltage.)

In the year 1919 the company, after the son of the founder, Henry Oertling, suffered from ill health and his own son Lewis had fallen in the war, was modified to a Limited Company. In 1925 it was purchased by Avery but kept the Oertling brand name. They subsequently became Avery Berkel which established the standard weighing apparatus in the vast majority of UK retail shops. In 2000 the whole group was merged into the conglomerate of the US company Avery Weigh-Tronix, Fairmont, MN, USA.

In 1946 Henry Morton Stanley and Albert W. Harrington founded Stanton Instruments Ltd. at London [8]. They produced all types of laboratory balances; early instruments have 'STANTON MORDEN' engraved on the beam. Success of the com-

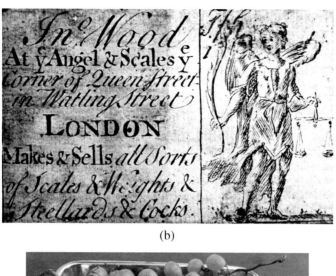

(b)

(c)

Fig. 8.3 (Continued)

pany depended heavily on Harrington's contacts with the science and research establishments, originating from his work during the war, whereas Stanley concentrated on issues of finance and long-term decision making. From 1951 they developed recording thermobalances and in 1953 they introduced the first British thermobalance. Later this was the leading product of Stanton-Redcroft, which was founded in 1965. In 1959 the main works moved to Copper Mill Lane, London-Wimbledon. In 1962 the mass flow thermobalance was introduced. In 1963 John Jamieson, Glasgow and Howard Rawson & Co. had been purchased. In 1966 the commercial DTA (differential thermal analysis) was introduced. In 1968 the entire group of Stanton companies was sold to Avery Ltd. and H.M. Stanley resigned from the board. In 1970 following the integration of the production into L. Oertling Ltd. which was at that time already a company of the Avery group, A.W. Harrington resigned from the board. The Stanton name gradually disappeared from Oertling/Avery range. In 1983

Stanton-Redcroft became part of Thermal Scientific plc which is now Rheometric Scientific.

The Avery Company can trace back its roots to a Birmingham shop of Charles de Grave 1670. In 1730 James Ford established the business as a maker of 'stilliards' [9–11]. The scalemaking business was passed to William Barton, Thomas Beach, and then in 1799 to Joseph Balden, who married Mary Avery. In 1813, as the Industrial Revolution gathered momentum, the workshop was transferred to the brothers William and Thomas Avery, who traded under the name W & T Avery. The business rapidly expanded and in 1885 they owned three factories: the Atlas Works in West Bromwich, the Mill Lane Works in Birmingham and the Moat Lane Works in Digbeth with almost 200 employees. In 1876 the first Avery weighbridge was designed by M.A. Brown. In 1891 the business became a public limited liability company with a board of directors and in 1894 became a public limited company with shares quoted on the London Stock Exchange. In 1895 the company bought the legendary Soho Foundry in Smethwick, a former steam engine factory owned by James Watt & Co. The steam engine business was gradually converted to pure manufacture of weighing machines. By 1914 the company had employed some 3000 employees. In the inter-war period the growth continued with the addition of specialised shops for cast parts, enamel paints and the product range diversified into counting machines, coin machines, testing machines, automatic packing machines and petrol pumps. During World War II the company also produced various types of heavy guns. The profound changes then led to the closure of the foundry, the introduction of load cells and electronic weighing with the simultaneous gradual disappearance of purely mechanical weighing machines. In the 20th century several acquisitions of other companies were made including the balance manufacturers Oertling Ltd. in 1925 and Stanton Redcroft Ltd. in 1968. After almost a century of national and international expansion the company was taken over by General Electric Company plc in 1979. Keith Hodgkinson, managing director at the time, completed the turn-around from mechanical to electronic weighing with a complete overhaul of the product range of retail scales and industrial platform scales. In 1993 GEC took over the Dutch food-slicer manufacturer Berkel and the Avery-Berkel name was introduced. In 2000 the business was in turn acquired by the US-American company Weigh-Tronix, who already owned Salter, and is today operating as Avery Weigh-Tronix.

In 1760 the brothers Richard and William Salter established a workshop for springs and pocket steelyards (spring balances) in a cottage in Bilston, England. Over the next two centuries, these springs were to be used in some of the most important industrial developments, including railways, cars and aeroplanes manufactured in the UK. In 1770 the company moved to West Bromwich, where it was one of the leading employers for more than 200 years. In the 1790s William's sons, John and George Salter took over the business, which also made bayonets. In 1838 George Salter took out the first of many weighing equipment patents, which related to various ways of indicating weight on a spring balance that are still used today. The 1880s was very busy time for Salters, with a product portfolio now including the first coin-operated public weighing machines as well as various food processing equipment, clothes washers, letter-clips, hat hooks and dynamometers. A factory was also opened in Berlin, in association with German firm Ubrig, which led to the development of automatic rotating display and vending machines. In 1895 Salter produced the first English typewriter—the British Empire. Rights to this business were sold in 1936. In 1920 Salter employed more than 1000 people. A repair company was opened in Australia. During World War II, the company manufactured more than 750 million springs for war purposes. In 1973 George Salter & Co. limited was bought by Staveley Industries, and in 2000 merged with Avery Weigh-Tronix.

8.1.2 France

The first balance which could be called a precision balance was made by Pierre Bernard Mégnié for Antoine L. Lavoisier (1743–1794) [12–14]. At the turn towards the 19th century Paris the instrument and globe maker Nicolas Fortin (1750–1831) and his son-in-law Hermann designed a sensitive equal-armed balance type for laboratory work. Later on the mechanic Louis-Joseph Deleuil (1795–1862) founded his own workshop for the production of optical and other physical instruments. He was highly respected for his perfected barometers, electrical batteries, photographic apparatus etc. Since 1830 he also improved the 'balance Fortin'. His son Jean-Adrien (1825–1894) took over the workshop and designed the 'Delieul balance' in collaboration with the chemist Henri-Victor Regnault [15]. For the production and certification of the 'kilogram des archives' as the basis of the new metric system, Deleuil arranged a specialised type of his balance in an iron vacuum compartment: that was the first vacuum balance [16] (Fig. 5.17). Figure 7.29 shows a laboratory balance constructed by Jean-Adrien Deleuil which had been presented in 1874 by Desains at the Académie des Sciences. The sensitivity was 5 mg at a maximum load of 3 kg. It could be used for weighing of voluminous bodies or of long-neck flasks with up to 25 cm diameter. These balances still exist and are exhibited in the Musée des Techniques du Conservatoire des Arts et des Métiers at Paris (France).

8.1.3 Austria

Physics cabinets had been established already in the medieval European universities as well as in monasteries and manorial courts. The golden age of physics cabinets was the 18th century. They stored artistic and rare instruments which were used both for demonstrations in lectures and in research. Such a physics cabinet also existed at the imperial court at Vienna. Instruments including balances and globes were bought or built by the 'Hofmechaniker' which had been clockmaker and optician. During the reign of Maria Theresia 1740–1780 on account of financial shortness the budget of the 'Hofmechaniker' was cancelled. Later on the 'Physikalisches Kabinett' was closed; most of the instruments were distributed to other locations.

Physics cabinets had been maintained also by monasteries functioning additionally as research institutes. Best known is the completely conserved instrument collection and the observatory of the Benedictine Monastery Kremsmünster. Many of the instruments were made by Georg Friedrich Brander (1713–1783) of Augsburg, who was considered at that time as one of the most famous mechanics. A hydrostatic balance of inclination type still exists in the collection [17]. Exhibited is a precision balance made in Vienna and a trade balance is also mentioned. Furthermore, the monastic pharmacy has several magnificent balances, one of them signed with 'Schindler M & M fecit'.

After years of travel the mechanic Johann Christoph Voigtländer (1732–1797) learned carpentry from his father at Leipzig. After years of travel he settled at Vienna and worked as a journeyman at Meinicke, which was the only 'Mechanicus' of Vienna producing physical and mathematical instruments. Then, in 1763 he got the 'Hof-Freyheit' (licence) for similar instruments, because all other master locksmiths of the town had not the necessary equipment. Furthermore money could be saved in reducing imports form Augsburg, Nuremberg or England because there was not any production in Austria at that time. Later on, Voigtländer concentrated his efforts on production of optical equipment and cameras.

In 1768 Josef Florenz (†1791) founded the Florenz Waagen- und Maschinenfabrik at Braunau-on-Inn, and produced balances and weights of all types. First the beam corresponded to the conical Ramsden-type; later on it was formed as flat, oblong rhombic bars of brass sheet, 3–4 mm thick. In 1972 the company was sold to the Carl Schenck AG at Darmstadt. Today the branch is called Schenck Process Austria GmbH with headquarters at Braunau [18]; main products are trade and industrial scales.

In the 19th century at Vienna, as at other places in Middle Europe, an efficient precision mechanical industry came into being [19]. In 1815 the 'k.k. Polytechnisches Institut' at Vienna was founded [20]. The first director Johann Josef Prechtl [21] established a mathematical-physical cabinet which was transformed later on to the institute's workshop. All kinds of scientific apparatus of high level, in particular geodetic and astronomical instruments but only few balances had been created. On account of basic governmental funding private workshops could hardly compete. After complaints of associations of engineering works and of private instruments makers in 1866 that state enterprise was privatised. Georg Christoph Starke (1794–1865), the former head of the workshop continued managing that privatised business, and after his death his son and Gustav Starke (1832–1917) founded together with his brother-in-law, Karl Kammerer the company Starke & Kammerer [22]. In 1937 the company became insolvent.

In 1823 the mechanic C.E. Kraft established his workshop of surveying and other instruments at Vienna. He became a significant figure in the making of precision balances, whose quality was "identical to what the Berlin mechanic Oertling" made.

Ludwig Seyss (1817–1891) ran a workshop, later extended to a factory, at Atzgersdorf near Vienna. Most important product of that company had been sorting machines for coins, but he produced also interesting precision balances. The firm was closed in 1976.

Most important manufacturer of precision balances and weights had been Albert Rueprecht (1833–1913), born at Halle (Saale), Germany. In 1848, after years of travel, he founded his workshop at Vienna, since 1904 called A. Rueprecht & Sohn. His balances with relative long beam had been in the tradition of those produced at Vienna; in particular rider weight displacement and locking mechanism had been similar to such made several decades earlier by J. Kusche, mechanic of the k.k. Polytechnisches Institut. He manufactured many types of balances including pharmaceutical, jewellers, and trade balances, decimal and hydrostatic balances, which of course were improved over the years. In 1878 he delivered a prototype balance for the BIPM, Paris [23]. The conception of that balance most probably was based on a design of Professor Friedrich Arzberger (1833–1905), Brünn and Vienna. (Arzberger's father Johann had a factory for mechanical instruments and was professor at Vienna's Polytechnic Institute). Friedrich Arzberger invented the first air-damping device. The balance was equipped with a weight change mechanism (made by H. Schorss). Locking, adding of weights and weight transposition were performed by means of rods 4 m in length. Reading of deflection was made by means of a reflecting prism and a telescope. With that balance about 1000 weighings had been made; that means that about 30 000 times the knife edges had been in contact with the supports. Such balances, improved in details, had been delivered until 1950s likewise to standardisation bureaus of other states also, which had been members of the metric convention. Variations were used for high-performance chemical tasks. In its zenith at the turn to the 20th century the workshop had about 20 workers. In 1902 more than 10 000 precision scales had been produced and delivered all over the world. In 1904 the founder's son Albert Jr., became associate and after the death of the founder Albert Jr.'s son was introduced. In 2009 in the Dorotheum at Vienna the estate of the 'well known Vienna balance manufacturer Ernst Albert Rueprecht & Sohn', was auctioned.

Another Vienna precision balance workshop of international importance was the 'Institut für Präcisionsmechanik' of Josef Nemetz (*1851), mentioned as 'k.k. Hoflieferant'. Besides telescopes, air and vacuum pumps, barometers, thermometers, hygrometers, he offered in particular precision balances up to a load of 75 kg for pharmacies, schools, and industry. Exceptional was a vacuum prototype balance after demands of Professor Stephan Kruspér and built in 1889 for the central Hungarian Office for Weight and Measures [24]. Though of similar performance in comparison to Rueprecht's prototype balances, the Kruper/Nemetz type hardly gained acceptance. The problems of weighing of mass prototypes in vacuum were connected with adsorption effects which had not been overcome at this time. Another vacuum balance he made for Dimitri Ivanowich Mendeleev at Petersburg (Russia). Nemetz developed devices for manual and automatic remote control of weights and his precision balances had convenience never known before and they can be regarded as being at the top of mechanical instruments. In 1923 Josef Nemetz retired and Jakob Hellmann (1868–1930) and later on his son Hans Hellmann (*1905) became owner of the workshop. In 1936 the company was sold off.

8.1.4 Germany

In comparison with other Western European countries the development of precision mechanics in Germany was poor. In the 18th century only Georg Friedrich Brander (1713–1783) of Augsburg was of some regional importance. He continued in the tradition of famous makers of watches and compasses and produced also other instruments like balances. Later on München (Munich) then became nursery of such industry. The interest in scientific instruments was growing and in the first decades of the 19th century in Germany new workshops for precision instruments were established at Munich, Hamburg, and Berlin [25]. Later on, mostly in close connection with local universities, the centres of instrument production moved to smaller towns like Göttingen, Freiberg, Dresden, Jena, Kassel, Pforzheim, and Celle [26]. By the influence of Justus Liebig (1803–1873) workshops in Giessen [27] went to the top. In particular analytical chemists required sensitive balances for their work. Likewise physicists needed balances, often of special design as well as instruments based on balance techniques, like seismometers. As is typical for an underdeveloped country German workshops produced first copies of foreign instruments (mostly English types) and then they improved them. They conquered the home market and then competed successfully with advanced types in other countries. At the end of the 19th century, Germany was leading in precision mechanics. Scientific instruments manufactured in German workshops enjoyed a high standing worldwide and contributed to the high esteem of the brand 'Made in Germany'. Scientific balances represented an important part of the spectrum of mechanical and optical instruments. The following survey is based essentially on investigations of H.R. Jenemann and S. Kiefer.

In the second half of the 19th century in chemical analysis the demands for precision weighing increased. Above all very small samples of organic products had to be analysed. For that purpose the measuring range of laboratory balances was extended into the microgram range and its resolution was increased. Knife-edges of such sophisticated chemical microbalances were made of improved hard metal and carefully machined. By soft unlocking these were settled on blocks of polished precious or half-precious stone. The notches for rider weights on the beam had been carefully dimensioned. To reduce beam's mass it was extensively punched, and besides brass, aluminium and light metal alloys were used. By means of mirror optics, magnifying glass or telescope the rider had been controlled and deflections of the balance observed. In this way the microgram range was made accessible; however attempts of extension into the sub-microgram range by 'ultra micro' balances failed.

8.1.4.1 Berlin

In 1711 the 'Académie Royale des Sciences et des Belles-Lettres' opened, and in 1744 renewed by Frederick the Great; this had a department for 'philosophie expérimentale'. One of the members was the famous physicist Johann Heinrich Lambert

(1728–1777), who invented an inclination balance with quadrant indication. Therefore at that time precision mechanic workshops should have existed.

In 1804 Johann Georg Tralles (1763–1822) became a member of the Académie. Tralles had been Swiss delegate of the commission which introduced during the French revolution the new system of measures and weights. For precision balances he favoured conical beams, but he remarked first also that a rhomboidal punched beam could be used. Such a relative short beam had been described already for an assay balance in 1781. Tralles designed also new types of hydrometers and hydrostatic balances. The Tralle hydrometer or alcoholometer is named after Johann Tralles; it is still in use. It was graduated so that it read the percentage of alcohol by volume at 15.6 °C. Tralles' hydrostatic balances consisted of three bodies swimming in water of constant temperature. The scale was below the swimming bodies and therefore so also was the centre of gravity. Because there are no mechanical bearings there was no friction. The balance was operated in the substitution mode. Such simple, sensitive and accurate scales with high load capacity had been widely used in the first half of the 19th century for trade and industry. A modern example is a kilogram prototype balance of the Physikalisch-Technische Bundesanstalt, Braunschweig (Germany) [28].

Tralles charged Karl Theodor Nathan Mendelssohn (1782–1852), son of the philosopher Moses Mendelssohn, to construct a balance of highest resolution and accuracy. His balance was equipped with a beam nearly 1 m in length [29] and consisting of two hollow truncated cones connected by a hollow cube as proposed by the English mechanic Jesse Ramsden. (Such a Ramsden balance is exhibited in the Science Museum of London.) The steel knife-edge was supported by an agate plate. The swing time was about 1 minute. In order to shorten the weighing time detent levers were arranged. The relative resolution was 10^{-6}; at that time Mendelson's balance was the most accurate worldwide. For more than a half century the conical beam was standard for precision balances.

All the balances made at that time had been unique examples. This changed in 1813, when the post councillor Heinrich Pistor (1778–1847) founded a mechanic workshop and began to produce instruments in series. His balances covered a range up to 1 kg load and had been designed according to English types and some of them may have been imported. Some of his balances were equipped with rider weights which had been introduced by the Swedish chemists Johann Gottlieb Gahn (1779–1848) and Jöns Jakob Berzelius (1779–1848). Furthermore he used a mirror as the indicator scale, thus avoiding parallax of readings. Several apprentices of Pistor later founded factories on their own and thus he may be regarded as the progenitor of the Berlin precision mechanic industry, though already factories existed for the production of trading scales and similar articles. Since that time in Germany the Berlin instrument workshops became market leaders. Pistor's factory collapsed in 1874.

Johann August Daniel Oertling (1803–1866) was a journeyman of Pistor. In 1826 he founded his own workshop 'August Oertling' and produced scientific instruments and 'chemical' balances with 30, 40 cm and even longer, rhombic lattice shaped beams for maximum loads of 100 g up to several kg. He hoped to improve sensitivity and accuracy in this way. The balances had now arrestment head with shaft and cam as well as the vertical arrestment rod at the centre of the column with two arms at the sides to fix the beam. The centre bearing consists of two small round agate planes, one in front and the other behind the beam, allowing contact to the steel centre knife running across the beam over only a small area. The upper edge of the beam is graduated at both sides into ten divisions, so that this scale serves to determine the position of the manually placed rider. Both ends of the beam transform into end pieces, into the eye on which the pans are hooked. These beam ends, to a degree similar to swan-neck ends of older balances, can be adjusted by means of small screws sideways and vertically, allowing to line up the three axes and (to a large degree) adjust both arms to the same length. Because Pistor's factory changed mainly to the production of other instruments, Oertling's improved balance models went to the top and became worldwide examples. He offered also conditioning apparatus for silk, which are precursors of thermogravimetric apparatus. After August Oertling's death in 1866 the workshop was continued by F. Oertling, probably his son. The Oertling factory was closed at the end of the 1880s, but his name survived as L. Oertling Ltd. at Orpington, Kent, UK. That company was founded by a relative of the Berlin Oertling.

Towards the middle of the 19th century several precision mechanical workshops (including balance workshops) had been founded. One of the most important was that of Ludwig Reimann (*1810) who opened his workshop in 1839. The name Reimann appears in several combinations with other workshops of A. Müller and Horn; obviously there was some cooperation. In 1872 his son Georg Reimann (*1849) followed as head of the factory. He produced several types of precision balances, spring balances, and scales for household use and industry. A special product had been the hydrostatic balance. He succeeded in incorporation of a thermometer into the sinking glass body. Until now the standardised sinking body according to Reimann is in use. For the laboratory balances Georg Reimann changed the design and took over Bunge's short arm beam, however with several significant differences. Furthermore he designed a special type of substitution balances, both for trade and laboratory. Such a balance was equipped with two scales one above the other, one for the sample and the second for weights. The always constant sum of the mass of sample + weight was compensated by a fixed counterweight on the opposite side of the balance beam. In 1910, the factory went bankrupt. During World War I the successor Hugo Rau produced—besides the standard program of balances—sorting machines for bullets and portioning machines for gun powder which had been developed earlier. Such a sorting machine was able to sort up to 1400 bullets by weight. Nevertheless, after the war, attempts of consolidation failed. In 1937 the company was closed.

More balance manufacturers are known from registers of exhibitions and registrations in the Chamber of Industry and Commerce: Alexander Bernstein, Böttcher & Halske, Friedrich Dopp, Hermann Fleischer, A. Hasemann, Hofmann & Eberhardt, W. Kleiner, Reinhold Löhmenn, J.F. Luhme & Co., E. Mentz, Müller, E. Petitpierre, Pintus & Co., Friedrich Plattner, W.J. Rohrbeck. Such factories had mostly been specified on the production of balances for special tasks, e.g. balances for production, analysis and portioning of pharmaceutical or pourable products like cereals, for post offices, for heavy loads, etc.

The most important designer of balances of the 19th century in Berlin was Paul Stückrath (1844–1916). After employment by Siemens and Reimann he opened his workshop in 1887 at Friedenau (now incorporated in Berlin) in which he produced precision balances, automatic coin balances for banks, geodesical and seismological instruments. One of his top products was a microgram balance with triangular aluminium beam. Instead of knife-edges it was equipped with two metal needle points for friction free movements. Indeed, such a design was realised already earlier. It was unfavourable in that the points were damaged quickly by load. The Stückrath point balance was made for maximum load of 1 gram, and deviations due to a mass change of 1 µg could be observed by means of a telescope. A device allowed placing of rider weights on the hanger. Remote control and exchange of weights and load was possible by means of long rods. Of quite different constructions had been prototype balances for the kilogram range equipped with knife edges partially for their use in controlled atmosphere or in a vacuum. Remote control and transposition weighing was performed with some meter long rods. Stückrath constructed also a balance for the determination of the mean density of Earth. For that purpose the Gravitational constant was determined in comparison to 100 000 kg lead mass in a casemate of Spandau's citadel during 1884–1896 [30]. The kilogram balance had one scale below and the other above the lead block; the difference in weight force was ~1.2 mg. Another product of Stückrath was a horizontal pendulum for the measurement of gravitational variations with time [31]. Some of Stückrath's balances and the pendulum are exhibited in the Deutsches Museum at Munich. Furthermore he produced pressure balances [32], sorting machines, automatic gun powder balances and pumps for liquefying of gases. Some balances were equipped for graphical record of weight changes; these may be regarded as forerunners of today's electro-mechanical balances [33]. After Stückrath's death his associate Lambert Lind, and in 1928 Lambert Lind Jr. took over the responsibility. At this time the workshop had about 20 employees; this was regarded as an optimal size for such businesses. In 1957 Lambert Lind died and in 1963 that last important scales workshop at Berlin was closed.

8.1.4.2 Dresden

In 1898 Hermann Göhring and Oskar Hebenstreit founded at Dresden the 'Maschinenfabrik Göhring & Hebenstreit which produced mainly engines for bakeries like waffle automats [34]. After World War II the factory was completely dismantled and moved to Russia. The proprietors moved to West-Germany and re-founded a factory at Mörfelden-Walldorf. Also at Radebeul a new factory was built and soon expropriated by the DDR government. At Dresden-Kadlitz existed the 'Fabrik für Ladeneinrichtungen Albin Lasch & Co.'. A department was engaged in repair of American scales. Wilhelm Richard Lehmann began the production of industrial and retail scales which worked very rapidly and therefore the company's name was changed to 'Rapido-Werke R. Lehmann & Co.'. In 1970 both companies merged. Main production of the 'VEB Wägetechnik Rapido' was now waffle-automats, mechanical and electronic scales and cash dispensers. In 1997 the production of balances was brought to an end.

8.1.4.3 Gießen

With transition from predominately speculative and metaphysical way of thinking and research of the philosophers to experimental research, precision balances became the most important instrument of chemists. For Justus Liebig (1803–1873) as the most important chemist of this time, the balance "was the incomparable instrument, promoting durability to each observation, tells the truth, detects errors, and indicates that we are on the right way."

At the end of the 19th century in Gießen some instrument factories may have existed. Obviously such workshops seem to be on a low level because Professor Georg Gottlieb Schmitt (1786–1837), who published a book in which balance theory was treated extensively [35], ordered balances by the artist Hauff at Darmstadt. Another supplier was master of the mint and mechanic Hector Rößler (1779–1863) of Darmstadt.

At the beginning of the 19th century Paris and London became centres of manufacturing of precision balances. With the development by Liebig of a centre of chemical science at Gießen, the number of workshops there increased which produced scientific instruments and in particular precision scales [27, 36, 37]. Balances from Gießen enjoyed high reputation in the scientific world and this is the case also today. Justus Liebig's balances are exhibited in the Liebig-Museum at Gießen. But already before Liebig scientific instruments including balances had been manufactured. Some balances of Liebig had been constructed by master joiner Konrad Ludwig Hoß (1795–1833). His successor was Carl Staudinger & Co. and the W. Spoerhase Company. In 1957 Mettler-Toledo Greifensee, Switzerland, the leading manufacture of laboratory balances [38–40], bought [40] and extended the Gießen factory to an application laboratory and department store of Germany, whereas the production of balances was stopped.

8.1.4.4 Göttingen

At Göttingen a development of fine mechanical arts took place in the 18th century. Friedrich Apel (1786–1851) was the first mechanic of some importance there. After physical mathematical studies he was engaged as mechanic at the university and ran besides a private workshop. His precision balances had been of Mendelssohn's and Ramsden types; some had been made after proposals of Wilhelm Eduard Weber (1804–1891) who taught there. In cooperation with scientists teaching at the university of Göttingen many improvements were made by Sartorius [41], a company already of industrial size from its start. Today it ranges as number two in the world market of laboratory scales.

8.1.4.5 Hamburg

Starting with the famous instrument workshop of Johann Georg Repsold (1770–1830) Hamburg became a centre of production of scientific instruments. Repsold was well known for his astronomical instruments; besides some precision balances had been manufactured. Carl Friedrich Gauss (1777–1855) used a Repsold balance comparing the pound standards of Hannover and Prussia. Also Carl August Steinheil (1801–1870) used such a balance, when in 1837 calibrating the standard kilogram of Bavaria made of rock crystal with the kilogram prototype at Paris.

In the construction of precision and chemical microbalances the workshop of Paul Bunge, opened in 1866, and his successors at Hamburg, were leaders. Paul Bunge (1839–1888) introduced the short balance beam [42–44] and created the last prototype of equal-armed laboratory scales, which were used until the 1970s [42]. Scaling down of the beam reduced its mass and therefore of the moment of inertia. Precursor to that measure was improved precision of the bearing elements which resulted in better reproduction of the weighing after locking. Optical devices were applied for observing the balance deviation. With some Bunge balances a relative resolution of 5×10^{-11} could be realised [45]. He also produced coin sorting machines.

With loss of his sight Bunge sold his workshop in 1886 to August Theodor Herzberg (1862–1946), who improved the Bunge high level balances with regard to easier and faster operation. In 1939 the workshop had 50 employees and was headed by Herzberg's son-in-law Friedich Wilhem Schmitt (1893–1969). In World War II the Bunge/Herzberg workshop was completely destroyed, but rebuilt afterwards. The introduction of micro-analytical methods, in particular by Friedrich Emich at Innsbruck and Graz, required adequate balances and the micro chemical and the ultra micro balance were created. After the achievement of the single armed and top scale balances the workshop produced also such types, based on earlier patent applications. However, analytical techniques changed and ultra-micro balances were hardly required. In addition mechanical microbalances were replaced by completely different electronic types. The small workshop could not compete with large companies like Sartorius and Mettler, and in 1974 balance production was stopped.

Bunges associate Wilhelm H.F. Kuhlmann (1867–1945) later opened his own workshop. A third workshop of that kind was established by Max Bekel (1857–1937). This workshop was continued by Georg Kurt Walter Eigenbrodt who diversified the production program to other fine mechanical work and industrial scales. In 1920 the workshop Kaiser & Sievers was opened. The company was headed by Friedrich Heinrich Gustav Kaiser (~1880–1949) who dealt with sales and marketing whereas his brother-in-law Georg Sievers (1879–1964) was a mechanic.

Towards the end of the 19th century all those fine-mechanical workshops closed or changed their production program.

8.1.4.6 Heilbronn

In 1875 Franz Schneider founded a precision mechanics workshop for repair of bicycles and sewing machines. In 1890 the company moved to Sontheim and began producing Roberval type household weighing scales, taps and components for the automotive industry. In 1901 the brothers Döft became associates. In 1920 Franz Schneider Jr. took the chair of the company. In 1936 the company was split and the brothers Döft specialised their work into the production of scales at Sontheim, whereas the production of fittings was moved to Nordheim.

Manager Krauth developed household and body scales based on the combination of the steelyard and Roberval principle [46]. He designated his household and baby scales 'microscales' because their sliding weight was equipped with a micrometer screw. Indeed its digital indication was only in grammes, but such scales were possible to be calibrated (Fig. 8.4). Most balances were sold during the preserving and Xmas baking season. With increasing prosperity, industrialising of food production, sale of ingredients divided in portions, weighing became unnecessary in most cases. Furthermore, working women had not time enough for cookery, and the need for kitchen scales decreased rapidly. Karl Döft died at the age of 80. He had neglected switching to electronics. His son Jens-Peter had not modernised the factory before his early death. In 1963 the company was closed.

8.1.4.7 München

Georg Reichenbach (1772–1826) established a workshop for scientific instruments at Munich and founded in 1804 the Mathematical-Mechanical Institute. Thus Munich became the centre of production of precision instruments. In 1819 he sent his foreman Traugott Lebrecht Ertel to Vienna in order to establish there a similar workshop.

8.1.4.8 Schwäbische Alb (Swabian Jura)

In a geographically narrow space of about 15 km around Onstmettingen (Figs. 8.4/8.5) near Stuttgart (Germany) in the middle of the 20th century a number of scales manufacturer workshops had been concentrated without parallel in Germany and worldwide [47]. In about 30 factories more than 3000 workers were engaged in the production of precision and assay balances for the chemical and pharmaceutical industry as well as household, trade and industrial scales. Also today three leading factories and three important balance museums exist.

The Schwäbische Alb is an idyllic landscape, however, with barren soil and harsh winters; thus agriculture is poor. To improve profits cottage industry began. By the end of the 17th century, the Black Forest area started producing clocks and, following their example, Onstmettingen began somewhat later (Fig. 8.5). But their approach was slightly different; they used metal instead of the customary wood of the

Fig. 8.4 Onstmettingen, Swabian Alb, Germany in 1822

Fig. 8.5 Onstmettinger desk clock

Black Forest. The profession of the worker was metal smith, but the whole family was engaged including children, even if it was only with menial work. The workshop was in the front room, later on in the converted barn.

The outstanding pioneer of the balance industry on the Schwäbische Alb is without a doubt the Pastor Philipp Matthäus Hahn (Fig. 8.6) [48, 49]. He was born on 25th November, 1739, in Scharnhausen as a son of a protestant vicar. The Family moved in 1756 to Onstmettingen. Phillipp Matthäus, at the age of 17, became a friend of Philipp Gottfried Schaudt who was of the same age. Both had been interested in science and performed experiments. Schaudt became teacher of the Onstmettingen School. During 1756–1760 Hahn studied theology at Tübingen and became home teacher and vicar at several villages in Württemberg. Similarly he

Fig. 8.6 Philipp Matthäus
Hahn

proceeded in studies of science and together with his friend and with help of Onstmettingen's clockmaker they constructed several small astronomical clocks and in
1768/69 a large one, the so-called 'Ludwigsburger Weltmaschine' (world engine).
Besides serving the parish he managed the production of clocks of every size and
scale and he constructed a four species calculator for up to 14 digits (Fig. 8.7). Especially he made mathematical calculations and drafts which then were realised by
Schaudt and the flourishing precision mechanical workshops in the Schwäbische
Alb, particularly of the brothers Sauter, who had been specialists in the treatment
of brass and steel [50]. Hahn was extremely successful in inventions, but in addition we have many notes of his sermons. He saw his religious activity as his most
important duty as we can see from his note dated August 10, 1773:

> "Oh I mused today already, with how many unnecessary impediments are you surrounded
> and did you entangle yourself? It is a thousand times more valuable to preach the gospel,
> interpret it and to work within it! These calculators, these astronomical machines, they are
> worldly! But in order to gain fame and honour to enter and spread the gospel, I will continue
> to carry the load."

Philipp Matthäus Hahn (1739–1790) [47, 49, 51] constructed a 'convenient
household scale' (Fig. 4.35), an inclination balance with quadrant-reading as his
most well-known work in his field. Furthermore he produced new types of hydrostatical balances (Fig. 8.8). In his workshop books we also find sketches of other
scales: a sol-scale, a quadrant scale with higher accuracy to weigh gold coins, a
double steel-yard, a more sensitive assay balance, a scale with sliding weight with
three ranges, and a more sensitive beam-scale with sliding weight [52]. Because he
had not enough time to make such instruments to order he passed them to his brothers who were barbers by profession but also skillful workmen and opened a new
market for the workshops of the region.

Fig. 8.7 Four species calculator of Philipp Matthäus Hahn

We may assume that after Hahn moved away in 1770, such work was continued, however, in particular by Philipp Gottfried Schaudt and the smiths and clockmaker families Sauter and Keinath, though we have no reports. Since the 1830s we have several publications in catalogues and newspapers on precision mechanic workshops which offered weapons, balances, and physical instruments, whereas manufacturing of clocks declined as a result of the meanwhile industrialised production in the Black Forest.

As usual in that time a young man took to the road to expand his knowledge and knowhow and gaining new ideas. August Sauter (1831–1874) reports that the mechanics from Onstmettingen went on travels, above all to Christian Ferdinand Oechsle in Pforzheim, which was at the beginning of the 19th century the leading maker for physical instruments in Baden Wurtemberg. Sources in Ebingen's town archives indicate that August Sauter travelled as far as Vienna and London where renowned mechanical shops existed. From there the mechanics brought back home what they had learnt, including their recognition that there was no future in clock making but rather in a market in need of physical instruments for the growing exact sciences, mainly in chemistry. During his travels he bought tools and established in 1856 his workshop as balance maker at Ebingen (Fig. 8.9). Successfully he exhibited his balances at exhibitions in London and Paris. Simultaneously he worked as a calibrator in his town. After his early death the workshop was sold to his employee Louis Armbruster. Twenty-two workers in a new factory building and up to 10 home-based workers expanded the production. After the death of Armbruster in 1903 Heinrich Cless developed a large concern with industrial production. A forge, a foundry and all workshops for finishing the product had been incorporated, with more than 200 workers in 1913. The Sauter Company exported its balances worldwide (Fig. 5.10). Armin Wirth developed in 1928 an automatic inclination balance with a five times revolving indicator (Fig. 8.10). In 1974, 150 000 scales of that type

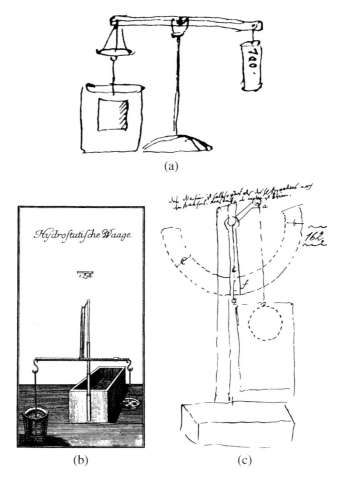

Fig. 8.8 Sketches in Hahn's catalog and workshop book II (1771–1773) for his hydrostatic balances

had been sold. When Heinrich Cless died in 1943, his son Oscar Cless followed. In 1971 the August Sauter KG was sold to the Mettler Toledo AG, Greifensee, Switzerland, which is now the biggest manufacturer of scientific instruments including weighing systems. The Mettler-Toledo (Albstadt) GmbH is a global centre of competence for the development and manufacture of weighing systems including hardware and software.

Gottlieb Kern was the son of an Onstmettingen mayor [53]. After apprenticeship with Sauters he went on travels, in particular to Pforzheim with master craftsman Oechsle. In 1844 he founded his workshop at Onstmettingen. He married Sauter's widow. The couple combined the Sauter and Kern workshops and began producing precision scales.

In 1852 Ludwig Bosch founded his metal workshop at Jungingen and produced different components and instruments including precision balances. In the 1990s

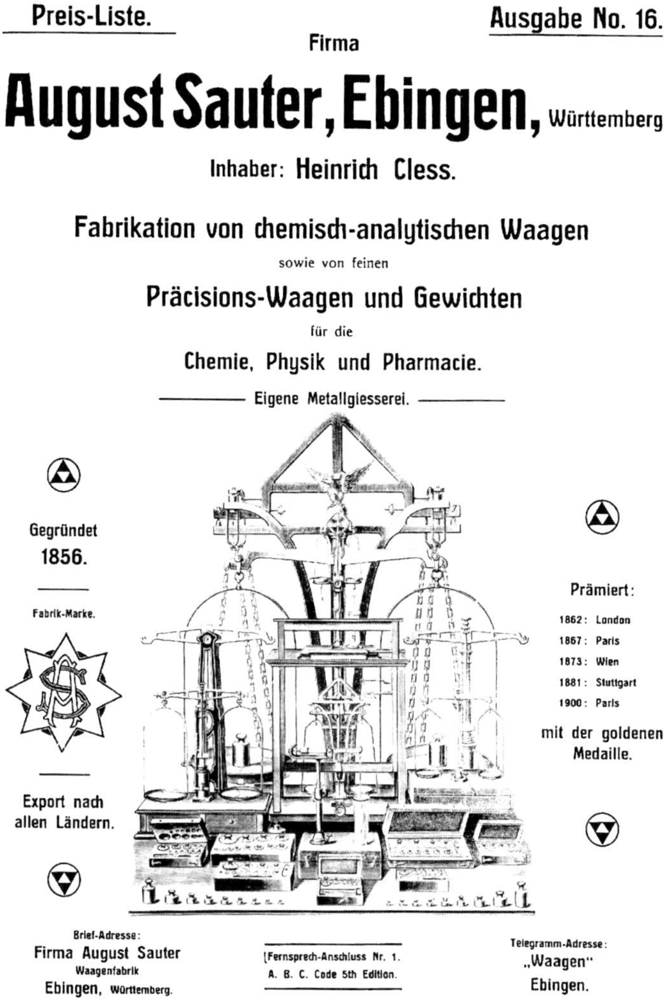

Fig. 8.9 (a) Price list of the Sauter Company, Balingen, Germany. (b) Letterhead of the Sauter Company, Balingen, Germany

balance production ceased. Today a company 'Bosch CNC-Komplettbearbeitung GmbH' works with lathe and milling machine.

Gustav Hartner (1853–1945) bought in 1893 the workshop of Ferdinand Rehfuß, who in 1879 started the production of trade and analytical balances with four workers at Ebingen. Hartner expanded and in 1911 he started in addition to analytical balances the large production of twist drills. During World War II the factory building was completely destroyed. Again rebuilt, priority was set on production of twist drills and today only precision drilling tools are produced, more than 10 000 different items.

(b)

Fig. 8.9 (Continued)

Fig. 8.10 Automatical
inclination balance Wirth of
the Sauter Company,
Balingen, Germany

Huber reported that in 1910 in Onstmettingen six factories of precision balances
existed, four at Ebingen, two at Jungingen [54]. They had 500 employees of whom
300 worked in the factories and 200 at home. Those factories needed many suppliers
like horn turner, brass and steel processors, and joiners. The yearly production was
about 50 000 balances and 1 million weights. About 80 % of scales and weights had
been calibrated by standardisation bureaus at Ebingen and Onstmettingen. In these
figures the production of the Bizerba Company is not included which was already
specialised on trading scales.

The economic crisis after World War I weakened the industry everywhere. After World War II the market conditions changed rapidly. Up to the 1950s precision weighing was time consuming and highly ritualised. In 1945 newly founded Mettler Company in Switzerland begun to produce consequently high-precision unequal-armed substitution balances [38, 39] which were much easier to operate and the weighing result was presented digitally. Far too late the Swabian manufacturers reacted. In addition, small companies were unable to rationalise their production by introducing assembly lines. They had also not enough capital to finance research, in particular for the beginning of the electronic age. Several factories closed, others changed their product. For example, Gustav Hartner, who already produced twist drills besides balances, confined himself successfully to the production of drilling tools. Until now the companies Sauter, Kern & Sohn, Bosch and Bizerba produce balances. Bosch and Sauter had been bought by Mettler-Toledo [55]. Now, Kern & Sohn claims to be the oldest German balance manufacturer. Haigis at Onstmettingen produces and certifies weights. Worldwide Bizerba is one of the leading manufacturers of scales for trade and industry.

8.1.5 Asia

In Asian countries, the economy was based on agriculture for several millennia. Weights dated to 2500 BC in the Huanghe valley (China) and so also balances would have been used [56]. However, measurements by volume are presumed to have preceded those by weight.

Earliest manifestations of the Iron Age are dated to 1000–800 BC. With the beginning of use of metals for fabrication of items for decorative and practical use, technology developed. Trade with countries to the West began with sailing ships and via the Silk Road. It may be concluded that measuring instruments were used, although there is no confirmation by historical records.

Early civilisations in Asia preceded those in Europe. Constructions such as the Great Wall of China, temples in Japan, Cambodia, Indonesia, India and Sri Lanka show that civil engineering was advanced in these countries. Constructions to enshrine relics of Buddha, 'stupas', began in India and Sri Lanka. Some of these were very high. The tallest stupa in Sri Lanka is second in height to the tallest pyramid in Egypt.

In Sri Lanka, the construction of artificial reservoirs or 'tanks' for storing water for agricultural and human use was prevalent from the third century BC. Elaborate system of canals channelled the water over long distances. Such activities would imply advanced skills in engineering. The gradient of some canals were very small— about one inch (2.5 cm) over a mile (1.6 km). It is implied that technical skills and instruments were used, and a knowledge of surveying and trigonometry existed, although much of the constructions involved large amounts of human labour [57].

Although historical events were recorded on parchment and in stone, technical skills passed down generations and were not recorded. This was often deliberate in order to protect intellectual property, there being no patent laws as is known today.

Applications of technology in Asia developed only from the 16th century, with colonisation by European powers. British colonial rule in South Asia and French colonial rule in South East Asia exposed these countries to the technical skills of the West. However, the colonial policy was to develop these countries as agricultural bases (especially for the supply of spices for European markets), and also as markets for industrial products of Europe.

8.1.6 America

When Spanish soldiers at the turn towards the 16th century conquered middle and the northern part of South America, accompanying monks observed the use of scales by the Native Americans. However, development of measuring instruments took place during the 18th century after immigration of European people in North America. The production of household and pharmaceutical scales of modest precision, but easy operation, began.

In ~1830 Thaddeus Fairbanks (1776–1886), in the USA, invented industrial scales with ground-levelled platform. The company still exists and its history is revealed in more detail in the next section.

The Ainsworth Company was founded in 1880 at the height of the legendary Colorado goldrush. The company founder, William Ainsworth, recognised the opportunity for a US manufacturer and entered the market with innovative analytical balance designs. In the middle of the 20th century Ainsworth offered a series of vacuum microbalances with maximum load up to 200 g. That company still exists and is described in the next section.

Denver Instrument, Bohemia, NY, developed and manufactured analytical precision instruments since 1880. In 2002, Denver was merged with Scaltec Instruments and the 'new' Denver Instrument was founded. A broad range of analytical and precision balances, moisture analyser and pH-meter were made. Denver now belongs to Sartorius, Göttingen, Germany.

The Ohaus Corporation began in 1907, when Gustav Ohaus decided to forgo a career in the gray iron foundry business for a business venture with his father, Karl, a German-trained scale mechanic. Together, they established a scale repair business in Newark, NJ. In 1914, Gustav and Karl Ohaus were incorporated as the Newark Scale Works, coinciding with their first production of grain testing equipment, and the issuing of their first patent. The onset of World War II increased the demand for laboratory equipment, and thus an additional Ohaus-owned factory opened in 1941. In 1947, the former Newark Scale Works was incorporated as Ohaus Scale Corporation, and soon after occupied a new factory in Union, NJ. In 1969, the Gustav Ohaus Award program was established to honour educators for their commitment to academics, and awarded to entrants for excellence in science teaching. In 1975 Ohaus established offices worldwide. In 1982, Ohaus introduced the first laboratory-quality portable balance. The first electronic moisture determination balance soon followed in 1984. In 1990 the laboratory balance production of Ohaus Corporation was brought into the Mettler-Toledo group of products.

From middle of the 10th century several big engineering and chemical works developed electronic balances for their own research but some of them offered for the public. We mention Beckman, Du Pont de Nemours, Instruments Division, General Electric and Perkin Elmer.

Lee Cahn (†1968) was the first introducing his vacuum microbalance to the public. In 1956 he founded the Cahn Instrument Company at Cerritos CA. The company merged with another business, and was sold in succession to several other companies. Today it belongs to Thermo Fisher Scientific. His Cahn balance was manufactured nearly unchanged until recently. Today service is made by Microtech Instruments and Precision Instrument Repair Company. Lee Cahn was the initiator of the Conferences on Vacuum Microbalance Techniques which started in 1960 and are still running.

8.2 The Balance Market

The profile and history of some of today's balance manufacturers is described, mainly based on published data of the companies.

8.2.1 Company Profiles and History

8.2.1.1 A&D

A&D Co., Ltd. was founded in 1977 to manufacture and distribute electronic measuring instruments. A head office was established in Nerima-ku, Tokyo and a factory in Kitamoto City, Saitama. In 1979 relocation of head office to Shibuya-ku, Tokyo, and in 1981 relocation of factory to Konosu City, Saitama, took place. In 1982 Kensei Industry Co. Ltd. was acquired and the electronic balance business started. In 1985 Litra Co., Ltd. was founded, a factory specialising in load cell and electronic balance manufacturing. In 1987 it entered into contract with Takeda Medical Inc. for sole representation and began distribution of digital blood pressure monitors. In 1987 it acquired Mercury Weighing and Control Systems Pty. Ltd. stocks and established it as the A&D subsidiary sales office in Oceania. In 1989 it merged with Takeda Medical Inc. In 1989 it established a research and development technical centre in Kitamoto City, Saitama, and transferred all research and development functions from the Konosu factory. This was followed by founding subsidiaries worldwide.

8.2.1.2 Acculab

Acculab has ceased business operations effective from January 2011. Acculab was part of the Sartorius Group and victim of the delevering strategy for Sartorius' reorganisation in 2010. Acculab offered low cost inexpensive weighing scales like the PP pocket scales and Econ digital scales for the elementary science classroom; these have been totally eliminated from the Sartorius scale portfolio. Acculab Vicon electronic scales have ceased production and replaced with the Sartorius AY Series. Acculab SVI bench scales have no direct replacement. ECL Excelleron light industrial weighing scales can be substituted with the higher quality Sartorius Midrics. Also, Acculab ALC and Atilon lab balance equivalent cross reference is the Sartorius AZ and ED series respectively.

8.2.1.3 Accuweigh

Accuweigh commenced operations in Adelaide during June 1992 by founding partners, Brenton Cunningham & Greg Brogan. Brenton's experience lay in engineering sales, management, marketing and administration, whilst Greg's skills lay in mechanical engineering, automation, and hands-on trouble shooting. Client enquiries for packaging and filling equipment led to the Brenton's close working relationship with Ross Waller, the owner of a small packaging machine manufacturer known as WeighPack Services. In 1997, Ross was admitted as a third partner due to his expertise in design, development and implementation of packaging equipment and complete automatic production lines. Soon after, Accuweigh acquired the Avery Scales agency for industrial products and picked up the exclusive agency for Salter Weightronix during 2001 which was renamed as Salter Australia Pty. Ltd. and further acquisitions were made. Accuweigh is now the largest weighing, filling, packaging and inspection equipment supplier in Australia with the ability to supply simple solutions up to complete turnkey systems.

8.2.1.4 Applied Weighing International Limited

Applied Weighing International Limited is the UK's leading manufacturer and supplier of process weighing equipment and load cells. They offer solutions for almost every kind of industrial weighing application, including in hazardous areas. The company claims "Whether your requirements are for just a single load cell or a complete batching system you can have confidence in our products, which will be based on thousands of successful installations over the last 20 years".

8.2.1.5 Ainsworth

The Ainsworth Company was founded in 1880 at the height of the legendary Colorado goldrush. The company founder, William Ainsworth, recognised the opportunity for a US manufacturer and entered the market with innovative analytical balance designs. In the past century, the Ainsworth product line has featured a variety of products including military compasses and watches. Balances and scales, however, have always been the foundation of the company. Through the years, mechanical balances, jeweller scales and postal scales have been successful under the Ainsworth name. The current product line includes a broad range of electronic precision analytical and toploading balances, portable balances, high-capacity industrial scales and infrared moisture analysers. Ainsworth now belongs to Denver Instrument / Sartorius

8.2.1.6 Avery Weigh-Tronix

Avery Weigh-Tronix [58] is a company specialising in weighing scales. Its headquarter stands on the site of the Soho Foundry overlooking Black Patch Park in Smethwick, Birmingham, England, with a US-based manufacturing and retail manufacturing plant. It was formed in 2000 when US weighing company Weigh-Tronix acquired the Avery Berkel group of businesses, bringing together the four brands of Avery, Berkel, Salter and Weigh-Tronix.

Avery Weigh-Tronix is one of the world's largest manufacturers of weighing equipment covering the full spectrum of industrial scale uses: counting scales, bench scales, checkweighers, deck scales, instrumentation, computer software, printers, remote displays, cargo scales, truck scales, lift truck scales, static or in motion rail scales, bin/tank/hopper scales, conveyor scales, crane scales, and almost all major components, including strain gauges, load cells and Weigh Bars With the addition of Avery Berkel products it has also food processing and slicing products. Salter/Berkel produces light commercial scale products, Dillon Overload protection and overhead weighing equipment, and Railweight train-in-motion weighing. In 2008 it was purchased by Avery Weigh-Tronix Illinois Tool Works Inc. which employs nearly 60,000 people in 825 business units across 52 countries, and is based in Glenview, Illinois.

8.2.1.7 Weigh-Tronix

Incorporated in 1971, Weigh-Tronix was one of the largest US suppliers of scale equipment applying Weigh Bar technology to build the first all-electronic deck scale. In 1986 National Controls, Inc. of Santa Rosa, CA, became part of Weigh-Tronix. The merger provided additional sophisticated load cell technology exhibited by its checkweighers, counting scales, retail and postal scales. It also added the respected Dillon force measurement equipment to the product family. In 1991 Weigh-Tronix became a subsidiary of Staveley Industries plc, joining the Staveley Weighing and Systems Group. This group integrated Weigh-Tronix, Chronos Richardson, and the Salter companies. In 1998 private investors purchased the Weigh-Tronix business from Staveley keeping the Salter Company as part of the Weigh-Tronix family. This group has manufacturing facilities in the US, UK, and Canada. In 2000 Weigh-Tronix expanded dramatically with the purchase of the Avery Berkel group of businesses from Marconi and the creation of Avery Weigh-Tronix. Avery Berkel brought an additional workforce of over 4,000 and created a strong presence in Europe, Africa, and the Far East.

8.2.1.8 Avery

In 1731 James Ford established a business as a maker of 'stilliards' in Birmingham, England. The company can trace its roots indirectly as far back as 1670 and Charles de Grave who had a shop near St. Paul's Cathedral in Birmingham. In 1760 James Ford retired and the business passed on to William Barton and then to Thomas Beach in 1782. In 1799 business passed on to Joseph Balden, who married Mary Avery. In 1813 as the Industrial Revolution gathered momentum, the business was transferred to William Avery, soon joined by his brother Thomas, who traded under the name W&T Avery. Business expanded and more premises had been acquired in Digbeth and London. The company now employed almost 200 people. In 1876 the first weighbridge was designed by Mr. A.W. Brown, an ex-Boulton & Watt apprentice. In 1895 the business of James Watt & Co. and the Soho Foundry site in Smethwick, West Midlands, was acquired. In 1918 W E T Avery died, last family member actively involved with the company. It now employs more than 3,000 people. In 1963 the first load cell weighbridges were made at Soho and in 1971 the first digital retail scales in volume production. In 1979 W&T Avery became part of the GEC Group of companies. Production of electronic weighing machines increased. In 1982 the first volume production retail scale with integral printer price look up began and in 1984 the first UK manufacture of retail scale with networking facility and data capture facilities. In 1993 the Berkel Company was acquired and combined with GEC Avery Ltd. to trade globally under the Avery Berkel name. In 2000 the Avery Berkel group was acquired by Weigh-Tronix, Inc.

8.2.1.9 Salter

Brothers Richard and William Salter established a business, manufacturing springs and pocket steelyards (spring balances) in a cottage in Bilston, England. Over the next two centuries, these springs were to be used in some of the most important industrial developments, including George Stephenson's Rocket and the first cars and aeroplanes manufactured in the UK. In 1770 the company moved to West Bromwich, where it was one of the leading employers for more than 200 years. On Richard's death in the 1780s the business was taken over by William's sons, John and George Salter, who also made bayonets. George Salter took out many weighing equipment patents, which related to various ways of indicating weight on a spring balance that are still used today. The product folio was widened to coin-operated public weighing machines as well as various food processing equipment, clothes washers, letter-clips, hat hooks and dynamometers. A factory was also opened in Berlin, in association with German firm Ubrig, which led to the development of automatic rotating display and vending machines. In 1895 Salter produced the first English typewriter—the British Empire. Rights to this business were sold in 1936.

In the 1920s Salter employed more than 1 000 people. A repair company was opened in Australia. During the war, the company manufactured more than 750 million springs for war purposes. After the war Salter continued to expand. Nevertheless in 1973 George Salter & Co. limited was bought by Staveley Industries and purchased then by Weigh-Tronix. In 2000 Salters merged with Avery Berkel following the latter's acquisition then by Weigh-Tronix.

8.2.1.10 Brecknell

Salter-Brecknell supplies a wide variety of weighing equipment, with capacities ranging from a fraction of a gram up to 20,000 pounds.

8.2.1.11 Berkel

At the turn of the nineteenth century the world's first food slicer was invented in Holland by Wilhelm van Berkel. At that time of increasing prosperity, satisfying the demands of the people for more meat and sausage, butchers were kept busy slicing with 16-inch long carving knives from 7 o'clock in the morning until 11 o'clock at night. This invention was set to revolutionise the butcher's trade, where quality of cut and the speed of the slicer would become the predominant benefits of the new machines. In 1898 Van Berkel had started production at factories based in Rotterdam, and soon slicers were in demand all over Europe. In 1908 a branch was established in London and in 1909 the U.S. Slicing Machine Co. Inc., now in La Porte, Indiana. In 1993, now part of the GEC Avery Company, slicers and food processing equipment are still sold under the Berkel brand throughout its companies and distribution network. With the merger between Weigh-Tronix and Avery Berkel in 2000, the Avery Berkel brand now signifies the premier retail scales in Europe and North America. In 2004 the Berkel brand of slicers was sold to Brevetti van Berkel of Italy. Avery Weigh-Tronix became the exclusive supplier of Berkel slicers in the UK, France, Austria and Ireland. Avery Berkel retail scales remain a key part of the Avery Weigh-Tronix family as a world leader in the grocery market.

8.2.1.12 Bähr Thermoanalyse

The first Bähr dilatometer was developed in 1980. All important thermal analysis systems such as DTA, TGA, STA, DSC and TMA followed. Offering a complete program for the high temperature viscosimetry Bähr is also world-leading in this field. In 2012 Bähr was purchased by TA Instruments.

8.2.1.13 BeWA-tec

The scales manufacter, Waagen Prümm', Th. Prümm GmbH & Co. KG was founded in 1839 by Theodor Prümm at Bergisch-Gladbach, Germany. Its first product had been calibrated decimal scales with maximum load up to 100 kg. Since 1985 the company was headed by Johanna Prümm who developed the factory to a sales

Fig. 8.11 Bizerba inclination type switchgear retail balance of Wilhelm Kraut, Bizerba, Balingen, Germany

and repair shop for precision laboratory, medical and industrial scales. In 2009 Jörg Bernard took the chair. He changed the firm into BeWA-tec, Bernard Waagen Automatisierungs- und Steuerungstechnik.

8.2.1.14 Bizerba

In 1866 the fitter Andreas Bizer (1839–1914) entered the Bizer Brothers weighing scales workshop in Balingen and received his first commission to build a weighbridge for the town. This weighbridge is exhibited now in the 'Museum für Waage und Gewicht'. With the introduction of the metric system a weights and measures inspector was established in Balingen with Andeas Bizer as inspector. In 1906 his daughter Elisabeth, married the later professor Wilhelm Kraut (1875–1957) who managed the establishing and repair of power stations in several towns including Balingen. In 1906 Andreas Bizer sold his business to his son-in-law. The company employed eight people and manufactured mainly table top scales (Fig. 8.11) and steelyards which were sold in Switzerland. In 1923 Wilhelm Kraut Jr. became manager of the company.

Together with his master craftsmen, Hauser und Schlaich, he developed the word's first pendulum quadrant scales, which quickly replaced the time-consuming method of balancing goods against weights. In 1927 the first mechanical shop scales with ticket printer were built. In 1928 the company became the largest scales producer in Germany with 800 employees. The younger brothers, Erwin and Arthur, entered the company as engineers, whereas the father still acted as 'buiding contractor' well into the 1950s. In 1949 the production of industrial and heavy duty industrial scales with printer was started. In 1942 Professor Wilhelm Kraut handed over his scales collection to the town of Balingen which established the 'Museum für Waage und Gewicht in the Zollernschloss (Zollern Castle). Following on from various postal scales and checkweighers, Bizerba launched new types of counting scales for bulk parts. In the 1940s the production of food slicers was started. In 1955 the first industrial scales with remote weight display and printer was presented and in 1967 the first calibrated invoicing machine. In the 1960s the heat seal label was introduced. In the 1980s the first digital evaluation and display system for industry was introduced comprising scales and label printer.

In 2011 Andreas Wilhelm Kraut took on the chairmanship. Bizerba now is a leading, worldwide-operating solutions-provider offering professional system solutions in scale, label, information and food service technologies, in the retail, food industry, producing and logistic segments. Sector-specific hardware and software, high-performance network-compatible management systems as well as a wide selection of labels, consumables and business services ensure transparent control of integrated business processes and high-level availability of Bizerba-specific performance features. Worldwide, Bizerba is present in over 120 countries—with 41 shareholdings in 23 countries, as well as 54 country distributors. The company's headquarters, with a staff of 3,000, is located in Balingen, Germany. Further manufacturing facilities are located in Meßkirch, Bochum, Vienna (Austria), Pfäffikon (Switzerland), Milan (Italy), Shanghai (China), Forest Hill (USA) and San Luis Potosi (Mexico).

8.2.1.15 Bruss

Bruss is a supplier of mobile weighing system built in lifting trucks, load boards and fork lift trucks. Bruss is a brand name of RAVAS GmbH mobile Wiegetechnik, Kleve, Germany.

8.2.1.16 Cardinal Detecto

Detecto Scale Company was started in 1900 in New York City by three immigrant brothers (The Jacobs Brothers) in the butcher supply business. The company produced bakers' dough scales, butchers' scales, clinical scales, counter scales and hanging scales. Cardinal/Detecto, officially began in 1950. Founder William H. Perry of Webb City, Missouri, an experienced scale industry man, started the new company as a result of customer need for a line of special-capacity scales. This need arose out of America's construction-era boom. In the development of the nation's highways, bridges and other structures, builder demand was at an all-time high for appropriate, new weighing systems. All concrete, stone and asphalt were batched and sold by weight; manufacture of the right weighing equipment called for someone with the right experience. In 1981 Detecto was acquired by Cardinal Scale Manufacturing Company based in Webb City, Missouri where its corporate offices and manufacturing facilities are still located today. Cardinal was a well-known name in the heavy-duty truck scale and floor scale market.

8.2.1.17 CAS

CAS is an international manufacturer of truck scales, crane scales and many other types of digital scales. For more than 25 years, CAS has manufactured high-quality scales that meet a wide range of industrial applications. Based in Korea, CAS is a global provider of industrial and retail scales, and the company operates two offices in the US. These versatile and reliable truck and crane scales are ideal for light to heavy-duty industrial applications.

8.2.1.18 Chatillon Ametek

The John Chatillon & Sons Corporation was formed in 1835 in New York. AMETEC acquired Chatillon in 1997. The Chatillon brand is now part of the AMETEC Test & Calibration Instrument business unit. Chatillon force measurement and weighing products are engineered and manufactured in Largo, Florida, USA and additionally in Mexico and China. Products are gauges, testers and test stands, software, sensors, grips and fixtures.

8.2.1.19 Chronos BTH

In 1881 the 'Hennefer Maschinenfabrik C. Reuther und Reisert' was established in Hennef, Germany and invented the first automatic weigher thus setting the standards for the future to come in automatic weighing. In 1982 the company became Chronos Richardson. At this time one remembers only the 'The Hennef Weigher-Walking-Way' [59, 60] and Chronos Richardson GmbH, engaged in the production of packing machines. In 2002 Chronos Richardson was acquired by Premier Tech, a Canadian privately owned corporation with more than 85 years tradition, with nearly 1 600 team members based in America, Europe and Asia. Premier Tech has been building its know-how and reputation in three groups—Horticulture and Agriculture, Industrial Equipment and Environmental Technologies. In 2009 Premier Tech acquired BTH BV, an innovative company in the packaging industry located in the Netherlands. The combination of Chronos Richardson and BTH operates under the name Chronos BTH both in Europe as well as in Asia. The new company, Chronos BTH, will offer some of the most advanced and innovative equipment and systems in the areas of bag packaging and palletising solutions as well as load security. The new entity will continue to operate from sites in the Netherlands, Germany, Italy, England, France and Thailand.

8.2.1.20 CI Precision

CI Precision is the new trading name for CI Electronics Ltd. and its sister company CI Systems Ltd. CI is a privately owned manufacturer of precision instruments and systems for use in quality assurance and research applications. The company was founded in 1965 to manufacture one of the first microbalances for research applications. That vacuum microbalance is often found in apparatus with other brands. Their areas of expertise include high precision weighing & handling, software design, electronic design, and the provision of GMP solutions for pharmaceutical applications. All of their design and implementation processes are closely controlled and they have particular expertise in the whole area of system validation.

8.2.1.21 Denver Instrument

Denver instrument as successor of Ainsworth can be traced back to 1880. English immigrant, William Ainsworth, recognised the opportunity for a US manufacturer and entered the market with innovative analytical balance designs. Around 1895, Ainsworth introduced the company's first short-beam analytical balance, which significantly reduced weighing time and improved reliability. The short-beam design was also one of the first commercial uses for aluminium. In the late 1880s, the only analytical balances available in America were also manufactured and serviced in Europe. The company's vertical orientation to manufacturing ensured total quality control of its finished products as all aspects of design and manufacture are performed in the company's modern production facilities in Arvada, Colorado. Denver Instrument's continued investment in state-of-the-art manufacturing technologies and machine tools gives the company's products a distinct advantage in the world market, without sacrificing the quality of the analytical instrument. In 2002, a new course was set. Denver was merged with Scaltec Instruments and the 'new' Denver Instrument was founded. A new generation of highly precise, easy-to-use laboratory instruments were created. A broad range of analytical—and precision balances, moisture analyser and pH-meter for all applications and budgets are available. Now, Denver, Scaltec, and Ainsworth belong to Sartorius, Göttingen.

8.2.1.22 Dillon

WC Dillon and Co., a family business, was founded in 1937 as a supplier of cable tensioning products, which were primarily marketed to the US armed forces. Willie Dillon invented a dynamometer to measure the amount of tension being placed in wires so that they would be neither over-stressed nor too loose. The company grew rapidly during the major wartime periods of the twentieth century. Ownership changed in the early 1980s, when WC Dillon's manufacturing operations were integrated with those of Santa Rosa, CA, based scale manufacturer National Controls Inc. (NCI), a division of Scope Inc. The Dillon brand was maintained. Dillon was acquired by Avery Weigh-Tronix, LLC in 1986. Dillon's primary manufacturing facility was relocated to Fairmont, MN, in 1988.

8.2.1.23 Ebinger Waagenbau GmbH

The company distributes and repairs balances of all kinds. In addition it designs solutions for apparatus and plants which include weighing tools. In Fig. 5.12c a medical scale with card reader is depicted.

8.2.1.24 ESIT Electronic Systemler

The Turkish Company started in 1980 as a manufacturer of electronic systems used in industrial applications, with a focal point primarily on electronics engineering, Esit Electronics majored on weight measurement and in 1987 started manufacturing electronic weighing equipment; in 2005 it had 200 employees. Esit weighing systems are based on strain-gauge load cells and comprise platform scales, truck and trailer scales, railway scales, axle weighing scales, monorail scales, tank weighing systems, conveyor belt scales, check weighers, filling, discharging and batching systems, bagging systems, and various custom weighing systems.

8.2.1.25 Fairbanks Scales

Thaddeus Fairbanks, a mechanic and builder, was a wagonmaker by trade. He built a foundry in 1823 to manufacture two of his inventions—the cast iron plough and a stove. In 1824, his brother Erastus formed together with Thaddeus, the E & T Fairbanks Company in St. Johnsbury, Vermont. Once in business together, the two brothers realised that the current weighing system yielded inaccurate results. So, Thaddeus decided to invent a new, more dependable weighing machine. On the basis of a steelyard and with additional levers according to the Roberval principle he constructed very accurate but easy to handle platform scales. He dug a pit for the levers, placing the platform level with the ground. This modification ended the task of having to hoist the entire load. On the morning he was to leave St. Johnsbury for the test marketing of the scale, he discovered a solution to his dilemma. By adding two short levers to his long ones, he established support points at all four corners of the platform. Now his scale was not only accurate, but very stable.

Erastus and Thaddeus were now joined by their younger brother, Joseph. With 10 employees, the company was making scales, ploughs and stoves. Within 50 years the company grow to more than 1000 people working in 40 buildings. In the 1830s manufacturing rights were sold to H. Poole and Sons in England. In 1846, trade began with China and Cuba. Following the end of the Civil War in 1865, the United States continued to prosper and grow—so did E & T Fairbanks & Company. Within two years of the war's end, Fairbanks was turning out 4 000 scales a month and meeting the needs of the expanding worldwide demand. By 1882, more than 80 000 Fairbanks scales were being produced annually. By 1897, the company held 113 patents for improvements and inventions in weighing. Fairbanks offered its customers 2,000 standard model scales, yet made as many as 10,000 different models and custom systems.

In 1916, Charles Hosmer Morse, a Fairbanks employee, acquired control of the company. During this time the Fairbanks-Morse company produced not only scales but diesel engines, electric engines and pumps for industrial use. In 1958, Fairbanks-Morse merged with Penn-Texas and was renamed Fairbanks-Whitney. New leadership was brought in four years later when George Strichman was appointed president. Renamed Fairbanks Weighing Division of Colt Industries, Fairbanks experienced a rebirth. A modern manufacturing plant replaced the deteriorating facilities in St. Johnsbury, Vermont, in 1966. And in 1975, a new factory was built in Meridian, Mississippi, producing a variety of products designed for heavy capacity weighing. It was in 1988 that Fairbanks came under the current management of F.A. 'Bill' Norden, president and major stockholder of Fairbanks Scales. He headed a group which acquired the company from Colt Industries. With new leadership came more changes. Finance, marketing and executive offices were moved from St. Johnsbury to the more central location of Kansas City, Missouri. In 1999, F.A. Norden was named Chairman of the Board and his son, Richard Norden, became Fairbanks' President and CEO.

Today, Fairbanks has more than 500 employees nationwide and maintains service centres, authorised distributors and sales offices in 49 states and in more than 25 countries. Manufactured are everything from precision and bench scales to heavy capacity truck scales and railroad track scales Also platform scales as described above often have a transformation of forces, e.g. Fairbanks scales. Scales achieving higher levels of accuracy are designed for diagnostic and sanitary capabilities.

8.2.1.26 Herbert Group Ltd.

Herbert & Sons was founded by John Wood at London, around 1740. John Wood died in March 1765 around the age of 50. After his death his business was taken on by Thomas Goulding, and later by his nephew Richard. In 1867, the Herbert family stepped in to take over the Woods scalemaking business.

An equal armed butcher scale 'Lion Quick Action Scale' became a best seller. In 1952 Herbert & Sons bought the Swift Scale Company and with it came a rather important new customer: Tesco Stores. In the 1960s to 1980s the application of electronics to retail weighing machines was pioneered and consolidated. The Lion 2000 was a hybrid design and the initial version of the machine was approved in June 1975. The machine incorporated many improvements and innovations such as a price-entry keyboard, rapid settling time and speedy weighing resulting in the display of computed price within a second. It was possible to totalise and display the sum of the prices. In October 1977 a joint venture company, Lion Electronics Ltd., was formed by Herbert & Sons and Gresham Lion. In 1979 load cell technology was introduced. In 1982 the labelling capability of the Sovereign range was enhanced by linking to printers incorporating thermal printing technology. It enabled the reliable and speedy printing of bar codes on variable weight items so that they too could be automatically identified at check-out. Printers were linked to the balance with fibre optics. Fully integrated scale and barcode printer with programmable preset price keys were developed in cooperation with Teraoka Seiko Co. Ltd., better known as Digi. Today are produced besides retail scales, wristband printers, labels, tags, Barcode scanners, mobile computers and tablet PCs with more than 225 employees.

8.2.1.27 Flintec

Founded in Sweden the company has focused upon the design and manufacture of strain gauges, load cells, measurement electronics and software for a diverse industry base from weighing machinery and now new markets in medical devices, industrial and agricultural machinery and process automation sectors. Flintec is part of the Swedish Indutrade Group. Flintec manufactures all key components in-house. Integrated manufacturing processes include strain gauge production, CNC machining, heat treatment, assembly and calibration. Manufacturing is carried out at two new ultramodern facilities totalling 12,000 square metres in Sri Lanka (Katunayake and Koggala).

8.2.1.28 Hiden Isochema

Hiden Analytical manufactures quadrupole mass spectrometers, instruments and apparatus for a diverse range of applications—precision gas analysis, plasma diagnostics by direct measurement of plasma ions and ion energies, SIMS probes for UHV surface science, catalysis performance quantification, and thermogravimetry—over a pressure range extending from 30 bar processes down to UHV/XHV. The first Intelligent Gravimetric Analyser (IGA) was originally developed in the Department of Physics and Space Research at the University of Birmingham (UK). Since 1989 Hiden Isochema designs and manufactures sorption instruments for research, development and production applications in surface chemistry and materials science. Through the next decade the IGA as well as other innovative products and services for sorption science were manufactured and developed by Hiden Analytical Ltd., from its manufacturing base in Warrington, England. Following the continued success of the IGA range, Hiden Isochema Ltd. was formed in January 2002 and now employs over 20 staff at its UK headquarters.

8.2.1.29 Hottinger Baldwin Messtechnik GmbH

HBM was founded in 1950 by Karl Hottinger at Vogtareuth, Germany. The first products developed and produced were for parameters like force, displacement, vibration and special inductive displacement transducers. In 1955 the company moved to Darmstadt, Germany, and started the production of wire strain gauges, corresponding sensors and sophisticated resistant bridges. HBM has become a leading international manufacturer in the field of test and measurement. In 1963 the company's name was changed to Hottiner Baldwin Messtechnik GmbH. In 1973, HBM founded a subsidiary in Marlborough, USA, and in 1997 a subsidiary along with production facilities in Suzhou, China. In 1977 a calibration laboratory was accredited. In 2008, HBM expanded its portfolio of acquisition instruments and software solutions by purchasing the companies nCode, LDS and SoMat. Recent product innovations include torque measurement systems and universal data acquisition devices.

8.2.1.30 IES Corporation

The National Lab Balance Repair Center is the largest American service company specialising in repair of laboratory and precision balances including vacuum microbalances. Since 1981, more than 33 000 lab balances have been repaired. IES does not carry out on-site calibrations or sell new equipment. A Field Service Handbook is offered free.

8.2.1.31 Itin Scale Company, Inc.

Itin is dealer of every type of balances. MINX is trademark of Itin for pocket scales.

8.2.1.32 Kern & Sohn GmbH

Mechanik Gottlieb Kern (1819–1886) founded his workshop in 1844. Indeed the root of that company is much older, because Gottlieb Kern married Auguste (1818–1900), the widow of Phillipp Matthäus Sauter. The couple combined the Kern and Sauter workshops (Fig. 8.12a). As reported in the previous chapter, the root of Sauter balance production can be traced back to Phillip Matthäus Hahn around 1770 (Fig. 8.12b). Kern & Sohn is now the oldest existing balance producing company [53]. His business became the core cell of internationally famous precision balance production in the Swabian Jura. Kern is still an independent, medium sized family business, in its 5th generation. Today's production program covers laboratory, industrial and medical balances, as well as accessories like test weights. Manging director is Martin Sauter.

8.2.1.33 Korona

In 1991, Korona introduced the personal scale into its range. Over time, the portfolio was extended to include numerous products linked to the theme of wellness. After a re-structuring of the range, Korona is starting to concentrate on its core competences again in 2009: the development of high-quality personal and kitchen scales.

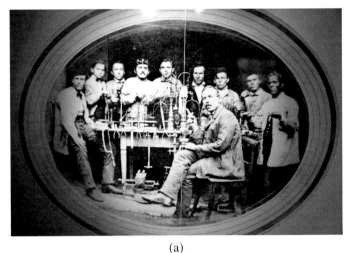

(a)

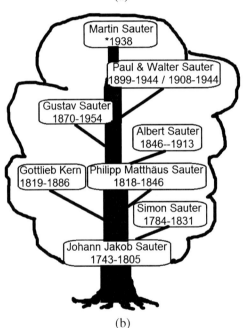

(b)

Fig. 8.12 (**a**) Workshop of Gottlieb Kern, Ebingen, Germany 1863. (**b**) Family tree of the company Kern & Sohn, Ebingen, Germany

8.2.1.34 Linseis

The production of Linseis is centered on thermal analysis, data logger and recorder. This includes thermobalances for high vacuum and pressures up to 150 bar, and temperature up to 2500 °C. A magnetic suspension balance for high vacuum investigations has been developed.

8.2.1.35 Magnescale

On January 8, 2010, the parent company Sony Corporation, Mori Seiki Co., Ltd. and Sony Manufacturing Systems Corporation signed a definitive agreement to transfer the measuring systems (Magnescale, Laserscale and digital gauge) business of SMS to Mori Seiki.

This new entity, Magnescale Co., Ltd. divested from SMS, is a wholly owned and consolidated subsidiary of Mori Seiki. Although, Magnescale Co. Ltd. will engage in the measuring systems business, Sony Manufacturing Systems America Inc. (SMSA) will continue to handle the measuring products in both sales and technical support. Magnescale offers a complete line of linear and rotary encoders for a wide variety of industrial applications. Included are digital scales in incremental and absolute feedback magnescale.

8.2.1.36 Mark-10 Corporation

Mark-10, founded in 1979, is a manufacturer of force and torque measuring instruments.

8.2.1.37 Mecmesin

Founded by Jim Oakley, an accomplished design engineer, in 1977, Mecmesin remains a privately owned company, based in Slinfold, West Sussex, UK, which has expanded worldwide, with regional offices in the US, China, Thailand and France. The focus of the company has always been to provide high quality test solutions. Products include force gauges, torque gauges, tensile and compression testing systems, torque testing systems and software.

8.2.1.38 Mettler-Toledo AG

Mettler-Toledo International Inc. is the world's leading manufacturer of precision weighing instruments and other precision analytical instruments for the laboratory and industrial/retail markets [55]. The company holds leading positions in most of its product categories, including weighing instruments, titrators, automatic lab reactors, thermal analysis systems, pH meters, electrodes, automated synthesis products, and moisture analysers. Mettler-Toledo is also the world leader in metal detection systems for the cosmetics, chemicals, food processing, pharmaceutical, and other industries.

Although parts of the company, including its Toledo line, established in 1901 as the Toledo Scale Company in the United States, predate the company, Mettler traces its official origins to 1945 and the design of the single-pan analytical balance by company founder Erhard Mettler. With the introduction of its scale, the company broke the two-pan weighing mould, using the so-called 'substitution principle' to achieve more accurate measurements. Large-scale production of the unit began in 1946. Mettler continued to refine its technology, introducing, in 1952, a microscale, capable of measurements to 0.0000001 grams.

The company's increasing sales brought it to its first expansion moves in 1952, with the opening of the Stäfa, Switzerland production facility. Two years later, the company's increasing international presence, in particular in the United States, led it to open its first international subsidiary, Mettler Instrument Corp., in Hightstown, New Jersey. This international expansion was followed by the company's move into Giessen, Germany, in 1957. In 1962 the company acquired Dr. Ernst Rüst AG, a maker of high-precision mechanical scales. The newly added division was renamed as Mettler Optic AG. Mettler also made its first moves at expanding its product categories, introducing a thermal-analyser in 1964. By then, Mettler also had moved into new facilities, with the construction of the first building of its Greifensee campus, which would remain the company's headquarters through to the end of the century. The company took a new step towards a diversified product line with the 1970 introduction of automated titration systems. At the same time, Mettler strengthened its core scales component with the acquisition of balance manufacturer Microwa AG. The following year, another acquisition, of August Sauter KG, of Albstadt-Ebingen, in Germany, added that company's specialised industrial and retail scales, as well as more than 500 employees to the Mettler payroll. By then, Mettler had taken its place among the world's leading manufacturers of specialised scales and other analytical apparatus. In 1973 Mettler's first fully electronic precision balance made its debut. This enabled the company to open an additional production facility, in Uznach, Switzerland.

In 1980 Mettler's shares were bought up by Ciba-Geigy. Mettler was added to Ciba-Geigy's industrial division. The purchase enabled the company's founder to retire, while ensuring Mettler the resources for future growth. By then, Mettler had begun to expand beyond its original laboratory market to produce balances and other equipment for the industrial and retail market, especially food retailing. In 1982 the company presented its first electronic precision balance for industrial applications, the Sauter MultiRange. A new product line was launched in 1985, when Mettler, in collaboration with Ciba-Geigy, introduced its first automated lab reactor. The company also moved into the Asian market, opening a joint venture operation in China, where it began production of laboratory equipment in 1987.

The second half of the 1980s was marked by three significant acquisitions. The first took place in 1986, when Mettler acquired fellow Swiss company Ingold Firmengruppe and its line of laboratory and industrial-use electrodes and sensors. Next, in 1987, Mettler took over Garvens Automation GmbH, near Hanover, Germany, and that company's dynamic checkweighers, dosage control, and other processing systems. In 1989 Mettler acquired the Toledo Scale Corporation, based in Worthington, Ohio. Toledo, which had been founded in 1901, was then the largest producer of industrial and food retailing scale systems in the United States. The acquisition would lead Mettler to capitalise on Toledo's strong brand recognition. Following the acquisition, Mettler changed its name to Mettler-Toledo AG.

Mettler-Toledo continued its product line expansion when it acquired the rheology and laboratory automation systems from Contraves AG in Zurich in 1990. Another acquisition made that year brought the laboratory balance production of Ohaus Corporation into the Mettler-Toledo group of products. On a larger scale, Mettler-Toledo inaugurated new production facilities in Albstadt and Giessen, in Germany, as well as new quarters in Schwerzenbach, in Switzerland.

In 1996 Mettler-Toledo regained its independence from Ciba-Geigy after a management buyout, assisted by AEA Investors, a New York-based investment group. The following year, AEA Investors brought Mettler-Toledo to the public, listing the company on the New York Stock Exchange. An important acquisition was the 1997 purchase of the United Kingdom's Safeline Ltd., the leading manufacturer of metal detection systems destined for the food, pharmaceutical, cosmetics, chemicals, and other industries requiring security testing procedures for their packaged products.

In 1998, the Illinois-based Bohdan Automation Inc. was purchased, a maker of laboratory automation and automated synthesis equipment. The move marked a further enhancement of Mettler-Toledo's position in the growing laboratory automation markets, especially in the synthesis reactor product category. At the end of 1998, Mettler-Toledo was able to boost this segment still further, as it announced the acquisitions of two drug and chemical compound automated discovery and development systems manufacturers, Applied Systems, of Annapolis, Maryland, and Myriad Synthesiser Technology, based in Cambridge, England. By then, Mettler-Toledo had elected a new company chairman, Robert Spoerry, who had been serving as company CEO since 1993. Mettler-Toledo continued on its acquisition march in 1999. In February of that year, the company announced its intention to acquire French industrial and retail scale manufacturer Testut-Lustrana. In 1999, Mettler-Toledo completed the acquisition of full control of its Chinese joint venture partnership, which, combined with company-owned production facilities in Shanghai, would enable the company to expand its sales throughout the Asia-Pacific region.

With 46 percent of sales in Europe and 43 percent of sales in the United States, Mettler-Toledo had become one of the most global of laboratory equipment manufacturers, with growing operations in the Asian region. By then, industrial products had risen to become the company's largest product area, with 49 percent of annual sales, as compared with just 38 percent for sales of laboratory products.

8.2.1.39 Michell Instruments GmbH

Michell Instruments produces moisture and humidity measurement instruments. With over 30 years experience, Michell designs and manufactures a wide range of transmitters, instruments and system solutions capable of measuring dew-point, humidity and oxygen in a vast range of applications and industries ranging from compressed air, power generation, process, oil and gas, pharmaceutical and many more. A quartz crystal sensor is applied for moisture trace analysis.

8.2.1.40 Netzsch Group

The Netzsch Group is a globally active family-owned enterprise with headquarters in the Bavarian town of Selb, Germany. Its international presence is backed by 127 sales and production centres in 23 countries on three continents. Its three business units are analysing & testing, grinding & dispersing, and pumps & systems. Netzsch Analysing & Testing was founded in 1952 as a department of Gebrüder Netzsch-Maschinenfabrik. First DTA and dilatometers were manufactured and in 1959, thermobalances. In 1970 combined TGA/DTG apparatus were offered. In 1993 cooperation with Bruker Optik GmbH began. In 2005 Dr. Thomas Denner and Walter H. Neumann took over from Dr. Wolf-Dieter Emmerich the management of the business unit Analysing & Testing. In 2006 Easy-to-use *Tarsus* for quality assurance was presented.

8.2.1.41 Microtech Instruments

Microtech Instruments started in 1993 as Baraboo Balance. The company provides repair and service for Cahn instruments.

8.2.1.42 Ohaus

For more than 100 years, Ohaus has had vast experience in the design, development and marketing of scales and balances in the laboratory, industrial, education, retail, food, and jewellery industries. The company has headquarters in Parsippany, New Jersey, with offices throughout Europe, Asia and Latin America. Ohaus now belongs to the Mettler group.

8.2.1.43 Perkin-Elmer

Perkin-Elmer design, manufacture and deliver advanced technology solutions that address the world's most critical health and safety concerns, including maternal and foetal health, clean water and air, and safe food and toys. The company was founded in 1937 by the banker Richard S. Perkin and the reporter Charles W. Elmer starting with optics and consulting business. Within the large number of instruments also a series of thermgravimetric apparatus had been developed.

8.2.1.44 Pesa

Pesa was founded in 1972 by merger of several companies working in the field of balance manufacturing. The company specialised in the development of electronic weighing cells based on an oscillating string. In general weighing cells and platforms are manufactured for Original-Equipment-Manufacturers (OEM) which can be incorporated in apparatus, machines and industrial plants. The digital output is evaluated by computer. Today Pesa is a member of the Bizerba group.

8.2.1.45 Porotec GmbH

In 1994 former employees of Carlo Erba Germany under the direction of Gerd
Schmidt founded the company POROTEC taking over the distribution of the microstructure lines for THERMO Electron (Carlo Erba) and the support of customers
in Germany, Switzerland, Austria and eastern Europe. Since 1995 POROTEC offers instruments for the characterisation of finely divided and porous solids. In its
laboratory measurements are made and instruments demonstrated concerning Hg-
porosimetry, sorption, density, particle size, temperature programmed desorption reduction oxidation and pulse-chemisorption. In 2000 Jürgen Adolphs took the chair.

In 1982 close contact with scientific researchers led to organisation of the first
'Workshop on the characterisation of finely divided and porous solids'. This 2 day
event has been repeated every 2 years at Bad Soden (Germany).

8.2.1.46 Precisa Instruments AG

In 1935 Precisa AG was founded at Dietkon, Switzerland by Ernst Jost. In 1942 began dedicated production of mechanical calculating machines in Zurich- Oerlikon.
In 1958 there were more than 1000 employees with a world network of agents and
service centres. In 1964 merger of Precisa AG with the typewriter manufacturer Hermes SA took place and the new company was named Hermes-Precisa International
AG. In 1978–1979 manufacture of a range calculating machine and new development of electronic precision-scales began. Manufacture of Hermes typewriters by
Olivetti started. During 1980–1996 Precisa precision and analytical balances became established in the world market with Swiss technology and production.

In 2005 a new production range in drying and ashing was introduced. In 2006
the company changed its name to Precisa Gravimetrics AG to emphasise the core
business.

8.2.1.47 Precision Instrument Repair Company

This company was founded in 1977. Its principals were former employees of the
Cahn Instrument Company in Cerritos, California. PIR is licensed to repair, calibrate and certify laboratory balances and scales, and this is the main business of the
company. It also repairs and restores antique balances of a collectible nature. They
also sell laboratory balances.

8.2.1.48 QCM Laboratory

QCM lab is a department of G&M Norden AB, which is concerned mainly with
manufacturing and dealing of computer components and of instruments based on
quartz crystal sensors. The QCM laboratory develops equipment, based on Quartz

Crystal Microbalance (QCM) technique. Balances have a resolution in the pictogram range and a mega-gravity field generator can produce accelerations up to one hundred million times the gravitational acceleration on the Earth. This opens new interesting fields in fundamental physics, materials science and biotechnology in the range of temperatures between $-200\ °C$–$500\ °C$ in gases and liquids.

8.2.1.49 Q-Sense

Q-Sense AB was founded in 1996 by a group of researchers at the Department of Applied Physics, Chalmers University of Technology in Gothenburg, Sweden, where QCM related research had been going on since the 1970s. The research group discovered and patented (1995) a whole new way of extracting mass and viscosity parameters for surface analysis in liquid, based on the quartz crystal microbalance (QCM). In 1999, when the first commercial system for measurements in liquid was launched, Q-Sense designed acoustic resonator based instruments for molecular binding events taking place on various surfaces. Applications are found in nanotechnology, biomaterials, detergents, drug discovery and biosensor development. The unique feature of QCM-D (dissipation) technology is the ability to provide real-time, structural information about the adsorption/interaction of molecules interacting with different surfaces. By measuring both the dissipation factor and the frequency, it became possible to analyse the whole chain of reactions that occur when a new substance is introduced to the surface of the quartz crystal. The new technique was called QCM-D (Quartz Crystal Microbalance with Dissipation monitoring).

At the beginning of 1998, the company began to develop its first commercial product, a first multi-frequency QCM that can also measure the viscoelastic properties of adsorbing layers in liquid. Today Q-Sense has a subsidiary in the US, a well established distributor network in Europe and Asia and a second generation system on the market, the Q-Sense E4.

8.2.1.50 Radwag Wagi Elektroniczne

Zaklad Mechaniki Precyzyjnej RADWAG is the biggest manufacturer of electronic weighing equipment in Poland and one of the biggest in EU. RADWAG manufactures and offers wide range of modern electronic balances, industrial and medical scales, weighbridges and checkweighers. This system has been recognised and certified with CE Metrology certificate, by European notified body NMi (Nederlands Meetinstituut). It authorises RADWAG to self declaration of conformity of their products with NAVI directive. They manufacture balances beginning from microbalances with very high precision ($d = 0.001$ mg, max 21 g) to weighbridges of high capacity ($d = 10$ kg, max 60000 kg).

8.2.1.51 RAVAS GmbH mobile Wiegetechnik

Ravas is manufacturer of mobile weighing systems built in lifting trucks, load boards and fork lift trucks etc. A brand name of RAVAS GmbH is Bruss.

8.2.1.52 Rigaku

Since its inception in 1951, Rigaku has developed hundreds of major innovations to their credit, the Rigaku Group of Companies are world leaders in the fields of protein and small molecule X-ray crystallography, general X-ray diffraction, X-ray spectrometry, X-ray optics, semiconductor metrology, automation, computed tomography, nondestructive testing and thermal analysis. Whether supplying the tools to create better semiconductor chips, enabling drug discovery, improving production line quality, or exploring the new frontier of nanotechnology, Rigaku products and services lead with innovation.

8.2.1.53 Rubotherm

Rubotherm develops, produces and sells measuring instruments and laboratory plants for process engineering. The main emphasis is placed on gravimetric measuring methods for recording mass changes under controlled environments. In the 1980's the company founders from the Ruhr-Universität Bochum (Germany) developed a modified method of the suspension balance whereby samples could be weighed in closed vessels without any contact. The principle is as follows: a sample is weighed from the outside by means of a new kind of magnetic suspension balance (which has been patented worldwide) and the suspension force is contactlessly transmitted from the pressurised measuring cell to a microbalance at ambient atmosphere. This means that mass changes of a sample can be recorded even under extreme conditions with the utmost accuracy. Since the mid-1980's these magnetic suspension balances have been applied to fundamental research work with enormous success. Numerous requests, including many from industry, induced the developers to set up the company Rubotherm GmbH in 1990, located in the Ruhr Technology Centre in Bochum.

8.2.1.54 Sartorius

The Sartorius Group is a leading international laboratory and process technology provider covering the segments of biotechnology and mechatronics. The company currently employs approximately 4 350 persons. In 1870 the Goettingen 'university mechanician' Florenz Sartorius founded the 'Feinmechanische Werkstatt F. Sartorius' and began with the production of short-beam analytical balances [41]. This already industrialised workshop has been developed to the mechatronic branch in which is primarily manufactured equipment and systems featuring weighing, measurement and automation technology for laboratory and industrial applications; the developments are described below.

In 1922 Richard Zsigmondy, chemist and physicist as well as Nobel Prize laureate in 1925, invented the membrane filter in 1918 and the 'cold ultrafilter'. This filter is the basis for commercial-scale production of Sartorius separation and filter technology products. Today the major areas of activity in its biotechnology segment focus on fermentation, filtration, fluid management, purification and laboratory applications. In the mechatronics segment, the company Sartorius has its own production facilities in Europe, Asia and America as well as sales subsidiaries and local commercial agencies in more than 110 countries.

On the recommendation of Otto Hahn (chemist and Nobel Prize winner in 1944), Sartorius initiated industrial-scale manufacture of hydrodynamic bearings (slide bearings). Hydrodynamic bearings are used increasingly by customers predominantly in high-speed applications, such as turbines, compressors and industrial pumps. As manufacturing of bearings was a technology separate from Sartorius's core activities the department was spun off in 2005 as a separate company with about 100 employees. In order to bring the product on the international market in 2007 Sartorius Bearing Technology GmbH was sold and integrated into the John Crane Inc. business of the British Smith Group plc which focuses on the design and manufacture of seals, the production remaining in Goettingen.

In 1955 the Gast microbalance for weighing under atmospheric conditions was presented. During 1961–1964 Theodor Gast (1916–2011) was scientific director of Sartorius. Under his guidance electronics was introduced in balance design. In 1963 the Gast electromagnetic suspension balance was presented. In 1964 the first laboratory balances with an analog output made their debut. In 1970 the first Sartorius electronic precision and analytical balances were presented. The resulting boom in orders led to the founding of numerous foreign subsidiaries, and contributed decisively to the significance of the Sartorius Group today. In 1971 a nanogram balance set the world record for the most precise weighing. This balance was used to weigh the moon rocks that astronaut Neil Armstrong brought back to Earth from his expedition. In 1972 a compact analytical balance with a digital, electronic weight display, stability control and digital output was launched. In 1975 a precision balance featuring microprocessor electronics premiered and in 1977 fully electronic analytical balance with microprocessor electronics had been launched. In 1979 manufacturing of toploading analytical balances began. In 1996 the world's first ultra-microbalance with a weighing capacity featuring a resolution one in 21 million and an accuracy of 0.1 µg made its debut. In 1997 the first monolithic weigh cell technology was presented. The monolithic weigh cell replaces a complicated weighing system made up of up to 150 different parts. This new mechatronic system is the basis for many successive generations of balances and scales.

In 1990 Sartorius AG went into public ownership. In 1998 capital was successfully raised. Boekels was acquired (leading in quality control in the processing industry and in metal detection). In 1999 Denver Instrument (laboratory balances and measurement technology) was acquired. In 2001 GWT (formerly known as Phillips

Wägetechnik and leading in high-capacity weighing technology) was incorporated. In 2004 the brands were integrated and new names were created for Group member companies: Sartorius Aachen GmbH & Co. KG (formerly Boekels), Sartorius Hamburg GmbH (formerly GWT, Global Weighing, Hamburg. Sartorius bought a 100 % stake in application specialist Omnimark Instrument Corporation based in Arizona, USA. In 2003 there was a change at the helm of the Executive Board: Dr. Joachim Kreuzburg became spokesman of the Executive Board and Group CEO.

In 2009 Sartorius signed a cooperation agreement with Metrohm Schweiz AG for marketing Sartorius Mechatronics laboratory equipment in Switzerland. The Executive Board was rearranged. In addition to Executive Board Chairman and CEO Dr. Joachim Kreuzburg, who is now also the Executive for Labour Relations, Reinhard Vogt and Jörg Pfirrmann were appointed as new members. Dr. Günther Maaz resigned from his position.

8.2.1.55 Schenck

In 1881 Carl Heinrich Johann Schenck founded his 'Eisengießerei und Waagenfabrik', in Darmstadt. As early as 1863, Carl Schenck had already co-founded the Mannheim Company of 'Schenck, Mohr & Elsässer', where the automotive pioneer Carl Benz was employed as a technical specialist. In 1881, Schenck left the Mannheim company in order to establish his own motor works. In 1902 the product range was extended to include conveyor technology, followed in 1907 by the start of development and production of balancing machines. At the end of the 1920s, the product range was again expanded to dynamometers and brake testing stands, although the iron foundry was closed.

After the Second World War, the company expanded enormously. Schenck products were now delivered to every part of the world, coming into use in many sectors of industry. Sales and production locations were also established in many countries. Manufacturing was extended in the field of measurement technology, automotive engineering, testing, conveying (gravimetric feeders, weighfeeders) and balancing technology. In 1972 incorporation of Florenz, Braunau, took place, Since 2000, Carl Schenck AG is a subsidiary of the globally-active Dürr Group. The operational business is today managed by the Balancing and Diagnosis Technology Division.

8.2.1.56 Scientec

Scientech, Inc. was founded in 1968 by Dr. Robert W. Zimmerer, a former physicist at the National Bureau of Standards located in Boulder, Colorado. Scientech's first product line was the original beamless electronic balance, which Dr. Zimmerer had invented. Scientech was also the first American balance manufacturer to utilise microprocessor technology in its line of electronic top-loading balances. This line included the first external keyboard for balances, which could select various functions such as parts counting, statistics, and weight averaging. Soon after introducing the balance line, Scientech marketed the first commercial laser power and energy measurement system based upon prototypes developed by the National Bureau of Standards.

8.2.1.57 Seca

In 1840, locksmith A.C.C. Joachims set up on his own company in Hamburg and began to build scales. He had acquired the necessary know-how from the Strasbourg mechanic and monk Quintenz who had invented the decimal scale in 1821. In the course of the next thirty years, the locksmith's workshop developed into a small factory. However, the business stagnated when Joachims died in 1874. Things started to improve again when Frederik Vogel, a young businessman, purchased the scales factory in 1888. He expanded the product range and introduced the brand name seca which he registered in 1897. Frederik Vogel passed the management over to his son Robert. When Robert Vogel died in 1966, his son Sönke joined the company. In 1970, Sönke Vogel focused production on medical measuring and weighing, on products which are needed for diagnosis and therapy purposes in medicine in order to determine the precise weight of a patient. Seca is now one of the market leaders for these products. The product range includes: baby scales, column scales, flat body scales, wheel scales, bed dialysis scales, and height measuring instruments. The company's headquarters are in Hamburg and it has branches in Austria, France, England, Switzerland, Japan, China, Mexico and the USA.

8.2.1.58 SII Seiko Instruments NanoTechnology Inc.

In 1877 Kintarō Hattor founded a workshop for watch repair. The workshop was developed to production first of wall clocks under the brand Seikōsha. In 1937 the production of watches was separated in a subsidiary, later named Seiko Instruments. High-tec quartz watches had been developed and besides scientific instruments, like thermoanalytical apparatus.

8.2.1.59 SETARAM

The production of SETARAM commenced with thermal analysis and calorimetry applications. In the 1850s Pierre Antoine Favre (1813–1880) demonstrated first measurements of heats of adsorption of gases on solids using a calorimeter [61, 62]. To measure adsorption at liquid air temperature James Dewar (1842–1923) built a diathermal phase-change, calorimeter. Albert Tian (1880–1972) invented the heat-flow microcalorimeter arranged in a multi-shielded thermostat with a thermopile of 42 junctions + 7 thermocouples and using Peltier power for heat compensation. His pupil Edouard Calvet (1895–1966) introduced the differential heat-flow microcalorimeter. He founded a special Institute built by the CNRS, in 1959, for development of microcalorimetry, now 'Jean Rouquerol-Laboratorie Chimie Provence',

Universites-Aix-Marseilles, France. The Tian-Calvet microcalorimeter, specially suited for studies of adsorption or immersion was commercialised first in 1948 by ARAM (measuring instrument repair department of Ugine, the steelmaking group) and in 1954 by the dealer DAM (Diffusion des Appareils de Mesure). The companies' commercialized also the Eyraud balance, an electrodynamic high vacuum thermobalance. In 1965 ARAM took over DAM and the company changed its name to SETARAM. It was bought by SFIM, an aeronautical company with 2000 employees that wished to diversify its activities. In 1997 SFIM sold its non-aeronautical activities, and Setaram joined the young high-tech manufacturing group KEP. In 2005 it signed an exclusive partnership with the Swiss company AKTS (Advanced Kinetics Software), and in 2006, an exclusive distribution partnership with MATHIS Instrument: TCi (thermal conductivity), becoming C-THERM company. In 2008 it acquired HY-ENERGY LLC (Newark, CA), a manufacturer of high performance gas sorption scientific instruments for the ever-expanding field of clean renewable energy research. In 2008 partnership with HEL, the leader in adiabatic calorimetry occurred.

8.2.1.60 Shimadzu

The Japanese company Shimadzu is one of the leaders in producing equipment for instrumental analysis. Its program includes UV chromatography, liquid chromatography, infrared chromatography, mass spectroscopy, cardiology and angiography, balances and instruments for particle size analysis.

8.2.1.61 Soehnle Professional GmbH & Co. KG

Soehnle was founded in 1868 at Murrhardt by Wilhelm Soehnle. Until 1920 all household balances were manufactured craftsmanlike, but then the so-called 'Reform-Balance' was built in series. In 1956 the first body scale was designed. Following this electronics was applied and in 1987 the first diet computer was presented. In 2001 Soehnle was merged into the Leifheit AG, a company centered on the production of household appliances. Today Soehnle deals in household, body and medical balances.

In addition Soehnle Professional produces transportation scales.

8.2.1.62 Surface Measurement Systems

(SMS) Ltd. develops and engineers innovative experimental techniques and instrumentation for physico-chemical characterisation of complex solids. The production comprises instruments for gravimetric dynamic vapour sorption and inverse gas chromatography.

8.2.1.63 TA Instruments

Headquartered in New Castle, DE, TA Instruments is supplier for apparatus for thermal analysis, rheology, and microcalorimetry. The TA group merged with several specialised companies, like Waters, Rheometric Scientific (rheology), VTI Corp (sorption analysis), and Bähr (thermal analysis).

8.2.1.64 Thass GmbH

Founded in 2002 by Jürgen Koch at Friedberg, Germany. The company provides apparatus for investigating surface properties and thermal analysis using instruments of Seiko Scienitfic NanoTechnology Inc. and Surfacee Electro Optics. Besides, laboratory and technical balances of Gibertini are dealt.

8.2.1.65 Taylor/Hanson

In 1851, George Taylor started a small company making precision thermomenters and barometers, and hygrometers. During the Industrial Revolution years, Taylor grew with the demands for precision and controlling instruments for the environment. During World War I it made altitude barometers for the fighter aircraft, and during World War II, compasses, besides the other products.

The other part of Taylor Precision Products was the Hanson Scale Company, founded in 1888 in Chicago, IL, making light industrial scales. Hanson set up a manufacturing plant in Ireland. Hanson Scale was eventually sold in the 1960's, but the family Hanssen then began Metro Corporation. Metro then became a key US brand in the bath scale market.

In 1996, Taylor Environmental Instruments was sold to Lawrence Zalulsky & Associates, run by the key personnel that had run Health-o-meter scales. In 1997, they purchased Metro Scale Corporation. Using the strong Taylor brand, Taylor was then used to brand bathroom scales. In 2002, Taylor Precision Products was acquired by the HoMedics Corporation, which was busy in the health and wellness product industry. In 2004, HoMedics purchased Salter Housewares Ltd., which were producing kitchen and bath scales in the United Kingdom and in the USA. Salter Housewares USA was merged into Taylor Precision Products in 2006, where the two product lines were combined to be a strong force in measurement products.

8.2.1.66 Teraoka Digi

The Japan Calculation Machinery Manufacturing Co. was founded in 1910. In 1928 Japan's first spring scale was manufactured. In 1934 the Teraoka Research Centre was established. An automated temperature compensating device for scales was developed. In 1938 the company name was changed to Teraoka Seikosho. In 1958 the

'Gold' spring scale was produced at the rate of 10 000 units per month. In 1958 the world's first price computing scale was presented. In 1973 the digital load cell price computing scale was introduced. In 1981 the Company name changed to Teraoka Seiko Co., Ltd. In 1984/85 the integrated fully automatic weigh/wrap/label system was marketed. In 1997 the latest postal scale system and electronic scales were delivered to post offices nationwide in Japan, and new postal systems were developed. In 2007 handy terminal type self shopping was released.

8.2.1.67   Thermo Fisher Scientific

Thermo Fisher Scientific Inc. is the world leader in serving science with approximately 37 000 employees. Instruments include some special balance types, as Shandon scales, and belt scales. In 2011 The Cahn program of thermogravimetric apparatus was discontinued and only repair was offered.

8.2.1.68   VTI

VTI is a manufacturer of gravimetric sorption analysis and thermogravimetric analysis (TGA) instruments. The company was founded in the 1970s by Dr. Augie Venero at Hialeah, FL. In 2008 VTI was purchased by TA Instruments.

8.2.1.69   Waters–TA Instruments

Waters Corp., headquartered in Milford, Massachusetts, holds worldwide leading positions in complementary analytical technologies—liquid chromatography, mass spectrometry, rheometry and microcalorimetry. Waters Corp. operates in two divisions: Waters Division and TA Instruments. TA Instruments rheometry and microcalorimetry instruments are used primarily in predicting the suitability of polymers and viscous liquids for various industrial, gravimetric sorption analysis, consumer goods and health care products.

8.2.1.70   WIPOTEC–OCS Checkweighers

Wipotec Weighing and Positioning Systems, Kaiserslautern, Germany, is the leading innovator and recognised system provider for the integration of intelligent weighing technology. The core business comprises the development, production and integration of ultra-fast precision weighcells and high tech weighing systems for high speed applications.

OCS Checkweighers GmbH is a daughter company located at Schwäbisch-Hall, Germany. OCS Checkweighers produces checkweighers, catchweighers, in-motion scales as well as X-ray scanners.

8.2.1.71 Yale–Columbus McKinnon

The brand 'Yale' goes back to Linus Yale Jr. who in 1858 invented the cylinder lock and founded together with Henry R. Towne the factory at Stamford (Connecticut), USA. Since 1857 the Weston Differential tackle block was produced. In 1920 it purchased C.L. Hunt & Co. which was producer of electro platform trucks. In 1952 it started the production of forklift trucks at Velbert (Germany). In the following years several companies with related production programs had been purchased or merged and worldwide places of business were founded. Since founding of the Hungarian company Columbus McKinnon Tigrip balances may be Incorporated in the hoists.

8.2.2 Tables of Manufacturers

To make the following tables several search machines have been used. It includes manufacturers and dealers, because often dealers use components of manufacturers for own designs. Only few manufacturers cover the whole program of balance types. Many of them buy the sensor or parts of it elsewhere and modify it for the special task. Most manufacturers are concerned with industrial mass and weight sensors. Several thousand companies have included mass sensors in the construction of apparatus and systems. So, it is clear that this survey cannot be complete.

Table 8.1 Manufacturer and dealer of balances, force gauges and apparatus, and systems supplied with mass sensors

Manufacturer	Address	Internet	Types
ABG-Packmat	Oldisstrasse 55, CH-7023 Haldenstein, Switzerland	www.abg-packmat.ch/	Counting Weighing
ABB Instrumentation & Analysis	Affolternstrasse 44, Zürich, Switzerland	www.abb.com/	Load cell
ADE	AD GmbH & Co. Hammer Steindamm 27-29, D-22089 Hamburg, Germany	http://www.ade-germany.de/	Kitchen retail
A&D Instruments Ltd.	3-23-14, Higashi-Ikebukuro, Toshima-ku, Tokyo 170-0013, Japan	www.aandd.jp/	All types
Acculab Scales	USA	acculab.balances.com/	Laboratory
Accuweigh	19 Yampi Way, Willetton, WA, 6155, Australia	www.accuweigh.com.au	Packaging
Acrison	20 Empire Blvd., Moonachie, NJ 07074, USA	www.acrison.com/	Weight feeder

Manufacturer	Address	Internet	Types
Adam Equipment Co. Ltd.	Bond Avenue, Denbigh East, Bletchley, Milton Keynes, MK1 1SW, UK	www.adamequipment.co.uk/	All types
ADE GmbH & Co.	Hammer Steindamm 27-29, D-22089 Hamburg, Germany	http://medicine.ade-germany.de/	Household, Retail, Medical
Ados Pesatura Electronica	Via Bersaglio, I-20, 22015 Gravedona (CO), Lombardy	www.ados.it/	Load cell Industrial
Affeldt Maschinenbau GmbH	Altendeich 14, D-25335 Neuendorf	www.affeldt.com/	Counting Weighing
Africaine de Fournitures Industrielles et Techniques (AFIT)	Route d' El jadida, RN 1. km 14, commune rurale Ouled Azouz 20190 Nouasseur, Morocco		Import
Ahlborn Mess- und Regelungstechnik GmbH	Eichenfeldstraße 1-3 83607 Holzkirchen	www.ahlborn.com/	Load cells
Ainsworth	Denver Instrument, 5 Orville Dr., Bohemia, NY 11716	www.ainsworthbalances.com/	Laboratory
AJT Equipment Ltd.	Premier Partnership Estate, Brockmoor, Brierley Hill, West Midlands DY5 3UP, UK	www.ajt-testing.com/	Load cell Crane
Alessandrini strumentatione SpA	Via Bosco 16, I-41030 S. Prospero (Mo), Italy		Laboratory
Alfa Laval	Lund, Sweden	www.alfalaval.com/	Force control
Alfamatic srl	Via Magenta, 25 S. Giorgio su Legnano 20010 (MI), Italy	www.alfamatic.net/	Laboratory
ALL-CON World Systems, Inc.	Seaford, DE 19973, USA	www.all-con.com/	Solid feeder
All-Fill	418 Creamery Way, Exton, PA 19341, USA	www.all-fill.com/	Bag weight filler
Allpax GmbH & Co. KG	Zur Seeschleuse 14, D-26871 Papenburg, Germany	http://www.allpax.de/	Industrial Dealer
Alluris GmbH & Co. KG	Basler Strasse 65, D-79100 Freiburg, Germany	www.alluris.de/	Force gauge
Alpha-Pack (Shenzhen) Co., Ltd.	Block 10, Changxin Industry Zone, Wan'an Road, Shayi, Shajing, Baoan District, Shenzhen, Guangdong, China	http://alpha-pack.en.made-in-china.com/	Packing

Manufacturer	Address	Internet	Types
Althen GmbH Meß- und Sensortechnik	Frankfurter Str. 150-152, D-65779 Kelkheim, Germany	http://www.althen.de/	Load cell
AMETEK GmbH Lloyd Instruments Chatillon	8600 Somerset Drive Largo, Florida 33773, USA	www.ametek.com/	Force gauge
AMT AG	Badstrasse 34, CH-5312 Döttingen, Switzerland	www.amt-ag.net	Film thickness
Andilog Technologies	BP 62001, 13845 Vitrolles CEDEX 9, France	www.andilog.com/	Load cell Spring
APE Engineering GmbH	Hansaring 18, D-63843 Niedernberg, Germany	www.ape-engineering.de/	Industry
Applied Measurements Ltd.	3 Mercury House, Calleva Park, Aldermaston, Berkshire, RG7 8PN, UK	www.appmeas.co.uk/	Torque sensor
Applied Weighing International Ltd.	Southview Park, Marsack Street, Caversham, Reading, RG4 5AF, UK	www.appliedweighing.co.uk/	Process equipment, Load cell
ARD Pesage	ZA Ste Julie, 34980 Montferrier, LEZ, France	www.ard-design.net/ www.net-pesage.fr	Industrial
Arodo bvba	B-2370 ARENDONK Belgium	www.arodo.com/	Weight solid filler
Arpège Master K	36 av. Des Frères Montgolfier, F-69686 Chassieu cedex	www.masterk.com/	Industry Laboratory
Ascell Sensor	Industrial Congost Cl Congost 56 Nave 3 08760 Martorell, Spain	www.ascellsensor.com/	Load cell Industry
ASGCO Manufacturing Inc.	301 Gordon Street, Allentown, PA 18102	www.asgco.com/	Conveyor
ATC Sensor Division Instronics	1550 Kingston Rd. #1116, Pickering, ON, L1V 6W9, USA	www.instronics.com/	Load cell Crane
Atexis	France	www.atexis.fr	Load cell
Atlas Copco	Sickla Industriväg 19, 131 34 Nacka, Sweden	www.atlascopco.com/	Torque calibration
Atlet	Mölnlycke, outside Gothenburg, Sweden	www.atlet.com/	Forklift
ATP Messtechnik GmbH	J.B. von Weiss Strasse 1, D-77955 Ettenheim, Germany	www.atp-messtechnik.de/	Dealer
Automat di Cirio Matteo	Via Asti, 49, 14053 Canelli (AT), Italy	www.automatimpianti.com/	Industrial
Automated Packaging Systems Ltd.	Böttgerstraße 2, D-38122 Braunschweig, Germany	www.autobag.com/	Packing

Manufacturer	Address	Internet	Types
Autoscales & Service Co. Ltd.	Truweigh House, Ordnance Street Blackburn, Lancashire BB1 3AE, UK	www.autoscales.co.uk/	
Avery Weigh-Tronix Berkel, Dillon, Salter	1000 Armstrong Drive Fairmont, MN 56031-1439, USA	www.averyweigh-tronix. com/ www.ipsc.com.pl www.wtxweb.com/	All types
Azar Machine Mfg. Ind.	Sahand Alley, Kuy-e Western Tohid Ind., Azarshahr Rd. 51978 Tabriz, Iran		Industry
AZO GmbH & Co.	Rosenberger Str. 28, D-74706 Osterburken, Germany	www.azo.de/	Filler
Bähr-Thermoanalyse GmbH →Ta Instruments	Postfach 1105, 32603 Hüllhorst, Germany	www.baehr-thermo.de	Thermal analysis
Bayerische Waagenbaustätte Althaus GmbH	Industriestraße 2, D-86919 Utting, Germany	http://www.bayernwaage.de/	Dealer
Balanças Marques de José Pimenta Marques, Lda	Parque Industrial Celeiros, 2ª. Fase – Celeirós, 4700-000 Celeirós BRG, Portugal	www.balancasmarques.pt/	Industry Load cell
Barcelbal–Balancas e basculas, Lda	Rua do Caires 197. Pavilhao B, Braga, Portugal	http://www.barcelbal.com/	Industry
Baron	USA	www.directindustry.com/	Wheel scales Fork lifter
Batarow Sensorik GmbH	Karow 17, D-18276 Lüssow	www.batarow.com/	Load cell
BCM Sensor Technologies	Industriepark Zone 4, Brechtsebaan 2, B-2900 Schoten-Antwerpen, Belgium	www.bcmsensor.com/	Load cell Platform
Beckman Coulter, Inc.	4300 N. Harbor Boulevard P.O. Box 3100, Fullerton, CA 92834-3100, USA	www.beckmancoulter.com/	Scientific Thermal analysis
Behn + Bates Maschinenfabrik GmbH & Co. KG	Robert-Bosch-Str. 6, D-48153 Münster, Germany	www.behnbates.de/	Bag weight filler
BEL Engineering srl	Via Carlo Carrà 5, Monza I-20900 (MB), Italy	www.belengineering.com/	Laboratory
Berthold Technologies GmbH & Co.KG	Calmbacher Str. 22, D-75323 Bad Wildbad, Germany	http://www.berthold.com/	Radiometric belt weigher

Manufacturer	Address	Internet	Types
Best Technology GmbH	Detmolder Straße 377, D-33605 Bielefeld	www.best-technology.eu/	Retail
Best Weight Scale	167 13th Concession, Scotland, Ont. N0E 1R0. South of Brantford, Canada	www.bestweighscale.ca/	Spring Load cell
BeWA-tec Bernard Waagen Automatisierungs- und Steuerungstechnik GmbH	Kempener Str. 47, D-51469 Bergisch Gladbach, Germany	www.waagen-pruemm.de/	Dealer, repair
Bernhardt Packing & Process	18 Z.I. de la Trésorerie 62126 Wimille, France	www.bernhardt.fr/	Bag weight filler
Bihl & Wiedemann	Floßwörthstr. 41, D-68199 Mannheim, Germany	www.bihl-wiedemann.de/	Weighing controller
Bilatron di Marcello & Fausto Fucili s.n.c.	7, Via Maestri Del Lavoro 05100 Terni (Terni), Italy	http://www.bilatron.it/	Industrial
Bilfinger Berger Industrial Services Swiss AG, Helios AG	Erlenstr. 56, CH-4106 Therwil, Switzerland	www.helios-msr.ch/	Industry
Bint SRL	Via Breno 7, I-20139 Milano, Italy	www.bint.it/	Force gauge
Biosoft	Great Shelford, Cambridge, GB-CB22 5WQ, UK	http://www.biosoft.com/	Laboratory
Bitzer Wiegetechnik GmbH	Benzstrasse 3, 31135 Hildesheim, Germany	www.bitzer-waage.de	Industry
Bizerba GmbH & Co. KG	Wilhelm-Kraut-Str. 65, 72336 Balingen	www.bizerba.com	All types
Bolder automation GmbH	In den Klostergärten 9, D-65549 Limburg	http://www.bolder-automation.com	Industry
Bondy A/S	Hasselunden 14, 2765 Smørum, Danmark	www.bondylmt.dk/	Industry
PJ Boner & Co. Ltd.	Unit 35, Western Parkway Business Centre Ballymount Drive, Dublin 12, Ireland	www.pjboner.com/	Industry Calibration
Bonso Electronics Co. Ltd.	132 Da Yang Road, Da Yang Synthetical Develop District, Fu Yong, Shenzhen, 518103, China	http://www.bonso.com/index.html	Household
Bosche Wägetechnik GmbH & Co. KG	Reselager Rieden 3 D-49401 Damme, Germany	www.bosche.eu/	All types

Manufacturer	Address	Internet	Types
Brabender Technologie KG	Kulturstrasse 55-73, D-47055 Duisburg, Germany	www.brabender-technologie. com/	Bag weight filler
Breitner Abfüllanlagen GmbH	Daimlerstrasse 43, D-74523 Schwäbisch Hall, Germany	www.breitner.de/	Weight liquid filler
Brosa AG	Dr. Klein Str. 1, D-88069 Tettnang	www.brosa.net/	Load cell
Brovind Vibratori S.p.A.	Via Valle Bormida, 5. I-12074 Cortemilia (CN), Italy	www.brovindvibratori.it/	Bulk weight filler
Bühler	Gupfenstrasse 5, CH-9240 Uzwil, Switzerland	www.buhlergroup.com/	Industial
Bruss	Aspastraße 25, D-59394 Nordkirchen, Germany	www.bruss-waagen.de/	Industry Fork lifter
Burimec spa	Via Nazionale 24, I-33042 Buttrio (Udine), Italy	http://www.burimec.it/	Laboratory
Burster Präzisions-messtechnik GmbH & Co. KG	Talstrasse 1-5, D-76593 Gernsbach, Germany	www.burster.de/	Load cell
Bushman AvonTec	W133 N4960 Campbell Dr. Menomonee Falls, WI 53051, USA	www.bushmanavontec.com/	Packing conveying
Byk Gardner	Abelstraße 45, D-46483 Wesel, Germany	www.byk.com/	Laboratory
Cairo Bearing Agency (S.B.A)	1, El-Sherifein St., Off Kasr El-Nil St., 11111 Down Town, Cairo, Egypt		Dealer
CamCorp	9732 Pflumm Road, Lenexa, KS 66215, USA	www.camcorpinc.com/	Hopper
Canada Crane & Construction Machinery	Gleichen T0J-1N0, AB Southern AB, Canada	http://www.cn-racoon.com	Crane
Cardinal Detecto	203 E. Dougherty, Webb City, MO 64870, USA	www.cardinalscale.com/ www.detecto.com/	Medical Industry
Carrelli Carmeccanica S.r.l.	I-25020 Flero (Brescia) via F. Lana, 5	www.carmeccanica.it/	Fork lift
Cassel Messtechnik GmbH	In der Dehne 10, D-37127 Dransfeld, Germany	www.cassel-inspection.com/	Checkweigher
Cavicchi Implanti srl	Via Matteotti, 35, I-40055 Villanova di Castenaso (Bo), Italy	www.cavicchiimpianti.com/	Batching Packing
Celmi s.r.l.	L.go Brugnatelli 13/16, I-20090 Buccinasco (MI), Italy	www.celmi.com/	Load cell Platform

Manufacturer	Address	Internet	Types
CEM Corporation	3100 Smith Farm Road, Matthews, NC 28106-0200	www.cem.com	Microwave dryer
Ceramic Instruments s.r.l.	Via Regina Pacis, 20/26, I-41049 Sassuolo (MO), Italy	www.ceramicinstruments.com/	Laboratory
Chantland MHS Co.	502 7th Street North, Dakota City, IA 50529, USA	www.chantland.com/	Bag weight filler
Chatillon Ametek Measurement & Calibration Technologies Division	8600 Somerset Drive, Largo, Florida 33773, USA	www.chatillon.com/	Force gauge Spring
Chronos Richardson Systems GmbH	Reisertstrasse 21 53773 Hennef, Germany	www.chronosrichardson.com	Packing
Chronos BTH	1, Avenue Premier, Rivière-du-Loup (Québec) G5R 6C1, Canada Meerheide 40, NL-5521 DZ Eersel, Netherlands	http://www.chronosbth.com	Industry
CI Precision	Brunel Rd. Churchfields Salisbury, Wiltshire SP2 7PX, UK	www.cielec.com/	Microbalance Thermal analysis
Citizon Scales (India) Pvt. Lted.	Shop No. 3, Pushpanjali Bldg., Gaushala Lane, Malad (East), Mumbai-400 097, India	www.citizenscales.com/	All types
Martin Christ	An der Unteren Söse 50, D-37520 Osterode, Germany	www.martinchrist.de/	Moisture
CM Bilance SNC di Marco Chiavelli e Franco Beggio	V. Piemonte 39, Due Carrare (PD),	www.cmbilance.com	All types
Cobos S.A.	Calle Calabria 236-240, E-08029 Barcelona, Spain		Laboratory
Cogo Bilance Srl	Via dell'industria 51, I-30010 Camponogara (VE), Italy	http://www.cogobilance.it/	Medical
Combi Scale	Florida, USA	www.combiscale.com/	Bag weight filler
Collischan GmbH & Co. KG	Saganer Straße 1-5, D-90475 Nürnberg, Germany	www.collischan.de/	Counting Weighing
Columbus McKinnon	140 John James Audubon Parkway, Amherst, New York 14228-1197, USA	www.cmworks.com/	Materials handling

Manufacturer	Address	Internet	Types
Container Consulting Sevice Inc. - CCS	455 Mayock Rd., Gilroy, CA 95020, USA	www.ccs-packaging.com/	Packing
CMC Cleveland Motion Controls Inc.	6 Enterprise Road Billerica, MA 01821, USA	www.cmccontrols.com/	Load cell
Coffee Service Sp. Zoo	ul. Niemcewicza 26/U7 02-306 Warszawa, Poland	www.coffee-service.eu/	Weight solid filler
Comef Aparatura Naukowo-Badawcza	ul. Gdańska 2, PL-40-719 Katowice, Poland	www.comef.com.pl	Instrument Dealer
Concetti S.p.A.	190 Frazione Ospedalicchio I-06083 Bastia Umbra (Perugia), Italy	www.concetti.com/	Solid filler
A J Cope & Son Ltd.	The Oval, London, E2 9DU, UK	http://www.ajcope.co.uk/	Laboratory
Collischan GmbH & Co. KG	Allersberger Str. 185, D-90461 Nürnberg	www.collischan.de	Checking Packaging
Coventry Scale Company		www.coventryscale.com	Load cell manufacturer Dealer of all types
Crane Electronics Ltd.	3 Watling Drive, Sketchley Meadows, Hinckley LE10 3EY, UK	www.crane-electronics.com/	Torque sensor
Dataq Instruments, Inc.	241 Springside Drive, Akron, OH 44333, USA	http://www.dataq.com/	Load cell
Daewon GSI Co. Ltd.	990-2, Geumsan-ri, Waegwan-eup, Chilgok-gun, Gyeongsangbuk-do 718-802, Korea	daewonmachinery.en	Hopper scale Packing
Denver Instrument	6542 Fig Street, Arvada, COL 80004	www.denverinstrumentusa.com/ www.denverinstrument.com/	Laboratory
Deprag CZ a.s.	T.G. Masaryka 113, 507 81 Lázně Bělohrad, Česká Republika	www.depragindustrial.de/	Torque meter
Dibal S.A.	Astintze Kalea, 24 48160 Derio, Spain	www.dibal.com	Platform Retail
Diffusion Service	ZA du Moulin, Avenue de Mocard, F-85130 La Verrie, France	www.diffusion-service.fr/	Indication
DigiWeigh	15830 El Prado Road, Unit B Chino, California 91708, USA	www.digiweighusa.com/	Household Laboratory Industry

Manufacturer	Address	Internet	Types
Dillon force gauges	1000 Armstrong Drive Fairmont, MN 56031, USA	www.dillon-force.com/	Crane, Overload protection
Dimatec Analysentechnik GmbH	Nünningstr. 22-24, D-5141 Essen, Germany	http://www.dimatec.de/	Labatory Industry
Dini Argeo, S.r.l.	Via Della Fisica 20, Loc. Spezzano, I-41042 Fiorano Modenese (MO), Italy	www.diniargeo.com/	Industry
Ditta Monti Maurizio & Figli S.r.l.	Via Sesto Fiorentino 27, I-00146 Roma (RM), Italy	www.paginegialle.it/ montibilance	
Nuova Dizma – Pavan Group	Via Monte Grappa, 8, I-35015 Galliera Veneta (PD), Italy	www.dizma.it/	Bag weight filler
Doga	8 avenue Gutenberg, ZA Pariwest, F-78111 Maurepas, France	www.dogassembly.com/	Torque gauge
Dogain Instruments, Inc Siltec	1765 Scott Blvd #105, Santa Clara, CA 95050	www.isiltec.com	Precision
Doric Instruments Vas Engineering, Inc.	4750 Viewridge Avenue, San Diego, CA 92123, USA	www.doric-vas.com/	Strain gauge
Doran Scales, Inc.	1315 Paramount Parkway, Batavia, IL 60510, USA	www.doranscales.com/	Retail Checkweigher
DS & M S.r.l., Dosing Systems & Machineries	Via Indipendenza 1/B, I-41100 Modena, Italy	www.dsem.it	Bag weight filler
D.R. Italia s.r.l.	Italy		Load cell
Dytran Instruments	21592 Marilla Street Chatsworth, CA 91311-4137, USA	www.dytran.com/	Load cell Piezoelectric
Easydur Italiana di Renato Affri	Via Monte Tagliaferro, 8, Induno Olona (VA), Italy	www.easydur.com/	Torque gauge
Easy Weigh	Lauriergracht 68-D, NL-1016 RL Amsterdam, The Netherlands	http://www.easy-weigh.com/	Retail
Ebinger Waagenbau GmbH	Obere Riedwiesen 15, D-74427 Fichtenberg, Germany	www.ebinger-waagenbau.de/	Systems Repair
E.D.C. S.r.L.	Via Enrico Cialdini 37, I-20161 Milano, Italy	www.edcnet.it	Torque gauge
Educated Design & Developmant, Inc. ED&D	901 Sheldon Drive, Cary, NC 27613, USA	www.productsafet.com/	Force gauge

Manufacturer	Address	Internet	Types
Eilersen Electric A/S	DK-2980 Kokkedal, Danmark	www.eilersen.com/	Load cell Platform
Eilon Engineering Weighing Systems Ltd. (Ron Crane Scales)	1 Etgar, ind. park Tirat Carmel 7036 Haifa, Israel 1243 Naperville Drive, Romeoville, IL 60446, USA	http://www.eilon-engineering.com/	Industry
Elbo Controlli	Via S. Giorgio, 21, 20821 Meda (MB), Italy	www.elbocontrolli.it/	Force gauge
Electra SAS	Galleria Nuovo Centro, 11 32043 Cortina d'Ampezzo, (BL)	www.electrasas.com/	Hopper
Elektrotermometriia PubJSC	Chervonoarmiiskyi Prov. 4, Room 22, Luhansk 91055, Ukraine		Dealer
Elimco	8. Cadde 68. Sokak No:16 Emek, Ankara 06510, Turkey	www.elimko.com.tr/	Belt weigher controller
Elpe - Etude de Ligne de Pesage et Ensachage	Z. I de l'Albanie, route d'Agremont, F-73490 La Ravoire, France	www.elpe-packaging.com/	Solid filler
Emde Industrie-Technik	Koppelheck, D-56377 Nassau/Lahn	www.emde.de/	Solid weight filler
Endevco Meggit Sensing Systems	30700 Rancho Viejo Rd., San Juan Capistrano, CA 92675, USA	www.endevco.com	Piezoelectric
Enerpac Actuant Corporation	N86 W12500 Westbrook, Crossing, Menomonee Falls, Wisconsin 53051, USA	www.actuant.com/	Hydraulic load cells
J. Engelsmann AG	Frankenthaler Str. 137-141, D-67059 Ludwigshafen, Germany	http://www.engelsmann.de/	Industry
Grupo Epelsa	Parque Tecnológico, Tecnoalcalá, 28805 Alcalá de Henares, Madrid, Spain	www.grupoepelsa.com/	Retail Industry
Epsilonlab	5, rue de l' Olympe, bd. 2 Mars résid. Houria rdc. 20100 Casablanca, Morocco		Laboratory
Epsilon Technology Corp.	3975 South Highway 89, Jackson, WY 83001, USA	www.epsilontech.com	Load cell
ERDE-Elektronik AB	Spikgatan 8, SE-235 32 Vellinge, Sweden	www.erde.se	Load cell
ScaleComp		www.scalecomp.com/	Weight transmitter
EL Erhard & Leimer	Albert-Leimer-Platz 1, 86391 Stadtbergen, Germany	www.erhardt-leimer.com/	Load cell Force gauge

Manufacturer	Address	Internet	Types
Erichsen GmbH & Co. KG	Am Iserbach 14, 58675 Hemer, Germany	www.erichsen.de/	Load cell
Escali scales	3203 Corporate Center Drive, Suite 150, Burnsville, MN 55306, USA	http://www.escali.com/	Household
ESIT Electronic Systemler, Imalatve Ticaret Ltd.STI.	Mühürdas Cad 91, TR 81300 Istanbul, Turkey	www.esitscale.com/	Industrial Load cells
ETH-Messtechnik GmbH	Hagstrasse 10, D-74417 Gschwend, Germany	www.eth-messtechnik.de/	Torque sensor
Euromec Strumenti srl	183, Via Romana Ovest I-55016 Porcari (Lucca), Italy	http://www.euromec.it/	Packing
Euro Press Pack Spa Unipersonale	Via M. Disma, 87, 16042 Carasco (GE), Italy	www.europresspack.it/	Load cells
Eurotech	Via Fratelli Solari 3/a, I-33020 Amaro (Udine), Italy	www.eurotech.com/	Weight liquid filler
Extech Instruments Corporation	9 Townsend West, Nashua, NH 03063, USA	www.extech.com/	Force gauge
Excell Precision Co. Ltd.	6 F, No.127, Lane 235, Pao-Chiao Road, Hsintien, Taipai Hsien, 231, Taiwan	www.excell.com.tw/	Platform Counting
Exonex S.R.L.	Via Francesco De Sanctis, 53, I-20141 Milano, Italien	www.econex.it/	Hydraulic force scale
Fairbanks Scales	821 Locust Street, Kansas City, MO 64106, USA	www.fairbanks.com/	Industrial Retail
Feige Filling GmbH	Rögen 6a, D-23843 Bad Oldesloe, Germany	www.feigeusa.com/	Filler
Femto Tools	Furtbachstrasse 4, CH-8107 Buchs/ZH, Switzerland	www.femtotools.com/	Femto force sensors
Filltech GmbH	Industriegebiet Nord, Anton-Boehlen-Str. 25, D-34414 Warburg, Germany	www.filltech.de/	Filler Platform
Flintab AB	Kabelvägen, 4, SE-55302 Jönköping, Sweeden	http://www.flintab.com/	All types
Flintec Tranducers CAD Library	Köpmangatan 1B, SE-72215 Västerås, Sweden	www.flintec.com/ www.indutrade.se	Load cell Systems Transmitter
Flomat Bagfilla International	Flomat Bagfilla International Ltd. Fairport House, 28 Charles Street, Higher Hillgate, Stockport SK1 3JR, UK	www.fairport.co.uk/pmh/ flomat/	Bag weight filler

Manufacturer	Address	Internet	Types
Fos&s Optical Fiber Sensors FBGS Group	Bell Telephonelaan 2H, B-2440 Geel, Antwerp, Belgium	www.opticalfibersensors.org/	Load cell
FMS Force Measuring Systems AG	Aspstrasse 6, CH-8154 Oberglatt, Switzerland	www.fms-technology.com/ www.fms-ag.ch/	Load cells
Oskar Frech GmbH + Co. KG	Schorndorfer Straße 32, D-73614 Schorndorf-Weiler, Germany	www.frech.com/	Industry
Fulcrum Inc. (Torbal)	100 Delawanna Ave. Suite 120, Clifton, NJ 07014, USA	http://www.torbalscales.com/	Torsion balance
Fuji Robotics– Fuji Yusoki Kogyo Co. Ltd.	2327-1. Higashi Takdomari, Onoda, Yamaguchi, Japan	www.fujirobotics.com/	Packing
Futec Advanced sensor technology, Inc.	10 Thomas, Irvine, CA 92618, USA	www.futek.com/	Tension force Load cell
Gamry Instruments	734 louis Drive, Warminster PA 18974, USA	www.gamry.com	
Gandus Saldatrici	Via Milano, 5, I-20010 Cornaredo (MI), Italy	www.gandus.it/	Weight filler
Garant Research and Production Centre	ul. Promishlennaya, 7. St. Petersburg 198095, Russia	www.gamry.com	Quartz crystal
Garvens	Mettler-Toledo AG, Im Langacher, CH-8606 Greifensee, Switzerland	www.mt.com/garvens	Industry
Gassner Wiege-und Messtechnik GmbH	Münchner Bundesstraße 123, A-5020 Salzburg, Austria	http://www.gassner-waagen.at/	Industry
GEA Colby Pty Ltd.	328 High Street Chatswood NSW Australia 2067	www.colbypowder.com/	Weight feeder
GEC General Dynamics Corporation	8000 Calender Rd, Arlington, TX, USA	www.gecscales.com/	Load cell, Air craft weighing
Gefran	Philipp-Reis-Str. 9a, D-63500 Seligenstadt, Germany	www.gefran.de/	Load cell Indicator
Gericke Holding Ltd.	Althardstrasse 120, CH-8105 Regensdorf, Switzerland	www.gericke.net/	Industry Bag weight filler
Gibertini Elettronica s.r.l.	via Bellini 37, I- 20026 Novate (MI), Italy	www.gibertini.com	Laboratory

Manufacturer	Address	Internet	Types
Gilson S.A.	3000 Parmenter Street, Middleton, WI 53562-0027, USA	http://www.gilson.com/	Laboratory
GIP Graf-Irle-Pitzer GmbH Waagen- und Maschinenbau KG	Essener Str. 27, D-57234 Wilnsdorf, Germany	www.gip-waagen.de	Industry
Götz Waagen und Industrieanlagen	Zusamstraße 3, D-86165 Augsburg, Germany	www.goetz-waagen.de/	Industry
Goglio	Via Dell'industria 7, I-21020 Daverio (VA), Italy	www.goglio.it/	Packing Belt weigher
GPS Sensors & Indicators Pvt. Ltd.	K-14, Udyog Nagar, Peera Garhi New Delhi 110041, India	www.loadcells.in/	Load cell
Grabher Indosa AG	Industriestrasse 24, CH-9434 Au (St. Gallen), Switzerland	www.indosa.ch/	Packing
Gradall Industries, Inc.	406 Mill Avenue SW, New Philadelphia, OH 44663, USA	www.gradall.com/	Excavator Crane
GROWA Stahlbau Berlin GmbH	An der Industriebahn 20-27, D-13088 Berlin, Germany		Industry
Gurley Precision Instruments GPI	514 Fulton Street, Troy, NY 12180, USA	www.gurley.com/	Precision Industrial
Heinrich Grifft	Blumenstr. 11, D-73728 Esslingen	www.waagen1.de/	All types Dealer
Häfner Gewichte GmbH	Hohenhardtsweiller Str. 4, D-74420 Oberrot, Germany	www.haefner.de	Weights
Haene elektronische Messgeräte GmbH	Heinrich-Hertz-Str. 29, D-40699 Erkrath, Germany	www.haehne.de/	Load cells
Hardy Process Solutions	9440 Carroll Park Drive, Suite 150, San Diego, CA 92121, USA	www.hardyinst.com/	Process Weighing Vibration
Hasco Hasenclever GmbH + Co. KG	Römerweg 4, D-58513 Lüdenscheid, Germany	www.hasco.com/	Load cell
Hasler Deutschland GmbH	Münsterstrasse 69, 49525 Lengerich, Germany	www.hasler-int.com	Weight filler Platform
Haver & Boecker OHG	Carl-Haver-Platz 3, D-59302 Oelde, Germany	www.haverboecker.de/	Bag weight filler
Health-o-meter	2381 Executive Center Drive, Boca Raton, FL 33431, USA	www.healthometer.com/	Medical
Hense Wägetechnik GmbH	Oststraße 18, D-59929 Brilon, Germany	http://www.hense-waegetechnik.de/	Industry

Manufacturer	Address	Internet	Types
Herbert Group Ltd.	18 Rookwood Way, Haverhill Suffolk CB9 8PD, UK	www.herbert.co.uk/	Retail Label Printer
Herweg Waagen- und Vibrationstechnik GmbH & Co. KG	Gewerbestr. 28, D-48249 Dülmen, Germany		Industry
Hiden Isochema Ltd.	231 Europa Bd., Gemini Business Park, Warrington WA5 7TN, UK	www.hidenisochema.com www.hidenanalytical.com/	Thermal analysis Sorption
Hilger u. Kern GmbH Industrietechnik	Käfertaler Straße 253, D-68167 Mannheim, Germany	http://www.hilger-kern.de/	Industry
Hirschmann Automation and Control GmbH	Stuttgarter Straße 45-51, 72654 Neckartenzlingen	hus.hirschmann.com/	Load cell
Hitachi	6F Sumitomo Fudosan Ueno Bldg, 9-3, Higashi-Ueno 6chome, Taitou-ku, Tokyo 110-0015, Japan	www.hitachi-c-m.com/	Crane Excavetor
Höfelmeier Waagen GmbH	Werner-von-Siemens Str. 33, D-49124 Georgsmarienhütte, Germany	www.hoefelmeyer.de/	Weighing systems
Honeywell	101 Columbia Road, Mailstop, M6/LM, Morristown, NJ 07962	honeywell.com/	Load cell
Horiba Jobin Yvon S.A.S.	16-18 rue du Canal, F-91165 Longjumeau, France	www.jobinyvon.com/ www.horiba.com/scientific/	Molecular weight
Horizon Systems	P.O. Box 5084, Herndon, VA 20170, USA	www.horizonsystemsinc. com/ www.horizon-systems.com/	Load cell Hopper
Hoto Instruments	USA	www.hoto-instruments.com/	Force testing
Hottinger Baldwin Messtechnik HBM	Im Tiefen See 45, D-64293 Darmstadt, Germany	www.hbm.com/	Load cell Crane
Hualong–Shanghai Hualong Test Instruments Co. Ltd.	Shanghai, China	www.hualong.net/	Drop weight tear tester
Huba Control AG	Industriestrasse 17, CH-5436 Würenlos	www.hubacontrol.com/	Piezoelectric transducer
Hu-Lift HanseLifter Hub- & Fördertechnik	Straubinger Straße 20, 28219 Bremen	www.hu-lift.de	Lifter

Manufacturer	Address	Internet	Types
Hyundai	140-2 Gye-dong, jongno-gu, Seoul, Korea 110-793	www.hyundaicorp.com/	Crane
ICA SpA.	Via del Litografo, 7, I-40138 Bologna, Italy	www.icaspa.it/	Weight solid filler, Packing
Idecon s.r.l.	Via dell'Industria, 242, I-48014 Castel Bolognese (RA), Italy	www.idecon.it/	Belt weigher
IES Corporation	3388 SE 20th Ave, Portland, OR 97202, USA	www.iescorp.com/	Repair
Iemmegi srl	Via A. Diaz 38, I 20020 Bienate di Magnago (MI), Italy	http://www.iemmegi.it/	Industry
Ilapak	CH-6916 Grancia, Lugano, Switzerland	www.ilapak.com/	Weight solid filler
Imada	3100 Dundee Rd., Suite 707, Northbrook, IL 60062, USA	www.imada.com/	Load cells Stroboscope
Imanpack	Via L. di Bolsena, 19, I-36015 Schio (VI), Italy	www.imanpack.it/	Bag weight filler
Imatec s.àr.l.	33-35 Avenue des Alliés, L-9012 Ettelbruck, Luxembourg	www.imatec.lu/	Drop weight tear tester, Packing
Indiana Scale Company Inc - Incell, InScale	1607 Maple Avenue, Terre Haute, IN 47804-3234, USA	www.inscale-incell.com/	Floor
IBR Industy Bilance Riunite S.P.A.	Via Dandolo, 5, 21100 Varese (VA), Italy	http://www.ibrspa.com/ibr/	Industry
Inelta Sensorsysteme GmbH & Co.	Haidgraben 9a, D-85521 Ottobrunn/München, Germany	www.inelta.de/	Load cell
Inficon	Hintergasse 15B, CH-7310 Bad Ragaz, Switzerland	www.inficon.com/	Quartz crystal
Ingersoll Rand plc	170/175 Lakeview Drive, Airside Business Park, Swords, Co. Dublin, Ireland	www.ingersollrand.com/	Packing, Transport Torque sensor
Instron	Werner von Siemens Strasse 2, 64319 Pfungstadt, Germany	www.instron.de/	Load cell
Instrument Specialists Inc	133 E Main Street, P.O. Box 260, Twin Lakes, WI 53181-0280, USA	www.instrument-specialists.com/	Thermal analyser
Intelligente Peripherien für Roboter GmbH - IPR	Industriestr. 29, D-74193 Schwaigern	www.iprworldwide.com/	Torque sensor

Manufacturer	Address	Internet	Types
Intercomp Crane Scales	3839 County Road 116, Medina, MN 55340, USA	http://www.intercompco.com/	Industry
Interface	7401 East Butherus Drive, Scottsdale, Arizona 85260, USA	www.interfaceforce.com/	Load cell Weight indicator
Isetron Industrie-Sicherheits-Elektronik GmbH	Elektronikring 8, D-26452 Sande, Germany	www.isetron.de/	Load cell
Ishida Europe	11 Kettles Wood Drive, Woodgate Business Park, Birmingham B32 3DB, UK	http://www.ishida.com/ www.ishidaeurope.com/	Industry Retail
Itin Scale Company, Inc.	Itin Scale Co., Inc., 4802 Glenwood Rd., Brooklyn, NY 11234, USA	www.itinscales.com/	Dealer
ITI Systems	Via G. Di Vittorio 30, I-20060 Liscate (Milano), Italy	www.itisystems-srl.com/	Industrial
IVDT Skidweigh	3439 Whilabout Terrace, Oakville, Ontario, Canada L6L 0A7	http://www.skidweigh.com/	Fork lift
Ixapack	Parc Economique de La Commanderie, Le Temple, F-79700 Mauléon	http://www.ixapack.com/	Belt weigher
Jakob Antriebstechnik GmbH	Daimler Ring 42, D-63839 Kleinwallstadt, Germany	http://www.jakobantriebstechnik.de/	Hydraulic transducer
Jergens & Acme	15700 South Waterloo Rd. Cleveland, OH 44110, USA	http://www.jergensinc.com/	Torque calibration systems
Jinan Testing Equipment TE	Floor 10th Tower A Century Fortune Plaza High Tech. Zone, Jinan 250101, China	http://www.testingequipmentie.com/	Drop weight tester
JÖST GmbH + Co. KG Herweg Waagen	Gewerbestraße 28-32, D-48249 Dülmen-Buldern, Germany	http://www.joest.com/	Industry
Albert Jordan Frankfurter Waagenfabrik	Fahrgasse 10, D-60311 Frankfurt am Main, Germany	www.waagenjordan.de	Dealer, all types
Jungheinrich	Am Stadtrand 35. D-22047 Hamburg, Germany	http://www.jungheinrich.de/	Palette
Amandus Kahl GmbH & Co. KG	Dieselstrasse 5-9, D-21465 Reinbek/Hamburg	www.akahl.de/	Industrial
Kalmar Industries Ltd. (Coventry Climax)	Coventry, UK	www.kalmarind.com/ www.cargotec.com	Cargo

Manufacturer	Address	Internet	Types
Kern & Sohn Gmbh	Ziegelei 1, 72336 Balingen-Frommern, Germany	www.kern-sohn.com/	All types
Kistler Werkzeuge GmbH	Hindenburgstr. 9/1, 89610 Oberdischingen	http://www.kistler.de/	Load cell
Kliklok-Woodman	5224 Snapfinger Woods Drive Decatur, Georgia 30035, USA	http://www.kliklok-woodman.com/	Weight solid filler
KPZ Waagen GmbH	Moosstrasse 3, CH 8803 Rüschlikon, Switzerland	http://www.kpzwaagen.ch/	Dealer
Krones AG	Böhmerwaldstraße 5, D-93073 Neutraubling, Germany	http://www.krones.com/en/	Packing
K-Tron Process Group	Routes 55 & 553 Pitman, NJ 08071	http://www.ktron.com/	Weight feeder
Kubota	2-47, Shikitsuhigashi 1-chome, Naniwa-ku, Osaka 556-8601, Japan	http://www.kubota-global.net/	Transport Industry Bathroom
Kulite Semiconductor Products, Inc	Leonia, NJ, USA	http://www.kulite.com/	Load cell
LAT - Labor- und Analysen-Technik GmbH - Raczek	Heinkelstr. 10, D-30827 Garbsen, Germany	http://www.raczek.de/	Laboratory Dealer
Laumas Elletronica	Via 1° Maggio n. 6, Piazza di Basilicanova, Parma, Italy	www.laumas.com/	Load cells Truck, wheel
Laxmi Engineers	D-12-B, Diamond Park, G.I.D.C., Naroda, Ahmedabad-382 330, Gujarat, India	http://www.laxmi-engineers.com/	Bag weight filler
L C Industrie Perrin	110, Traverse de la Malvina, F-13013 Marseille	www.lcipx.com/	All types
Le Barbier	Route du Pont 7, F-76600 Le Havre	www.lebarbier.com/	Industy Laboratory
LECO Instruments	3000 Lakeview avenue, St. Joseph, MI 49085, USA	www.leco.com	Thermo-gravimetry
Leifheit AG	Leifheitstraße 1, D-56377 Nassau, Germany	www.leifheit.de	Household
Leon Engineering	Leon Engineering Tsoka 8, Thesi Xiropigado, Mandra Attikis 19600, Greece	www.leon-engineering.com/	Industrial Truck
L.F. Spa	Via Voltri 80, I-47521 Cesena (FC), Italy	http://www.lfspareparts724.com/	Kitchen

Manufacturer	Address	Internet	Types
Liebel	Weissenbrunner Hauptstr. 8D-91227 Leinburg	www.liebel-waegetechnik. de/	Weight solid filler
Liebisch Labortechnik	Eisenstr. 34, D-33649 Bielefeld	www.liebisch.com/	Dealer Laboratory
Lifesource			
Link Engineering	43855 Plymouth Oaks, Blvd. Plymouth, MI 48170-2539, USA	www.linkeng.com/	Load cells
Linseis Messgeräte GmbH	Viellitzer Str. 43, D-95100 Selb, Germany	www.linseis.net/	Thermal analysis
Lloyd Instruments Chatillon, AMETEK	Steyning Way, Bognor Regis West Sussex, PO22 9ST, UK	www.lloyd-instruments.co. uk	Force gauge
Load Controls Inc.	53 Technology Park road, Sturbridge, MA 01566, USA	www.loadcontrols.com/	Load control
Loadstar Sensors	Loadstar Sensors, Inc., 48501 Warm Springs Blvd, Suite 109 Fremont, CA 94539, USA	http://www.loadstarsensors. com/	Load cells
LOC	5, avenue du Vert Galant, F-95310 Saint Ouen L'aumone, France	http://www.loc.fr/en/	Fork lift
Lorenz Messtechnik GmbH	Obere Schloßstraße 131, D-73553 Alfdorf, Germany	http://www.lorenz-messtechnik.de/	Load cell
LSI	9633 Zaka Road, Houston, TX 77064, USA	http://www.loadsystems. com/	Load cell Crane
Magna-Lastic Devices MDI	7401 W Wilson Ave, Harwood Heights, IL 60706-4548, USA	mdi-sensor.com	Torque sensor
Magnescale	Hinagawa Intercity Tower A-18F, 2-15-1, Konan, Minato-ku, Tokyo 108-6018, Japan	www.magnescale.com/	Industry
Magtrol GmbH	Gutleutstrasse 322, D-60327 Frankfurt am Main, Germany	http://www.magtrol.de/	Load cell
Malvern Instruments	Enigma Business Park, Grovewood Road, Malvern, Worcestershire, UK, WR14 1XZ	http://www.malvern.com/	Laboratory
Mantracourt Electronics Ltd. ME	The Drive Farringdon Exeter, Devon EX5 2JB, UK	www.mantracourt.co.uk/	Indicator Load cell
Marel Norge AS Marel Food Systems	Vestvollveien 10, 2019 Skedsmokorset, Norway	www.marel.com/norway/	Retail Floor

Manufacturer	Address	Internet	Types
Mark-10	11 Dixon Avenue, Copiague, NY 11726, USA	www.mark-10.com/	Force Torque
H.P. Maroc Automatisme et Environnement	23, rue Mouatamid Ibn Abbad-ex Vuillanier km 7, (r.p.1), Aïn Sebaâ, 20580 Casablanca, Morocco		Industry
Jakob Maul GmbH	Jakob-Maul-Str. 17, D-64732 Bad König, Germany	www.maul.de	Office balances
Chr. Mayr GmbH & Co. KG	Eichenstraße 1, D-87665 Mauerstetten, Germany	http://www.mayr.com/en/	Torque sensor
Maxcess International Magpowr	Oklahoma City, OK 73126, USA	http://magpowr.maxcessintl.com/	Load cell Tension Torque
MCPI	Rue des Cols verts, F-74966 Meythet Cedex, France	www.mcpi.fr/	Dosing Packing
Measurement Specialities - MSI Sensors	1000 Lucas Way, Hampton, VA 23666, USA	www.meas-spec.com/	Load cell
MEA Testing Systems Ltd.	4 Hagavish St., Poleg Industrial Park, Netanya 42504, Israel	http://www.meatesting.com/	Torque gauge
Mecmesin	Slinfold, West Sussex, UK	www.mecmesin.com/	Force gauges
Mediseal GmbH Körber Medipack	Flurstrasse 65, 33758 Schloss Holte, Germany	www.mediseal.de/	Packing
Megatron Elektronik AG & Co.	Hermann-Oberth-Str. 7, D-85640 Putzbrunn, München	www.megatron.eu/	Load cell
Merobel	Senouches, F-45210 Ferrières	http://merobel.redex-andantex.com/	Load cells
Merrick industries	10 Arthur Drive, Lynn Haven, FL 32444	http://merrick-inc.com/	
Metra	Obninsk, Russia	http://www.metra.ru/	Load cell
Metrolog	Rua 7 de Setembro, 2671, Centro 13560-181 São Carlos, SP, Brazil	http://metrolog.net/	Load cell
Mettler-Toledo AG	Im Langacher, CH-8606 Greifensee, Switzerland	www.mt.com/	All types
Michell Instruments Ltd.	48 Lancaster Way Business Park, Ely, Cambridgeshire CB6 3NW, UK	www.michell.com/	Quartz oscillator
Micro Photonics	4972 Medical Center Circle, PO Box 3129, Allentown, PA 18106-0129, USA	www.microphotonics.com/	Force gauge
Microtech Instruments	14732 Benfield Avenue, Norwalk, CA 90650, USA	microtech@microbalance.com	Cahn balance repair

Manufacturer	Address	Internet	Types
Minebea Co. Ltd.	Japan	http://www.minebea.co.jp/	
Mobile Industries Inc.	3610 Mavis Road, Mississauga, Ontario, Canada LSC 1W2	http://www.mobilept.com/	Palette
MLU-Monitoring für Leben und Umwelt Ges.m.b.H.	Babenbergerg. 12, A-2340 Mödling, Austria	http://www.mlu.at/	Laboratory
Mobtaker Electronic Tehran Eng. Ltd.	No.122/1, 2nd floor, Dr. Hoshyar St., Azadi Ave., 13418 Tehran, Iran		Import
Maschinenfabrik Möllers GmbH	Sudhoferweg 93-97, D-59269 Beckum, Germany	http://www.moellers.de/en/	Packing
Montalvo Corporation	50 Hutcherson Drive, Gorham, Maine 04038, USA	http://www.montalvo.com/	Load cell
Moretto S.p.A.	Via dell'Artigianato, 3, I-35010 Massanzago (PD), Italy	http://www.moretto.com/	Weight solid filler
Motorman S/A	Polígono Industrial 'La Ferreria', c/ Ceràmica 1, 08110 Montcada i Reixac (Barcelona)	http://www.motorman.es/	Industry Crane
MTS Systems Norden AB	Södra Långebergsgatan 16, 42132 Västra Frölunda, Sweden	http://www.mts.com/	Thermo-balance
Multipond Wägetechnik GmbH	Traunreuter Straße 2-4, D-84478 Waldkraiburg	http://www.multipond.com/	Weight solid filler
MyWeigh	3315-B W. Buckeye RD, Phoenix AZ 85009, USA	http://www.myweigh.com/	Household Industry
National Instruments	Austin, TX, USA	www.ni.com/	Industry
NBC ELETTRONICA GROUP SRL	2/c, v. Tavani 23014 DELEBIO (SO), Italy	ww.nbc-el.it	Load cell
Netzsch-Gerätebau GmbH	Wittelsbacherstraße 42, 95100 Selb/Bayern	www.netzsch.com www.netzsch-thermal-analysis.com/	Thermal analysis
Newtec	Stærmosegårdsvej 18, 5230 Odense M, Danmark	http://www.newtec.com/	Packing
Nexen	Nexen Group, Inc. 560 Oak Grove Parkway, Vadnais Heights, Minnesota 55127, USA	http://www.nexengroup.com/	Load cell
Noax Technologies AG	Am Forst 6, D-85560 Ebersberg, Germany	http://www.noax.com/	Weighing terminal

Manufacturer	Address	Internet	Types
Noshok	1010 West Bagley Road, Berea, OH 44017, USA	http://www.noshok.com/	Hydraulic load cell
Novatech Measurements Ltd.	83 Castleham Road, St Leonards on Sea, East Sussex, TN38 9NT, UK	http://www.novatechloadcells.co.uk/	Load cell Platform
Nova Weigh	Walkers Road, North Moons Moat, Redditch, Worcestershire	http://www.novaweigh.co.uk/	Industry
E H Oakley & Co. Ltd.	43 Edison Road, Rabans Lane Industrial Area, Aylesbury, Buckinghamshire HP19 8TE, UK	http://www.oakleyweigh.co.uk/	Dealer All types
OCS Checkweighers GmbH	Max-Planck-Str. 7, D-74523 Schwäbisch-Hall, Germany	www.ocs-cw.com	Industry
Ocme S.r.l.	Via del Popolo, 20/A 43122 Parma, Italy	www.ocme.com	Packing
Oelikon	Churerstrasse 120, CH-8808 Pfäffikon, Switzerland	http://www.oerlikon.com/	Industry
Ohaus Corporation	7 Campus Drive, Suite 310, Parsippany, NJ 07054, USA	us.ohaus.com/	All types
OK International	OK International Corp., 73 Bartlett St., Marlboro, MA 01752, USA	www.oksealer.com/	Packing
Omag di Affri, Sas (di Affri Davide & C.	Via Monte Tagliaferro 8, I-21056 Induno Olona (VA), Italy	www.omagaffri.com/	Fork lift
Omega Electronic Scale Co. Ltd.	No. 27, Tamarind Lane, Rajbahadur Buildngg, Near Mumbai, India	www.indiamart.com /company/630784/	All types
Oussama s.a.r.l.	5, rue Ibn El Mouataz -ex Saint Quentin, 20300 Casablanca, Morocco		Import
Outillage Général pour l'Industrie (Ogin)	21, bd de la Gironde, 20500 Casablanca, Morocco		Import
Pack-Line di Buda Mara	Via Bellaria, 33/A, I-47030 S. Mauro P.li (FC), Italy	http://www.pack-line.com/	Load cell Mass indicator
Panalytica	ul. Rydygiera 8, PL-01-793 Warszawa, Poland	www.panalytica.pl/	Dealer
Parimix	18, avenue du Bois-Préau, F-92502 Rueil-Malmaison, France	http://www.parimix.com/	Dosage

Manufacturer	Address	Internet	Types
Pattyn Packing Lines nv	Industriepark 'Blauwe Toren' Hoge Hul 4-6-8, BE-8000 Brugge, Belgium	www.pattyn.com/	Bulk weigher Packing
Pavone Sistemi S.r.l.	via Dei Chiosi, 18, 20040 Cavenago Brianza (MI) Italy	http://www.pavonesistemi.it/en	Industrial
Payper S.A.	Polígono Industrial el Segre, parcela, 25191 Lleida, Spain	http://www.payper.com/	Packing
PCB Piezotronics, Inc.	3425 Walden Avenue, Depew, New York 14043-2495, USA	http://www.pcb.com/	Piezo cells
PCE Ibérica S.L.	Calle Mayor, 53, 02500 Tobarra (AB), Spain	http://www.pce-iberica.es/	Laboratory Forcce gauge
PD LiftTec AB	Mörsaregatan 11 A, 25466 Helsingborg, Sweden	http://www.pdlifttec.se/	Transport
Peekel Instruments B.V.	Industrieweg 161, NL-3044 AS Rotterdam, The Netherlands	http://www.peekel.com/	Load cell
Penko Engineering B.V.	Schutterweg 35, NL-6718 XC Ede, The Netherlands	http://penko.com/	Industial
Pennsylvania Counting Scales	1042 New Holland Ave., Lancaster BA, 17601, USA	www.pennsylvaniascale.net/	Counting scales
Perkin-Elmer	761 Main Ave, Norwalk CT 06859-0012, USA	las.perkinelmer.com/	Thermal analysis
Perkin-Elmer Life and Analytical Sciences	940 Winter Street, WalthamMassachusetts 02451, USA		Microbalance
Jambu Pershad & Sons	6275/22, Near Nigar Cinema G.P.O. Box No. 58 Nicholson Road, Ambala Cantt, 133001, Haryana, India	www.japson.com/	Laboratory
Pesa Waagen AG	Leutschenbachstrasse 45, CH-8050 Zürich	www.pesawaagen.com/	Oscillating string
Pesatura Servizi Industriali s.r.l. – PE.S.I.	Via dell'inudustria 8, I-36051 Creazzo (VI), Italy	www.pesisrl.it	Industrial
PESCALE	Kapuzinerweg 3, D-72406 Bisingen, Germany	www.pescale.de	Dealer
Pesola AG	Rebmattli 19, CH-6340 Baar, Switzerland	http://www.pesola.com/	Spring Load cell
PFM Packing Machinery Spa	Via Pasubio 29, I-36036 Torrebelvicino (Vicenza), Italy	www.pfm.it/en/	Packing

Manufacturer	Address	Internet	Types
Pfreundt GmbH	Robert-Bosch-Str. 5, D-46354 Suedlohn, Germany	http://www.pfreundt.org/	Industry
Plus Trade	100, bd Abdelmoumen, imm. Cimr 14°ét. 20340 Casablanca, Morocco		Import
Pneumatic Scale	4485 Allen Road, Stow, Akron, OH 44224, USA	http://www.psangelus.com/	Filler
Porotec GmbH	Niederhofheimer Str. 55a, D-65719 Hofheim, Germnay	www.porotec.de	Dealer Thermal analysis
Pramac S.p.A.	Località Il Piano, 53031 Casole d'Elsa (SI), Italy	http://www.pramac.com/	Transport
Precia Molen	BP 106 – F-07001 Privas, France	www.preciamolen.com/	All types
Precisa Gravimetrics AG	Moosmattstrasse 32 CH-8953 Dietikon, Switzerland	precisa.ch/Precisa/en www.biltekas.com	Laboratory
Precision Instrument Repair Company PIR	13414 Woodruff Ave. Bellflower, CA 90706, USA	precisioninstrumentrepair. com	Dealer Laboratory
Precision Weighing Balances – Scaleman - Siltec - CAS	30 South Cross Road, Bradford, MA 01835-8232, USA	www.balances.com/ http://scaleman.com/ siltec.balances.com/	Dealer
Premier Tech Chronos	1, Avenue Premier, Rivière-du-Loup (Québec) G5R 6C1, Canada	http://www.premiertechieg. com/	Packing
Prins UK ltd	Unit 140 Hartlebury Trading Estate Kidderminster, Worcestershire DY10 4JB	http://www.prinsuk.com	Industrial
PT Ltd.	5E Marken Place, Glenfield, North Shore City, Auckland, New Zealand	http://www.ptglobal.com/	Load cell Platform
QCM Laboratory G&M Norden AB	Nettovägen 11, 175 41 Järfälla, Sweden	www.qcmlab.com/ www.wedoo.se/	Quartz crystal
Q-sense AB	Hängpilsgatan 7, SE-426 77 Västra Frölunda, Sweden	www.q-sense.com	Quartz crystal
Qingdao Just Scale Manufacturing Co., Ltd.	Donggu Zhen Chengyang Qingdao 266000 Qingdao, China	http://www.mthq-just.com/	Industry
Radpak Ltd. Engineering Company	PL-87-800 Wloclawek, ul. Okrezna 2, Poland	http://www.radpak.net/	Packing

Manufacturer	Address	Internet	Types
Radwag Wagi Elektroniczne	Bracka 28, PL-26-600 Radom, Poland	www.radwag.eu/ www.radwag.pl/	All types
Ravas GmbH mobile Wiegetechnik	Dinnendahlstrasse 27, D-47533 Kleve, Germany	www.ravas.com	Industrial
Ravas Europe BV	NL-5300 CA Zaltbommel, Netherlands		
RDP Electronics Ltd.	RDP Electronics Ltd., Grove Street, Heath Town, Wolverhampton WV10 0PY, UK	http://www.rdpe.com/	Load cell
RDS Technology Ltd.	Cirencester Rd GL6 9BH Stroud, UK	www.rdstec.com/	Transport
Rein Kraftmessgeräte	Gottlieb-Daimler-Str. 62, D-89150 Laichingen, Germany	http://www.kraftmessgeraet.de/	Load cell Spring
Reliant Technology LLC	257 El Camino Verde Street, Henderson, NV 89074, USA	http://www.extensometry.com/	Load cell
REMA Holland BV	Galjoenweg 47, D-6222 NS Maastricht, Netherlands	http://www.rema.eu/	Floor Platform
Rembe GmbH – Safety Control	Gallbergweg 21, 59929 Brilon, Germany	http://www.rembe-safety-control.de/	Industry
Research Equipment (London) Ltd.	72 Wellington Road, Twickenham, Middlesex TW2 5NX, UK	http://www.research-equipment.com/	Laboratory
R.G.S Implanti	Via Mavore 1640/C, I-4109 Zocca (MO), Italy	http://www.rgsimpianti.com/	Transport
Rheometric Scientific Ltd. (TA Instruments)	Surrey Business Park, Weston rd. Kiln Lane, Epson, Surrey KT 17 1JF, UK	www.tainstruments.com/	Thermal Analysis
Rice Lake Weighing Systems	230 West Coleman Street, Rice Lake, WI 54868 USA	http://www.ricelake.com/	All types
Rigaku Corporation	4-14-4, Sendagaya, Shibuya-Ku,Tokyo 151-0051, Japan	rinttyo@rigaku.co.jp	Thermal analysis
Rigaku Europe	Unit B6, Chaucer Business Park Watery Lane, Kemsing Sevenoaks, Kent TN15 6QY, UK		
Rinstrum Pty Ltd.	41 Success Street, Acacia Ridge, Brisbane 4110, QLD, Australia	www.rinstrum.com/	Load cell Retail
Roga-Instruments	Steinkopfweg 7, D-55425 Waldalgesheim, Germany	http://roga-messtechnik.de/	Load cell

Manufacturer	Address	Internet	Types
Romanian Business Consult RBC	Bulevardul Ficusului 42, Bucureşti 013975 Sector 1, Romania	http://www.rbc.com.ro/	Industry
Ron Crane Scales	1243 Naperville Drive	www.ron-crane-scales.com/	Crane scale
Eilon Engineering	Romeoville, IL 60446, USA	www.eilon-engineering.com/	Force gauge
Ronchi Mario SpA	Via Italia, 43, I-20060 Gessate, Milano, Italy	http://www.ronchi.it/en/	Packing
Rovema GmbH	Industriestrasse1, D-35463 Fernwald, Germany	http://www.rovema.de/	Packing
Roure Tectosa S.L.	C/Poeta Maragall, 16, 08130 Santa Perpétua de Mogoda, Barcelona (España)	http://www.rouretectosa.com/	Packing
Rubotherm	Konrad-Zuse-Str. 4. D-44801 Bochum, Germany	www.rubotherm.de/	Magmetic suspension balance
Rudrra Sensor	3rd Floor, Karnavati Complex, Opp. Baroda Express Way, C.T.M., Ahmedabad-380 026. Gujarat, India	www.rudrra.com/	Load cell
Salter Brecknell	1000 Armstrong Drive. Fairmont, MN 56031, USA	www.brecknellscales.com/	Industry
Salter Housewares	HoMedics House, Somerhill Business Park, Five Oak Green Road, Tonbridge, Kent, UK	www.salterhousewares.com/	Household
Sany	Changsha, China	www.sany.comm www.sanygroup.com	Crane
Sartorius AG	Weender Landstrasse 94-108, D-37070 Göttingen, Germany	www.sartorius.com	All types
Sauter GmbH	Tieringerstr. 11-15, D-72336 Balingen, Germany	www.kern-sohn.com	Force gauge
Scaime	Technosite Altéa, F-74105 Annemasse, France	http://www.scaime.com/	Load cell
Scales Galore	4802 Glenwood Rd. Brooklyn, NY 11234, USA	http://www.scalesgalore.com/	Dealer
Scan-Sense AS	Bekkeveien 163, 3173 Vear, Norway	http://www.scan-sense.com/	Load cell
Carl Schenck AG	Landwehrstraße 55, D-64293 Darmstadt	www.schenck.net/	Industrial
Schenck Process		www.carlschenck.de	

Manufacturer	Address	Internet	Types
Schenck Accurate	746 E. Milwaukee Street, P.O. Box 208, Whitewater, WI, USA	http://www.accuratefeeders.com/	Solid weight feeder
Hans Schmidt & Co. GmbH Control Instruments	Schichtstrasse 16, D-84478 Waldkraiburg, Germany	www.hans-schmidt.com/	Force gauge
Schneider Leichtbau	Raiffeisenstraße 24, D-35236 Breidenbach-Oberdieten, Germany	www.schneider-logistics.de/	Palette
Albert Schumann GmbH - GF.Sh.-Renani	Schulstr. 12, 24969 Sillerup, Germany	www.schumann-gmbh.de/	Dealer
Schwer & Kopka GmbH	Herknerstraße 4, D-88250 Weingarten, Germany	http://www.sk-gmbh.de/	Load cell
Scientech	5649 Arapahoe Avenue, Boulder, Colorado 80303-1399	www.scientech-inc.com/	Laboratory Laser
Seca Gmbh & Co. kg.	Hammer Steindamm 9-25, D-22089 Hamburg	www.seca.com	Medical balances
SII Seiko Instruments NanoTechnology Inc.	1-8, Nakase, Mhama-ku, Chiba-shi, Chiba 261, Japan	www.seikoinstruments.com/	Thermo-gravimetry
J.P. Selecta	Autovía A-2, Km 585, 1-08630 Abrera (Barcelona), Spain	http://www.grupo-selecta.com/	Laboratory
Semat Technical Ltd.	Unit 1 Executive Park Hatfield Road, St. Albans, Hertfordshire AL1 4TA, UK		Laboratory
Seneca	Via Germania, 34, 35127 Padova, Italy	http://www.seneca.it/	Load cell amplifier
Sennebogen Maschinenfabrik GmbH	Hebbelstraße 30, D-94315 Straubing, Germany	www.sennebogen.com	Crane
Sensor Developments Inc.	1050 W. Silver Bell Rd. Orion, MI 48359-1327, USA	http://www.sendev.com/	Tension force sensors
Sensortechnics GmbH	Boschstr. 10, D-82178 Puchheim, Germany	http://www.sensortechnics.com/	Load cell
Sensy SA	Z.I. de Jumet, Allée Centrale. BE-6040 Jumet, Belgium	http://www.sensy.com/	Load cells
Sentran Corporation	2547 Aerial Way SE, Salem, Oregon, USA 97302	http://sentrancorp.com/	Load cell Transmitter

Manufacturer	Address	Internet	Types
Serac INc	300 Westgate Drive, Carol Stream, IL 60188	http://www.serac-usa.com/	Weight liquid filler
SETARAM	7 rue de l'Oratoire, F-69300 Caluire, France	www.setaram.com/	Calorimeter Thermogravi-metry
Setra Sensing Solutions	159 Swanson Road, Boxborough, MA 01719, USA	http://www.setra.com/	Weighing counting
Sharp Pack Weight Machines	Upavan Industrial Area, Thane (W) 400606, Maharastra, India	http://www.sharppackmachines.com/	Bag weight filler
Sherborne Sensors Ltd.	1 Ringway Centre, Edison Road Basingstoke, Hampshire RG21 6YH, UK	http://www.sherbornesensors.com/	Load cells
Shimadzu Corporation	3 Kanda-Nishikicho 1-chome Chiyoda-ku, Tokyo 101, Japan	www.shimadzu.de www.laborildam.com	All types
Shimpo NanoTechnology Inc.	1701 Glenlake Avenue, Itasca, Illinois 60143-1072, USA	http://www.shimpoinst.com/	Force gauge
Shinko Denshi Co. Ltd. Vibra	3-9-11 Yushima, Burkyo-ku, Tokyo 113 0034, Japan	www.vibra.co.jp/global/	Laboratory
SII Nano Technology Inc.	1-8, Nakase, Mihama-Ku261-8507 Chiba, Japan	http://www.siint.com/	Thermal analysis
Siemens Process Instrumentation	Wittelsbacherplatz 2, D-80333 München, Germany	http://www.automation.siemens.com/	Industry Coriolis mass flow
Simex Sp. z o.o.	ul. Wielopole 7, 80-556 Gdańsk, Poland	http://www.simex.pl/	Weight indicator
Sipesa Sistemas de Pesaje	C/Aumet, s/n Pol. Ind. Celà, 17460 Celrà (Griona), Spain	www.sipesa-pesaje.com	Industrial
Sipi Srl	Via Lazaretto, 10, I-21013 Gallarate Varese, Italy	http://www.sipi.it/	Industrial Precision
Soehnle (Leifheit) Soehnle Professional GmbH & Co. KG	Manfred-von-Ardenne-Allee 12, D-71522 Backnang, Germany	www.soehnle.com/ www.leifheit.de/ www.soehnle-professional.com	All types
Soemer Messtechnik GmbH	Kaiser-Otto-Str. 2, D-57368 Lennestadt, Germany	www.soemer.de/ waegetechnik/	Load cells Systems
Solotop Oy	Laippatie 14 A, FIN-00880 Helsinki, Finland	http://www.solotop.fi/	Industry Kitchen
Speedshield	1663-1665 Centre Rd, Springvale VIC 3171, Australia	http://www.speedshield.com/	Fork lift

Manufacturer	Address	Internet	Types
Stanford Research Systems SRS	1290-D Reamwood Avenue, Sunnyvale, CA 94089, USA	http://www.thinksrs.com/	Quartz crystal
Statec Binder GmbH	Industriestrasse 32, 8200 Gleisdorf, Austria	http://www.statec-binder.at/i	Force sensor Packing
Stiavelli – Divisione Packaging di Pavan	Via G. Amendola 27, Loc Torrichio, I-51010 Uzzano, Italy	http://www.stiavelli.it/	Bag weight filler
Strain Measurement Devices	55 Barnes Park North, Wallingford, CT 06492, USA	http://www.smdsensors.com/	Load cell
Strainsert company	12 union Hill Road, West Conshohocken, PA 19428, USA	http://www.strainsert.com/	Load cell
Sunshine Sensors	A wing, 4th floor, Narnarayan Complex, Nr. Swastik Cross Road, C.G. Road, Navrangpura, Ahmedabad, 380009, Gujarat, India	www.sunshineloadcell.com	Load cell
Sunward	Sunward Intelligent Industrial Park, Xingsha, Changsha, Hunan, China	http://www.sunward.com.cn/	Crane Fork lift
Surya Lab Expotech	250, Pradeep Vihar, Village-Ibrahimpur, Gali No:4, Block-A110036, Delhi, India	www.kompass.in/surya-lab-expotech/	Laboratory
SureTorque	1461 Tallevast road, Sarasota, FL 34243, USA	http://suretorque.com/	Torque gauge
Surface Measurement Systems Ltd.	5 Wharfside, Rosemont Road, Alperton, Middlesex, HA0 4PE, UK	www.thesorptionsolution.com/ www.cm-labo-gastro.com.pl	Thermal analysis Sorption
Surya Lab Expotech	250, Pradeep Vihar. Village-Ibrahimpur, Gali No:4, Block-A110036, Delhi, India		Export
Systec GmbH	Ludwig-Erhard-Str. 6, D-50129 Bergheim-Glessen	http://www.systecnet.com/	Industy
TA Instruments	109 Lukens Drive, New Castle, DE 19720-276, USA	www.tainstruments.com/	Calorimetry Thermal analysis Rheology
Tamtron Oy	Vestonkatu 11, 33730 Tampere, Finland	http://www.tamtron.fi/	Industry
Tanita	Tokyo, Japan	www.tanita.com/	Household Medical

Manufacturer	Address	Internet	Types
Tassinari Bilance srl	P.le Gaetano e Giovanni Tassinari 2, I-44047 Sant Agostino, Italy	http://www.tassinaribilance.fe.it/	Industry Medical
Taylor Precision Products	2311 W. 22nd Street, Oak Brook, IL 60523, USA	www.taylorusa.com/	Household
Tech-Weigh Electronics Co. Ltd.	C-1-603, Lurun Mingshang Plaza, No. 84, Yingxiongshan Road, Jinan, China	http://www.tech-weigh.com/en/	All types
Tecsys s.a.r.l.	108, rue Rahal Ben Ahmed, ex Dinant, 6°ét. 20300 Casablanca, Morocco		Laboratory Load cell
Tehničar-Servag doo	Medulićeva 14, 10000 Zagreb, Croatia	www.tehnicar-servag.hr/	Import
Tekscan	307 West First Street, South Boston, MA 02127-1309, USA	http://www.tekscan.com/	Load cell
Teldyne Test Service	513 Mill Street, Marion, MA 02738	http://www.teledyne-ts.com/	Load cell
TD Models & Scientific Co.	34 Industrial Estate, Ambala Cantt-133006, Haryana, India		Models Eduction
Tehničar-Servag doo	Medulićeva 14, 10000 Zagreb, Croatia	www.tehnicar-servag.hr/	Laboratory
Teraoka Digi	3-12 Kugahara 5-chome, Ohta-Ku, Tokyo 146-8580, Japan	www.digisystem.com/	Household Retail
Testing Machines Inc. TMI	40 McCullough Drive, New Castle, DE 19720, USA	http://www.testingmachines.com/	Testing
Thayer Scale Hyer Industries Inc.	Pembroke, MA, USA	http://www.thayerscale.com/	Solid flow control
Thass - Thermal Analysis & Surface Solutions GmbH	Pfingstweide 21, D-61169 Friedberg, Germany	www.thass.org/ http://www.thass.net	Thermal analysis
Thermo Fisher Scientific, Inc.	81 Wyman Street, Waltham, MA 02454, USA	www.thermofisher.com/	Laboratory
Thermo Cahn, Inc.	5225 Verona Road, Madison, WI 53711, USA	www.thermoscientific.com/	Cahn balances
TheScaleShop.com	USA	www.thescaleshop.com/	Dealer
Tohnichi Mfg CO Ltd.	2-12 Omori-Kita, 2-Chome Ota-ku, Tokyo, Japan	http://www.tohnichi.be/	Torque
Toshiba TEC Europe Retail Information Systems	Toshiba America, Inc., 1251 Avenue of the Americas, Suite 4110, New York, NY 10020, USA	http://www.toshibatec-eu.co.uk/ http://www.toshiba.com/	Platform

Manufacturer	Address	Internet	Types
Tractel	Tractel Greifzug GmbH, Scheidtbachstr. 19-21, D-51469 Bergisch Gladbach, Germany	http://www.tractel.com/	Crane
Transfer Multisort Elektronik Sp. z o.o. - TME	ul. Ustronna 41, PL-93-350 Lodz, Poland	www.tme.eu/de/	Dealer
Trilogica	Im Mühlahl 23, D-61203 Reichelsheim, Germany	www.trilogica.de/	Thermal analysis
T-Scale	282, Sect. 3, Hoping W. Rd., Taipei, Taiwan 99, Shunchang Road Kunshan Hi-Tech industrial Park, Kunshan, Jiangsu, China	http://www.taiwanscale.com/	Household Industry
U-Therm International (UK) Ltd. - Lab-Kits	RM. 905, Workingberg Commercial Bldg., 41-47 Marble Road, Hong Kong	www.lab-kits.com/	Laboratory
Utilcell, Tecnicas de Electronica y Automatismos SA	C/Espronceda 176-18008018 Barcelona, Spain	http://www.utilcell.es/	Load cell
Varlik Makina	Ömerli Mahallesi Seciye Sok. No:2 Arnavutköy/İstanbul, 34555, Turkey	http://www.varlikmakina.com.tr/	Bag weight filler
Velteko s.r.o	K Borovičkám 1716, CZ-258 01 Vlašim, Czechia	http://www.velteko.de/	Bag weight filler
Vendee Concept	57 rue alexander Fleming, Z.I. Sud Belle Place, 85000 La Roche Sur Yon, France	http://www.vendeeconcept.com/	Weighing maschine Filling
Velomat Messelektronik GmbH	Schwarzer Weg 23 b, D-01917 Kamenz, Germany	http://www.velomat.de/	Load cell
Vetek Weighing AB	Industrivägen 3, S-760 40 Väddö, Sweden	www.vetek.com	
VinSyst Technologies	Citizen House, MIDC, Andheri (East), Mumbai-400 093, India	http://www.vinsyst.com/	Check-weigher Packing
Vishay Precision Group	3 Great Valley Parkway, Suite 150, Malvern, PA 19355, USA	www.vishaypg.com/	Load cell Platform
VTI Corp. (TA Instruments)	7650 West 26th Ave Hialeah, FL 33016, USA	www.vticorp.com/ www.tainstruments.com/	Sorption analyzer
VWR Scientic Producte	1310 Goshen Parkway, West Chester PA 19380, USA	http://www.vwr.com/	Laboratory
Waagenbau Harald Riedel	Industriestraße 28, D-01640 Coswig, Germany	www.waagenbau-riegel.de/	Industry

Manufacturer	Address	Internet	Types
Waagen Friederichs GmbH	Vangerowstr. 33, D-69115 Heidelberg, Germany	www.waagen-friederichs.de/	Dealer
Waagen Jordan	Fahrgasse 10, D-60311 Frankfurt am Main, Germany	www.waagenjordan.de/	Dealer
Waagen-Kissling GmbH	Bahnhofstraße 17A, D-64668 Rimbach	http://www.waagen-kissling.de/	Dealer
Waagen-Schmitt	Hammer Steindamm 27-29, D-22089 Hamburg, Germany	www.waagen-schmitt.de/	Dealer
Waagen-Steitz	Lutrinastraße 12, D-67655 Kaiserslautern, Germany	mail@waagen-steitz.de	Dealer
Waagen Merry GmbH	Buchenhofener Str. 23-25, Halle 11, D-42329 Wuppertal, Germany	http://www.waagen-merry.de/	Dealer
WamGroup S.p.A.	Via Cavour, 338, I-41030 Ponte Motta / Cavezzo (MO), Italy	http://www.wamgroup.com/	Bag weight filler
Wånelid AB	Smedstorpsgatan 10, S-53221 Skara, Sweden	www.wanelid.com/	All types
Waters - TA Instruments	109 Lukens Drive, New Castle, DE 19720-2795, USA	www.waters.com/ www.tainstruments.com/	Thermo-analysis Sorptometry
WeightPack Systems Inc.	Las Vegas, USA		Packing
Wesco Industrial Products Inc.			Industry
Wipotec	Adam-Hoffmann-Str. 26, D-67657 Kaiserslautern, Germany	www.wipotec.com/	Weigh cell Platform
Xilin	North Floor 3, West Lake Ming Building, 296 Qingchun Road, HangZhou 310006, Zhejiang, China	http://www.xilin.com/	Fork lift
Yale – Columbus McKinnon	Am Lindenkamp 31, 42549 Velbert	www.cmco.eu, www.yale.de	Rotary weight filler Crane
Yamato Scale Co. Ltd.	5-22 Saenba-cho 66 Akashi-shi, Hyogo 673-8688, Japan	www.yamato-scale.co.jp	Bag weigher Retail
Yueqing Handpi Instruments Co., Ltd. Handpi	No. 40 Minfeng Road, Yueqing Zhejiang, China	http://handpi.en.china.cn/	Force gauge
Zapadpribor Private Business	Henerala Kurmanovycha St. 9b, Floor 3, Lviv 79052, Ukraine	zapadpribor.en.ecplaza.net/	Laboratory

Manufacturer	Address	Internet	Types
Zhejiang Hero Time Machinery Co., Ltd.	Zhuji City, Zhejiang Province, China	zjhaotai.en.alibaba.com/	Transport Crane
Zwick Roell	Zwick GmbH & Co. KG August-Nagel-Str. 11 D-89079 Ulm, Germany	www.zwick.de	Industry
Zwick/Roell -Amsler	Herrgass 21, CH-8413 Neftenbach, Switzerland	www.zwick.de/	Testing apparatus
Zwiebel	Zwiebel - Saint Jean Saverne, F 67701 Saverne, France	http://www.zwiebel-weights.com/en/	Laboratory Household

Table 8.2 Manufacturer of vacuum balances, crystal oscillators, gravimetric sorption, moisture and thermogravimetric measuring apparatus

Manufacturer	Address	Internet	Types
Bähr Thermoanalyse GmbH	Altendorfstr. 12, 32609 Hüllhorst, Germany	www.baehr-thermo.de	Thermal analysis
Beckman Coulter, Inc.	4300 N. Harbor Boulevard, P.O. Box 3100, Fullerton, CA 92834-3100, USA	www.beckmancoulter.com/	Scientific Thermal analysis
C3 Prozess- und Analysentechnik GmbH	Peter-Henlein-Str. 20, D-85540 Haar	www.c3-analysentechnik.de	Quartz crystal
CEM	Carl-Friedrich-Gauß-Str. 9, D-47475 Kamp-Lintfort, Germany	www.cem.de/	Microwave Moisture
CI Preccision	Brunel Rd. Churchfields, Salisbury, Wiltshire, SP2 7PX, UK	www.cielec.com/	Vacuum Balance Thermal analysis
Denver Instrument	6542 Fig Street, Arvada, COL 80004	www.denverinstrumentusa.com/ www.denverinstrument.com/	Laboratory
Hiden Analytical	420 Europa Bd., Gemini Business Park, Warrington WA5 5UN, UK	www.hidenisochema.com www.hidenanalytical.com/	Thermal analysis Sorption
Inficon	Hintergasse 15B, CH-7310 Bad Ragaz, Switzerland	www.inficon.com/de/	Quartz crystal
LECO Instruments	3000 Lakeview avenue, St. Joseph, MI 49085, USA	www.leco.com	Thermo-gravimetry
Linseis GmbH	Viellitzer Str. 43, D-95100 Selb, Germany	www.linseis.net/	Thermal analysis Suspension

Manufacturer	Address	Internet	Types
Mettler-Toledo AG	Im Langacher, CH-8606 Greifensee, Switzerland	www.mt.com/	All types
Michell Instruments Ltd.	48 Lancaster Way Business Park, Ely, Cambridgeshire CB6 3NW, UK	www.michell.com/	Quartz oscillator
Netzsch Gerätebau	Wittelsbacher Str. 42, D-95100 Selb, Germany	www.netzsch.com www.netzsch-thermal-analysis.com/	Thermal analysis
Perkin-Elmer	761 Main Ave, Norwalk, CT 06859-0012, USA	las.perkinelmer.com/	Thermal analysis
Perkin-Elmer Life and Analytical Sciences	940 Winter Street, Waltham Massachusetts 02451, USA		
POROTEC GmbH	Niederhofheimer Str. 55a, D-65719 Hofheim, Germany	www.porotec.de	Thermal analysis Sorptometry
Precisa Gravimetrics AG	Moosmattstrasse 32, CH-8953 Dietikon, Switzerland	precisa.ch/Precisa/en www.biltekas.com	Moisture analysis
QCM Laboratory G&M Norden AB	Nettovägen 11, 175 41 Järfälla, Sweden	www.qcmlab.com/ www.wedoo.se/	Quartz crystal
Q-sense AB	Hängpilsgatan 7, SE-426 77 Västra Frölunda, Sweden	www.q-sense.com	Quartz crystal
Radwag Wagi Elektroniczne	Bracka 28, PL-26-600 Radom, Poland	www.radwag.eu/ www.radwag.pl/	Dealer Suspension
Rheometric Scientific Ltd. (TA Instruments)	Surrey Business Park, Weston rd. Kiln Lane, Epson, Surrey KT 17 1JF, UK	www.tainstruments.com/	Thermal analysis
Rigaku Corporation	4-14-4, Sendagaya, Shibuya-Ku,Tokyo 151-0051, Japan	rinttyo@rigaku.co.jp	Thermal analysis
Rigaku Europe	Unit B6, Chaucer Business Park Watery Lane, Kemsing Sevenoaks, Kent TN15 6QY, UK		
Rubotherm	Universitätsstr. 142, D-44799 Bochum, Germany	www.rubotherm.de/	Suspension balance
Sartorius AG	Weender Landstrasse 94-108, D-37070 Göttingen, Germany	www.sartorius.com	Moisture
SII Seiko Instruments NanoTechnolgy Inc.	1-8, Nakase, Mhama-ku, Chiba-shi, Chiba 261, Japan	www.seikoinstruments.com/	Thermo-gravimetry
SETARAM	7 rue de l'Oratoire, F-69300 Caluire, France	www.setaram.com	Thermal analysis
Shimadzu Corporation	3 Kanda-Nishikicho 1-chome Chiyoda-ku, Tokyo 101, Japan	www.shimadzu.de	Laboratory Industry

Manufacturer	Address	Internet	Types
Shimadzu Deutschland GmbH	Albert-Hahn-Str. 6-10, D-47269 Duisburg		Moisture
Shinko Denshi Co. Ltd. Vibra	3-9-11 Yushima, Bunkyo-ku, Tokyo 113-0034, Japan	http://www.vibra.co.jp/	All types
Surface Measurement Systems	5 Wharfside, Rosemont Road, Alperton, Middlesex, HA0 4PE, UK	www.thesorptionsolution. com/ www.cm-labo-gastro.com.pl	Thermal analysis Sorption
TA Instruments	109 Lukens Drive, New Castle, DE 19720-2795, USA	www.tainstruments.com/	Thermal analysis Rheology
THASS - Thermal Analysis & Surface Solutions GmbH	Pfingstweide 21, D-61169 Friedberg, Germany	www.thass.org/	Thermal analysis
Trilogica	Im Mühlahl 23, D-61203 Reichelsheim, Germany	www.trilogica.de/	Thermal analysis
VTI Corp. (TA Instruments)	7650 West 26th Ave Hialeah, FL 33016, USA	www.vticorp.com/ www.tainstruments.com/	Sorption analyzer
Waagenbau Dohmen GmbH	Am Weiweg 6, D-52146 Würselen, Germany	www.waagenbau-dohmen.de	Car scales
Waagencenter Koch GbR	Dorfstraße 7, D-94486 Osterhofen, Germany	http://www.waagencenter. com	Dealer
Waagen Kissling GmbH	Bahnhofstraße 17A, D-64668 Rimbach, Germany	www.waagen-kissling.de/	Dealer
Hermann Waldner GmbH & Co. KG	Anton-Waldner-Straße 10-16, D-88239 Wangen, Germany	http://www.waldner.de/	Laboratory Process
Waters - TA Instruments	109 Lukens Drive, New Castle, DE 19720-2795, USA	www.waters.com/ www.tainstruments.com/	Thermo-analysis Sorptometry
WSK Mess- & Datentechnik GmbH	Güterbahnhofstraße 1, D-63450 Hanau, Germany	http://www.tabase.de/	Software

References

1. G. Clifton, G.L.E. Turner (eds.), *Directory of British Scientific Instrument Makers 1550–1851* (1998)
2. G.L.E. Turner, *Scientific Instruments, 1500–1900: an Introduction* (1998)
3. A. McConnell, *Jesse Ramsden (1735–1800): London's Leading Scientific Instrument Maker (Science, Technology and Culture, 1700–1945)* (Ashgate, Farnham, 2007)
4. J.T. Stock, D.J. Bryden, A Robinson balance by Adie & son of Edinburgh. Technol. Cult. **13**(1), 44–54 (1972)
5. J.T. Stock, Thomas Charles Robinson and his balances. J. Chem. Educ. **45**(4), 254 (1968)
6. H. Magellan, Lettre sur les balances d'essai. Observations sur la Physique, sur l'Histoire. Naturelle et sur les Arts **17**, 43–49 (1781)

7. H.R. Jenemann, Die Berliner Werkstätten. Hofheim (1996). Translator Thomas Allgeier
8. T. Allgeier, Standton Instruments Ltd. http://stanton-instruments.co.uk/ (2010)
9. Wikipedia, W & T Avery Ltd. http://en.wikipedia.org/wiki/W_%26_T_Avery_Ltd (2011)
10. L. Sanders, *A Short History of Weighing* (Avery, Birmingham, 1947)
11. L.H. Broadbent, *The Avery Business* (Avery, Birmingham, 1949)
12. M. Daumas, Les appareils d'experimentation de Lavoisier. Chymia **3**, 45–61 (1950)
13. M. Daumas, Les instruments scientifiques aux 17. et 18. siècles. Paris (1953)
14. M. Daumas, *Scientific Instruments of the Seventeenth and Eighteenth Centuries and Their Makers* (B.T. Batsford, London, 1972)
15. A. de Lapparent, Biographie de Victor Regnault, in *(Ecole Polytechnique) Livre du Centenaire* (Gauthier-Villars et fils, Paris, 1897), p. 326 et suiv
16. E. Robens, C. Eyraud, P. Rochas, Some comments on the history of vacuum microbalance techniques. Thermochim. Acta **235**, 135–144 (1994)
17. G.F. Brander, Beschreibung einer neuen hydrostatischen Waage. Augsburg (1771)
18. C. Schenck, Schenck Process Austria GmbH. http://www.schenckprocess.at/ (2011)
19. H.R. Jenemann, Zur Geschichte der Herstellung von Präzisionswaagen hoher Leistung in Wien. Blätter für Technikgeschichte **49**, 7–85 (1987)
20. H. Sequenz (ed.), 150 Jahre Technische Hochschule in Wien, 1815–1965. Wien (1965)
21. J.J. Prechtl (ed.), *Technologische Enzyklopädie oder alphabetisches Handbuch der Technologie, der technischen Chemie und des Maschinenwesens*. Stuttgart (1855)
22. E. Dolezal, Christoph und Gustav Starke, Begründer der mathematisch-mechanischen Werkstätte der Firma Starke & Kammerer zu Wien, ihr Leben und Wirken. Z. Instrum.kd. **61**, 58–64 + 82–94 (1941)
23. BIPM (ed.), *Le Bureau International des Poids et Mesures 1875–1975, Cent ans de Métrologie* (BIPM, Paris, 1975)
24. S. Krusper, Eine Vakuumwaage neuer Konstruktion. Z. Instrum.kd. **9**, 81–86 (1889)
25. H.R. Jenemann, Paul Bunge und die Fertigung wissenschaftlicher Waagen in Hamburg. Z. Unternehm.gesch. **31**, 117–140 (1986), see also 165–183
26. H.R. Jenemann, Die Waagenkonstruktionen von Georg Westphal, in *125 Jahre 1860–1985 Westphal-Mechanik/Westphal-Augenoptik*, ed. by W. Mechanik-Augenoptik (Westphal Mechanik-Augenoptik Celle, 1985), pp. 35–45
27. H.R. Jenemann, Zur Geschichte der Präzisionsmechanik und der Herstellung feiner Waagen in Gießen. Mitt. d. Oberhessischen Geschichtsvereins Gießen, NF **66**, 5–54 (1981)
28. M. Gläser, Methods of mass determination, in *Comprehensive Mass Metrology*, ed. by M. Kochsiek, M. Gläser (Wiley/VCH, Berlin, 2000), pp. 184–231
29. H.R. Jenemann, Zur Geschichte des langarmigen Balkens von Präzisionswaagen. Mass Gewicht, Beih. Z. Metrol. **29**(3), 672–687 (1994)
30. J.H. Poynting, *The Mean Density of the Earth* (Charles Griffin, London, 1894)
31. O. Hecker, Das Horizontalpendel. Z. Instrum.kd. **16**, 2–6 (1896)
32. H.F. Wiebe, Über die Genauigkeit der Druckmessung mit der Stückrathschen Druckwaage. Z. Instrum.kd. **30**, 205–217 (1910)
33. H.R. Jenemann, Über die Grundlagen und die geschichtliche Entwicklung elektromechanischer Wägesysteme. Teil I–III. CLB, Chem. Labor Betr. **36**, 393–396 (1985), see also 500–504, 629–632
34. Wikipedia, Hebenstreit-Rapido. http://de.wikipedia.org/wiki/Hebenstreit-Rapido (2011)
35. G.G. Schmidt, *Sammlung physisch-mathematischer Abhandlungen*, vol. 1 (Friedrich Heyer, Gießen, 1793)
36. C. Beyer, Waagen und Wägen. Pharmazie in unsrer Zeit **14**(6), 161–174 (1985)
37. H.R. Jenemann, *Die langarmigen Präzisionswaagen im Liebig-Museum zu Gießen* (Mettler, Gießen, 1988)
38. H.R. Jenemann, Substitutionswägung – heute und vor 200 Jahren. Wägen + Dos. **11**, 24–31 (1980)
39. H.R. Jenemann, Die mechanische Analysenwaage: Einführung und Durchsetzung des Substitutionsprinzips in der Zeit von 1945–1960. Wägen + Dos. **18**, 57–63 (1987), see also 101–103

40. Mettler-Toledo, M. Kochsiek, *Dictionary of weighing terms* (Mettler-Toledo, Greifensee, 1990)
41. H.R. Jenemann, Die Göttinger Präzisionsmechanik und die Fertigung feiner Waagen, in *Göttinger Jahrbuch*. Göttingen (1988), pp. 181–230
42. H.R. Jenemann, Zur Geschichte des kurzen Balkens an der gleicharmigen Balkenwaage. Technikgeschichte **52**, 113–137 (1985)
43. P. Bunge, Neue Construction der Wage. Repertorium für Physikalische Technik + für Mathematische und Astronomische Instrumentenkunde **3**, 269–271 (1867)
44. P. Bunge, Analytische Wage. Repertorium für Experimentalchemie, für Physikalische Technik, Mathematische und Astronomische Instrumentenkunde **3**, 382–384 (1867)
45. H.R. Jenemann, E. Robens, History of vacuum macrobalances, in *Microbalance Techniques*, ed. by J.U. Keller, E. R. (Multi-Science Publishing, Brentwood, 1994), pp. 13–23
46. R. Gaul, Mikro-Waagenfabrik Döft, Heilbronn-Sontheim. Bad Rappenau (2012)
47. S. Kiefer, E. Robens, Die Waagenindustrie in der Schwäbische Alb. Mass Gewicht **9**, 135–144 (2010)
48. S. Kiefer, Philipp Matthäus Hahn und seine Rolle als Begründer des Präzisionswaagenbaus in Süddeutschland, in *Unsichtbare Hände – zur Rolle von Laborassistenten, Mechanikern, Zeichnern u.a. Amanuenses in der in der physikalischen Forschungs- und Entwicklungsarbeit 978392818685*, ed. by K. Hentschel (Verlag für Geschichte der Naturwissenschaft & Technik GNT-Verlag, Diepholz, 2008)
49. H.R. Jenemann, Die wägetechnischen Arbeiten von Philipp Matthäus Hahn, in *Philipp Matthäus Hahn, 1739–1790*, ed. by C. Väterlein, Stuttgart (1989)
50. H.R. Jenemann, Der Mechaniker-Pfarrer Philipp Mathäus Hahn und die Ausbreitung der Feinmechanik in Südwestdeutschland. Z. Württemb. Landesgesch. **46**, 117–161 (1987)
51. H.R. Jenemann, Waagen, in *Philipp Matthäus Hahn, 1739–1790, Katalog*, ed. by C. Väterlein, Stuttgart (1989), pp. 357–366
52. P.M. Hahn, in *Die Kornwestheimer Tagebücher*, ed. by M. Brecht, R.F. Paulus. Berlin, New York (1979)
53. *Kern-Themen, Kern-Topics* (Kern & Sohn, Balingen, 2012)
54. F.C. Huber (ed.), *Festschrift zum 50jährigen Bestehen der württembergischen Handelskammern*. Stuttgart (1910)
55. H.R. Jenemann, Die frühe Geschichte der Mettler-Waage, in *Siegener Abhandlungen zur Entwicklung der materiellen Kultur*. St. Katharinen (1992)
56. H.R. Jenemann, The development of the determination of mass, in *Comprehensive Mass Metrology*, ed. by M. Kochsiek, M. Gläser (Wiley/VCH, Berlin, 2000), pp. 119–163
57. Mahavamsa, The Great Chronicle of Ceylon (1912)
58. Avery-Weigh-Tronix, The History of Weighing. http://www.wtxweb.com/. Avery Weigh-Tronix (2008)
59. W. Euler, Hennefer Waagen-Wanderweg. chronos begleitheft-online.pdf (2008). Hennef: METAP Metrology/Bürgermeister von Hennef
60. E. Gareis, W. Euler, The Hennef Weigher-Walking-Way. chronos begleitheft-online.pdf (2008). Hennef: METAP Metrology/Bürgermeister von Hennef
61. F. Rouquerol, J. Rouquerol, K. Sing, *Adsorption by Powders & Porous Solids* (Academic Press, San Diego, 1999)
62. J.C. Moreno-Piraján, *Adsorption Calorimetry: Fundamentals and Applications in Characterization of Catalysts* (Universidad de Los Andes, Bogotá, 2010)

Appendix

A.1 Prefixes and Numbers

Table A.1 Prefixes and numbers

Multiple	SI prefix	Symbol	Designation		
			Germany, UK, etc.	France	USA, Italy
10^{33}			Quintilliarde		decillion
10^{30}			Quintillion	quintillion	nonillion
10^{27}			Quadrilliarde		octillion
10^{24}	yotta	Y	Quadrillion	quatrillion	septillion
10^{21}	zetta	Z	Trilliarde, thousand trillions	Mille trillions	sixtillion
10^{18}	exa	E	Trillion	trillion	quintillion
10^{15}	peta	P	Billiarde	mille billions	quadrillion
10^{12}	tera	T	Billion	billion	trillion
10^{9}	giga	G	Milliarde	milliard	billion
10^{6}	mega	M	Million	million	million
10^{3}	kilo	k	Thousand	mille	thousand
10^{2}	hecto	h	Hundert, hundred	cent	hundred
10^{1}	deca	da	Zehn, Ten	dix	ten
10^{0}			Eins, one	un	one
10^{-1}	deci	d	Zehntel, tenth	dixième	tenth
10^{-2}	centi	c	Hundertstel, hundredth	centième	hundredth
10^{-3}	milli	m	Tausendstel, thousandth	millième	thousandth
10^{-6}	micro	μ	Millionstel, millionth	millionième	millionth
10^{-9}	nano	n			
10^{-12}	pico	p			
10^{-15}	femto	f			
10^{-18}	atto	a			
10^{-21}	zepto	z			
10^{-24}	yocto	y			

E. Robens et al., *Balances*, DOI 10.1007/978-3-642-36447-1,
© Springer-Verlag Berlin Heidelberg 2014

A.2 Books and Reviews

A.2.1 Literature on Mass, Weights and Balances

Table A.2 Recent books and reviews on mass, weights, and balances

	1981	*Les anciens systèmes de mesures: projet d'enquêtes métrologique; table ronde du 17 Oct. 1981 (Caen).*	Paris: Institut d'Histoire Moderne et Contemporaine
	2006	*Virtual Laboratory.* http://vlp.mpiwg-berlin. mpg.de/technology/search?-max=10&- title=1&-op_varioid=numerical&varioid=2.	Berlin: Max-Planck-Institut für Wissenschaftsgeschichte
Balek, V. Tolgyessy J.	1984	*Emanation thermal analysis and other radiometric emanation methods. Wilson and Wilson's comprehensive analytical chemistry,* vol. 12, Part C.	Elsevier, Amsterdam
Benedetti-Pichler, A.A.	1959	*Waagen und Wägungen,* in: F. Hecht, M.K. Zacherl: *Handbuch der Mikrochemischen Methoden.* Bd. 1, Tl. 2.	Wien: Springer
Beyer, Christian	1985	*Waagen und Wägen.* In: *Pharmazie in unsrer Zeit,* **14**(6): p. 161–174.	
Boeckh, August	1978	*Metrologische Untersuchungen über Gewichte, Münzfüße und Maße des Alterthums in ihrem Zusammenhange.* Reprint.	Berlin
Connor, R.D.	1987	*The weights and measures of England.*	London: HMSO
Crawforth, M.A.	1979	*Weighing coins—English folding gold balances of the 18th and 19th centuries.*	London: Cape Horn Trading
Crawforth, M.A.	1984	*Handbook of old weighing instruments.*	Chicago: International Society of Antique Scale Collectors
Crosby, Alfred W.	1997	*The measure of reality: quantification and Western society; 1250–1600.* Reprint.	Cambridge: Cambridge Univ. Press
Cruz, A.	2007	*Weights and measures in Portugal.*	Caparica: Instituto Português da Qualidade
Dilke, Oswald	1999	*Mathematik, Masse und Gewichte in der Antike.*	Stuttgart: Reclam
Doursther, Horace	1976	*Dictionnaire universel des poids et mesures anciens et modernes: contenant des tables des monnaies de tous les pays.* 3. ed. Reprint.	Amsterdam: Meridian Publ.
Dubler, Anne-Marie	1975	*Masse und Gewichte im Staat Luzern und in der alten Eidgenossenschaft.*	Luzern: Luzerner Forschungsstelle fuer Wirtschafts- und Sozialgeschichte im Staatsarchiv

Table A.2 (Continued)

Earnest, Charles M. (Ed.)	1988	*Compositional analysis by thermogravimetry.*	Philadelphia, PA: ASTM
Garnier, Bernard	1989	*Introduction à la métrologie historique.*	Paris: Economica
Gläser, Michael Kochsiek, Manfred (Eds.)	2010	*Handbook of Metrology.*	Weinheim: Wiley-VCH
Graham, J.T.	1981	*Scales and Balances.*	Aylesbury: Shire Publications
Guerra, François-Xavier (Ed.)	1998	*Los espacios públicos en Iberoamérica: ambigüedades y problemas; siglos XVIII–XIX.*	México: Centro Francés de Estudios Mexicanos y Centroamericanos
Gyselen, Rika	1990	*Prix, salaires, poids et mesures.*	Paris: Groupe pour l'Etude de la Civilisation du Moyen-Orient
Haas, M.	2012	*Scales and Weights. A collection of historical scales and weights from different periods of the past 3000 years.*	http://www.s-a-w.net/
Haeberle, K.E.	1966	*10 000 Jahre Waage.*	Bizerba, Balingen
Hager, Claus	2006	*Württembergische Stein- und Metallgewichte 1557–2000.*	Stuttgart: Koch
Hase, Wolfgang	1994	*Damit mußten sie rechnen … auch auf dem Lande: zur Alltagsgeschichte des Rechnens mit Münze, Maß und Gewicht.*	Cloppenburg: Museumsdorf Cloppenburg
Haustein, Heinz-Dieter	2001	*Weltchronik des Messens: Universalgeschichte von Maß und Zahl, Geld und Gewicht.*	Berlin: de Gruyter
Haustein, Heinz-Dieter	2004	Quellen der Meßkunst: zu Maß und Zahl, Geld und Gewicht.	Berlin: de Gruyter
Haustein, Heinz-Dieter	2007	*Universalgeschichte des Messens.*	Berlin: Directmedia Publishing
Heit, Alfred	1996	*Bibliographie zur historischen Metrologie.*	Auenthal-Verl.
Hellwig, Gerhard	1982	*Lexikon der Masse und Gewichte.*	Gütersloh: Lexikothek Verl.
Hocquet, Jean-Claude	1992	*Anciens systèmes de poids et mesures en Occident.*	Aldershot: Variorum
Hultsch, Friedrich	1971	*Griechische und römische Metrologie.* Reprint 1882.	Graz: Akad. Druck- u. Verl.-Anst.,
Jammer, Max	1974	*Der Begriff der Masse in der Physik.*	Darmstadt: Wiss. Buchgesellschaft
Jenemann, H.R.	1979	*Die Waage des Chemikers.*	DECHEMA, Frankfurt am Main
Jenemann, Hans R.	1992	*Die frühe Geschichte der Mettler-Waage.*	St. Katharinen: Scripta-Mercaturae
Jenemann, Hans R.	1994	The early history of balances based on electromagnetic or electrodynamic force compensation. In: U. Keller, E. Robens (eds.): *Microbalance Techniques.*	Brentwood: Multi-Science Publishing,

Table A.2 (Continued)

Jenemann, Hans R.	1997	*Die Waage des Chemikers. The Chemist's Balance.*	Frankfurt am Main: DECHEMA
Kahnt, Helmut	1987	*Alte Maße, Münzen und Gewichte: ein Lexikon.*	Mannheim: Bibliograph. Inst.
Kind, Dieter	2000	*Für alle Zeiten, für alle Menschen – vor 125 Jahren wurde die Meterkonvention unterzeichnet.* In: *Physikalische Blätter.* 56, 6, 63–66.	Weinheim: Wiley-VCH
Kisch, Bruno	1966	*Scales and Weights. A Historical Outline.*	New Haven: Yale University Press
Klimpert, Richard	1972	*Lexikon der Münzen, Masse, Gewichte, Zählarten und Zeitgrößen aller Länder der Erde.* 2. ed. Reprint 1896.	Graz: Akad. Dr.- u. Verl.-Anst.
Kochsiek, M.	1977	*Über die Luftauftriebskorrektion bei der Weitergabe der Masseeinheit.* PTB-Bericht Me-15.	Braunschweig: PTB
Kochsiek, M. (Ed.)	1989	*Massebestimmung hoher Genauigkeit.*	Braunschweig: PTB
Kochsiek, M. (Ed.)	1989	*Handbuch des Wägens.* 2. ed.	Braunschweig: Vieweg
Kochsiek, M. Gläser, M. (Eds.)	1996 2008	*Massebestimmung.*	Weinheim: VCH
Kochsiek, M. Gläser, M. (Eds.)	2000	*Comprehensive Mass Metrology.*	Weinheim: Wiley-VCH
Kochsiek, M. Meißner, B.	1986	*Wägezellen, Prinzipien, Genauigkeit, praktischer Einsatz für eichfähige Waagen. PTB-Bericht MA-4.*	Braunschweig: PTB
Kottmann, Albrecht	1992	*Die Kultur vor der Sintflut: das gleiche Zahlendenken in Ägypten, Amerika, Asien und Polynesien.*	Heiligkreuztal: Verl. Aktuelle Texte
Kretzschmar, Gunter	2003	*Alte Maße und Gewichte in der Westlausitz.*	Elstra: Elstraer Heimat- und Geschichtsverein
Kula, Witold	1986	*Measures and men.*	Princeton, N.J.: Princeton Univ. Press
M., A. (Ed.)	1980	*Wägetechnische Probleme bei der Massebestimmung.* PTB-Bericht. Vol. Me-26.	Braunschweig: PTB
Rochhaus, Peter	1997	*Alte Maße und Gewichte im Erzgebirge: ein Abriß.*	Annaberg-Buchholz: Adam-Ries-Bund
Rochhaus, Peter	1999	*Rechnen mit alten deutschen Maßen und Gewichten: tabellarische Übersichten und Rechenbeispiele zu 100 ausgewählten Orten in Deutschland.*	Annaberg-Buchholz: Adam-Ries-Bund
Rochhaus, Peter	2002	*Alte Maße und Gewichte im Erzgebirge.*	Annaberg-Buchholz: Adam-Ries-Bund
Rottleuthner, Wilhelm	1985	*Alte lokale und nichtmetrische Gewichte und Masse und ihre Groessen nach metrischem System.*	Innsbruck: Wagner

Table A.2 (Continued)

Siefert, Kurt	2003	*Alte Maße und Gewichte: mit Maß- und Gewichtssystemen; alte Ortsmaße.*	Beerfelden-Gab.: Siefert
Skinner, F.G.	1967	*Weights and Measures—Their ancient origins and their development in Great Britain up to AD 1855.*	London: Science Museum
Spichal, Reinhold	1990	*Jedem das Seine: Markt und Maß in der Geschichte am Beispiel einer alten Hansestadt.*	Bremen: Brockkamp
Trapp, Wolfgang	1998	*Kleines Handbuch der Maße, Zahlen, Gewichte und der Zeitrechnung. 3. ed.*	Stuttgart: Reclam
Tuor, Robert	1977	*Mass und Gewicht im Alten Bern, in der Waadt, im Aargau und im Jura.*	Bern: Haupt
Vieweg, R.	1962	*Maß und Messen in kulturgeschichticher Sicht.*	Wiesbaden: Franz Steiner
Vieweg, R.	1966	*Aus der Kulturgeschichte der Waage.*	Bizerba, Balingen
Von Alberti, Hans-Joachim	1957	*Mass und Gewicht: geschichtliche und tabellarische Darstellungen von den Anfängen bis zur Gegenwart.*	Berlin: Akad.-Verl.
von Graffenried, Charlotte	1990	*Akan: Goldgewichte im Bernischen Historischen Museum.*	Bern: Benteli
von Hippel, Wolfgang	1994	*Mass und Gewicht: im Gebiet von Bayerischer Pfalz und Rheinhessen (Departement Donnersberg) am Ende des 18. Jahrhunderts.*	Mannheim: Inst. für Landeskunde und Regionalforschung der Univ. Mannheim
von Hippel, Wolfgang	1996	*Mass und Gewicht: im Gebiet des Großherzogtums Baden am Ende des 18. Jahrhunderts.*	Mannheim: Inst. für Landeskunde und Regionalforschung
von Hippel, Wolfgang	2000	*Maß und Gewicht im Gebiet des Königreichs Württemberg und der Fürstentümer Hohenzollern am Ende des 18. Jahrhunderts.*	Stuttgart: Kohlhammer
Witthöft, Harald	1979	*Umrisse einer historischen Metrologie zum Nutzen der wirtschafts- und sozialgeschichtlichen Forschung: Maß und Gewicht in Stadt und Land Lüneburg, im Hanseraum und im Kurfürstentum/Königreich Hannover vom 13. bis zum 19. Jh.*	Göttingen: Vandenhoeck & Ruprecht
Witthöft, Harald	1984	*Münzfuß, Kleingewichte, pondus Caroli und die Grundlegung des nordeuropäischen Maß- und Gewichtswesens in fränkischer Zeit.*	Ostfildern: Scripta-Mercaturae
Witthöft, Harald (Ed.)	1986–	*Handbuch der historischen Metrologie.*	St. Katharinen: Scripta Mercaturae
Ziegler, Heinz	1997	*Studien zum Umgang mit Zahl, Maß und Gewicht in Nordeuropa seit dem Hohen Mittelalter.*	St. Katharinen: Scripta Mercaturae
Zimmermann, Albert	1983–	*Mensura – Mass, Zahl, Zahlensymbolik im Mittelalter.*	Berlin: de Gruyter

A.2.2 Literature on Thermal Analysis and Sorptometry

Table A.3 Recent books and reviews on the application of balances in thermal analysis and in controlled environments. The survey is based on a compilation of H.K. Cammenga and S.M. Sarge: *Monographien auf dem Gebiet der Thermischen Analyse.* http://www.gefta.org/literatur.html. 2011, GEFTA

Anderson, H.C.	1966	*Thermal Analysis.*	Dekker, New York
Behrndt, K.	1956	Die Mikrowaagen in ihrer Entwicklung seit 1886. *Z. angew. Physik* 8, 9, 453–472.	
Blazek, A.	1973	*Thermal Analysis.*	Van Nostrand Reinhold, London
Brown, M.E.	1988	*Introduction to Thermal Analysis. Techniques and Applications.*	Chapman and Hall, London
Calvet, E. Prat H.	1963	*Recent Progress in Microcalorimetry.*	Pergamon Press, Oxford
Charsley, E.L. Warrington, S.B. (Eds.)	1992	*Thermal Analysis—Techniques and Applications.*	Roy. Soc. of Chemistry, Cambridge
Cunningham, B.B.	1949	*Microchemical methods used in nuclear research. Nucleonics* 5, 5 62–85.	
Czanderna, A.W. Wolsky, S.P.	1980	*Microweighing in Vacuum and Controlled Environments.*	Elsevier, Amsterdam
Daniels, T.C.	1973	*Thermal Analysis.*	Kogan Page, London
Dodd, J.W. Tonge, K.H.	1987	*Thermal Analysis.* John Wiley for Analytical Chemistry by Open Learning.	Wiley, London
Eder, F.X.	1983	*Arbeitsmethoden der Thermodynamik, Bd. 2: Thermische und kalorische Stoffeigenschaften.*	Springer, Berlin
Emich, F.	1919	*Methoden der Mikrochemie.* In: E. Abderhalben (ed.): *Handbuch der biolochemischen Arbeitsmethoden.*	Berlin: Urban & Schwarzenberg
Emich, F.	1921	*Methoden der Mikrochemie.* In: E. Abderhalben (ed.): *Handbuch der biologischen Arbeitsmethoden.*	Berlin: Urban & Schwarzenberg
Eyraud, C. Robens, E. Rochas, P.	1990	*Some comments on the history of thermogravimetry. Thermochimica Acta* 160, 25–28.	
Ford, J.L. Timmins, P.	1989	*Pharmaceutical Thermal Analysis—Techniques and Applications.*	Ellis Horwood, Chichester
Garn, P.D.	1965	*Thermoanalytical methods of investigation.*	Academic Press, New York
Gorbach, G.	1936	Die Mikrowaage, in: *Mikrochemie*, 20, 2/3, 254–236.	
Grewer, T. Klais, O.	1988	*Exotherme Zersetzung – Untersuchung der charakteristischen Stoffeigenschaften.*	VDI-Verlag, Düsseldorf

Table A.3 (Continued)

Haines, P.J.	1995	*Thermal Methods of Analysis—Principles, Applications and Problems.*	Chapman & Hall, London
Harwalker, V.R. Ma, C.-Y. (Eds.)	1990	*Thermal Analysis of Foods.*	Elsevier Applied Science, London
Hatakeyama, T. Quinn, F X.	1994	*Thermal Analysis—Fundamentals and Applications to Polymer Science.*	Wiley, Chichester
Heide, K.	1982	*Dynamische thermische Analysenmethoden*, 2. Aufl.	VEB Deutscher Verlag für Grundstoffindustrie, Leipzig
Hemminger, W.F. Cammenga, H.K.	1989	*Methoden der Thermischen Analyse.*	Springer, Berlin
Hemminger, W. Höhne, G.	1979	*Grundlagen der Kalorimetrie.*	Verlag Chemie, Weinheim
Hemminger, W. Höhne, G.	1984	*Calorimetry—Fundamentals and Practice.*	Verlag Chemie, Weinheim
Hemminger, W. Höhne, G. Flammersheim, H.-J.	1996	*Differential Scanning Calorimetry.*	Springer, Berlin
Huggle, Ursula	1998	Maße, Gewichte und Münzen: historische Angaben zum Breisgau und zu angrenzenden Gebieten.	Bühl/Baden: Konkordia-Verlag
Jespersen, N.D. (Ed.)	1982	*Biochemical and clinical applications of thermometric and thermal analysis. Wilson and Wilson's comprehensive analytical chemistry*, vol. 12, Part B.	Elsevier, Amsterdam
Keattch, C J. Dollimore, D.	1975	*An Introduction to Thermogravimetry*, 2nd ed.	Heyden, London
Keattch, C.J.	1977	*The History and Development of Thermogravimetry.*	University of Salford
Keller, U. Robens, E. (Eds.)	1994	*Microbalance Techniques.*	Brentwood: Multi-Science Publishing
Koch, E.	1977	*Non-isothermal Reaction Analysis.*	Academic Press, London
Kopsch, H.	1995	*Thermal Methods in Petroleum Analysis.*	VCH, Weinheim
Kubaschewski, O. Alcock, C.B. Spencer, P.J.	1993	*Materials Thermochemistry*, 6th ed.	Pergamon Press, Oxford
LeNeindre, B. Vodar, B. (Eds.)	1975	*Experimental Thermodynamics*, vol. 2.	Butterworths, London
Liptay, G.	1971 1977	*Atlas of Thermoanalytic Curves.*	Akadémiai Kiadó and Heyden, Budapest/London
Lodding, W.	1967	*Gas effluent analysis.*	Dekker, New York
Lu, C. Czanderna, A.W.	1984	*Applications of Piezoelectric Quartz Crystal Microbalances.*	Elsevier, Amsterdam

Table A.3 (Continued)

Mackenzie, R.C. (Ed.)	1970 1972	*Differential Thermal Analysis*; vol. 1, vol. 2.	Academic Press, London
Mathot, V.B.F. (Ed.)	1994	*Calorimetry and Thermal Analysis of Polymers.*	Carl Hanser, München
McCullough, J.P. Scott, D.W. (Eds.)	1968	*Experimental Thermodynamics*, vol. 1.	Butterworths, London
Mikhail, R.S. Robens, E.	1983	*Microstructure and thermal analysis of solid surfaces.*	Wiley, Chichester
Paulik, F.	1995	*Special Trends in Thermal Analysis.*	Wiley, Chichester
Paulik, J. Paulik, F.	1981	*Simultaneous thermoanalytical examinations by means of the derivatograph. Wilson and Wilson's comprehensive analytical chemistry,* vol. 12, Part A.	Elsevier, Amsterdam
Pope, M.I. Judd, M.D.	1977	*Differential Thermal Analysis. A Guide to the Technique and Its Applications.*	Heyden, London
Ramachandran, V.S.	1969	*Applications of differential thermal analysis in cement chemistry.*	Chem. Publ. Co., New York
Ramachandran, V.S. Garg, S.P.	1959	*Differential thermal analysis as applied to building science.*	Central Building Research Institute, Roorkee
Robie, R.A. Hemingway, B.S.	1995	*Thermodynamic properties of minerals and related substances at 298.15 K and 1 bar (10^5 Pascals) pressure and at higher temperature.*	United States Geological Survey Bulletin, 2131
Schultze, D.	1972	*Differentialthermoanalyse.*	Verlag Chemie, Weinheim
Šesták, J.	1984	*Thermophysical Properties of Solids. Their Measurements and Theoretical Thermal Analysis. Wilson and Wilson's comprehensive analytical chemistry,* vol. 12, Part D.	Elsevier, Amsterdam
Skinner, H.A. (Ed.)	1962	*Experimental Thermochemistry*, vol. 2.	Interscience Publishers, New York
Slade, Jr. P.E. Jenkins, L,T. (Eds.)	1966	*Techniques and methods of polymer evaluation.*	Dekker, New York
Slade, Jr. P.E. Jenkins, L.T. (Eds.)	1970	*Thermal characterization techniques.*	Dekker, New York
Smothers, W.J. Chiang, Y.	1966	*Differential thermal analysis: theory and practice.*	Chemical Publ. Co., New York
Smothers, W.J. Chiang, Y.	1966	*Handbook of differential thermal analysis.*	Chemical Publ. Co., New York
Smykatz-Kloss, W.	1974	*Differential Thermal Analysis. Application and Results in Mineralogy.*	Springer, Berlin
Smykatz-Kloss, W.	1991	*Thermal Analysis in the Geosciences.*	Springer, Berlin
Speyer, R.F.	1994	*Thermal Analysis of Materials.*	Dekker, New York

Table A.3   (Continued)

Todor, D.N.	1976	*Thermal analysis of minerals.*	Abacus Press, Turnbridge Wells
Turi, E.A. (Ed.)		*Thermal Characterization of Polymeric Materials.*	Academic Press, New York
Utschick, H.	1996	*Anwendungen der Thermischen Analyse.*	Ecomed, Landsberg
Wendlandt, W.W. Collins, L.W.	1976	*Thermal Analysis.*	Dowden, Hutchinson & Ross, Stroudsberg
Wendlandt, W.W.	1986	*Thermal Analysis*, 3rd ed.	Wiley, New York
Widmann, G. Riesen, R.	1987	*Thermal Analysis: Terms, Methods, Applications.*	Hühitg, Heidelberg
Wiedemann, H.G. Bayer G.	1978	*Trends and Applications of Thermogravimetry*, in *"Topics in Current Chemistry"*, vol. 77.	Springer, Berlin
Wiederholt, E.	1984	*Differenzthermoanalyse (DTA) im Unterricht.* Praxis Schriftenreihe Chemie, Bd. 37.	Aulis, Köln
Wolsky, S.P. Zdanuk, E.J.	1969	*Ultra Micro Weight Determination in Controlled Environments.*	Wiley, New York
Wunderlich, B.	1982	*Audio Course Thermal Analysis.* 10 C-90 audio cassettes and textbook.	Rensselaer Polytechnic Institute, Troy, NY
Wunderlich, B.	1990	*Thermal Analysis.*	Academic Press, Boston

Table A.4   *Hans Jenemann's scales library*, stored in the Philipp-Matthäus-Hahn-Museum, Onstmettingen, Germany

1885	*Beschreibung und Erläuterung zu den Bildlichen Darstellungen von aichfähigen Gattungen von Maaßen, Werkzeugen, Gewichten und Waagenkonstruktionen*	Berlin: W. Moeser
1933	*75 Jahre sächsisches Eichwesen.*	Dresden: Vereinigung sächsischer technischer Eichbeamten
1934	*Waagen und Gewichte – Warum eine Sartorius-Analysenwaage?* Katalog.	Göttingen: Sartorius
1939	*Saalburg Jahrbuch.* Vol. 9.	Frankfurt am Main: F.B. Auffarth
1941	*Catalogue du Musée Section K Poids et Mesures Métrologique.*	Paris: Conservatoire Nationale des Arts et Métiers
1967	*150 Jahre Hessische Eichbehörde.*	Darmstadt: Hessische Eichbehörde
1970	*1870–1970 Sartorius.*	Göttingen: Sartorius

Table A.4 (Continued)

	1975	*Die SI-Basiseinheiten Definition Entwicklung Realisierung.* PTB-Mitteilungen.	Braunschweig: Physikalisch- Technische Bundesanstalt
	1984	*Internationales Wörterbuch der Metrologie.*	Berlin: Beuth
		L'artz & l'argent.	Paris: Banque Nationale de Paris
Ahrens, D.	1993	*I. Angewandte Metrologie in Geschichte und Gegenwart – II. Die Geometrie von Trierer Kirchengebäuden.* Trierer Museums-Seminare, ed. G. Gaffga.	Trier: Städtisches Museum Simeonsstift
Albertazzi, C., Dazzani, C., eds.	1992	*Collezzione Albertazzi–Dazzani.*	Campogallione: Museo della Bilancia
Amman, J.	1568	*Das Ständebuch.*	Frankfurt am Main, Leipzig: Insel-Verlag
Appius, M.	1972	*Feingerätebau am Beispiel der Waagen und Kraftmesser.*	Buchs: Interstaatliche Ingenieurschule Neu-Technikum Buchs
Balhorn, R., Buer, B., Gläser, M., Kochsiek, M.	1983	*Massebestimmung. Teil 1: Weitergabe der Masseneinheit – Waagen, Gewichtstücke und Prüfräume.* PTB-Bericht. Vol. MA-24.	Braunschweig: PTB
Bamberger, F.	1832	*Beschreibung und Abbildung der in neuerer Zeit erfundenen und verbesserten Maschinen zum Wägen sowie kleinerer, sehr empfindlicher Wagen insbesondere der Brückenwagen, Federwagen, hydraulischen Wagen, Wagen für Chemiker und Apotheker etc.*	Quedlinburg: Gottfr. Basse.
Bauerreiß, H.	1913/ 1914	*Zur Geschichte des spezifischen Gewichts im Altertum und Mittelalter.*	Erlangen: Universität Erlangen
Bayer-Helms, F., German, S., Becker, G., Schrader,H.-J., Schley, U., Thomas, W., Drath, P., Bauer, G.	1975	*Die SI-Basiseinheiten, Definition, Entwicklung, Realisierung.* PTB-Bericht.	Braunschweig: PTB
Behrendsen, O.	1900	*Die mechanischen Werkstätten in Göttingen.*	Leipzig: Kiepert'sche Verlagsbuchhandlung
Ben Chanaa, M., ed.	1995	*Proceedings of the 26th International Conference on Vacuum Microbalance Techniques Marrakesh 1995.*	Marrakesh: Faculty of Sciences Semlalia
Bevan, S.C., Gregg, S.J., Parkyns, N.D., eds.	1973	*Progress in Vacuum Microbalance Techniques.* Progress in Vacuum Microbalance Techniques. Vol. 2.	London: Heyden & Son

Table A.4   (Continued)

Block, W.	1928	*Messen und Wägen*. Chemische Technologie, ed. A. Binz.	Leizig: Otto Spamer
Biétry, L., Kochsiek, M.	1982	*Mettler Wägelexikon*.	Greifensee: Mettler
Block, W.	1912	*Maße und Messen*. Aus Natur und Geisteswelt, ed. B.G. Teubner.	Leipzig: B.G. Teubner
Arlaud-Chabalian, M.-C.	1992	*La balance historique, technique et iconographie*.	Cahiers de Métrologie, **10**: pp. 77–88
Böttcher, P., ed.	1910	*Kollektiv-Ausstellung der Deutschen Präzisionsmechanik und Optik, Brüssel*.	Leipzig: Hermann Hönnicke
Bosch, L., Jenemann, H.R.	1977	*Festschrift zum 125jährigen Jubiläum der Firma Gebr. Bosch*.	Jungingen: Gebr. Bosch
Braunbeck, J., Hasenauer, W., Lewisch, R., eds.	1980	*Von der Elle zum Atommaß – Messen: gestern, heute, morgen*.	Wien: Bundesamt für Eich- und Vermessungswesen
Brander, G.F.	1771	*Beschreibung einer neuen hydrostatischen Wage nebst zweyen hiezu gehörigen Abhandlungen*.	Augsburg: Eberhard Kletts sel. Wittib
Brandes, Gmelin, Horner, Muncke, Pfaff, eds.	1825	*Johann Samuel Traugott Gehler's Physikalisches Wörterbuch, Erste Abtheilung A und B*. Vol. 10.	Leipzig
Brandes, P., Kochsiek, M.	1982	*Maßnahmen zur ‚Funktionsfehlererkennbarkeit (FFE)' an einer elektromechanischen Waage*. PTB-Bericht. Vol. Me-37.	Braunschweig: PTB
Brauer, E.	1880	*Die Konstruktion der Waage*. Neuer Schauplatz der Künste und Handwerke.	Weimar: Bernhard Friedrich Voigt
Brauer, E.	1880	*Atlas zur Konstruktion der Waage nach wissenschaftlichen Grundsätzen und nach Massgabe ihres Special-Zweckes*.	Weimar: B.F. Voigt
Brauer, E., Lawaczek, F.	1906	*Die Konstruktion der Wage*.	Osnabrück: Reinard Kuballe
Child, E.	1940	*The tools of the chemist*.	New York: Reinhold
Crawforth, M.A.	1979	*Weighing coins—English folding gold balances of the 18th and 19th centuries*.	London: Cape Horn Trading
Crawforth, M.A.	1984	*Handbook of old weighing instruments*.	Chicago: International Society of Antique Scale Collectors
Czanderna, A.W., ed.	1971	*Vacuum Microbalance Techniques—Proceedings of the Wakefield Conference 1969*. Vol. 8.	New York: Plenum Press
Czanderna, A.W., Wolsky, S.P.	1980	*Microweighing in Vacuum and Controlled Environments*.	Amsterdam: Elsevier

Table A.4 (Continued)

Darius, J., Mills, A., Millburn, J.R., Lawrence, J., eds.	1983– 1994	*Bulletin of the Scientific Instrument Society.*	Leicester: The Scientifc Instrument Society
Daumas, M.	1972	*Scientific Instruments of the Seventeenth and Eighteenth Centuries and their Makers.*	London: B.T. Batsford
de Clercq, P.R., ed.	1985	*Nineteenth-Century scientific instruments and their makers.*	Leiden, Amsterdam: Rodopi
de Saussure, H.B.	1783/ 1900	*Versuch über die Hygrometrie.* Ostwald's Klassiker der exakten Wissenschaften ed. A.J. von Oettingen. Vol. 115 + 119.	Neuchâtel/Leipzig: Engelmann
Dettmer, H., ed.	1977	*Messen und Wiegen – Alte Meßgeräte aus Westfalen.*	Münster: Landesverband Westfalen-Lippe
Domke, J., Reimerdes, E.	1912	*Handbuch der Aräometrie.*	Berlin: Julius Springer
Dove, H.W.	1835	*Über Maass und Messen.* 2 ed.	Berlin: Sandersche Buchhandlung
Düwel, K., et al., eds.	1987	*Untersuchungen zu Handel und Verkehr der vor- und frühgeschichtlichen Zeit in Mittel- und Nordeuropa.* Abhandlungen der Akademie der Wissenschaften in Göttingen Philologisch-Historische Klasse. Vol. 3/156.	Göttingen: Vandenhoek & Rupprecht
Duval, C.	1953	*Inorganic Thermogravimetric Analysis.*	Amsterdam: Elsevier
Felgenträger, W.	1932	Feine Waagen, Wägungen und Gewichte.	Berlin: Julius Springer
Fleckenstein, J.O., ed.	1979	*Travaux du II.Congrès International de la Métrologie Historique, Edinbourg 1977.*	München: Forschungsinstitut des Deutschen Museums
Franken, N.	1994	*Aequipondia – Figürliche Laufgewichte römischer und frühbyzantinischer Schnellwaagen.*	Alfter: VDG
Gaude, W.	1979	*Die alte Apotheke.* 2 ed.	Leipzig: Koehler & Amelang
Garnier, B., Hocquet, J.C., Woronoff, D., eds.	1989	*Introduction a la metrologie historique.*	Paris: Economica
Garnier, B., Hocquet, J.-C., eds.	1989	*Genèse et diffusion dy système métrique.*	Caen: Édition-Diffusion du Lys
Garnier, B., Hocquet, J.C., eds.	1990	*Le médieviste et la métrologie historique, 2. partie.* Cahiers de Métrologie, ed. J.C. Hocquet.	Caen: Centre de Recherche d'Histoire Qauntitative

Table A.4 (Continued)

Gast, T., Robens, E., eds.	1972	*Progress in Vacuum Microbalance Techniques. Proceedings of the 9th Conference on Vacuum Microbalance Techniques 1970 Berlin.* Progress in Vacuum Microbalance Techniques.	London: Heyden & Son
Giesecke, P., Preußer, T.	1991	*Von Waagen und Waagenbauern.*	Darmstadt: Carl Schenk
Gläser, M.	1989	*100 Jahre Kilogrammprototyp.* PTB-Bericht. Vol. MA-15.	Braunschweig: PTB
Gorbach, G.	1936	*Die Mikrowaage.*	Mikrochemie, 1936. **20**(2/3): p. 254–236
Grossi, L., ed.	1995	*La massa e la sua misura—Mass and its measurement—Proceedings of the International Congress, Modena 1993.*	Bologna: CLUEB
Haeberle, K.E.,	1966	*10 000 Jahre Waage.*	Balingen: Bizerba
Häussermann, U.	1962	*Ewige Waage.*	Köln: DuMont Schauberg
Haim, E., ed.	1929	*Festschrift zum 60.Geburtstag von Hofrat Prof. Dr. Fritz Pregl.*	Mikrochemie, Wien: Haim
Haim, E., ed.	1930	*Festschrift zum 70. Geburtstag von Hofrat Prof. Dr. Friedrich Emich.*	Mikrochemie, Wien: Haim
Hasenauer, W., Lewisch, R., eds.	1982	*60 Jahre Bundesamt für Eich- und Vermessungswesen.*	Wien: Bundesamt für Eich- und Vemessungswesen
Hein, W.-H.	1974	*Christus als Apotheker.*	Frankfurt am Main: Govi
Hein, W.-H.	1992	*Christus als Apotheker.* 2 ed.	Frankfurt am Main: Govi
Hess, E., ed.	1970	*PTB-Prüfregeln Feinwaagen 9.01-70.*	Braunschweig: Deutscher Eichverlag
Hess, E.	1983	*Waagen – Bau und Verwendung.*	Berlin: Deutscher Eichverlag
Hoffmann, D., Witthöfft, H., eds.	1996	*Genauigkeit und Präzision in der Geschichtee der Wissenschaften und des Alltags.* PTB-Texte.	Braunschweig: Physikalisch-Technische Bundesanstalt
Hoppe-Blank, J.	1975	*Vom metrischen System zum Internationalen Einheitensystem – 100 Jahre Meterkonvention.* PTB-Bericht. Vol. ATWD-5.	Braunschweig: PTB
Hürlimann, A., Reininghaus, A., eds.	1996	*Mäßig & Gefräßig.*	Wien. MAK-Österreichisches Museum für angewandte Kunst
Ibel, T.	1908	*Die Wage im Altertum und Mittelalter.*	Erlangen: Friedrich-Alexander-Universität

Table A.4 (Continued)

Jammer, M.	1964	*Der Begriff der Masse in der Physik.*	Darmstadt: Wissenschaftliche Buchgesellschaft
Jenemann, H.R.	1979	*Die Waage des Chemikers.*	Frankfurt am Main: DECHEMA
Jenemann, H.R.	1988	*Die langarmigen Präzisionswaagen im Liebig-Museum zu Gießen.*	Gießen: Mettler
Justi, E., Vieweg, R.	1958	*Elektrothermische Kühlung und Heizung. – Maß und Messen in Geschichte und Gegenwart.* Arbeitsgemeinschaft für Forschung des Landes Nordrhein-Westfalen, ed. L. Brandt.	Köln: Westdeutscher Verlag
Jewell, B.	1978	*Veteran scales and balances.*	Tunbridge Wells: Midas Books
Kamper, P., Rekittke, H.J., Dehn, M.H., Schmidt, W., Frahm, H.	1974– 1995	*Wägen und Dosieren.*	Mainz: Keppler-Kirchheim
Karpin, E.V.	1960	*Wägemaschinen.*	Leipzig: VEB Fachbuchverlag
Keller, J.U., Robens, E., eds.	1994	*Microbalance Techniques—Proceedings of the 25th Conference on Vacuum Microbalance Techniques 1993 Siegen.*	Brentwood: Multi-Science Publishing
Kisch, B.	1966	*Scales and Weights. A Historical Outline.*	New Haven: Yale University Press
Kissel, O.R.	1984	*Die Justitia.*	München: C.H. Beck
Kochsiek, M., ed.	1984	*Massebestimmung hoher Genauigkeit.* PTB-Bericht.	Braunschweig: PTB
Kochsiek, M.	1985	*Handbuch des Wägens.* 2 ed.	Vieweg: Braunschweig
Kochsiek, M., Meißner, B.	1986	*Wägezellen, Prinzipien, Genauigkeit, praktischer Einsatz für eichfähige Waagen. PTB-Bericht.* Vol. MA-4.	Braunschweig: PTB
KG., H., ed.	1971	*Eins zwei drei vier fünf sechs sieben – eine Geschichte des Zählens und des Zählers.*	Aldingen: J. Hengstler
K. Normal-Eichungskommission, ed.	1885	*Beschreibung und Erläuterung zu den bildlichen Darstellungen der aichfähigen Gattungen von Maaßen, und Meßwerkzeugen, Gewichten und Waagenkonstruktionen.*	Berlin: W. Moeser
K. Normal-Aichungs-Kommission, ed.	1911	*Bildliche Darstellungen der aichfähigen Gattungen von Maaßen, Meßwerkzeugen, Gewichten und Waagenkonstruktionen.*	Berlin: W. Moeser

Table A.4 (Continued)

K. Normal-Eichungskommission, ed.	1913–1914	*Bildliche Darstellungen der eichfähigen Gattungen von Meßgeräten.*	Berlin: W. Moeser
K. Normal-Eichungskommission, ed.	1914	*Bildliche Darstellungen der eichfähigen Gattungen von Meßgeräten – Beschreibung und Erläuterung 2. Teil.*	Berlin: W. Moeser
Kramm, D., Beringer, K., eds.		*Das rechte Maß.* Schriftenreihe des Bayrischen Staatsministeriums für Wirtschaft und Verkehr-	München: Bayrisches Staatsministerium für Wirtschaft und Verkehr
Kristóf, J., Nowák, C., eds.	1999	*Proceedings of the 7th European Symposium on Thermal Analysis and Calorimetry—ESTAC 7, Balatonfüred.* Journal of Thermal Analysis and Calorimetry.	Budapest: Akadémiai Kiadó
Kruhm, A.	1934	*Die Waage im Wandel der Zeit.*	Frankfurt am Main: W. Kramer
Laitinen, H.A., Ewing, G.W., eds.	1977	*A history of Analytical chemistry.*	York, Pa.: Div. of Analytical Chemistry of the American Chemical Society
Lohrengel, J.	1988	*Organisation und Eichzeichen der Eichbehörden seit 1871.* PTB-Bericht. Vol. TWD-32.	Braunschweig: PTB
Luppi, G., Grossi, L., Vaughan, D., Menna, E., Dragoni, G., eds.	1993	*Bilance a bracci uguali—equal-arm balances.*	Modena: Museo della Bilancia
Machabey, J.A.	1949	*Mémoire fur l'histoire de la balance et de la balancerie.*	Paris: Imprimerie Nationale
Massen, C.H. Van Beckum, H.J.	1970	ed. *Vacuum Microbalance Techniques—Proceedings of the Eindhoven Conference 1968.* Vol. 7.	New York: Plenum Press
Maushart, M.-A.	1993	*Die Entwicklung der Feinwaagenindustrie auf der Schwäbischen Alb am Beispiel der Firma Kern & Sohn.*	Stuttgart: Universität
Meißner, B., Volkmann, C.U.	1981	*Prüfung von Dehnungsmeßstreifen-Wägezellen.* PTB-Bericht. Vol. Me- 30.	Braunschweig: PTB
Mendelejew, D.L.	1931	*Olythnoe issledwanie kolebanija wesow (Experimentelle Untersuchung über Waagen).*	Leningrad: Gosudarstwennoe Nautschno-Technitscheskoe Izdatelstwo – Lenchimsektor
Menzel-Rogner, H.	1944	*Nikolaus von Cues: Der Laie über Versuche mit der Waage.* 2 ed. Philosophische Bibliothek. Vol. 220.	Leipzig: Felix Meiner

Table A.4 (Continued)

Mettler, ed.		*Mettler.*	Stäfa: Mettler
Mettler	1956	*Service manual—Balance d'analyse.*	Greifensee: Mettler
Mettler, ed.	1656–1966	*Mettler News.* Vol. 1–36.	Zürich: Mettler
Mettler-Toledo	1989	*Mettler Wägefibel.*	Greifensee: Mettler
Mettler-Toledo, Kochsiek, M., eds.	1990	*Dictionary of weighing terms.*	Greifensee: Mettler-Toledo
Mettler-Toledo, Kochsiek, M., eds.	1991	*Grundlagen der Massebestimmung.*	Greifensee: Mettler-Toledo
Mez, L.	1975	*Womit der Apotheker einst hantierte.*	Basel: Gute Schriften
Milano, E., Luppi, G.	1991	*Libra—la bilancia nei codici estensi.*	Modena: Il Bulino edizione d'arte
Munz, A.	1977	*Philipp Matthäus Hahn.*	Sigmaringen: Jan Thorbecke
Padelt, E.	1931	*Waagen, Eigenschaften, Wartung, Instandsetzung.* RKW-Veröffentlichungen.	Berlin: Beuth
Paret, O.	1939	*Von römischen Schnellwaagen und Gewichten*, in *Saalburg Jahrbuch.*	Frankfurt am Main: F.B. Auffarth
Payen, J., ed.	1980	*L'industrie Francaise des instruments de précision—Catalogue 1901–1902.*	Paris: Alain Brieux
Payen, J., ed.	1984	*Catalogue de l'exposition collective Allemande d'instruments d'optique et de mécanique de précision 1900.*	Paris: Alain Brieux
Pelly, I., Shoval, S., Steinberg, M., eds.	1988	*The 9th International Congress on Thermal Analysis—ICTA.* 1988, Israel Academy of Sciences and Humanities: Jerusalem.	Jerusalem: Jewish University
Perrier, G.	1949	*Wie der Mensch die Erde gemessen und gewogen hat.*	Bamberg: Bamberger Verlagshaus Meisenbach
Place, F.	1869	*Theorie und Konstruktion der Neigungswaage (Zeigerwaage).*	Weimar: Bernhard Friedrich Voigt
Pressouvre, L., Chabalian, M.C.	1984	*La balance, historique, technique et iconographie du XIe au XVe siècles (1450).*	Paris: Université de Paris I
Probst, R.	1983	*Untersuchung eines hydrostatischen Wägeverfahrens hoher Genauigkeit.* PTB-Bericht. Vol. Me-45.	Braunschweig: PTB
Raudnitz, M., ed.		*Die Konstruktion der von Hand bedienten Waagen.* Handbuch des Waagenbaus, Vol. 1. 19335.	Leipzig: Bernhard Friedr. Voigt
Reimpell, J.	1955	*Handbediente Waagen.* 5 ed. Handbuch des Waagenbaus, ed. M. Raudnitz and J. Reimpell. Vol. 1.	Berlin: Bernhard Friedr. Voigt

Table A.4 (Continued)

Reimpell, J., Krackau, E.	1966	*Selbstanzeigende und selbsttätige Waagen.* Handbuch des Waagenbaus, ed. M. Raudnitz and J. Reimpell. Vol. 2.	Berlin-Hamburg: Bernhard Friedr. Voigt - Verlag Handwerk und Technik
Reimpell, J., Bachmann, W.	1966	*Elektromechanische Waagen.* Handbuch des Waagenbaus.	Berlin-Hamburg: Bernh. Friedr. Voigt - Verlag Handwerk und Technik
Robens, E.	1980	*Errors in thermogravimetric experiments resulting from adsorption on the counterweight,* in *Thermal Analysis,* H.G. Wiedemann, Editor, p. 213–218.	Basel: Birkhäuser
Rutishauser, H., Reichmuth, A.	1988	*Die neue Analysenwaage AT von Mettler.*	Greifensee: Mettler
Saint-Paul, R., ed.	1989	*L'aventure du mètre.*	Paris: Musée National des Techniques— CNAM
Saint-Paul, R., ed.	1990	*Inventaire des Poids.*	Paris: Musée National des Techniques— CNAM
Sawelski, F.S.	1977	*Die Masse und ihre Messung.*	Thun: Harri Deutsch
Scheel, K.	1911	*Grundlagen der praktischen Metronomie.* Die Wissenschaft.	Braunschweig: Vieweg
Schember, C., ed.	1888	*Brückenwaagen.* Katalog.	Wien: C. Schember & Söhne
Schenk-Behrens, K.W., ed.	1981– 1990	*Waagen Auktionen Essen.*	Essen: Schenk-Behrens
Schmidt, G.G.	1793	*Sammlung physisch-mathematischer Abhandlungen.* Vol. 1.	Gießen: Friedrich Heyer
Schostin, N.A.	1990	*Otscherki Istorii Russkoj Metrologii.*	Moskwa: Isdatelstwo Standastow
Schröter, G.	1991	*Eichgesetze und Waagen – Ein Leitfaden.* 3 ed.	Greifensee: Mettler-Toledo
Schulz, W.	1990	*Das gesetzliche Meßwesen in der Bundesrepublik Deutschland.* PTB-Bericht. Vol. TWD-36.	Braunschweig: PTB
Sessler, A.	1960	*Die Waage in Handel und Industrie.*	Stuttgart: Deva Fachverlag
Sheppard, T., Musham, J.F.	1975	*Money, scales and weights.*	London: Spink & Son
Skinner, F.G.	1967	*Weights and Measures—Their ancient origins and their development in Great Britain up to AD 1855.*	London: Science Museum
Spieweck, F., Bettin, H.	1991	*Methoden zur Bestimmung der Dichte von Festkörpern und Flüssigkeiten.* PTB-Bericht. Vol. W-46.	Braunschweig: PTB
Steinle, P., ed.	1968	*1868 Söhnle.*	Murrhardt: Gebrüder Söhnle

Table A.4 (Continued)

Steinle, P., ed.	1993	*125 Jahre Soehnle-Waagen – Der Soehnle-Waagenstammbaum – Von der Waagenwerkstatt zum Marktführer.*	Murrhardt: Soehnle Waagen GmbH & Co
Stock, J.T.	1969	*Development of the chemical balance.* A Science Museum Survey.	London: Her Majesty's Stationery Office
Strecker, A.	1990	*Eichgesetz Einheitengesetz und Durchführungsverordnungen.* 3 ed.	Braunschweig: Deutscher Eichverlag
Tamás, K.	1980	*Mérlegtechnikai Kézikönyv.*	Budapest: Franklin Nyomda
Tauchnitz, O.	1913	*Automatische Registrierwagen.*	München: Oldenbourg
Terrien, J., ed.	1975	*Le Bureau International des Poids et Mesures 1875–1975.*	Sèvres: BIPM
Testut, C.	1946	*Mémento du pesage.*	Paris: Hermann
Trapp, W.	1983	*Kurze Geschichte des gestzlichen Meßwesens.* PTB-Bericht. Vol. ATWD-20.	Braunschweig: PTB
Trapp, W.	1994	*Geschichte des gesetzlichen Meßwesens und ausführliches Literaturverzeichnis zur historischen Metrologie.* PTB-Bericht. Vol. TWD-43.	Braunschweig: PTB
Unshelm, G., Schmitz, D., eds.	1986– 1995	*Maß und Gewicht.*	Solingen: Maß & Gewicht
Vangroenweghe, D., Geldof, T.	1989	*Apothecaries' weights—Pondera medicinalia.*	Brugge: Studiecentrum voor Apothekersgewichten
Vieweg, R.	1962	*Maß und Messen in kulturgeschichticher Sicht.*	Wiesbaden: Franz Steiner
Volkmann, C.U., ed.	1983	*Neuere Entwicklungen und eichtechnische Probleme bei elektromechanischen Waagen.* PTB-Bericht. Vol. ME-41.	Braunschweig: PTB
Walther, L.	1927	*Waagenbau und Reparatur.* Colemans Fachbibliothek für das Schlossergewerbe.	Lübeck: Charles Coleman
Westphal, J., ed.	1984	*Messen Prüfen Eichen – Eichwesen in Schleswig-Holstein.*	Kiel: Minister für Wirtschaft und Verkehr des Landes Schleswig-Holstein
Wiedemann, H.G., ed.	1972	*Thermal Analysis—Proceedings of the 3rd International Conference on Thermal Analysis, Davos 1971.*	Basel: Birkhäuser
Wiedemann, H.G., Hemminger, W., eds.	1980	*Thermal Analysis—Proceedings of the 6th International Conference on Thermal Analysis 1980 Bayreuth—ICTA 80.*	Basel: Birkhäuser
Wittenberger, W.	1957	*Chemische Laboratoriumstechnik.* 5 ed.	Wien: Springer
Zingler, J.	1928	*Theorie der zusammengesetzten Waagen.*	Berlin: Springer

Table A.5 Hans Jenemann's publication list

1. Jenemann, Hans R.: Eine kurze Entwicklungsgeschichte der wissenschaftlichen Waage. In: Bosch, Ludwig und Jenemann, Hans R.: „Festschrift zum 125jährigen Jubiläum der Firma Gebr. Bosch". Jungingen 1977, 29–66.

2. Jenemann, Hans R.: Zur Geschichte der Entstehung der Substitutionswägung zur genauen Massenbestimmung (Borda'sche Wägung). „Fresenius Zeitschrift für Analytische Chemie" 291 (1978), 1–9.

3. Anonym (Redaktioneller Beitrag des Chefredakteurs): Freizeit-Historiker im Dienste der Präzisionswaage. „Wägen und Dosieren" 9 (1978), 203–207.

4. Jenemann, Hans R.: „Die Waage des Chemikers – Betrachtung zu einer Darstellung im Dechema-Haus". Frankfurt am Main 1979.

5. Anonym („Je" als Verfasser): Hans Meier- Persönlichkeiten prägen Waagen und Waagen prägen Persönlichkeiten. „Wägen und Dosieren" 10 (1979), 256–257.

6. Jenemann, Hans R.: Zur Entwicklungsgeschichte der Neigungswaage. Wägen und Dosieren 11 (1980), 210–215; 248–253.

7. Jenemann, Hans R.: Die Pharmazie und die Entwicklung der analytischen Waage. „Deutsche Apotheker Zeitung" 120 (1980), 2057–2064.

8. Jenemann, Hans R.: Substitutionswägung – heute und vor 200 Jahren. Wägen und Dosieren 11 (1980) 24–31.

9. Jenemann, Hans R.: Philipp Mathäus Hahn und die Verwirklichung arbeitssparender und bequemer Waagen. „Blätter für Württembergische Kirchengeschichte" 80/81 (1980/1981), 142–174.

10. Jenemann, Hans R.: Zur Geschichte der Präzisionsmechanik und der Herstellung feiner Waagen in Gießen. „Mitteilungen des Oberhessischen Geschichtsvereins, Gießen", N. F. 66 (1981), 5–54.

11. Jenemann, Hans R.: Auf dem Weg zur modernen Analysenwaage. „CR-Magazin," Farbbeilage zu: „Chemische Rundschau" (Zürich), Nr. 14 (1 April 1981), 4–10.

12. Jenemann, Hans R.: Zur Geschichte der Substitutionswägung und der Substitutionswaage. „Technikgeschichte" 49 (1982), 89–131.

13. Jenemann, Hans R.: Zur Geschichte der Waage in der Wissenschaft. In: Giebeler, E.H.W. u. K.A. Rosenbauer (Hg.): „Historia scientiae naturalis". Darmstadt 1982, 97–119.

14. Jenemann, Hans R.: Über die Grundlagen der Ausführung von Wägungen im Laboratorium. „CLB Chemie für Labor und Betrieb", 33 (1982), 315–320; 356–358.

15. Jenemann, Hans R.: Zur Geschichte der mechanischen Laboratoriumswaage. „Physikalische Blätter" 38 (1982), 316–321.

16. Jenemann, Hans R.: Early History of the Inclination Balance. Equilibrium, Quaterly Magazine of ISASC, International Society of Antique Scale Collectors (1983) 571–578; 602–610.

17. Jenemann, Hans R.: Eine römische Waage mit nur einer Schale und festem Gegengewicht. „Archäologisches Korrespondenzblatt" 14 (1984), 81–96.

18. Jenemann, Hans R.: Entwicklung und heutiger Stand der mechanischen Analysenwaage „CLB Chemie für Labor und Betrieb" 34 (1983), 560–564; 35 (1984), 74–76, 188–189, 296–300, 390–393.

19. Jenemann, Hans R.: Nachwort zu: Bunge, P.: „Beschreibung der Präcisionswaagen", Hamburg 1884; Repr. Mainz 1984, 19–24.

20. Jenemann, Hans R.: Robert Hooke und die frühe Geschichte der Federwaage. „Berichte zur Wissenschaftsgeschichte" 8 (1985), 121–130.

Table A.5　(Continued)

21. Jenemann, Hans R.: Zur Geschichte des kurzen Balkens an der gleicharmigen Balkenwaage. Technikgeschichte 52 (1985) 113–137.

22. Jenemann, Hans R.: Die Entwicklung der mechanischen Präzisionswaage. In: M. Kochsiek (Hg.): „Handbuch des Wägens". Braunschweig, Wiesbaden 1985, 547–587.

23. Jenemann, Hans R.: Die Waage in ihrer jahrtausendlangen Geschichte – Die alten Ägypter wogen die Seele. „Chemische Rundschau" 38 (1985), 13.

24. Jenemann, Hans R.: Die Waagenkonstruktionen von Georg Westphal. In: Westphal Mechanik-Augenoptik (Hg.): 125 Jahre 1860–1985 Westphal-Mechanik / Westphal-Augenoptik. Celle 1985, 35–45.

25. Jenemann, Hans R.: Über Ausführung und Genauigkeit von Münzwägungen in spätrömischer und neuerer Zeit. „Trierer Zeitschrift" 48 (1985), 163–194.

26. Jenemann, Hans R.: Die Waage – mehr als eine Blackbox zum Knöpfedrücken: „Labor 2000 „Sonderpublikation Labor-Praxis" 85/86 (1985), 56–76.

27. Jenemann, Hans R.: Zur Geschichte der Federwaage. „Wägen und Dosieren" 17 (1986), 177–182.

28. Jenemann, Hans R.: Über die Grundlagen und die geschichtliche Entwicklung elektro-mechanischer Wägesysteme. CLB – Chemie für Labor und Betrieb 36 (1985) 393–396, 500–504, 629–632; 37 (1986) 169–172, 344–345, 631–633; 38 (1987) 240–246.

29. Jenemann, Hans R.: Paul Bunge und die Fertigung wissenschaftlicher Waagen in Hamburg. Zeitschrift für Unternehmensgeschichte 31 (1986) 117–140, 165–183.

30. Jenemann, Hans R.: Die mechanische Analysenwaage: Einführung und Durchsetzung des Substitutionsprinzips in der Zeit von 1945–1960. Wägen und Dosieren 18 (1987) 57–63, 101–103.

31. Jenemann, Hans R.: Der Mechaniker-Pfarrer Philipp Mathäus Hahn und die Ausbreitung der Feinmechanik in Südwestdeutschland. „Zeitschrift für Württembergische Landesgeschichte" 46 (1987), 117–161.

32. Jenemann, Hans R.: La Balance électro-magnétique de A.C. Becquerel. „Le Système métrique—Bulletin de la Société métrique de France" 1987 4ème trimestre.

33. Jenemann, Hans R.: Über die Aufhänge- und Arretierungsvorrichtung der ägyptischen Waage der Pharaonenzeit: „Berichte zur Wissenschaftsgeschichte" 11 (1988), 67–82.

34. Jenemann, Hans R.: Zur Geschichte der Herstellung von Präzisionswaagen hoher Leistung in Wien. „Blätter für Technikgeschichte" 49 (1987), 7–85.

35. Jenemann, Hans R.: Die Göttinger Präzisionsmechanik und die Fertigung feiner Waagen. Göttinger Jahrbuch 36 (1988) 181–230.

36. Jenemann, Hans R.: Die langarmigen Präzisionswaagen im Liebig-Museum zu Gießen, hg. von: Mettler Instrumente GmbH, Gießen. Gießen 1988.

37. Jenemann, Hans R.: Die mechanische Analysenwaage: Weiterentwicklung und Vollendung in der Zeit von 1960 bis 1975. „Wägen und Dosieren" 19 (1988), 138–141, 188–190, 225–228; +20 (1989), 18–20.

38. Jenemann, Hans R.: Entwicklung der Präzisionswaage. In: M. Kochsiek (Hg.): „Handbuch des Wägens." 2. Aufl., Braunschweig, Wiesbaden 1989, 745–779.

39. Jenemann, Hans R.: Die Geschichte der Mikrochemischen Waage. In: "Acta Metrologiae Historicae II—Travaux du 4, Congrès International de la Métrologie Historique", Linz 1988, 337–374. (mit Abbildungen)

40. Jenemann, Hans R.: Zur Geschichte der Waagen mit variablem Armlängenverhältnis im Altertum. Trierer Zeitschrift 52 (1989), 319–352.

Table A.5 (Continued)

41. Jenemann, Hans R.: Teilmetrische Mass- und Gewichtseinheiten in den deutschen Ländern in der Zeit vor 1872. „Le Système Métrique—Bulletin de la Société métrique de France" 1989, 1er trimestre, 431–438.

42. a. Jenemann, Hans R.: Abschnitt „Waagen". In: „Philipp Matthäus Hahn, 1739–1790, I, Katalog" hg. von Ch. Väterlein, Stuttgart 1989, 357–366.

42. b. Jenemann, Hans R.: Die wägetechnischen Arbeiten von Philipp Matthäus Hahn. In: „Philipp Matthäus Hahn, 1739–1790, II, Aufsätze" hg. von Ch. Väterlein, Stuttgart 1989, 479–499.

43. Jenemann, Hans R.; Robens, Erich: Indicator Systems and Suspension of the old Egyptian Scales, „Thermochimica Acta" 152 (1989), 249–258.

44. Jenemann, Hans R.: Zur Geschichte der Dichtebestimmung von Flüssigkeiten, insbesondere des Traubenmostes in Oechsle-Graden. Schriften zur Weingeschichte Nr. 98 (1990).

45. Basedow, Arno M.; Jenemann, Hans R.: Waage und Wägung. In: Friedrich Ehrenberger (Hg.): „Quantitative organische Elementaranalyse" (1991), 79–107.

46. Jenemann, Hans R.: Zur Geschichte der Bestimmung der Dichte fester und flüssiger Körper. In: „Acta Metrologiae Historicae III—Traveaux du 5. Congrès International de la Métrologie Historique" (Linz 1989)". St. Katharinen 1991, 95–161.

47. Jenemann, Hans R.: Zehntausend Jahre Waage? „Maß und Gewicht – „Zeitschrift für Metrologie, Heft 21 (März 1992), 470–487, und Heft 22 (Juni 1992), 509.

48. Jenemann, Hans R.: Die Werkstatt von Paul Bunge: 100 Jahre Präzisionswaagenherstellung in Hamburg. „Beiträge zur deutschen Volks- und Altertumskunde" 26 (1988/91), 169–188, Tafel 21–24.

49. Jenemann, Hans R.: Die frühe Geschichte der Mettler-Waage, Band 11 der Siegener Abhandlungen zur Entwicklung der materiellen Kultur, St. Katharinen 1992.

50. Jenemann, Hans R.: Die Geschichte der Mikrowa-Waagenfabrik, Anhang zu: „Die frühe Geschichte der Mettler-Waage" (s. Nr. 49).

51. Jenemann, Hans R., Basedow, Arno, M. Robens, Erich: Die Entwicklung der Makro-Vakuumwaage, PTB-Bericht TWD-38, Braunschweig 1992.

52. Jenemann, Hans R.: Eine römische Waage vom Typ Besmer. „Archäologisches Korrespondenzblatt" 22 (1992), 525–535.

53. Jenemann, Hans R.: Zur Geschichte des langarmigen Balkens von Präzisionswaagen. „Maß und Gewicht – Zeitschrift für Metrologie", Heft 29 (März 1994), 672–687.

54. Jenemann, Hans R.: Zur Geschichte des Reitergewichtes und der Reiterverschiebung an analytischen Waagen, in: „Acta Metrologiae Historicae IV—Travaux du 6. Congrès International de la Métrologie Historique" (Lille 1992); „Cahiers de Métrologie," ed. Jean Hoquet 11–12 (1993–1994), 169–202.

55. Jenemann, Hans R.: The early history of balances based on electromagnetic and elektrodynamic force compensation. In: Keller, J.U.; Robens, E. (eds.): Microbalance Techniques—Proceedings of the 25th Conference on Vacuum Microbalance Techniques" (September 2–4, 1993, Siegen, Germany). Brentwood, Essex 1994, 25–53.

56. Jenemann, Hans R., Robens, Erich: History of the Vacuum Macrobalance. In: „Microbalance Techniques—Proceedings of the 25th Conference on Vacuum Microbalance Techniques" (September 2–4, 1993, Siegen, Germany). Ed.: Keller, J.U.; Robens, E. Brentwood, Essex, 1994, 13–28.

57. Jenemann, Hans R.: Die wägetechnischen Arbeiten von Carl August von Steinheil. PTB-Bericht TWD-42 der Physikalisch-Technischen Bundesanstalt, Braunschweig, Braunschweig 1994.

Table A.5 (Continued)

58. Jenemann, Hans R.: Einige Aspekte zur Geschichte der Waage. „Beiträge zur Geschichte von Technik und technischer Bildung", Hg.: Hiersemann, Lothar, FH. Leipzig. 10 (1994), 28–72.

59. Jenemann, Hans R.: Die frühe Geschichte der Waagen mit elektromagnetischer und elektrodynamischer Kraftkompensation. „Wägen und Dosieren" 26 (1995), 12–18.

60. Jenemann, Hans R.: Die Geschichte der Anzeigevorrichtung an der gleicharmigen Balkenwaage. „Maß und Gewicht – Zeitschrift für Metrologie" Heft 33 (März 1995), 771–796.

61. Jenemann, Hans R.: Über altägyptische Kleinwaagen und artverwandte Wägeinstrumente. *Technikgeschichte* 62 (1995) 1–28.

62. Jenemann, Hans R.: Die Geschichte der Waage im Mittelalter. *NTM Internationale Zeitschrift für Geschichte und Ethik der Naturwissenschaften, Technik und Medizin* 3 (1995), 145–166.

63. Jenemann, Hans R.: The early history of balances based on electromagnetic or electrodynamic force compensation. In: L. Grossi (ed.): *La massa e la sua misura—Mass and its Measurement—Proceedings of International Congress; Modena (Italy), 15–17 September 1993*. CLUEB, Bologna 1995, p. 9–20.

64. Jenemann, Hans R.: Die Besmer-Waage im Altertum. *Jahrbuch des Römisch-Germanischen Zentralmuseums*, Mainz 41 (1994), 199–229.

65. Jenemann, Hans R.: Die Geschichte der Anzeigevorrichtung an der Präzisionswaage. *Maß und Gewicht—Zeitschrift für Metrologie*, (Juni 1996) 38, 895–918.

66. Jenemann, Hans R.: Das Kilogramm der Archive vom 4. Messidor des Jahres 7: Konform mit dem Gesetz vom 18. Germinal des Jahres 3? In: Genauigkeit und Präzision – in der Geschichte der Wissenschaften und des Alltags, hrsg. von D. Hoffmann und Harald Witthöft. PTB-Texte Bd. 4, Braunschweig 1996, 183–213.

67. Jenemann, Hans R.: Über die Zahlenmystik an der Großen Pyramide zu Giseh. Int. Zs. f. Gesch. u. Ethik der Naturwiss., Technik u. Med. 4 (1996) 249–268.

68. Jenemann, Hans R.: Die Entwicklung der Massebestimmung. In: M. Kochsiek, M. Gläser (Hrsg.): Massebestimmung. VCH, Weinheim 1997, S. 109–159.

69. E. Robens, H. R. Jenemann: "Remarks on the notion "Microbalance" ". In: M'bark Ben Chanaa: *Proceedings of the XXVIth International Conference on Vacuum Microbalance Techniques*. Faculté des Sciences Semlalia, Université Cadi Ayyad, Marrakech 1995, p. 7–12.

70. Jenemann, Hans R., Translated in English by Arno Basedow: Die Waage des Chemikers—The Chemist's Balance. DECHEMA, Frankfurt am Main 1997.

71. Jenemann, Hans R., Die Geschichte der Dämpfung an der Laboratoriumswaage. Berichte der Wissenschaftsgeschichte 20 (1997), 1–17.

72. Gast, Theodor, Jenemann, Hans R., Robens, Erich: Vakuumwaagen: Teil 1: Einleitung und Makrowaagen. Vakuum in Forschung und Praxis (1997) 4, 262–266.

73. E. Robens, Th. Gast, H.R. Jenemann: Vakuumwaagen. Teil 2: Mikrowaagen. *Vakuum in Forschung und Praxis* 10 (1998) 3, 203–206.

74. H.R. Jenemann, E. Robens: A history of the balance in society. *J. Thermal Analysis and Calorimetry* 55 (1999) 339–346.

75. Th. Gast, H.R. Jenemann, E. Robens: The damping of balances. *J. Thermal Analysis and Calorimetry* 55 (1999) 347–355.

76. Jenemann, Hans R.: Precision Balance. Equilibrium, Quaterly Magazine of ISASC, International Society of Antique Scale Collectors (1983?) 493–494, 563–564, 589–590, 620, 642–643, 664–665, 700–702, 718–720, 750–752, 790–792, 843–844, 872–873.

77. Jenemann, Hans R.: Substitution Weighing. Equilibrium, Quaterly Magazine of ISASC, International Society of Antique Scale Collectors (1983?) 873–874.

Table A.5 (Continued)

78. Jenemann, Hans R.: The Development of the Determination of Mass. In: M. Kochsiek, M. Gläser (eds.): *Comprehensive Mass Metrology*. Wiley-VCH, Weinheim 2000, p. 119–163. ISBN 3–527-29614-X.

79. Jenemann, Hans R.: Die Digitalisierung der Anzeigevorrichtung an der Analysen- und Präzisionswaage. „Maß und Gewicht – Zeitschrift für Metrologie", Heft 58 (Juni 2001), 1391–1408. ISSN 0933-4246.

80. Jenemann, H.R.: Die Berliner Werkstätten. Übersetzt ins Englische von Thomas Allgeier.

Table A.6 International Conferences on Vacuum Microbalance Techniques (ICVMT). (For abbreviations see the list at the end of the References)

#	Year	Place	Organiser	Proceedings
1	1960	Fort Monmouth, NJ, USA	M.J. Katz	M.J. Katz (ed.), VMT, Vol. 1, 1961
2	1961	Washington, DC, USA	R.F. Walker	R.F. Walker (ed.), VMT, Vol. 2, 1962
3	1962	Los Angeles, CA, USA	K.H. Behrndt	K.H. Behrndt (ed.), VMT, Vol.3, 1963
4	1964	Pittsburgh, PA, USA	P.M. Waters	P.M. Waters (ed.), VMT, Vol. 4, 1965
5	1965	Princeton, NJ, USA	K.H. Behrndt	K.H. Behrndt (ed.), VMT, Vol.5, 1966
6	1966	Newport Beach, CA, USA	A.W. Czanderna	A.W. Czanderna (ed.), VMT, Vol. 6, 1967
7	1968	Eindhoven, Netherlands	C.H. Massen, J.A. Poulis	C.H. Massen, H. van Beckum (eds.), VMT, Vol. 7, 1970
8	1969	Wakefield, MA, USA	A.W. Czanderna	A.W. Czanderna (ed.), VMT, Vol. 8, Plenum, New York 1971
9	1970	Berlin, Germany	Th. Gast, E. Robens	Th. Gast, E. Robens (eds.), PVMT, Vol. 1, 1972
10	1972	Uxbridge, UK	S.C. Bevan, S.J. Gregg, N.D. Parkyns	S.C. Bevan, S.J. Gregg, N.D. Parkyns (eds.), PVMT, 1973
11	1973	New York, NY, USA	A.W. Czanderna	A.W. Czanderna (ed.), JVST 11 (1974) 396–439
12	1974	Lyon, France	C. Eyraud, M. Escoubes	C. Eyraud, M. Escoubes (eds.), PVMT, Vol. 3, 1975
13	1975	Philadelphia, PA, USA	W. Kollen	W. Kollen (ed.), JVST 13 (1976) 541–560
14	1976	Salford, UK	D. Dollimore	D. Dollimore (ed.),TA 24 (1978) 204–431
15	1977	Boston, MA, USA	P. Ficalora	P. Ficalora (ed.), JVST 15 (1978) 745–821
16	1978	Kiel, Germany	O.T. Sørensen, H.-J. Seifert	O.T.Sørensen (ed.), TA 29 (1979) 198–360
17	1979	New York, NY, USA	A.W. Czanderna	A.W. Czanderna (ed.), JVST 17 (1980) 90–124
18	1981	Antwerpen, Belgium	R. de Batist, A. van den Bosch	E. Robens (ed.), TA 51 (1981) 1–95

Table A.6 (Continued)

#	Year	Place	Organiser	Proceedings
19	1982	Baltimore, MD, USA	R. Vasovsky	R. Vasovsky (ed.), JVST 20 (1983)
20	1983	Plymouth, UK	S.A.A. Jayaweera	S.A.A. Jayaweera (ed.), TA (1984)
21	1985	Dijon, France	N. Gérard	N. Gerard, S.A.A. Jayaweera (eds.), TA 103 (1986)
22	1987	Rabat, Morocco	L. Belkbir	L. Belkbir, S.A.A. Jayaweera (eds.), TA 152 (1989)
23	1989	Middlesbrough, UK	S.A.A. Jayaweera	W. Hemminger, S.A.A. Jayaweera, E. Robens (eds.), TA 1993/94
24	1991	Hammamet, Tunisia	M. Jemal	W. Hemminger, M. Jemal, E. Robens (eds.), TA 1993/94
25	1993	Siegen, Germany	J.U. Keller, E. Robens	J.U. Keller, E. Robens (eds.), VMT'94
26	1995	Marrakech, Morocco	M. Ben Chanaa	M'bark Ben Chanaa (ed.), VMT'95
27	1997	Lublin, Poland	P. Staszczuk	P. Staszczuk (ed.), JTAC 1999
28	1999	Kyiv, Ukraine	V.A. Tertykh	V.A. Tertykh (ed.), JTAC
29	2001	Middlesbrough, UK	S.A.A. Jayaweera	S.A.A. Jayaweera, E. Robens (eds.), JTAC 71 (2003)
	2003	(cancelled)		E. Robens, S.A.A. Jayaweera (eds.), JTAC 76 (2004)
30	2005	Wroclaw, Poland	G.W. Chadzyński	G.W. Chadzyński, E. Robens, S.A.A. Jayaweera (eds.), JTAC 86 (2006) 1
31	2007	Izmir, Turkey	D. Balköse	D. Balköse, E. Robens, S.A.A. Jayaweera, (eds.), JTAC 94 (2008) 3
32	2009	Kazimierz, Poland	P. Staszczuk, D. Sternik, A. Deryło-Marczewska	P. Staszczuk, D. Sternik (eds.), JTAC (2010)
33	2011	Zamość, Poland	Z. Rzączyńska	JTAC (2013)
34	2014	Kyiv, Ukraine	V. Gun'ko, V. Tertykh, Irina Laguta	JTAC (2015)

JTAC:	*Journal of Thermal Analysis and Calorimetry*, Kluwer Academic Publishers, Dordrecht & Akadémiai Kiadó, Budapest.
JVST:	*Journal of Vacuum Science and Technology*, American Institute of Physics, New York.
PVMT:	*Progress in Vacuum Microbalance Techniques*, Vol. 1–3, Heyden, London 1972–1975.
TA:	*Thermochimica Acta*, Elsevier, Amsterdam. ISSN 0040-6031.
VMT:	*Vacuum Microbalance Techniques*, Vol. 1–8, Plenum Press, New York 1961–1971. SBN 306–38408-6.
VMT'94:	J.U. Keller, E. Robens (eds.): *Vacuum Microbalance Techniques '94*, Multiscience Publishing, Brentwood 1994.
VMT'95:	M'bark Ben Chanaa (ed.): *Proceedings of the XXVIth International Conference on Vacuum Microbalance Techniques*. Faculté des Sciences Semlalia, Université Cadi Ayyad, Marrakech 1995.

Name Index[1]

---

[1]The name index includes company names only in case they refer directly to its founders. Thus company names registered in Tables 8.1 and 8.2 in general are not included.

E. Robens et al., *Balances*, DOI 10.1007/978-3-642-36447-1,
© Springer-Verlag Berlin Heidelberg 2014

Subject Index

E. Robens et al., *Balances*, DOI 10.1007/978-3-642-36447-1,
© Springer-Verlag Berlin Heidelberg 2014

Printed by Books on Demand, Germany